Lecture Notes in Engineering

Edited by C. A. Brebbia and S. A. Orszag

26

Shell and Spatial Structures: Computational Aspects

Proceedings of the International Symposium
July 1986, Leuven, Belgium

Edited by:
G. De Roeck, A. Samartin Quiroga,
M. Van Laethem and E. Backx

Springer-Verlag
Berlin Heidelberg New York
London Paris Tokyo

Editors
G. De Roeck
Katholieke Universiteit Leuven
Civil Engineering Department
de Croylaan 2
3030 Leuven
Belgium

A. Samartin Quiroga
E.T.S. Ingenieros de Caminos
Canales y Puertos
Departamento Analisis de Estructuras
Avda, Los Castros, Santander
Spain

M. Van Laethem
Katholieke Universiteit Leuven
Civil Engineering Department
de Croylaan 2
3030 Leuven
Belgium

E. Backx
Parkdreef 23
3030 Heverlee
Belgium

ISBN 3-540-17498-2 Springer-Verlag Berlin Heidelberg New York
ISBN 0-387-17498-2 Springer-Verlag New York Berlin Heidelberg

Library of Congress Cataloging-in-Publication Data
Shell and spatial structures.
(Lecture notes in engineering; 26)
1. Shells (Engineering)--Congresses.
2. Space frame structures--Congresses.
3. Structures, Theory of--Data processing--Congresses.
I. Roeck, G. de (Guido).
II. Series.
TA660.S5S465 1987 624.1'776 87-4291
ISBN 3-540-17498-2 (U.S.)

Printed in Germany

Printing: Mercedes-Druck, Berlin
Binding: B. Helm, Berlin
2161/3020-543210

FOREWORD

In recent years powerful engineering workstations for a reasonable price become a valuable tool for the design of complicated constructions such as shell and spatial structures. This availability causes an increasing use of advanced numerical techniques for the static and dynamic analysis of these structures, also in the non-linear range. The I.A.S.S. Working Group n° 13 concerned with "Numerical Methods in Shell and Spatial Structures" and the Department of Civil Engineering of the Katholieke Universiteit Leuven have taken the initiative to organise an International Symposium, providing a forum for discussion and exchange of views between researchers, specialists in numerical analysis on one hand and designers, practising engineerings on the other hand.

These Proceedings contain the papers presented at the Symposium, held in Leuven, July 14-16 1986.
The papers are organised in five sections :

1. Shell structures
2. Spatial structures
3. Dynamic analysis
4. Non-linear analysis
5. Presentation and interpretation of results

The papers covering more than one domain are classified following the main subject.

We hope that researchers as well as practising engineers will find a lot of useful information in the book.

G. De Roeck
A. Samartin
M. Van Laethem
E. Backx

Table of Contents

Shell Structures

Different methods of numerical analysis of shells
W.C. Schnobrich 1

Some finite element methods for linear thin shell problems
M. Bernadou 18

Computer aided design of shell structures
H. Sardar Amin Saleh 29

Direct integration of elastic thin shallow shell's governing equations and lower order method
Fan Jashen 38

Applications of axisymmetric thin shell finite elements to the analysis of a radiating flexural shell sonar transducer
B. Hamonic, J.C. Debus, J.N. Decarpigny, D. Boucher, B. Tocquet .. 43

Finite element analysis of cylindrical shells under concentrated loading
P.L. Gould, M. Hourani, J.S. Lin, T.G. Harmon, K.J. Han 54

A curved stiffener element for shell structures of general geometry
S.-C. Chang, J.H. Lee 64

Membrane stresses in hyperbolic paraboloid sectorial shell with cantilevered free edge
I. Sajtos 75

Approximate analysis of saddle Cable Net Roofs
Ding Shen-Si 91

Solutions of the equations of the unified approach in the analysis of wall beams, plates, arches and shells
P. Hernandez, E. Ramirez, J. Soler 101

Bond slip equations of reinforced concrete shells
M. Kurata 111

Spatial Structures

Analysis methods for spatial structures
M. Papadrakakis 121

Analysis of large scale space frames by minimization of total potential energy
H. Tabar-Heydar 149

Influence of Boundary conditions upon stress distribution in single-layer diagonal grids
Z.S. Makowski, D. Maranzano, G.A.R. Parke 159

Extensible cable network analysis through a nonlinear optimization code
J.P. Coyette, P. Guisset 173

Practical formex analysis of space structures
H. Nooshin, P.G. King 185

Traglast und Formänderung von Raumfachwerkplatten in anschaulicher Näherung
M. Herzog 200

Dynamic analysis

Practical dynamic analysis of translational shells
A. Samartin, J. Martinez 208

Dynamic experiments and earthquake observation of reticulated single-layer domes
T. Tanami, Y. Hangai 221

Atmospheric turbulence and its consequences on the dynamic behaviour of shell structures
J. Monbaliu 231

Numerical analysis of concrete pedestal of antenna building at Lessive, Belgium
D. Van Gemert, C. Vanoverschelde, P. Wouters, J. Decock 241

On the nonlinear behaviour of R/C frame structures under seismic loads
W.B. Kratzig, K. Meskouris 251

Dynamic analysis of liquid filled tanks including plasticity and fluid interaction - Earthquake effects
J. Eibl, L. Stempniewski 261

Static and dynamic analysis of tensegrity systems
R. Motro, S. Najari, P. Jouanna 270

Non-linear analysis

Stability of thin-walled beams. A general theory
Y.B. Yang 280

Geometrically and phisically non-linear analysis of plane frames sensitive to imperfections
M.A. Gizejowski 290

Numerical analysis in the vicinity of critical points by the generalized inverse
Y. Hangai 299

Nonlinear computation of a barrel vault
M. Van Laetham, J. De Coen 306

A finite element model for the nonlinear analysis of reinforced concrete shell structures
M. Cervera, A.J. Kent, E. Hinton 315

The behaviour of a steel cylinder under the influence of a local load in the elastic and elasto-plastic area
J. Spiekhout, A.M. Gresnigt, G.M.A. Kusters 329

Large displacement analysis of thin shells
S. Jingju, A. Peterson, H. Petersson 337

Nonlinear analysis of tension fabric structure
A. Yoshida, S. Ban, H. Tsubota, K. Kurihara 350

Non linear analysis with model investigation of buried pipelines
D.H. Jiang, M.C. Chang, C.J. Hsiao, C.C. Lee, Y.F. Lee, W.P. Lin 360

An unconventional class of elements for nonlinear shell analysis
M.K. Nygärd, P.G. Bergan 369

Shakedown and limit analysis of shells - A variational and numerical approach
P. Morelle, G. Fonder 381

Displacement fields and large deformation analysis of 3-D beams
G. Ying Qiao, J.L. Batoz 406

Presentation and interpretation of results

An algorithm for graphical computer results in shells
A. Samartin, J. Cardona 418

Analysis of shell structures on microcomputers
J.P. Rammant, E. Backx, L. Knoops 425

Using microcomputers for the interactive generating of spatial meshes and results evaluation
J.F. Stelzer 435

The use of a CAD-system for pre- and postprocessing of finite element calculations
G. De Roeck, C. Vanoverschelde 445

Mathematical model for calculation of rebuild of R/C mushroom slab
D. Van Gemert, C. Vanoverschelde, M. Vanden Bosch 454

Design and analysis of the Winter Garden in New York City
P. Lew, A. Gutman, L. Zborovsky, L. Petrella 460

Experimental and numerical verification of the structural behaviour of a foamed hyperbolic paraboloid shell
G. De Roeck, C. Vanoverschelde 467

Numerical analysis of tensile structures with minimum surface based on survey of soap film models
G. Valente, M. Moscetti 478

DIFFERENT METHODS OF NUMERICAL ANALYSIS OF SHELLS

W.C. SCHNOBRICH

UNIVERSITY OF ILLINOIS
URBANA ILLINOIS

Advances in the capabilities of computerized numerical methods over the past two plus decades has pushed those techniques to such a level that the analyst now has the tools which allow him to investigate shells while including in his analysis consideration of the complex details present in the real structure. Of particular importance is the ability to model the actual or near actual manner of supporting those structures. Any discussion of shell analysis nowadays should focus on these numerical methods. The majority of these methods are based upon an energy formulation. The major exceptions being the numerical integration schemes described initially by Goldberg [15], Zarghamee [41], Cohen [10], and Svalbonas [40]. Also the early finite difference methods reported by Radkowski [36], Septeoski [39] and Budiansky and Radkowski [5] represent simple direct implementation of these mathematical techniques.

The three basic numerical techniques which have been used with some regularity to solve shell problems are Numerical Integration, Finite Difference and Finite Element. In the sixties there was considerable activity expended on each technique. The discussion to follow will be directed first to shells of revolution, followed by application to general shells. Each of the procedures will be reviewed and evaluated.

FINITE DIFFERENCE METHODS

The finite difference method was first employed by Radkowski [36] for the analysis of layered shells of revolution subject to axisymmetric loading. The two second order equations of the Reissner type were transformed to algebraic equations using standard finite difference molecules. The analysis included (Y) branched shells. Septeoski studied the effects of grid size and error accumulation that might develop during the Gaussian elimination process. Budiansky and Radkowski included nonsymmetric loading using a Sanders shell theory. Soon half-station difference equations were found to produce systems of equations which often yielded significant increases in accuracy. For shells of revolution this approach was popularized by Bushnell [6] who used it in an energy formulation to establish the BOSOR program. The half station finite difference energy method when employed with displacements as variables uses points of definition of the tangential displacements intermediate between points of definition of normal displacements. The result is a system of algebraic equations that are equivalent to a finite element model using linear functions for the tangential displacements with quadratic normal displacements. Externally the result is implemented just like any displacement based finite element system. For axisymmetric shells these difference schemes prove very efficient.

Two dimensional difference schemes prove much more difficult. One of the earliest applications of a difference based variational method to the solution of general shell problems is that of Johnson [22]. In his paper the inplane displacement control points are described at different points than are the normal displacements, Fig.[1]. The need to use such an arrangement for the definition of displacements had been observed earlier in the development of analog models under the direction of Newmark [38] capable of reproducing governing equations equivalent to proper difference equations, Fig.[2]. Such difference equations proved to be quite efficient in achieving sound solutions while a direct mathematical application displayed serious defficiencies. The principal drawback of any difference procedure was the need to use special equations close to and on the boundaries of the shell. This condition made the development of general purpose programs significantly more complex. The result has been the disappearance over the years of the difference technique as a method for analyzing general shells even though with the formulation as a variational method went a along way toward obliterating the differences between finite element and finite difference techniques.

Many practical shells are reinforced by eccentric stiffness. To include their influence first the shell equations are cast as layered orthotropic systems. Then the stiffeners may be smeared out with their flexural and extensional contributions being added to the properties in the appropriate direction. Such a manner of handling the stiffeners is acceptable in the computation of buckling levels and frequencies if the stiffeners are at a reasonably closely spaced interval (i.e., two or three stiffeners per half wavelength of mode). Such a smearing process is inadequate for stress analyses; however, because of the local nature of the stress concentrations attendant with the stiffeners. For such problems the actual stiffener and its attachment method must be considered if stresses of any accuracy are to be computed. For shells of revolution the details of incorporating discrete ring stiffeners are summarized by Bushnell [7]. The techniques reduce the steps involved to the application of a coordinate transformation as is done in a finite element analysis. When the shell must be analyzed as a two dimensional system little can be found in the technical literature directed to the analysis by finite differences with eccentric stiffeners. The analog approach [32] organized in an energy basis could be used yielding basically the same equations as derived by an element analysis.

Nonlinear applications of the finite differences focused primarily on buckling problems. Bushnell's summary covers these aspects quite well. Applications with nonlinear material behavior; however, was not so successful. Use of the half station concept meant that different stress components were described at different locations. Therefore to establish when the stress at a specific location reached a controlling value required some form of averaging to describe the shear stresses at the same location as the extensional stresses. This is analogous to the situation that results when using selective integration as a form of reduced integration with isoparametric based finite elements.

NUMERICAL INTEGRATION

For shells of revolution with their possibility of reducing the equations down to a system involving only one dimension, results in the numerical integration method being one of the more efficient solution techniques. Goldberg [15] was one of the early developers of the method with several pioneering papers on its use. Cohen [10] analyzed orthotropic shells of revolution using nonshallow shell theory of Novozhilov and a Runge-Kutta method of forward integration. Shell branching was not included however unsymmetric loads were expanded in Fourier series and each harmonic analyzed separately. The nature of the edge effect and its rapid exponential decay from the edge, demonstrates itself by the effect that any presence of a far edge contribution term in the initial edge terms results in that far edge term growing beyond bounds very quickly. To circumvent this problem early integration papers subdivided the shell into a series of short segments, integrations were carried over the individual segments and compatibility equations used to reunite the segments. The consequence was a system of a significant number of simultaneous algebraic equations so that the numerical integration approach lost most of its advantage being more like finite differences or finite elements than an integration process. Zarghamee [41], Carter [9] and Goldberg [16] described a suppression scheme which allowed the integration process to continue without subdivision. This reinstated much of the efficiency.

FINITE ELEMENT MODELS

The final procedure, the finite element technique has become the dominant technique for shells. The volume of literature is so vast that a review of all papers presenting a significant contribution to the field would be a major undertaking in its own right. Only a very selective review is practical. The early reviews by Gallagher [13] [14] summarize the activity up to 1976. Although they cover twenty years of activity significant work has been done in the subsequent years.

For shells of revolution early programs which modeled the shells as conical frustra were quickly shown to be inadequate because of the spurious bending moments triggered by the slope discontinuities at the nodal circles[24] when the shell is subject to pressure or other distributed loads. Doubly curved elements solved this problem [23]. Bushnell's very comprehensive summary chronicles the status of analysis capabilities for shells of revolution. He has encapsulated this status in a chart which shows the level of generality and complexity that can be achieved by the various existing programs. Linear analysis for symmetric and nonsymmetric loads for branched segmented shells, axisymmetric material and geometric nonlinear analyses, vibrations with nonlinear prestress, are all capable of being accomplished within many existing programs. A summary of the theory and assumptions that form the basis of the techniques can be found in the Bushnell paper[7].

Although using axisymmetric curved shell elements is the preferred modeling to be used for shells of revolution many programs do not provide such elements, then use of an isoparametrically formulated axisymmetrical solid element can be substituted as an acceptable alternative. Eight node isoparametric elements perform quite satisfactorily for steel pressure vessels and for concrete shell geometries. Aspect ratios of 10 to 1 prove no problem for the analysis. When using such solid elements the presence of ring beams, stiffeners, branching zones, etc.,require no special elements. They all remain simple axisymmetric solid elements. Multiple elements through the thickness allows the three dimensional aspects of these regions to be recognized. When removed from those regions thin shell behavior again applies. Modeling by axisymmetric solids can still be used however some economies can be gained if the analysis reverts back to shell elements if that possibility exists. Transition elements then allow the remainder of the structure to be modeled by shell elements if desired for economy. Bushnell terms this mixture of solid and shell elements "hybrid" bodies of revolution. The transition elements have solid degrees of freedom at one edge, shell degrees of freedom for the remainder.

Regions where several shells intersect is a particular example that represents an area where the three dimensional nature of the problem may be present, Fig. [3]. When those shells are of different radia then the problem is no longer axisymmetric in a global sense, multidimensional behavior applies. Two dimensional shell elements can be used however at the intersection zone the shell reference frame becomes confused. The intersection zone is three dimensional. The concept of using three dimensional isoparmetric elements in the intersection region then transitioning through special elements back to shell elements as a means of addressing this problem is the technique to mix these different elements. One of the first applications of transition elements was to this problem, Bhakrebah[3]. This concept has subsequently been applied to the reinforced intersection regions of two cylindrical shells [37]. The transition element should be three dimensional on one face, while shell degrees of freedom should be specified at all remaining nodes of the element Fig. [4]. If solid element nodes are used for all but one face of the transition element, the thickness direction of the intermediate nodes experiences a problem from the zero strain assumption of the adjacent shell nodes. Transition elements should therefore be solid, shell, shell in makeup.

Openings, pipe connections, local supports, local irregularities, etc., generate stress fields in the shell which can not be adequately investigated within the constraints of a rotational shell solution. These problems require a two dimensional solution. This however can be very demanding computationally if the entire problem is analyzed with two dimensional shell elements. Outside the local disturbance one would like to be back to an axisymmetric process. Again by using transition elements that are two dimensional thin shell elements on one side and axisymmetric rotational elements on the other side, Gould (0) was able to provide a smooth transition between these two different dimensioned domains. This transition element forms a compatibility between the trigonometric "Fourier" displacement field of the rotational element with the polynomial field of the two dimensional element. If the program does not have such a capability, the problem must be solved in two pieces with the axisymmetric case treated first then the region with the local changes or irregularities cut out and treated separately with two dimensional elements. The loading around the boundary of this region is the general stress field observed in the axisymmetric solution at those boundary locations.

Computational abilities have been pushed beyond the limits of linear analysis . With layered shell elements the inclusion of nonlinear material and geometric behavior is possible, admittedly at significant computational costs. For metal shells simple plasticity theory is normally adequate [34]. For concrete shells the cracking phase is important. The chosen material model must be capable of allowing each layer to experience cracking in the direction deemed appropriate by the stress field. This should also include the possibility for the initial crack direction to change as the loading increases if the steel placement does not produce an isotropic condition. The rotating crack model [28] has developed to meet this requirement. Inclusion of mathematically described creep characteristics into the material model is also possible at an even greater computational cost [25]. Because the load driving the shell into nonlinear behavior is not a sustained load the creep analysis can be performed within the linear phase. The most comprehensive discussion of finite element procedures on nonlinear problems is the Hughes papers [19].

GENERAL SHELL ELEMENTS

Three basic element types have evolved for general shell analysis: a) flat shell elements from the superposition of plane stress elements with plate bending elements, b) curved shell elements formed from inserting assumed displacement fields into the selected thin shell theory, and c) degenerate elements generated by applying constraints or modifying continuum elements to comply with thin shell assumptions without resorting to a specific shell theory.

Flat elements were the first to be employed as they required no additional development work, simply a superposition. These are primarily low order elements in the form of triangles. Sometimes four triangles are assembled into a quadrilateral element. The spurious bending problem present in the conical frustra axisymmetric elements is not activated to any major degree for general shell problems as the orthogonal direction does not contain the closed rings that are present in shell of revolution cases. Thus the geometric approximation associated with their use does not have an overriding influence on the results even for distributed load situations. The major detrimental factor in the use of flat elements then is the excess stiffness present in low order triangular elements. Use of higher order elements is needed to overcome this excess stiffness. However then the error present in the geometric approximation associated with flat elements has a constraining influence on the minimum number elements that must be used. In spite of this there remains interest in this class of element [12].

Curved elements derived from a selected shell theory were the main focus in the sixties. This activity was motivated to overcome the shortcomings of the flat shell element approach. The early elements used shallow shell theory [35]. These shallow elements were superseded by some elements capable of handling deep shells [26]. Although nominally sucessful the curved shell element technology introduced some new problems. Failure to maintain compatibility along interelement boundaries and generation of strain under rigid body motion were the main undesirable characteristics that evolved and whose correction schemes occupied most investigators. It was argued that higher order polynominal displacements would result in a good approximation to the rigid body motion requirement which description is in a trigonometric form. A study on the result of the polynomial approximation can be found in Morris [31]. Rectangular elements were reasonably successful when applied to cylindrical shells [4]; however, for general shells most triangular elements proved to be excessively stiff unless very high order elements were used. As far as rigid body motion generating strain the fix often used was to restrict the opening angle present in any element to an angle less than ten degrees. This did not represent too severe a restriction as accuracy requirements would often demand such a level of grid refinements. A second fix was to define the displacements for the element with enough generalized coordinates that the stipulation of zero strain under rigid body motion could be enforced as a condition to determine one of the generalized coordinates. Unfortunately a consequence of this process was that the resulting element which starts as a conforming element becomes nonconforming. This is the lesser of the two evils. Another characteristic which caused some worry was the inherent violation of compatibility between shell elements and eccentrically placed support or stiffening beams. With out of plane displacements set as cubic or higher order while inplane displacements are described by linear shapes this compatibility problem can not be resolved.

The advent of the degenerate shell elements began with the Ahmad [1] paper. This element received resonable success for moderately thick shells. This model introduced normal rotations as independent unknowns. The process was cast in a Mindlin type theory which includes shear deformations. When applied to thin shells where Kirchhoff type conditions are to prevail however the system was not that sucessful. For thin shells excess stiffness characteristics, a phenomena termed membrane and/or shear locking was soon observed. Reduced integration was one technique found to soften any excessive stiffness [42]. Two forms of reduced integration evolved. In one the terms involving shear are integrated to a lower order than are the terms associated with extension (selective integration). For the second the same reduced number of integration points or patterns are used while integrating all strain contributions, usually one order less than that necessary to maintain the order of the displacement assumptions in the computation of the element stiffness. The correction however introduced a new problem that of excess zero energy modes. These underintegrated elements seem to have more deformation states which produce zero energy than just those shapes associated with rigid body motions. This is a consequence of computing energy on the basis of the stress states at a limited number of points. If those points happen to be null points then zero energy is assumed. For Ahmad defined elements this does not represent too severe a restriction as these zero energy modes are constrained out if several elements are connected together. For the more recent Lagrangian elements however the number of zero modes is rather large [33]. Two special procedures have been developed to suppress these modes: the stabilization method by Belytschko [2] and a-posteriori method applied after the fact by Oden [21]. One inexpensive suppression technique is to overlay one element with a fully integrated element. The material properties for this element are chosen to be very small, therefore this element contributes little to the overall stiffness. The superiority of the Lagrangian element for shell problems which experience very sharp gradients is demonstrated for cylindrical and spherical shapes under concentrated forces applied at select points on the shells [2]. However these elements which have these spurious zero energy modes are deemed unsuitable for use in general purpose programs because of the possibility of being used by an investigator unfamiliar with shell behavior and not aware of some of the special behavior characteristics of the elements.

In order to verify the theoretical convergence of these element based solutions, the displacement functions must satisfy certain conditions, i.e. those set down by Irons in his 'patch test'. The patch test places the requirements that four characteristic displacement modes be satisfied by the system. These modes entail rigid body motion, constant inextensional bending, and constant membrane strain fields. The interpretation of the patch test to cylindrical shell problems was addressed by Morley [29] then to spherical shells again by Morley [30]. For spherical shells Gallagher notes that one has to resort to membrane strain and curvature changes which are only approximately constant within the accuracy of the classical shell theory.

One important aspect of the element method is the ability to include support or stiffener beams in the analysis. However even with the simple two noded elements there provides to be an incompatibility between the shell element and the line element modeling the beam [28]. A number of attempts were made to overcome this compatibility problem [11] [27]. Miller [30] presented one of the closest attempts at establishing a compatible shell beam system while working within Kirchhoff conditions. Because of the efficiency of the eight noded Ahmad element, however, the need is for a beam element to matches that shell element. Fortunately for most problems the satisfaction of compatibility between shell and beam is not necessary, in fact, is probably not desirable. Quite reasonable solutions have been achieved using nonconforming shell-beam systems. Thus using two beam elements, spanning from corner to midside and midside to corner, along the edge of a eight noded Ahmad element is an acceptable model. These beam elements can even be straight members however use of curved arch members would be preferable.

One advantage of using the degenerate element formulation is the presence of rotational degrees of freedom as nodal variables is the possibility of developing conforming stiffener or beam elements. Using a Timoshenko beam theory Milford [28] derived a curved line element to be used with a Mindlin based Lagrangian shell element. This three noded element matches the three nodes of the nine noded Lagrangian element. A further advantage of this form of stiffener is the inclusion of the geometry in the isoparametric form of element description.

APPLICATIONS

To illustrate the level of complexity one can achieve several shells are analyzed with the Ahmad and Lagrangian elements. In his paper Belytschko applies the Lagrangian shell element to a number of shell problems selected on the basis of the difficulties they present to the achievement of sound solutions. Belytschko proposes his problem class as the obstacle course any sound shell element must be capable of solving. To succeed a good shell element must have the ability to handle inextensional bending modes ,rigid body motion without straining and produce complex membrane stress states. The nine noded Lagrangian shell element with reduced integration handled the problems without demonstrating a breakdown on any of the problems. The element seems best in class. Those problems will not be repeated here.Rather several additional problems to see the importance of beam eccentricity on hypar shells solutions were chosen. First a gable roof configuration Fig. [5] is loaded by gravity loads. Constant cross-section beams are used for simplicity. The shell is supported at its four corners with tension ties spanning between those corners to provide horizontal restraint to the corners. Because of symmetry only one quadrant of the shell was analyzed with a 6x6 then an 8x8 mesh. As for many shell problems the critical elements are not the shell but the perimeter beams. The ridge member is a particularly important member. Figure [6] shows the computed axial force and bending moments in that member for the case when the beam is placed above the shell and when its placed below. The eccentricity is seen to have an overriding effect upon the results. With this large a difference in internal forces in the beams the shell must also adjust its stress field to reflect its changing role in the composite system. Figure [7] shows the normal membrane stresses near the ridge line. With the reduction of axial force in the ridge beam the significant increase in compression stresses in the shell reestablishes the necessary total compression force at the ridge. This in effect maintains the global T-C couple which provides the bending moment necessary to satisfy basics statics, this moment being the center moment in a simple beam. The large flexural moments existing in the ridge beam is a consequence of flatness of the shell near midspan. In this region there is no support to carry the dead weight of the ridge beam, which due to its straightness it must do by bending.

The second example is the often used inverted umbrella. Again the most important results are the stresses in the beams. Also only one quadrant is analyzed. The edge members selected included shear deformation, i.e. Timoshenko beams. The three noded beam has the advantage of meshing with the eight and nine node shell elements. Stresses in the shell are shown in Fig. [8] while the axial force in the edge beam is in Fig. [9]. Bending in the edge beam is shown in Fig. [10]. Conformity between the shell and beam is retained.

A last example is an interior cylindrical shell which is continuous in both lateral and longitudinal directions. Diaphragm support conditions are provided by the support frames which span transverse to the longitudinal direction of the shell. Of interest here are the longitudinal stresses Fig. [11] which should reflect a stress picture similar to the fiber stresses in a continuous beam. Due to the ability of the cross section to distort these fiber stresses are distorted some from the usual linear variation. A second stress that is important is the membrane shear at the support Fig. [12]. This shear is markedly different from the parabolic distribution that one would expect. An explanation for this variation is the fact that at the support no displacements are permitted, thus contrary to conditions assumed for a beam, there is no ability for movement along the edge. In other words no there is provision for the system to breath through the depth as in a beam.

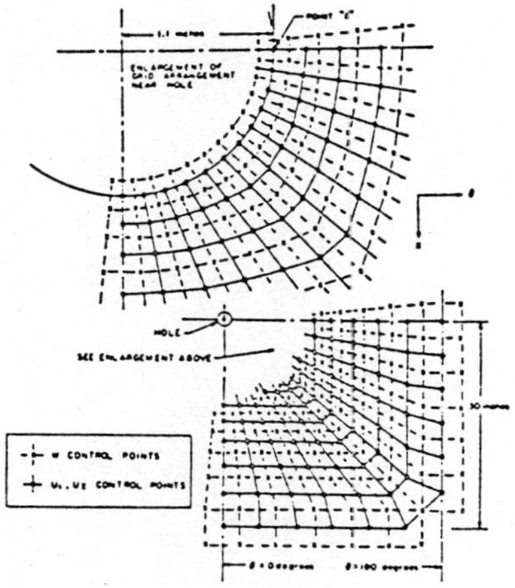

Fig. 1 Finite Difference Grid: Johnson

DESIGNATION	FUNCTIONS DEFINED AT THE POINTS
o	$W, N_x, N_y, M_x, \& M_y$
□	N_{xy} & M_{xy}
▲	U, ϕ_x & Q_x
x	V, ϕ_y & Q_y

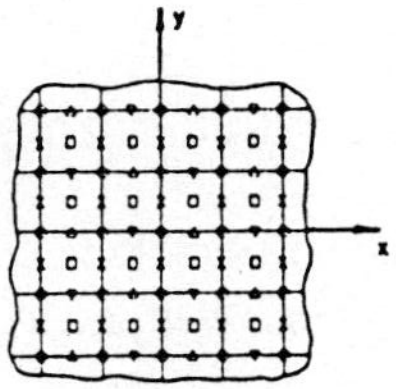

(a) Principal Curvature Lines

DESIGNATION	FUNCTIONS DEFINED AT THE POINTS
o	$W, N_x, N_y, N_{xy}, M_x, M_y$ & M_{xy}
Δ	$U, V, \phi_x, \phi_y, Q_x$ & Q_y

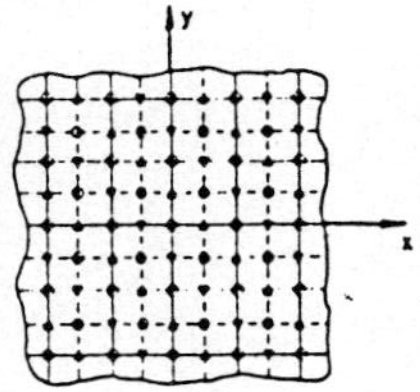

(b) General Shell

Fig. 2 Points of Definition of Displacements and Stress Resultants

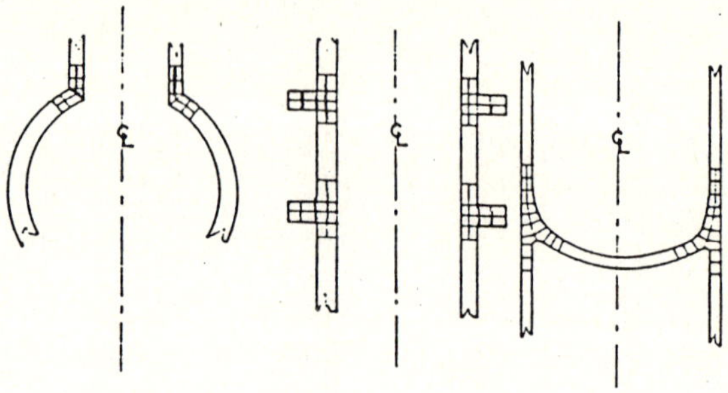

Fig. 3 Examples of Hybrid Regions

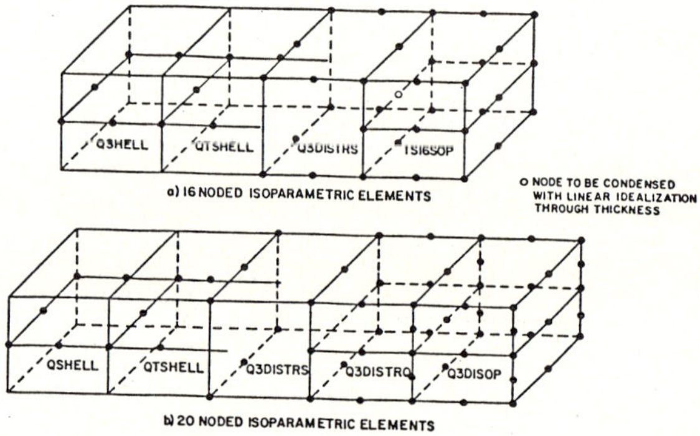

Fig. 4 Transition Elements

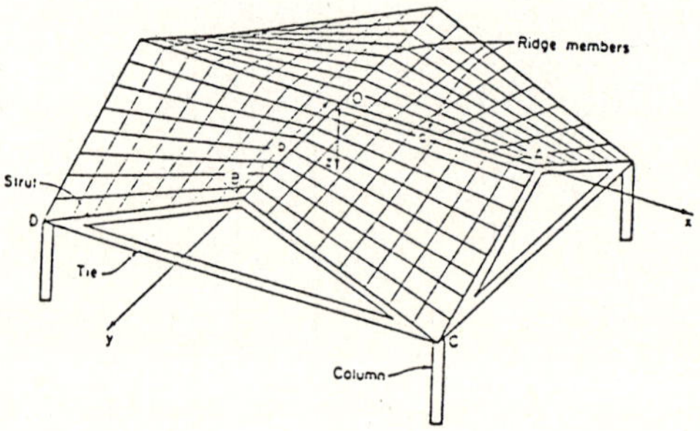

Fig. 5 Gable Roof

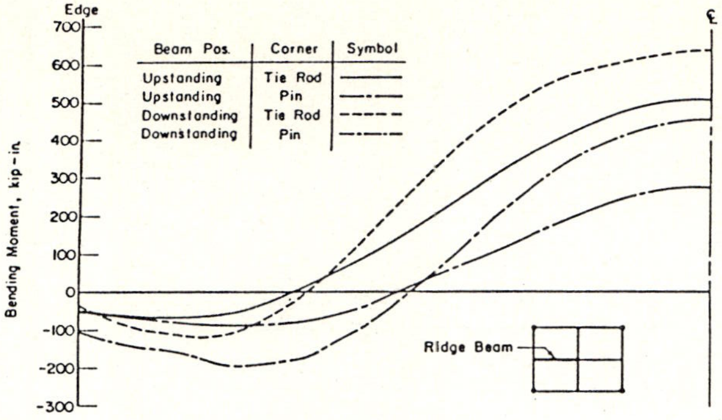

(a) Moment

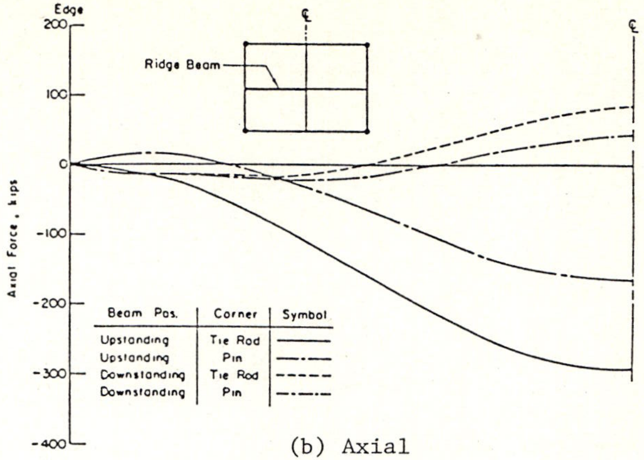

(b) Axial

Fig. 6 Forces in Ridge Beam

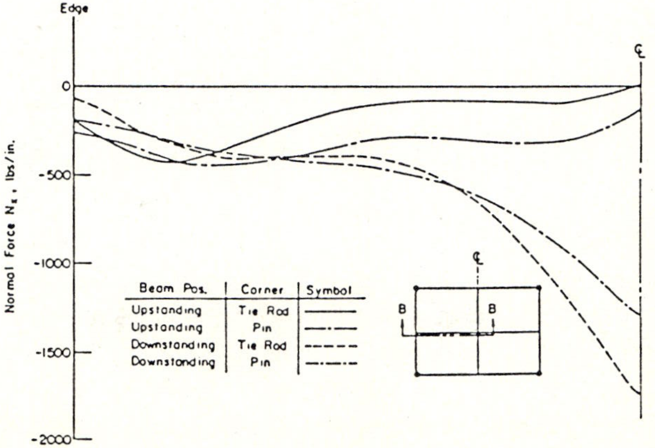

Fig. 7 Membrane Stresses Near Ridge

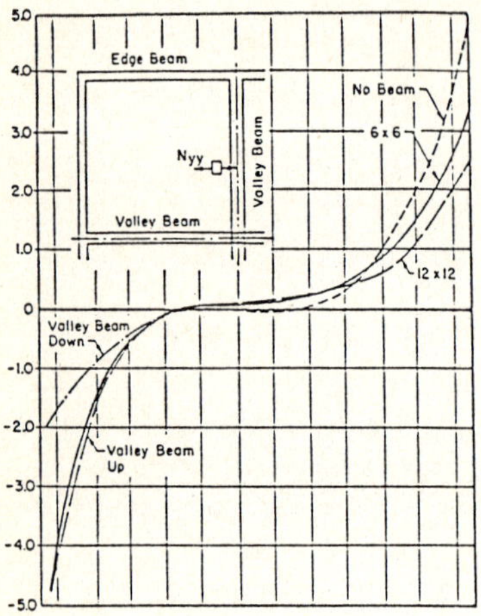

Fig. 8 Membrane Forces Near Valley Beam

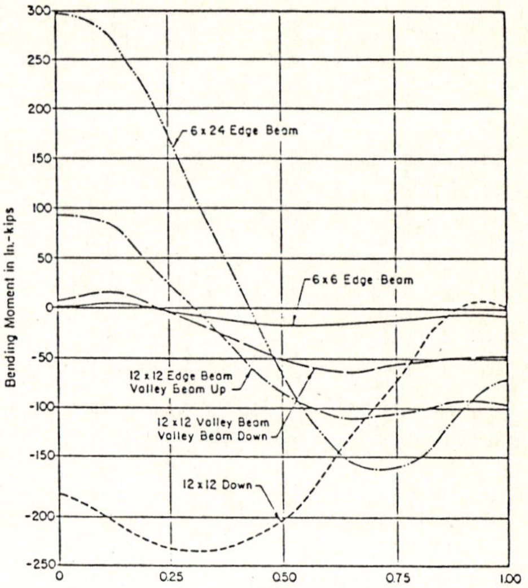

Fig. 10 Bending of Edge Beam

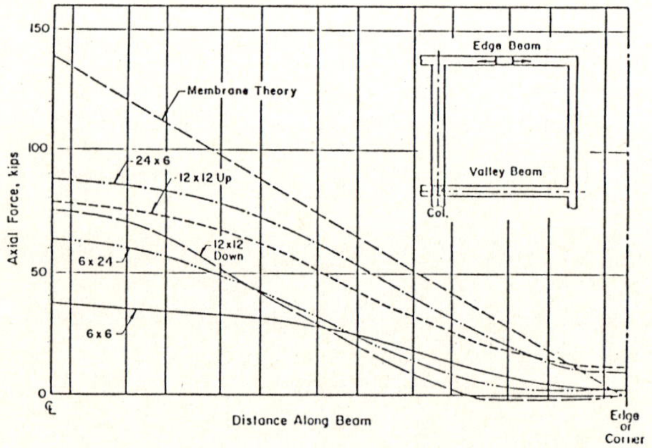

Fig. 9 Axial Force in Edge Beam

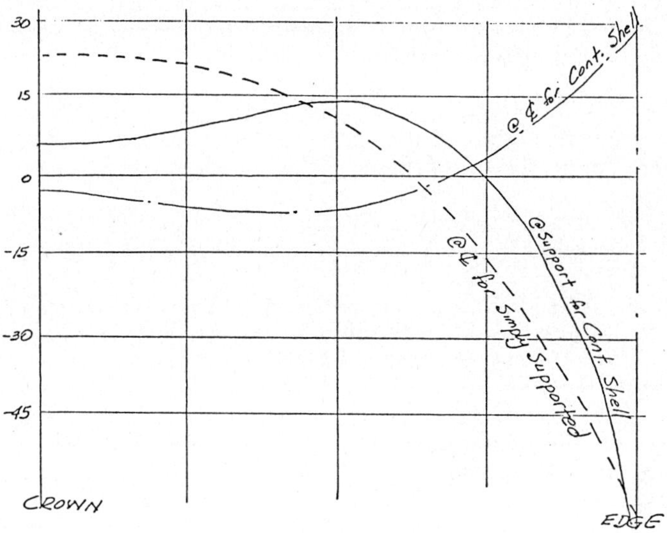

Fig. 11 Longitudinal Stresses in Continuous Cylinder

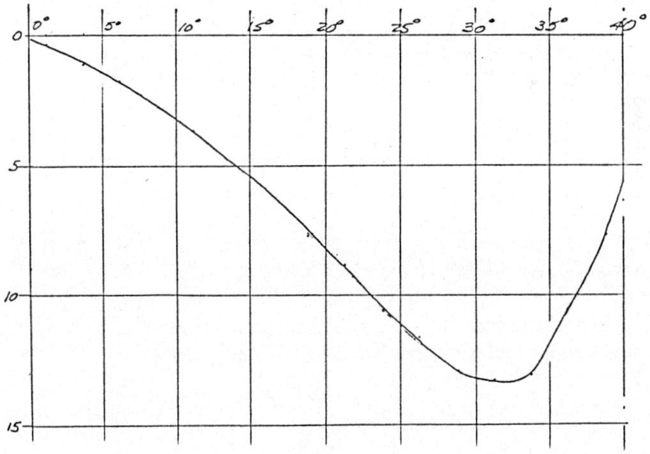

Fig. 12 Membrane Shear at Support Cylinder

REFERENCES

1. Ahmad, S., Irons, B. M. and Zienkiewicz, O. C., "Analysis of Thick and Thin Shell Structures by Curved Finite Elements," *Int. Journal of Numerical Methods in Engineering,* Vol. 2, 419-451 (1970).

2. Belytschko, T., Stolarski, H., et al., "Stress Projection for Membrane and Shear Locking in Shell Finite Elements," *Computer Methods for Applied Mechanics Engineering,* Vol.51, 221-258 (1985)

3. Bhakrebah, S. A. and Schnobrich, W. C., "Finite Element Analysis of Intersecting Cylinders," *Structural Research Series No. 400,* University of Illinois (1973).

4. Bogner, F.K., Fox, R.L., and Schmit, L.A., "The Generation of Interelement, Compatible Stiffness and Mass Matrices by the *Use of Interpolation Formulas" Proc. Conf. on Matrix Methods in Structural Mechanics* Wright-Patterson A.F. Base 397-444 (1965)

5. Budiansky, B. and Radkowski, P. P., "Numerical Analysis of Unsymmetrical Bending of Shells of Revolution," *AIAA Journal,* 1, 1833-1842 (1963).

6. Bushnell, D., "Finite-difference Energy Models Versus Finite Element Models: Two Variational Approaches in One Computer Program," *Numerical and Computer Methods in Structural Mechanics,* Ed. Fenves, S. J., et al., Academic Press, 291-336 (1973).

7. Bushnell, D., "Computerized Analysis of Shells--Governing Equations," *Computers and Structures,* Vol. 18, No. 3, 471-536 (1984).

8. Cantin, G. "Rigid Body Motions in Curved Finite Elements" *AIAA J.* Vol. 8, 1252-1255 (1970)

9. Carter, R. L., Robinson, A. R. and Schnobrich, W. C., "Free Vibrations of Hyperboloidal Shells of Revolution," *Journal of Engineering Mechanics Division of ASCE,* Vol. 95, No. 5, 1033-1052 (1969).

10. Cohen, G. A., "Computer Analysis of Asymmetric Deformation of Orthotropic Shells of Revolution," *AIAA Journal,* 2, 932- (1964).

11. Chu, T. C. and Schnobrich, W. C. "Finite Element Analysis of Translational Shells" *Computers and Structures* Vol. 2 ,197-222 (1972)

12 Dhatt, G. Marcotte, L. and Matte, Y. "A New Triangular Discrete Kirchhoff Plate/Shell" *Int. J. Num. Methods in Engr* Vol. 23, 453-470 (1986)

13. Gallagher, R. H., "Analysis of Plate and Shell Structures," *Proceedings, Symposium Applied Finite Element Methods in Civil Engineering,* Ed. Rowan, W. H., et al., Vanderbilt University and ASCE, 155-205 (1969).

14. Gallagher, R. H., "Problems and Progress in Thin Shell Finite Element Analysis," *Finite Elements for Thin Shells and Curved Members*, Ashwell, D. G. and Gallagher, R. H., Eds. Wiley, 1-14 (1976).

15. Goldberg, J. E. and Bogdanoff, J. L., "Static and Dynamic Analysis of Conical Shells under Symmetric and Unsymmetric Conditions," *Proceedings of 6th Symposium on Ballistic Missiles and Aerospace Technology*, Vol. 1, 219-238 (1961).

16. Goldberg, J. E., Setlur, A. V. and Alspaugh, D. W., "Computer Analysis of Non-Circular Cylindrical Shells," *Symposium on Shell Structures in Engineering Practice*, IASS, Vol. 2, 451-464 (1965).

17. Gould, P. L., *Finite Element Analysis of Shells of Revolution.* Pitman Publishing, (1985).

18. Han, K. J. and Gould,P. L., "Line Node and Transitional Shell Element for Rotational Shells," *International Journal of Numerical Methods in Engineering*, Vol. 18, 879-895 (1982).

19. Hughes, T.J.R. and Liu W.K. "Nonlinear Finite Element Analysis of Shells: Part I Three-Dimensional Shells Part II Two Dimensional Shells" *Computer Methods in Applied Mechanics and Engineering* Vol. 26,331-362 (1981) and Vol. 27 167-181 (1981)

20. Irons, B. M., "The SemiLoof Shell Element," *Finite Elements for Thin Shells and Curved Members*, Ashwell and Gallagher (Eds.), 197-222 (1976).

21. Jacquotte O. P. and Oden J. T. "Analysis of Hourglass Instabilities and Control in Under-integrating Finite Element Methods" *Computer Methods in Applied Mechanics and Engineering* Vol. 44 , 339-363 (1984)

22. Johnson, D. E. "A Difference Based Variational Method for Shells," *International Journal of Solids and Structures*, 6, 699-724 (1970).

23. Jones, R. E. and Strome, D. R., "Direct Stiffness Method Analysis of Shells of Revolution Utilizing Curved Elements," *AIAA Journal*, 4, 1519-1525 (1966).

24. Jones, R. E. and STrome, D. R., "A Survey of Analysis of Shells by the Displacement Method," *Proceedings of Conference on Matrix Methods in Structural Mechanics*, 205-229 (1966).

25. Kabir, A. F. and Scordelis, A. C. "Analysis of R C Shells for Time Dependent Effects" *Bulletin IASS* Vol. XXI (1979)

26 Key, S. W. and Beisinger,Z. E. "The Analysis of Thin Shells by the Finite Element Method" *High Speed Computing of Elastic Structures, IUTAM Symposium Liege* Vol. 1 ,209-252 (1971)

27. Kohnke, P. C. and Schnobrich, W. C. "Analysis of Eccentrically Stiffened Cylindrical Shells" *J. Str. Div. ASCE* Vol. 98, 1493-1510 (1972)

28. Mehrain, M., "Finite Element Analysis of Skew Composite Girder Bridges," Ph.D. thesis, Univ. of California, Berkeley, CA.

29. Milford R. V. and Schnobrich W. C. "The Application of the Rotating Crack Model to the Analysis of Reinforced Concrete Shells" *Computers and Structures* Vol. 20 , 225-234 (1985)

30. Miller, R. A., "Reduction of the Error in Eccentric Beam Modelling," *Int. J. for Num. Methods in Engr.,* Vol. 15, 575-582 (1981).

31. Morley, L. S. D. and Merriefield B. C. "Polynomial Comparsion Solutions in the Sander-Koiter Theory of Circular Cylindrical Shells" *Finite Elements for Thin Shells and Curved Members,* Ashwell, D. G. and Gallagher, R. H., Eds. Wiley, 1-14 (1976).

32. Morley, L. S. D. and Morris, A. J. "Conflict Between Finite Elements and Shell Theory" *Finite Element Methods in the Commercial Environment* Vol. 2 (1978)

33. Morris A. J. "A Summary of Appropriate Governing Equations and Functionals in Finite Element Analysis of Thin Shells" *Finite Elements for Thin Shells and Curved Members,* Ashwell, D. G. and Gallagher, R. H., Eds. Wiley, 15-41 (1976).

34. Noor, A. K. and Schnobrich, W. C., "On Improved Finite Difference Discretization Procedures," *Variational Methods in Engineering,* Southampton (1972).

35. Parisch, H., "A Critical Survey of the 9-Node Degenerated Shell Element with Special Emphasis on Thin Shell Application and Reduced Integration," *Computer Methods in Applied Mechanics,* Vol. 20, 323-350 (1979).

36. Parrisch, H. "Large Displacements of Shells Including Material Nonlinearities" *Computer Methods in Applied Mechanics,* Vol. 27, 183-214 (1981).

37. Pecknold, D. A. and Schnobrich, W. C. "Finite-Element Analysis of Skewed Shallow Shells" *J Str. Div. ASCE* Vol. 95, 715-744 (1969)

38. Radkowski, P. P., Davis, R. M. and Bolduc, M. R., "Numerical Analysis of Equations of Thin Shells of Revolution," *ARS Journal,* 32, 36-41 (1962).

39. Rajkotia, D. P. and Schnobrich, W. C., "A Stress Analysis of Circular Cylindrical Shell Intersections Including the Influences of Reinforcement, Cyclic Plasticity and Fatigue," *Structural Research Series No. 483,* University of Illinois (1980).

40. Schnobrich, W C. and Pecknold, D. A., "The Lumped Parameter or Bar Node Model Approach to Thin Shell Analysis," 1st *Symposium on Numerical and Computer Methods in Structural Mechanics,* Urbana (1971).

41. Septeoski, W. K., Pearson, C. E., Dingwell, I. W., "A Digital Computer Program for the General Axially Symmetric Thin Shell Problem," *Journal of Applied Mechanics,* No. 29, 655-661 (1962).

42. Svalbonas, V., *Numerical Analysis of Stiffened Shells of Revolution,* NASA CR-2273 (1973).

43. Zarghamee, M. and Robinson, A. R., "A Numerical Method for Analysis of Free Vibrations of Spherical Shells," *AIAA Journal,* Vol. 5, 1256-1261 (1967).

44. Zienkiewicz, O. C., Taylor, R. L., and Too, J. M., "Reduced Integration Techniques in General Analysis of Plates and Shells," *International Journal of Numerical Methods in Engineering,* Vol. 3, 275-290 (1971).

SOME FINITE ELEMENT METHODS FOR LINEAR THIN SHELL PROBLEMS

M. Bernadou
INRIA, Domaine de Voluceau
B.P. 105, Rocquencourt
78153 Le Chesnay Cedex, France

1 - Summary

This paper is essentially based upon some of our works on numerical analysis of thin shell problems according to Koiter's equations.

In paragraph 2, we record the general variational formulation of linear thin shell problems and the corresponding result of existence and uniqueness of a solution.

The paragraph 3 is devoted to the basic results concerning the approximation of the linear model by conforming finite element methods. In particular, we recall some convergence results and error estimates for an approximation of high degree of precision. Such results are very interesting because :

i) they lead to reliable methods whose convergence is mathematically proved ;

ii) they give sufficient conditions that the quadrature scheme has to satisfy ;

iii) they allow to "optimize" the different approximations in order to get efficient methods.

As an illustration of point iii), we discuss in paragraph 4 some aspects of the implementation of such an optimal high precision method when applied to the solution of the classical bench-mark of a clamped hyperbolic paraboloid roof under uniform pressure loading.

Some other approximation methods, whose mathematical studies are in progress, are briefly formulated in paragraph 5, essentially flat plate element methods, mixed methods and Discrete Kirchhoff Triangle (D.K.T.) methods.

2 - The continuous problem ; existence and uniqueness of a solution

2.1. Notations

Let Ω be a bounded open subset in a plane $\mathscr{E}^2$, with boundary Γ. Then, the middle surface S of the shell is the image of the set $\bar{\Omega}$ by a mapping $\vec{\phi} : \bar{\Omega} \subset \mathscr{E}^2 \to \mathscr{E}^3$, where $\mathscr{E}^3$ is the usual Euclidean space. Subsequently, we will assume that $\vec{\phi} \in (\mathscr{C}^3(\bar{\Omega}))^3$ and that all points of $S = \vec{\phi}(\bar{\Omega})$ are regular, in the sense that the two vectors $\vec{a}_\alpha = \vec{\phi}_{,\alpha}$, $\alpha = 1,2$, are linearly independent for all points $\xi = (\xi^1,\xi^2) \in \bar{\Omega}$. With the covariant basis $(\vec{a}_\alpha)$ of the tangent plane, we associate the contravariant basis $(\vec{a}^\alpha)$, which is defined through the relations $\vec{a}^\alpha.\vec{a}_\beta = \delta^\alpha_\beta$, where δ^α_β is the Kronecker's symbol. We also introduce the normal vector $\vec{a}_3 = \vec{a}^3 = \vec{a}_1 \times \vec{a}_2 / |\vec{a}_1 \times \vec{a}_2|$.

The thickness e of the shell $\mathscr{C}$ can be viewed as an application $e : (\xi^1,\xi^2) \in \bar{\Omega} \to \{x \in \mathbb{R} \; ; \; x > 0\}$. Then, the **shell** $\mathscr{C}$ is the set

$$\left.\begin{aligned} \mathscr{C} = \{M \in \mathscr{E}^3 \; ; \; \vec{OM} = \vec{\phi}(\xi^1,\xi^2) + \xi^3 \vec{a}_3 \,, \; (\xi^1,\xi^2) \in \bar{\Omega} \,, \\ -\tfrac{1}{2}\, e(\xi^1,\xi^2) \le \xi^3 \le \tfrac{1}{2}\, e(\xi^1,\xi^2)\} \,. \end{aligned}\right\} \qquad (2.1)$$

2.2. V.riational formulation of the continuous problem

Subsequently, we use the linear modelization of KOITER [1,2] who reduce the study of the deformation of a thin shell to the determination of the displacement field $\vec{u} = u_i \vec{a}^i$ of the particles of the middle surface (we use once more the convention for summation on repeated indices : Greek letters take their values in the set {1,2} while Latin letters take their values in the set {1,2,3}). In this way, KOITER uses the two following basic hypotheses : i) conservation of the normals ; ii) plane stresses.

In what follows, we assume that the shell is (i) loaded by a distribution of forces whose resultant is $\vec{p}$ on the middle surface S and whose resultant moment is $\vec{0}$ on S ; (ii) clamped on the part Γ_0 of its boundary $\Gamma = \partial\Omega$; (iii) free on the complementary part $\Gamma_1 = \Gamma - \Gamma_0$ of its boundary.

Then, the strain energy 1/2 $a(\vec{v},\vec{v})$ is associated to the following bilinear form

$$a(\vec{u},\vec{v}) = \int_\Omega e E^{\alpha\beta\lambda\Gamma} \{\gamma_{\alpha\beta}(\vec{u})\gamma_{\lambda\Gamma}(\vec{v}) + \frac{e^2}{12}\, \bar{\rho}_{\alpha\beta}(\vec{u})\bar{\rho}_{\lambda\Gamma}(\vec{v})\} \sqrt{a}\; d\xi^1 d\xi^2 \qquad (2.2)$$

where e is the thickness of the shell,

$$E^{\alpha\beta\lambda\Gamma} = \frac{E}{2(1+\nu)}\,[a^{\alpha\lambda}a^{\beta\Gamma} + a^{\alpha\Gamma}a^{\beta\lambda} + \frac{2\nu}{1-\nu}\, a^{\alpha\beta}a^{\lambda\Gamma}] \,, \qquad (2.3)$$

E, ν are the Young's modulus and Poisson's coefficient, and where the strain tensor $(\gamma_{\alpha\beta})$ and change of curvature tensor $(\bar{\rho}_{\alpha\beta})$ are given by

$$\left.\begin{aligned} \gamma_{\alpha\beta}(\vec{v}) &= \tfrac{1}{2}\,(v_{\alpha|\beta} + v_{\beta|\alpha}) - b_{\alpha\beta} v_3 \,, \\ \bar{\rho}_{\alpha\beta}(\vec{v}) &= v_{3|\alpha\beta} - b^\lambda_\alpha b_{\lambda\beta} v_3 + b^\lambda_{\alpha|\beta} v_\lambda + b^\lambda_\alpha v_{\lambda|\beta} + b^\lambda_\beta v_{\lambda|\alpha} \,. \end{aligned}\right\} \qquad (2.4)$$

In these expressions $a^{\alpha\lambda} = \vec{a}^\alpha.\vec{a}^\lambda$, $(b_{\alpha\beta})$ denotes the second fundamental form of the surface, the covariant derivatives $v_{\alpha|\beta}$, $v_{3|\alpha\beta}$, $b^\lambda_{\alpha|\beta}$ are defined by $v_{\alpha|\beta} = v_{\alpha,\beta} - \Gamma^\lambda_{\alpha\beta} v_\lambda$, $v_{3|\alpha\beta} = v_{3,\alpha\beta} - \Gamma^\lambda_{\alpha\beta} v_{3,\lambda}$ and $b^\lambda_{\alpha|\beta} = b^\lambda_{\alpha,\beta} + \Gamma^\lambda_{\beta\nu} b^\nu_\alpha - \Gamma^\nu_{\alpha\beta} b^\lambda_\nu$, and $(\Gamma^\lambda_{\alpha\beta})$ are the Christoffel symbols of the surface, i.e., $\Gamma^\lambda_{\alpha\beta} = \vec{a}^\lambda.\vec{a}_{\alpha,\beta}$.

Finally, the potential energy of exterior forces can be written in the following form

$$f(\vec{v}) = \int_\Omega \vec{p}.\vec{v}\,\sqrt{a}\,d\xi^1 d\xi^2 = \int_\Omega p^i v_i\,\sqrt{a}\,d\xi^1 d\xi^2 . \tag{2.5}$$

The **admissible displacement space** is defined by

$$\vec{V} = \{\vec{v} = v_i \vec{a}^i \ ; \ v_\alpha \in H^1(\Omega) \ , \ v_3 \in H^2(\Omega) \ , \ \vec{v}|_{\Gamma_0} = \vec{0} \text{ and } \partial_n v_3|_{\Gamma_0} = 0\} . \tag{2.6}$$

Then, we have proved in [3], the existence and uniqueness of a solution for the corresponding variational problem.

$$\text{Find } \vec{u} \in \vec{V} \text{ such that } a(\vec{u},\vec{v}) = f(\vec{v}) \ , \quad \forall \vec{v} \in \vec{V} . \tag{2.7}$$

3 - The approximate problem

The definition of the middle surface S through the application $\vec{\phi}$ is very interesting with respect to the approximation by finite element methods. Indeed, according to (2.2) and (2.5), the problem is defined on the plane reference domain Ω instead to be defined on the middle surface S. And now, the geometry of the shell appears only through variable coefficients. More precisely, according to [4], the variational formulation (2.7) can be written

$$\int_\Omega {}^tU\,[A_{IJ}]\,V\,d\xi^1 d\xi^2 = \int_\Omega {}^tF\,V\,d\xi^1 d\xi^2 \tag{3.1}$$

where the column matrix V (and similarly for the matrix U) is given by

$${}^tV = [v_1\ v_{1,1}\ v_{1,2}\ v_2\ v_{2,1}\ v_{2,2}\ v_3\ v_{3,1}\ v_{3,2}\ v_{3,11}\ v_{3,12}\ v_{3,22}] \tag{3.2}$$

and where the square matrix $[A_{IJ}]_{12\times 12}$ is only dependent of e, E, ν and of the first, second and third partial derivatives of the mapping $\vec{\phi}$.

In order to approximate problem (3.1), we have

i) to construct a finite dimensional subspace $\vec{V}_h$ of space $\vec{V}$. For this, we assume that Ω is a polygonal domain and we introduce finite element spaces

$$\left.\begin{aligned} V_{h1} \subset V_1 &= \{v \in H^1(\Omega) \ , \ v|_{\Gamma_0} = 0\} , \\ V_{h2} \subset V_2 &= \{v \in H^2(\Omega) \ , \ v|_{\Gamma_0} = \frac{\partial v}{\partial n}|_{\Gamma_0} = 0\} , \\ \vec{V}_h &= V_{h1} \times V_{h1} \times V_{h2} . \end{aligned}\right\} \tag{3.3}$$

ii) to introduce a numerical integration scheme, i.e., for any element K of the triangulation $\mathcal{T}_h$:

$$\int_K \phi(\xi)\,d\xi \sim \sum_{\ell=1}^{L} \omega_{\ell,K}\,\phi(b_{\ell,K}) . \tag{3.4}$$

Then, from (3.1), we obtain

$$\left.\begin{aligned} a_h(\vec{u}_h,\vec{v}_h) &= \sum_{K\in\mathcal{T}_h} \sum_{\ell=1}^{L} \omega_{\ell,K}\,({}^tU_h\,[A_{IJ}]\,V_h)\,(b_{\ell,K}) , \\ f_h(\vec{v}_h) &= \sum_{K\in\mathcal{T}_h} \sum_{\ell=1}^{L} \omega_{\ell,K}\,({}^tF\,V_h)\,(b_{\ell,K}) . \end{aligned}\right\} \tag{3.5}$$

Thus, the approximate problem can be stated as follows :

$$\underline{\text{Find}}\ \vec{u}_h \in \vec{V}_h\ \underline{\text{such that}}\ a_h(\vec{u}_h, \vec{v}_h) = f_h(\vec{v}_h)\ ,\ \forall \vec{v}_h \in \vec{V}_h\ ; \tag{3.6}$$

The mathematical analysis of such an approximation is complicated owing to the presence of variable coefficients and to the coupling between tangential components and normal component of the displacement. All proofs concerning such studies can be found in [4,5].

The figure 3.1 gives some illustrations of these results in case of an approximation of high degree of precision. In [4,5], we give analogous results for approximations of low degree of precision using for instance the H.C.T. elements.

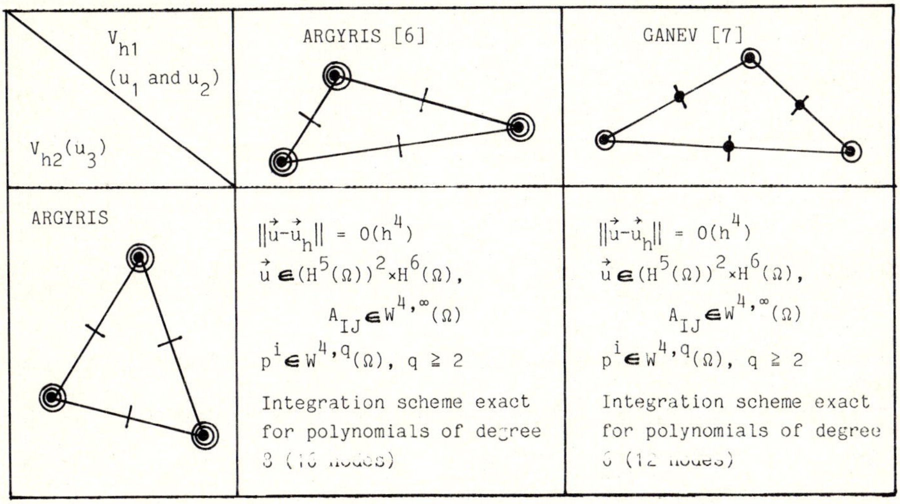

Figure 3.1 : Examples of high degree of precision

The space V_{h2} is construct from ARGYRIS finite elements. For the space V_{h1}, we can take $V_{h1} = V_{h2}$, but this is expensive and useless. Indeed, in this case, the mathematical analysis shows that the optimal choice to construct V_{h1} is to use the GANEV's element [7]. For the same error estimate, for the same regularity on the data, that means less degrees of freedom and less integration nodes. This is illustrated by the example considered in paragraph 4.

All these methods have been implemented and compared on the example of an arch dam [4]. They are also implemented in MODULEF [8].

4 - TEST OF ARGYRIS/ARGYRIS AND ARGYRIS/GANEV METHODS ON A CLAMPED HYPERBOLIC PARABOLOID ROOF UNDER UNIFORM PRESSURE LOADING

As an illustration of previous considerations, we report hereunder some results obtained in the approximation of the classical bench-mark of a clamped hyperbolic paraboloid roof by the methods given in figure 3.1. This bench-mark has been considered by different authors [9] [10] [11] ; it is described in figure 4.1.

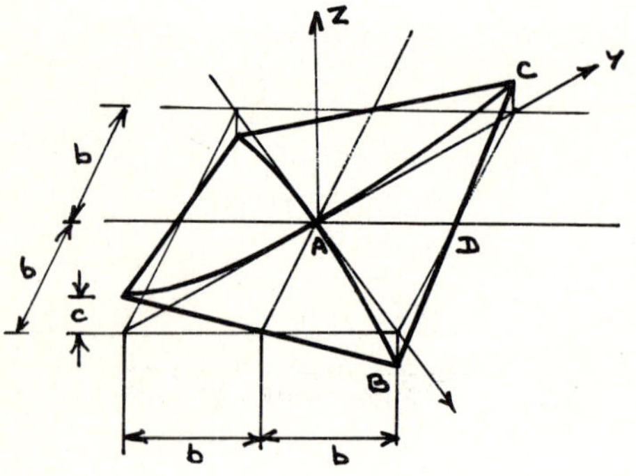

Equation :

$Z = c(Y^2-X^2)/2a^2$

where

$a = 50$ cm

$c = 10$ cm

$h = 0.8$ cm

$E = 2.85.10^4$ kp/cm^2

$\nu = 0.4$

$q = 0.1$ kp/cm^2

Figure 4.1 : The hyperbolic paraboloid roof

The mesh in use is described in figure 4.2 and takes into account the symmetry conditions. In figure 4.2, we also indicate the number of degrees of freedom, the number of unknowns, the number of integration nodes and the computing time corresponding to both methods : ARGYRIS/ARGYRIS or ARGYRIS/GANEV. For more details see [12].

V_{h1} (u_1 and u_2)	ARGYRIS	GANEV
V_{h2} (u_3)	ARGYRIS	ARGYRIS
Number of d.o.f.	693	651
Number of unknowns	483	471
Number of integration nodes	576	432
Computing time (on BULL DPS8/70)		
. Rigidity matrix	693s	412s
. Second member	14s	9s
. Resolution	153s	107s
Total	860s	528s

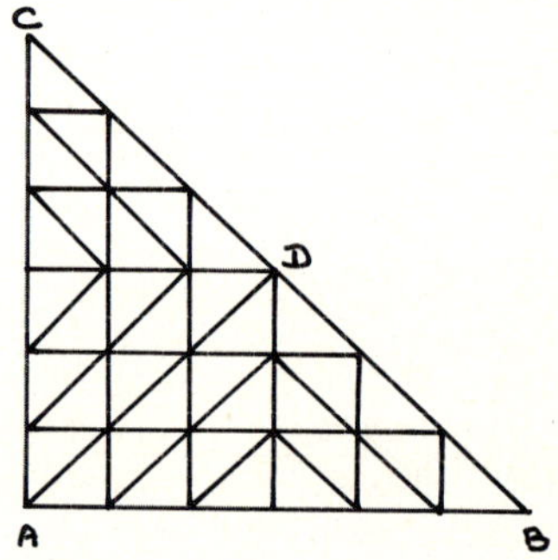

Figure 4.2 : The mesh used and some characteristic values of the implementation

Thus the ARGYRIS/GANEV method allows us to win about 38% of CPU time : 25% on the number of integration nodes plus about 20% on the obtention of elementary rigidity matrices (51 degrees of freedom for ARGYRIS/GANEV instead of 61 in case of ARGYRIS/ARGYRIS).

Let us add that the obtention of basis functions for the GANEV element has been realized by using the symbolic computation language MACSYMA [13]. It is quite sure that the systematic use of such language for all the implementation could significantly improve the total computing time.

5 - Some other approximation methods

Many approximation methods for shells are directly inspired from similar methods for plates by adding a certain approximation of the geometry. The conforming approximation method reported in paragraph 3 enters in this description since the use of numerical integration can be considered as an approximation of the geometry. But, by contrast to plate problems, very few numerical analysis results are avalaible for shells. In this paragraph, we are going to describe some approximation methods used in mechanical engineering, whose numerical analysis is in progress and seems to be promising. For more details see [14].

5.1. Flat plate element methods

Hereunder we describe one of the most representative flat plate element methods due to CLOUGH-JOHNSON [15]. The main steps of this method are :

Step 1 : The middle surface $\overline{\mathcal{S}}$ is replaced by a continuous faceted surface $\overline{\mathcal{S}}_h$ whose vertices are on $\overline{\mathcal{S}}$.

Step 2 : On each plane facet, we define the corresponding plate energy which is approximated, facet by facet, through conforming finite element methods : P_1-triangles for membrane components of the displacement and reduced HCT-triangle for the deflection.

Step 3 : In order to get an approximate energy which is consistent (i.e. sufficiently closed) to the exact energy of the shell, we introduce the following compatibility conditions on the discrete space :

i) the displacement is continuous at the vertices of the surface $\mathcal{S}_h$;

ii) the tangential components of the rotation vector, taken with respect to the initial middle surface $\mathcal{S}$, are continuous at the vertices of the surface $\mathcal{S}_h$.

By using such a method, Clough and Johnson have obtained good numerical results, specially for cylindrical shells.

For our part, we have tried to justify such methods from numerical analysis point of view. At the present time, the results are not completely satisfying. Indeed, when applying such kind of methods to one-dimensional general arch problems, we have proved convergence and error estimates : see [16] who generalize similar results obtained in case of circular arch problems by [17] and [18]. By contrast, until now, we cannot complete the proof concerning the general linear thin shell problem : it still remains to prove, on each facet, the consistency between the bending part of the energies computed on the exact middle surface and on the corresponding flat facet. Perhaps, a possible lack of consistency for such terms can explain some nonconvergent results reported by some engineers. In this way, the comments of [19, page 329] and [20, page 281] are very interesting.

5.2. Mixed finite element methods

Previous methods present two kind of difficulties that the mixed finite element methods can circumvent :

i) the stress field is derived from the knowledge of the displacement field, that means discontinuity at the interelements ;

ii) the use of finite element of class $\mathscr{C}^1$ whose implementation is not straightforward.

By contrast, the major drawback of mixed methods is the computation of the discrete solution which generally requires some specialized duality techniques.

For thin shell problems, the discrete mixed formulation of Hermann-Miyoshi type can be stated as follows :

Find $\vec{u}_h \in (X_{ho})^3$, $\underset{\approx}{n}_h \in (Y_h)^4_S$, $\underset{\approx}{m}_h \in (X_h)^4_S$ such that

$$\left.\begin{aligned} &\int_\Omega \frac{1}{e} G_{\alpha\beta\lambda\Gamma}\, n_h^{\lambda\Gamma}\, \eta_h^{\alpha\beta} \sqrt{a}\, d\xi^1 d\xi^2 + \int_\Omega \frac{12}{e^3} G_{\alpha\beta\lambda\Gamma}\, m_h^{\lambda\Gamma}\, \tau_h^{\alpha\beta} \sqrt{a}\, d\xi^1 d\xi^2 \\ &\qquad = \int_\Omega \{\gamma_{\alpha\beta}(\vec{u}_h)\, \eta_h^{\alpha\beta} + f_{\alpha\beta}(\underline{u}_h)\, \tau_h^{\alpha\beta} - u_{3h|\alpha}\, \tau_h^{\alpha\beta}|_\beta\} \sqrt{a}\, d\xi^1 d\xi^2 \\ &\forall \underset{\approx}{\eta}_h = (\eta_h^{\alpha\beta}) \in (Y_h)^4_S , \quad \underset{\approx}{\tau}_h = (\tau_h^{\alpha\beta}) \in (X_h)^4_S \end{aligned}\right\}$$

$$\left.\begin{aligned} &\int_\Omega \{\gamma_{\alpha\beta}(\vec{v}_h)\, n_h^{\alpha\beta} + f_{\alpha\beta}(\underline{v}_h)\, m_h^{\alpha\beta} - v_{3h|\alpha}\, m_h^{\alpha\beta}|_\beta\} \sqrt{a}\, d\xi^1 d\xi^2 \\ &\qquad = \int_\Omega \vec{p}\vec{v}_h \sqrt{a}\, d\xi^1 d\xi^2 , \quad \forall \vec{v}_h \in (X_{ho})^3 \end{aligned}\right\}$$

where $(n^{\alpha\beta})$ and $(m^{\alpha\beta})$ denote respectively the symmetrical tensors of stress resultants and stress couples, $\rho_{\alpha\beta}(\underline{v}) = \rho_{\alpha\beta}(\vec{v}) - v_{3|\alpha\beta}$, ${}_4\rho_{\alpha\beta} = \bar{\rho}_{\alpha\beta} - \frac{1}{2}(b^\lambda_\alpha \gamma_{\lambda\beta} + b^\lambda_\beta \gamma_{\lambda\alpha})$ is the modified change of curvature tensor, $\underline{v} = v_\alpha \vec{a}^\alpha$, $(\ )_S$ denotes a quadruplet whose second and third components are equal, and where

$$G_{\alpha\beta\lambda\Gamma} = \frac{1+\nu}{2E}\{a_{\alpha\lambda}a_{\beta\Gamma} + a_{\alpha\Gamma}a_{\beta\lambda} - \frac{2\nu}{1+\nu} a_{\alpha\beta}a_{\lambda\Gamma}\}$$

$$X_h = \{v_h \in \mathscr{C}^0(\bar{\Omega}) | v_{h|T} \in P_1 , \forall T \in \mathscr{T}_h\} , \text{ so that } X_h \subset H^1(\Omega)$$

$$X_{ho} = \{v_h \in X_h | v_h = 0 \text{ on } \Gamma\} ,$$

$$Y_h = \{v_h | v_{h|T} \in P_0 , \forall T \in \mathscr{T}_h\} , \text{ so that } Y_h \subset L^2(\Omega).$$

For numerical analysis of such problems we refer to [21] [22] [23] [24].

5.3. Approximation by D.K.T. methods

The D.K.T. (Discrete Kirchhoff Triangle) methods are now very popular when approximating linear or nonlinear plate problems. These methods are based on

MINDLIN plate models which take into account the effects of transverse shear deformations and which are well adapted to moderately thick plates.

In case of shell approximations, the original idea of D.K.T. method appears in [25]. Next, many numerical experiments were done for different kinds of shells : for a summary of contributions in this way, we refer to [26] and to the bibliography of this paper.

Hereunder, we give a short description of the formulation of a D.K.T. method in the context of the Koiter's formulation (paragraph 2), and we refer to [27] for more details.

We start with a set of thin shell equations including the effect of transverse shear deformations. That means we take into account the components β_α of the rotation of the normal $\vec{a}_3$ as additional unknowns. Then, the corresponding variational formulation can be written as

$$\left.\begin{aligned}
&\underline{\text{Find}}\ (\vec{u},\underline{\beta})\in V^5\ ,\ \underline{\beta} = \beta_\alpha \vec{a}^\alpha\ ,\ \underline{\text{such that}}\\
&\int_\Omega eE^{\alpha\beta\lambda\Gamma}\{\gamma_{\alpha\beta}(\vec{u})\gamma_{\lambda\Gamma}(\vec{v}) + \frac{e^2}{12}\chi_{\alpha\beta}(\vec{u},\underline{\beta})\chi_{\lambda\Gamma}(\vec{v},\underline{\delta})\}\sqrt{a}\, d\xi^1 d\xi^2\\
&\quad + \int_\Omega \frac{Ea^{\alpha\beta}}{2(1+\nu)}\{\phi_\alpha(\vec{u}) + \beta_\alpha\}\{\phi_\beta(\vec{v}) + \delta_\beta\}\sqrt{a}\, d\xi^1 d\xi^2\\
&\quad = \int_\Omega \vec{p}.\vec{v}\sqrt{a}\, d\xi^1 d\xi^2 + \int_{\Gamma_1}(\vec{N}.\vec{v} - M^\alpha\delta_\alpha)\, d\gamma\ ,\ \forall(\vec{v},\underline{\delta})\in V^5\ ,
\end{aligned}\right\}$$

where

$$\left.\begin{aligned}
&V = \{v | v\in H^1(\Omega)\ ;\ v_{|\Gamma_0} = 0\}\ ,\\
&\chi_{\alpha\beta}(\vec{u},\underline{\beta}) = \frac{1}{2}[(\beta_{\alpha|\beta}+\beta_{\beta|\alpha}-b^\lambda_\alpha(u_{\lambda|\beta}-b_{\lambda\beta}u_3)-b^\lambda_\beta(u_{\lambda|\alpha}-b_{\lambda\alpha}u_3)]\\
&\phi_\alpha(\vec{u}) = u_{3|\alpha} + b^\lambda_\alpha u_\lambda\ .
\end{aligned}\right\}$$

Note that this formulation gives back the usual Koiter's formulation when the normal remains normal, i.e., when $\beta_\alpha + \phi_\alpha = 0$ or, equivalently, when $\beta_\alpha = -(u_{3|\alpha}+b^\lambda_\alpha u_\lambda)$.

The finite element spaces used to approximate $\vec{u}$ and $\underline{\beta}$

The unknowns $\vec{u}$ and $\underline{\beta}$ are approximated in finite dimensional spaces V_{h1} or V_{h2} constructed by using the P_2-Lagrange elements (for u_1, u_2, β_1 and β_2) and P'_3-Hermite (for u_3).

Constraints imposed to the degrees of freedom

In order to define the D.K.T. approximate of $\vec{u}$ and $\underline{\beta}$, consider a triangle whose vertices and midsides are respectively denoted a_i and b_i, i = 1,2,3. Previous choice of finite element spaces involves that there are 33 degrees of freedom by triangle. By imposing 12 constraints, we reduce this number up to 21 effective degrees of freedom, i.e., $u_{\alpha h}(a_i)$, $u_{\alpha h}(b_i)$, $u_{3h}(a_i)$, $\beta_{\alpha h}(a_i)$, α = 1,2 and i = 1,2,3. These

12 constraints are :

i) the Kirchhoff-Love condition imposed at the vertices ;

ii) the Kirchhoff-Love condition "tangentially" imposed at the midsides ;

iii) the "normal" component of vector $\underline{\beta}_h(b_i)$ with respect to the side $a_{i-1}a_{i+1}$, is equal to the mid-sum of normal components at the adjacent vertices a_{i-1} and a_{i+1}.

Then, the D.K.T. approximated problem can be stated as follows : Find $(\vec{u}_h, \underline{\beta}_h) \in \overset{\approx}{V}_h$ such that

$$\left.\begin{aligned} &\int_\Omega eE^{\alpha\beta\lambda\Gamma} \{\gamma_{\alpha\beta}(\vec{u}_h)\gamma_{\lambda\Gamma}(\vec{v}_h) + \frac{e^2}{12} \chi_{\alpha\beta}(\vec{u}_h, \underline{\beta}_h)\chi_{\lambda\Gamma}(\vec{v}_h, \underline{\delta}_h)\} \sqrt{a}\, d\xi^1 d\xi^2 \\ &\quad = \int \vec{p}.\vec{v}_h \sqrt{a}\, d\xi^1 d\xi^2 , \ \forall(\vec{v}_h, \underline{\delta}_h) \in \overset{\approx}{V} , \end{aligned}\right\} \quad (4.1)$$

where the space $\overset{\approx}{V}_h$ is defined as follows :

$$\left.\begin{aligned} &\overset{\approx}{V}_h = \{(\vec{v}_h, \underline{\delta}_h) \text{ such that } \vec{v}_h \in V_{h1} \times V_{h1} \times V_{h2} , \ \underline{\delta}_h \in V_{h1} \times V_{h1} \\ &\quad \text{and } \vec{v}_h, \underline{\delta}_h \text{ satisfy the constraints i) ii) and iii)}\}. \end{aligned}\right\}$$

The question of existence and uniqueness of a solution for this problem is in progress as well as the study of the effect of the use of numerical integration schemes to compute the integrals in (4.1). Meanwhile let us mention that :

i) closed related methods give good results for some classical bench-marks in thin shell problems [26] ;

ii) Kikuchi [28] has proved the convergence for plate problems and has got error estimates such that

$$\sum_{i,j=1}^{2} \|\partial_{hij} u_h - \partial_{ij} u\|_{L^2} \leq Ch \|f\|_{L^2} ,$$

where u and u_h denote respectively the solutions (plate deflections) of the continuous and D.K.T. approximate problems.

References

[1] KOITER W.T. : "On the nonlinear theory of thin elastic shells", Proc. Kon. Ned. Akad. Wetensch, B69, 1-54, 1966.

[2] KOITER W.T. : "On the foundations of the linear theory of thin elastic shells", Proc. Kon. Ned. Akad. Wetensch, B73, 169-195, 1970.

[3] BERNADOU M. ; CIARLET P.G. : Sur l'Ellipticité du Modèle Linéaire de Coques de W.T. Koiter, in Computing Methods in Applied Sciences and Engineering (R. Glowinski and J.L. Lions Editors), Lecture Notes in Economics and Mathematical Systems, Vol. 134, Springer-Verlag, Berlin, pp. 89-136, 1976.

[4] BERNADOU M. ; BOISSERIE J.M. : The Finite Element Method in Thin Shell Theory : Application to Arch Dam Simulations, Birkhäuser, Boston, 1982.

[5] BERNADOU M. : "Convergence of conforming finite element methods for general shell problems", Int. J. Engng. Sci., 18, 249-276, 1980.

[6] ARGYRIS J.H. ; FRIED I. ; SCHARPF D.W. : The TUBA Family of Plate Elements for the Matrix Displacement Method, Aero. J. Royal Aeronaut. So., 72, pp. 701-709, 1968.

[7] GANEV H.G. ; DIMITROV Tch.T. : Calculation of Arch Dams as a Shell Using IBM-370 Computer and Curved Finite Elements, in "Theory of Shells", W.T. KOITER and G.K. MIKHAILOV, Eds., North-Holland Publishing Co, Amsterdam, pp. 691-696, 1980.

[8] BERNADOU M. ; GEORGE P.L. ; LAUG P. ; VIDRASCU M. : "MODULEF : A Library of Computer Procedures for Finite Element Analysis", in Structural Analysis Systems, Ed. by A. Niku-Lari, Vol. 3, Pergamon Press, Oxford, pp. 155-173, 1986.

[9] CONNOR J.J. ; BREBBIA C. : Stiffness Matrix for Shallow Rectangular Shell Elements, Proc. ASCE, Vol. **93**, N° EM5, pp. 43-65, Oct. 1967.

[10] BATOZ J.L. : Analyse Non Linéaire des Coques Minces Elastiques de Formes Arbitraires par Eléments Triangulaires Courbes, Doctoral Thesis, Dept. of Civil Engng., Université Laval, Québec, 1977.

[11] ADINA System Verification Manual, Report AE 83-5, June 1983, Adina Engineering, Inc., 71 Elton Avenue, Watertown, Mass. 02172, U.S.A.

[12] BERNADOU M. ; HASSIM A. : Approximation de Problèmes Généraux de Coques dans un Espace d'Eléments Finis d'Argyris-Ganev, Rapport de Recherche INRIA (To appear).

[13] MACSYMA, Reference Manual, The Mathlab Group Laboratory for Computer Science, M.I.T., Version 10, Second Printing, Vol. 1, December 1983.

[14] BERNADOU M. : Some finite element approximations of thin shell problems, in "Finite Element Methods for Plate and Shell Structures", Edited by T.J.R. Hughes and E. Hinton, Pineridge Press, 1986.

[15] CLOUGH R.W. ; JOHNSON C.P. : "A finite element approximation for the analysis of thin shells", Internat. J. Solids and Structures, 4, 43-60, 1968.

[16] BERNADOU M. ; DUCATEL Y. : "Approximation of general arch problems by straight beam elements, Numer. Math., 40, 1-29, 1982.

[17] JOHNSON C. : "On finite element methods for curved shells using flat elements" in : Numerische Behandlung von Differentialgleichungen, pp.147-154, International Series of Numerical Mathematics, Vol. 27, Basel, Stuttgart : Birkhäuser Verlag, 1975.

[18] CIARLET P.G. : "The Finite Element Method for Elliptic Problems", North Holland, Amsterdam, 1978.

[19] ZIENKIEWICZ O.C. : "The Finite Element Method in Engineering Science", Third Edition, Mc Graw-Hill, 1977.

[20] IRONS B. ; AHMAD S. : "Techniques of Finite Elements", John Wiley and Sons, 1980.

[21] BERNADOU M. ; BOUZIANE OUARITINI A. ; THOMAS J.M. : "A mixed finite element method for thin shell problems" (to appear).

[22] BOUZIANE OUARITINI A. : "Méthodes d'éléments finis mixtes pour des problèmes de coques minces", Thèse 3ème cycle, Univ. de Pau et des Pays de l'Adour, Oct. 1984.

[23] DESTUYNDER P. ; LUTOBORSKI A. : "A penalty duality method for the BUDIANSKY SANDERS shell model", Comp. Meth. Appl. Mech. Engng., 35, 127-151, 1982.

[24] MIYOSHI T. : "Finite element method for mixed type and its convergence in linear shell problems" Kumamoto J. Sci. (Math.), 10, 35-38, 1973.

[25] WEMPNER G.R. ; ODEN J.T. ; KROSS D. : "Finite element analysis of thin shells", J. Engng. Mech. Div., ASCE, 94, 1273-1294, 1971.

[26] BATOZ J.L. ; BENTAHAR M. ; DHATT G.S. : "Les éléments DKT et DKQ et l'analyse des plaques et coques minces", in Conference : Tendances Actuelles en Calcul des Structures, Sophia-Antipolis, 1-3 février 1982, 41 pages.

[27] BERNADOU M. ; MATO EIROA P. : "Approximation de problèmes linéaires de coques minces par une méthodes d'éléments finis de type D.K.T., Rapport de Recherche. INRIA, (to appear).

[28] KIKUCHI F. : "On the discrete Kirchhoff approach for plate bending problems", Theoretical and Applied Mechanics, 31, 3-21, 1981.

COMPUTER AIDED DESIGN OF SHELL STRUCTURES

H. Sardar Amin Saleh   Katholic University Leuven

INTRODUCTION

How far can we go in the automatic design of shell structures and its numerical processing by computer? It is our intension to design any kind of shell structure in civil engineering with the graphical and numerical processing power of the computer. By design we mean to allow the user to define any shape of shell, like cylinders, one-leaf hyperboloid, ellipsoids, hyperbolic paraboloids, conoids, cones, torusses and helicoids, and to optimise it with regard to its strenght and dimensions (Fig.1-8). In other words to analyse it with a finite element method and to evaluate the result (graphically) in order to adjust the original design in a convergent way.

DESCRIPTION

It is our aim to make a proto-type package, and for that purpose we developed already a great part. The preprocessing part has been developed for hyperbolic paraboloid shell structures. This choice is arbitrary, we could as well have chosen a cylindrical or a spherical or any other type of shell.
Starting our program, we can choose any type of shell (Fig.1-8) and for a choosen type (e.g. Hypar) we get a menu on the display to have an idea of the possibility of the program, consisting of five parts as are illustrated in the data flow diagram (Fig.A).

1. Definition and visualisation of hyperbolic paraboloid.
2. Finite element model + area/volume calculation.
3. Preparation of SAP-IV input file.
4. SAP-IV analyse.
5. Postprocessing + extraction of feed-back information.

At any time we have the possibility to:
- start a new project.
- repeat any part.
- continue with next part or stop.

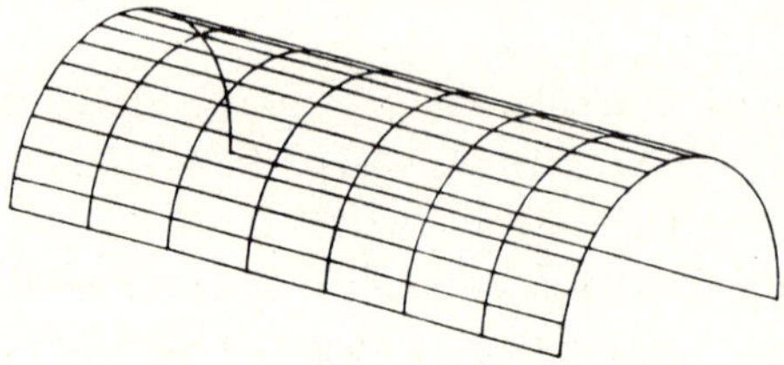

Fig. 1 Cylinder

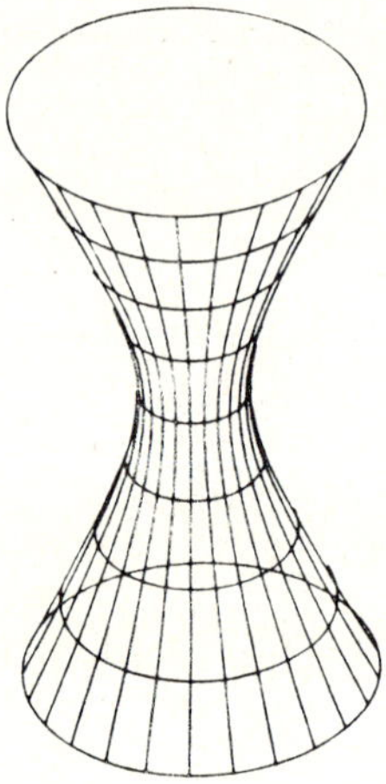

Fig. 2
One Leaf Hyperboloid

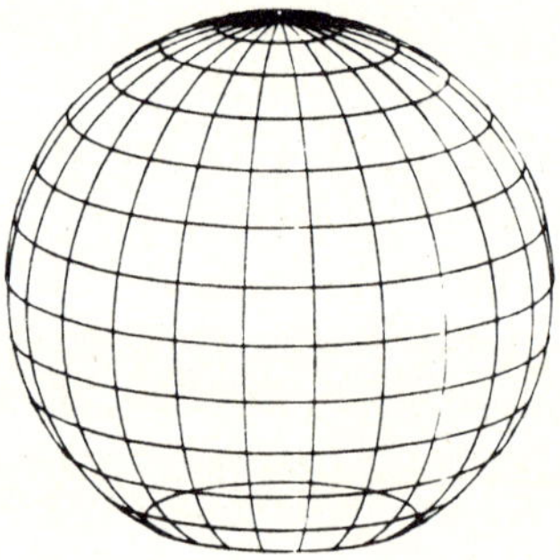

Fig. 3
Ellipsoid

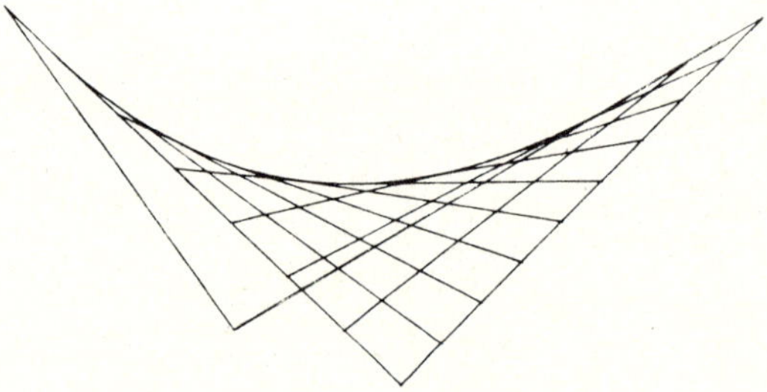

Fig. 4 Hyperbolic Paraboloid

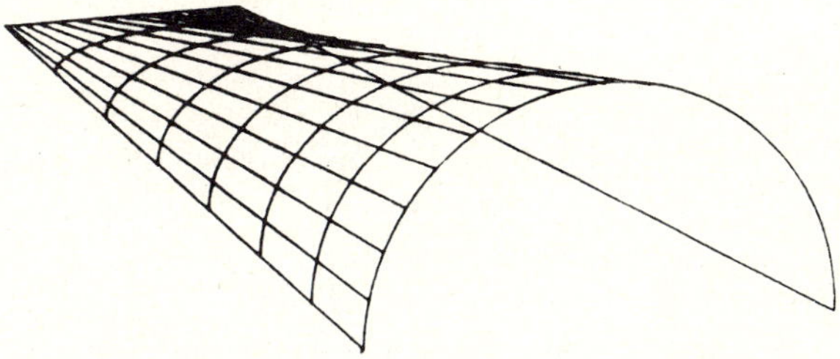

Fig. 5 Conoid

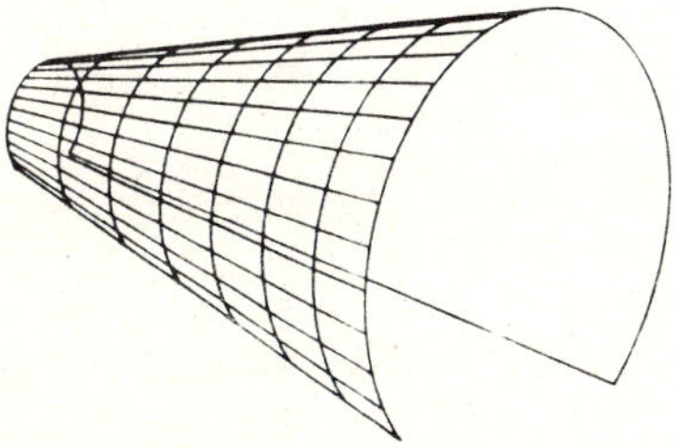

Fig. 6 Cone

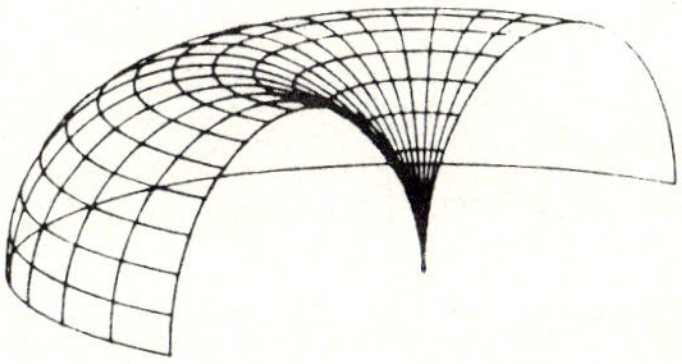

Fig. 7 Torus

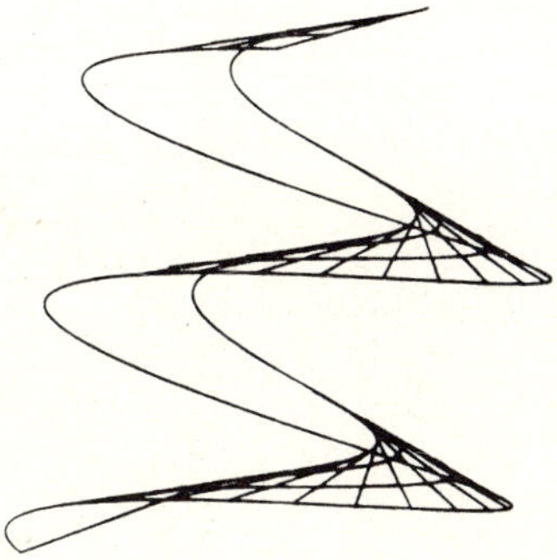

Fig. 8 Helicoid

DATA FLOW DIAGRAM

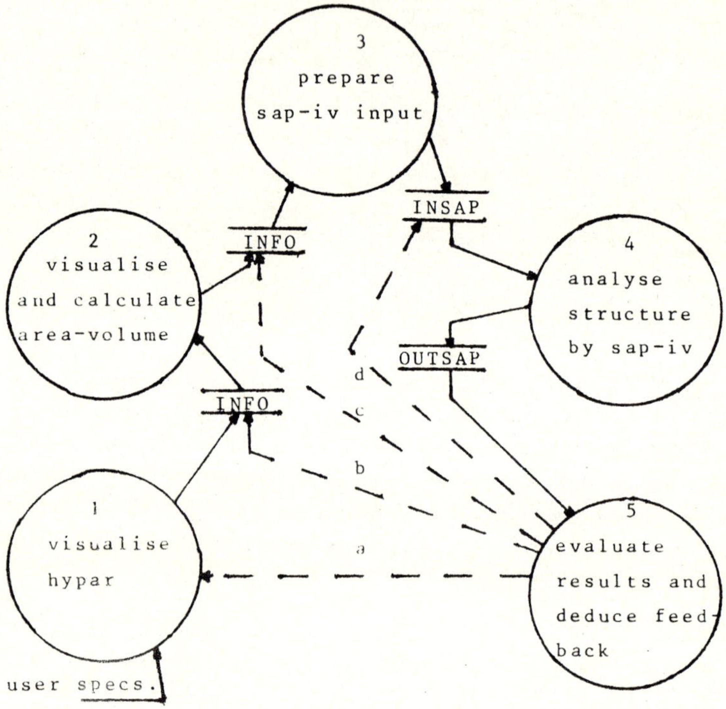

Fig. A

a........normal feed back way
b,c,d....internal ways

The preprocessing is divided in (the first) three parts, mainly for economic use of computer-resources and also for reason of splitting up the problem in to subproblems, afterwords we can always melt it back to one single part. Let's take a closer look to the different parts:

1. Definition and visualisation of hyperbolic paraboloid

Any hypar can be defined by four corner points, in three dimensions, (a,b,c), (k,l,m), (p,q,r), (u,v,w) see Fig.C. The user has to specify also the number of lines (=number of intervals + 1) in each of the two directions "s" and "t" to describe the hypar (Fig.B).
To give an idea about the used method we have to refere to LAGRANGE Interpolation Formula. The formula is the transfinite bilinear interpolation of wich an exact description can be found in "surface for computer aided design of space forms" from S.A. Coons. We picted out some qualitative considerations in order to have some feeling about the geometric model.

Consider a surface F in three-dimensional space F represents a relation from the (s,t)-surface in to 3-dimensional space F (s,t). Restricting the (s,t) surface to a square (0,0), (1,0), (0,1) and (1,1), relating with a finite element that can serve to assemble a shell surface (Fig.B).

We are looking for a formula that allows us to calculate points of the surface based on the knowledge of sides of the square, i.e. F(s,0), F(s,1), F(0,t) and F(1,t). Required formula is:

$$R(s,t) = R1(s,t) + R2(s,t) - R3(s,t) \quad \ldots\ldots (1)$$

With

$$R1(s,t) = (1-s)F(0,t) + s.F(1,t) \quad \ldots\ldots (2)$$

$$R2(s,t) = (1-t)F(s,0) + t.F(s,1) \quad \ldots\ldots (3)$$

$$R3(s,t) = (1-s)(1-t)F(0,0) + (1-s)t.F(0,1) + s(1-t)F(1,0) + s.t.F(1,1) \quad \ldots\ldots (4)$$

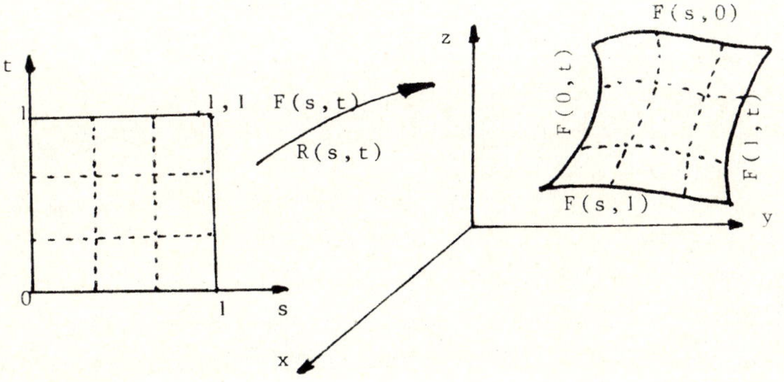

Fig. B

So, formula 2 is merely a linear interpolation formula between the points F(0,t) and F(1,t). These points are infact two arbitrary points of the sides F(0,t) and F(1,t) so that the formula is a transfinite interpolation formula between the given sides F(0,t) and F(1,t).
It is obvious in analogy, that formula 3 is an interpolation formula between the given sides F(s,0) and F(s,1).
Formula 4 is a bilinear interpolation formula between the four corner points of our surface, i.e. F(0,0), F(1,0), F(0,1) and F(1,1) required interpolation formula is given by combination of R1, R2 and R3 as stated in formula 1. One can easily check that F and R are identical at the sides. However this method of describing surfaces is only exact for ruled surfaces, so far instance for the ellipsoids we use parameter equations instead.

As the visualisation is meant to provide the user a good view, the shell is displayed in axometric projection (4/5/6) with hidden line removal (boundary lines always drawn) based on a simple algorithm using the scalar product of the view vector and the normal vector in a point on the surface. Infact the same algorithm is used to draw the edges-lines where the visible part becomes invisible to avoid dead-end lines. The result is shown in Fig.C.

In order to have a better view on the shell, rotation can be performed by any angle around any axis in 3-D. Of course at this stage it is quite easy to adjust any coordinate of the four corner points or to change the mesh-density. Finally we store all the information necessary for the following parts in a file named INFO.

In preprocessing part we have some checks based on ideas of Artificial Intelligence (AI) to perform a good predesign. For instance when we give unnormal value of thickness for a certain span for our shell, the AI program (depending on knowledge data base) does not accept the value and it propose a reasonable value for the thickness. At the end of this paper we give more detail about AI in our program.

2. Finite element model + Area/Volume calculation

Once a reasonable concept is set up, we can prepare for finite element modeling. Shells with ruled surfaces can be mathematically modelled with a trans finite linear interpolation method which is also used to devide the structure in finite shell elements.
We redraw our design ed shell with complete (no hidden line removal) element modeling. This model can be completely visualised together with the projection of the hypar on a two dimensional plane to indicate the finite element generation with or without the automatic node and/or element numbering (Fig.D).
To have an idea about the area covered by the shell, the surface area of the shell and the volume between zero-level and the surface area we have the options to calculate any combination of these items. As all nodal data and shell-element data is generated in this part and visually checked, the data check option of the F.E. program (SAP-IV) is made superfluous regarding geometry.

3. Preparation of SAP-IV input

Once part two has been finished we can write the data input file, named INSAP, for SAP-IV analysis interactiveley, with almost all the possibilities of SAP-IV features. Geometrical information is retrieved automatically while the user still has to provide material characteristics and load cases, although most of them are default for stell and concrete.
As a special case we have the possibility of calculating any plate by this program.

4. SAP-IV analysis

When we have executed all three parts we are able to submit the structure to SAP-IV, feeding it with the data file INSAP, for analysing and evaluating the results.

5. Postprocessing + Extraction of feed-back information

The output of SAP-IV analysis is directed to different files holding information about displacements of the nodes, the moments and stresses in the elements. As far as displacements are concerned we are able to produce graphs of the deformed structure, eventually superposed to the original structure. Such a case is shown in Fig.E, where a hypar has been submitted to a vertical distributed load. Of cource the deformations can be "overscaled" to allow for a better view. Also we visualise bending moment components, principle and membrane stresses in contour-maps.

Our intention is to integrate design rules and norms in a knowledge data base in order to deduce some information regarding adjustment of our design. The program will indicate to the user the dangerous parts in the design but is left to the user to adjust his consept in a sensible way. The implementation of Artificial Intelligence to deduce some possible actions, based on a knowledge data base, is our aim to help the user in selecting the possibile changes, by compairing values with build-in standards or experience rules, and to have an easy feed-back for a good design. For this part a test program using the very high level programming language PROLOG is in progress. The AI-interface works on a VAX 750/Unix 4.2BSD. The benefit from AI in this project is the modularity of the knowledge and the separation from its use so that it can be adapted very easily. The design rules and norms are given to the expert system controller cricket. It asks the user for specific data, it searches for possible solutions by using the knowledge base, and it can give explanations in a kind of natural language (for a good solution).
A problem is caused by the fact the presently prepared preprocessor is written in FORTRAN and the postprocessing in PROLOG. So we have some difficulties to make a link between both parts in two directions. The PROLOG should use numerical results from the FORTRAN program and the results of the PROLOG program have to be used by the FORTRAN program to improve the design. There is a possible solution by using the FORTRAN language for AI if we can make a good knowledge database. Also we can create a file in FORTRAN containing the necessary information for the expert system and use this file for interactive processing of AI with PROLOG language.

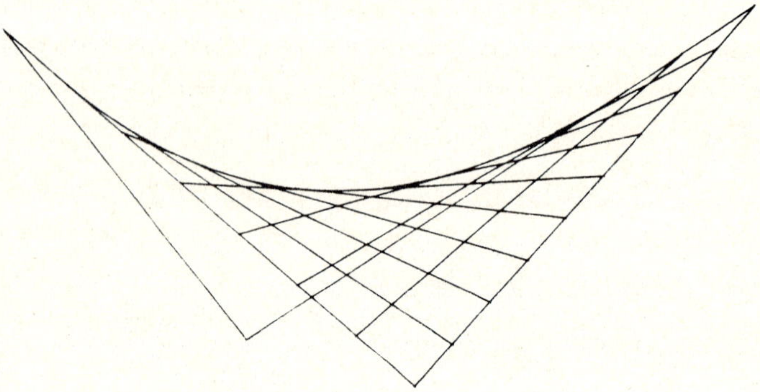

Fig. C

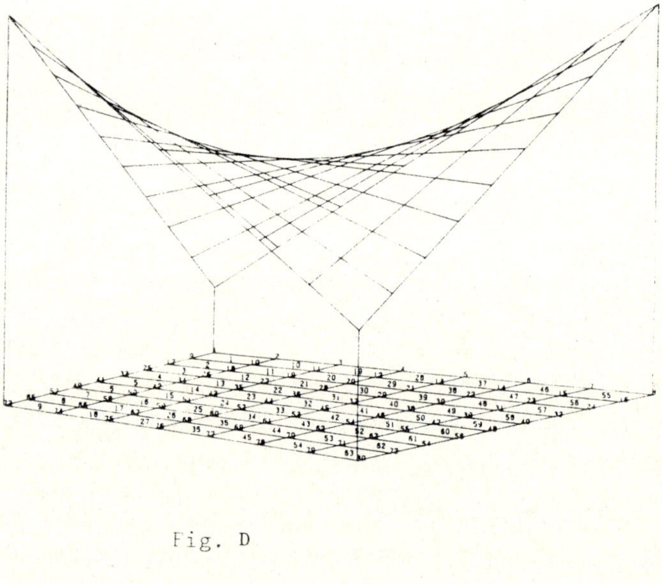

Fig. D

However this second method is not so elegant because the user has to change from one computer to the other.

ACKNOWLEDGEMENT

The outher is most greatful to Prof. Dr.ir. M. VAN LAETHEM and Prof.Dr.ir.Arch. H. H. NEUCKERMANS for their valuable guidance in my work. My special great thanks are also due to Dr.ir. E. BACKX for his continuous help in computer. I express my thanks to Dr.ir. BRUYNOOGHE and ir. KREKELS for their help in Artificial Intelligence.

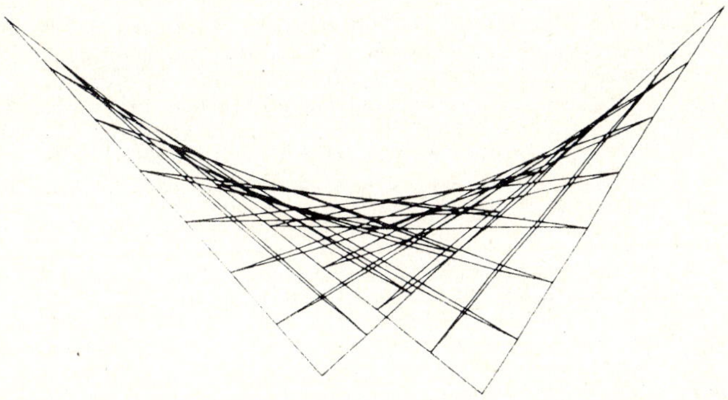

Fig. E

REFERENCES

1. E. BACKX; "Projectie van 3-D naar 2-D", Studiedag: Grafische Werken met computer, v.v.T.J. 15/1/1982.

2. H. CALLENS, P. HANDEKIJN, H. LECLUSE, L. MISSIAEN; "Projectwerk Schaalkonstrukties"; Begeleider E. BACKX, maart 1983, K.U.L.

3. J. ENCARNACAO, E.G.SCHLECHTENDAHL: "Computer Aided Design", Fundamentals and system Architectures, 1983.

4. R. KOWALSKI, "Logic For Problem Solving", 1979

5. W.F. CLOCKSIN, C.S. MELLISH: "Programming in Prolog", c.1981.

6. W.M. NEWMAN, R.F. SPROULL: "Principles of Interactive Computer Graphics", second edition c.1979.

DIRECT INTEGRATION OF ELASTIC THIN SHALLOW SHELL'S GOVERNING EQUATIONS AND LOWER ORDER METHOD

Fan Jashen
Associate Professor
Yunnan Institute of Technology
Kunming, Yunnan Province, China

Abstract

In this paper, we deduce the governing equations of an elastic thin shallow shell to be a Poisson equation of order two and a two fold Helmholtz equation of order four in partial differential equations, then we obtain the solutions of them directly from the corresponding partial differential equations' solutions. This method is very convenient in use for the case of round bottom shells, as for rectangular bottom ones, we introduce another simple approach, i.e. to lower the orders of governing equation in complex variable form. Although we discuss only spherical shell, yet it is the 1st asymptotic solutions of any other thin shallow shells with positive constant curvature such as elpar shell, elliptic shell, etc.

1. Solutions of elastic thin spherical shallow shell

The governing equations of an elastic thin spherical shallow shell are [1] :

$$D\nabla^2\nabla^2 W + \frac{1}{R}\nabla^2\phi = q \qquad (1)$$

$$\frac{1}{Eh}\nabla^2\nabla^2\phi - \frac{1}{R}\nabla^2 W = 0 \qquad (2)$$

$$\left(\nabla^2 = \frac{\partial^2}{\partial x^2} + \frac{\partial^2}{\partial y^2} = \frac{\partial^2}{\partial r^2} + \frac{1}{r}\frac{\partial}{\partial r} + \frac{1}{r^2}\frac{\partial^2}{\partial\theta^2}\right)$$

Here in (1) and (2), D is the flexural rigidity of the shell, R is its radius, W is its deflection, q is the normal load intensity exerted on the shell, E is the Young's modulus of it, h is its thickness, ϕ is the stress function of the shell's middle surface, it is denoted as:

$$N_x = \frac{\partial^2\phi}{\partial y^2}, \quad N_y = \frac{\partial^2\phi}{\partial x^2}, \quad N_{xy} = -\frac{\partial^2\phi}{\partial x\partial y} \qquad (3)$$

The N_x, N_y and N_{xy} correspond to the membrane normal stress in x,y directions and the membrane shear stress respectively.

It is mentioned in [2] that the solutions of a elastic thin spherical shallow shell's governing equations may be used as the 1st asymptotic solutions of any other shells with positive constant Gaussian curvature such as elpar, elliptic, ... shells. Therefore, our work possesses more general meaning than that for spherical one.

From eq. (2), we have:

$$\nabla^2\left(\nabla^2\phi - \frac{Eh}{R}W\right) = 0$$

$$\therefore \nabla^2\phi = \frac{Eh}{R}W + H \qquad (4)$$

Now in (4), H is a certain harmonic function. Substitute (4) in (1), we have:

$$\nabla^2\nabla^2 W + \frac{Eh}{DR^2} W = \frac{1}{D}(q-H)$$

The above equation may be divided as:

$$[\nabla^2 + (\sqrt{i}\lambda)^2][\nabla^2 + (\sqrt{-i}\lambda)^2] W = \frac{1}{D}(q-H), \quad (\lambda^4 = \frac{Eh}{DR^2}) \qquad (5)$$

Let:

$$\nabla^2 W + (\sqrt{i}\lambda)^2 W = F \qquad (6)$$

Then (5) is deduced into:

$$\nabla^2 F + (\sqrt{-i}\lambda)^2 F = \frac{1}{D}(q-H) \qquad (7)$$

Now (6) and (7) are all of Helmholtz equation in partial differential equations, it is derived from the theory of plane wave propagation, its solution is known [3]. Apply the solution twice, we obtain the following expression:

$$W(x,y) = \frac{1}{16D}\iint_S H_0^{(2)}[\sqrt{i}\lambda\sqrt{(x-\xi)^2+(y-\eta)^2}]\Big\{\iint_S H_0^{(2)}[\sqrt{-i}\lambda\sqrt{(\xi-\alpha)^2+(\eta-\beta)^2}][q(\alpha,\beta)-H(\alpha,\beta)]d\alpha d\beta\Big\}d\xi d\eta, \quad (8)$$

$$W(r,\theta) = \frac{1}{16D}\iint_S H_0^{(2)}[\sqrt{i}\lambda\sqrt{r^2+e^2-2re\cos(\theta-\psi)}]\Big\{\iint_S H_0^{(2)}[\sqrt{-i}\lambda\sqrt{e^2+f^2-2ef\cos(\psi-\Omega)}][q(f,\Omega)-H(f,\Omega)]f\,df\,d\Omega\Big\}e\,de\,d\psi, \quad (8)'$$

Once W is solved, the solution of ϕ is readily obtained as a Poisson equation [4]:

$$\phi(x,y) = -\frac{1}{2\pi}\iint_S [\ln\frac{1}{\sqrt{(x-\xi)^2+(y-\eta)^2}}][\frac{Eh}{R}W(\xi,\eta)+H(\xi,\eta)]d\xi d\eta + H' \qquad (9)$$

$$\phi(r,\theta) = \sum_{n=0}^{\infty} r^n(a_n\cos n\theta + b_n\sin n\theta) - \frac{1}{2\pi}\iint_S [\ln\frac{1}{\sqrt{r^2+e^2-2re\cos(\theta-\psi)}}][\frac{Eh}{R}W(e,\psi)+H(e,\psi)]e\,de\,d\psi \qquad (9)'$$

Now in (9), H' is another harmonic function differing from H. The constants in H or H' and a_n or b_n in (8) and (8)' or (9) and (9)' are determined by the boundary conditions of W and ϕ.

But in carrying out the exact and analytic integrals of this four formulae, it is very troublesome as the defined region is rectangular, thus another method suits it will be introduced later. Now we give an example as S is a circle, i.e. the application of (8)' and (9)'. The following formulae used here are adopted from [5]:

$$-\frac{1}{2}\ln[r^2+e^2-2re\cos(\theta-\psi)] = -\ln e + \sum_{n=0}^{\infty}\frac{1}{n+1}(\frac{r}{e})^{n+1}\cos(n+1)(\theta-\psi) \qquad (10)$$

$$H_0^{(2)}[\sqrt{i}\lambda\sqrt{r^2+e^2-2re\cos(\theta-\psi)}] = J_0(\sqrt{i}\lambda e)H_0^{(2)}(\sqrt{i}\lambda r) + 2\sum_{n=1}^{\infty} J_n(\sqrt{i}\lambda e)\cos n(\theta-\psi) \qquad (11)$$

$$J_n(\sqrt{i}\lambda e) = ber_n(-\lambda e) - i\,bei_n(-\lambda e), \quad H_0^{(2)}(\sqrt{i}\lambda e) = ker_n(-\lambda e) - i\,kei_n(-\lambda e) \qquad (12)$$

Besides, we assume:

$$H(r,\theta) = \sum_{n=0}^{\infty} r^n(A_n\cos n\theta + B_n\sin n\theta) \qquad (13)$$

Here in (13), A_n and B_n are undetermined constants. Substitute (10)-(13) into (8)', after computation, we have:

$$W(r,\theta) = \sum_{n=0}^{\infty}[\bar{a}_n ber_n(\lambda r) + \bar{b}_n bei_n(\lambda r)]\sin n\theta, \text{ (in case of odd function of } \theta) \qquad (14)$$

$$W(r,\theta)=\sum_{n=0}^{\infty}[\bar{a}_n' ber_n(\lambda r)+\bar{b}_n' bei_n(\lambda r)]\cos n\theta,$$ (in case of even function of θ)

The real constants $\bar{a}_n$ and $\bar{b}_n$ or $\bar{a}_n'$ and $\bar{b}_n'$ can be determined by the boundary conditions of W. Now in deriving (14), we have applied the assumptions that $W(0,\theta)\neq\neq\infty$ and $\left.\frac{\partial W}{\partial r}\right|_{r=0}\neq\infty$. Any load condition may be decomposed into the aboved two cases plus together.

Substitute (13) and (14) into (9)' , we can readily obtain the solution of $\phi(r,\theta)$ by using the expressions (10) and (12), the constants in H' may also be determined by the two boundary conditions of ϕ.

The purpose of this section is to show the close connection between the theory of elasticity and partial differential equations. To note, all the governing equations of elastic problems are partial differential equations, yet scarcely were the later's solutions applied directly to the problems in theory of elasticity, hence we give a concrete example to illustrate how to fill this gap other than to solve a problem.

2. The solutions of shells with rectangular bottom

The governing equation of elastical thin shallow spherical shell in complex form is [2] :

$$\left.\begin{aligned}&\nabla^2\nabla^2\chi+i\frac{\sqrt{12(1-\nu^2)}}{Rh}\nabla^2\chi=\frac{q}{D}\\&\chi=W+i\frac{\sqrt{12(1-\nu^2)}}{Eh^2}\phi\end{aligned}\right\}\quad(15)$$

Let :

$$\alpha=\frac{\sqrt{12(1-\nu^2)}}{Eh^2}\quad,\ \bar{\lambda}^2=\frac{\sqrt{12(1-\nu^2)}}{Rh}i\qquad(16)$$

In many cases, the solution of the plate equation is known while the bottom of the shell possesses the same shape and the same boundary conditions with the plate together with the same load condition. Let the solution of the plate be W_1 , i.e. :

$$D\nabla^2\nabla^2W_1=q\ ,\quad D=\frac{Eh^3}{12(1-\nu^2)}\qquad(17)$$

Then we take the W in (15) to be:

$$W=W_1+W_2\qquad(18)$$

Substitute (18) into (15) and use the simplified notations in (16), we have:

$$\left.\begin{aligned}&\nabla^2\chi_1+\bar{\lambda}^2\chi_1=\bar{H}-\bar{\lambda}^2W_1\\&\chi_1=W_2+i\alpha\phi\end{aligned}\right\}\quad(19)$$

Here in (19), $\bar{H}$ is a harmonic function. The partial differential equation in (19) is of order two which is two orders lower than that in (15) and is much easier to solve.

Example: A rectangular bottomed spherical thin shallow shell, its bottom's edges are axb with a and b along x and y coordinates respectively. It is exerted by uniformly distributed pressure q_0 , all edges are simply supported.

The boundary conditions are :

When x=0, x=a; y=0, y=b:

$$W_1=\nabla^2W_1=\phi=\nabla^2\phi=0\qquad(20)$$

$$W_1(x,y)=\frac{16q_0}{\pi^6 D}\sum_{m=1,3,\cdots}^{\infty}\sum_{n=1,3,\cdots}^{\infty}\frac{\sin\frac{m\pi x}{a}\sin\frac{n\pi y}{b}}{mn\left(\frac{m^2}{a^2}+\frac{n^2}{b^2}\right)} \qquad (21)$$

Since the boundary value of the harmonic function $\bar{H}$ is zero, it is identically zero everywhere, thus we assume:

$$\chi_1(x,y)=\sum_{m=1,3,\cdots}^{\infty}\sum_{n=1,3,\cdots}^{\infty}(\alpha_{mn}+i\beta_{mn})\sin\frac{m\pi x}{a}\sin\frac{n\pi y}{b} \qquad (22)$$

Substitute (22) into (19) and equate the real and imaginary parts of both sides, we obtain at once:

$$\left.\begin{aligned} W_2(x,y)&=\sum_{m=1,3,\cdots}^{\infty}\sum_{n=1,3,\cdots}^{\infty}\alpha_{mn}\sin\frac{m\pi x}{a}\sin\frac{n\pi y}{b}\\ \phi(x,y)&=\sum_{m=1,3,\cdots}^{\infty}\sum_{n=1,3,\cdots}^{\infty}\frac{\beta_{mn}}{\alpha}\sin\frac{m\pi x}{a}\sin\frac{n\pi y}{b}\end{aligned}\right\} \quad (23)$$

$$\left.\begin{aligned} \alpha_{mn}&=\frac{16E\alpha q_0}{Rh\pi^{10}Dmn\left(\frac{m^2}{a^2}+\frac{n^2}{b^2}\right)^2\left[\frac{R}{E\alpha h}\left(\frac{m^2}{a^2}+\frac{n^2}{b^2}\right)+\frac{E\alpha h}{R}\right]}\\ \beta_{mn}&=\frac{16q_0}{\pi^8 Dmn\left(\frac{m^2}{a^2}+\frac{n^2}{b^2}\right)\left[\frac{R}{E\alpha h}\left(\frac{m^2}{a^2}+\frac{n^2}{b^2}\right)^2+\frac{E\alpha h}{R}\right]}\end{aligned}\right\} \quad (24)$$

Now in (23) and (24), α is defined in (16). The above results in this section are the same as the known results but obtained with less labor.

3. Conclusion

Equation (19) is a Helmholtz equation, its solution is readily obtained as:

$$\chi_1(x,y)=-\frac{i}{4}\iint_S H_0^{(2)}\left[\bar{\lambda}\sqrt{(x-\xi)^2+(y-\eta)^2}\right]\left[\bar{H}(\xi,\eta)-\bar{\lambda}^2 W_1(\xi,\eta)\right]d\xi d\eta \qquad (25)$$

Thus we see, the known plate solution lowers the order of the corresponding shell equation from 4 to 2, therefore we may conclude whatever semi-analytical or direct integration method is applied, this method of lowering order save us much labor.

References

[1]. V.Z.Vlasov, General theory of shells, Nat. The.-Tec. Publishing House, 1949, (in Russian).

[2]. Long Yu-qiu, et al, Some problems of stress concentration in shallow shells of eliptic paraboloid, Acta Mechanica Sinica, vol.8, No.2, 1965.

[3]. A.N.Chexonnov, et al, Mathematical physics equations, Chap.7, Nat. The.-Tec. Publishing House, 1953, (in Russian).

[4]. E.T.Copson, Partial differential equations, Cam. Uni. Press, 1975, 135.

[5]. A.Erdelyi, Higher transcendental functions, vol.II, Mc Graw-Hill Book Co. Inc., 1953.

[6]. Hu Hai-chang, A simplified method for calculating simple-supported rectangular-bottomed shallow spherical shells used as roofs, Acta Mechanica Sinica, vol.5, No.1, 1962.

APPLICATION OF AXISYMMETRIC THIN SHELL FINITE ELEMENTS TO THE ANALYSIS OF A RADIATING FLEXURAL SHELL SONAR TRANSDUCER

B. Hamonic, J.C. Debus, J.N. Decarpigny
Institut Supérieur d'Electronique du Nord
41, boulevard Vauban, 59046 Lille Cedex, France

D. Boucher, B. Tocquet
Groupe d'Etudes et de Recherches de
Détection Sous-Marine, Le Brusc,
83140 Six Fours les Plages, France

Introduction

In SONAR projectors, the electrical energy is converted into acoustical energy using piezoelectric transducers. Classical piezoelectric transducers are length expander vibrators that are built with a ceramic stack tightly bound between a heavy tailmass and a light tapered headmass [1]. However, the use of these transducers for low frequency applications is difficult, due to size requirements. Thus, new flexural shell transducers are currently developed. They are characterized by a large volume velocity and a drastic reduction of their resonance frequencies as soon as they are flooded, due to added mass effect [2].

To design these new transducers , a finite element modeling is required, which has to take account of the assembling of three dimensional and shell elements, the piezoelectric effect, the fluid-structure interaction and the radiation damping. These points are included in the ATILA code [3-7] that has been developed in our laboratory for the French Navy [8]. This paper describes the analysis of a test axisymmetrical structure (figure 1) and proves the ability of the finite element method to handle the previous aspects and to provide complete and accurate physical information.

The first part of this paper sums up the main equations. The second part describes the shell elements. The third part gives the results of an in-air modal analysis. Finally the fourth part presents the final results for a flooded transducer and compares them to measurements.

I. Description of the method

To model a radiating piezoelectric transducer following the method that has been retained in the ATILA code, the finite element mesh includes the structure as well as a part of the nearfield fluid domain (figure 2). The unknown fields are the displacement field $\underline{u}$ in the whole structure, the electrical potential ϕ in the ceramic material and the pressure field p in the fluid. Kinematic and dynamic continuity equations between displacement and pressure fields are enforced on the interface, due to the variational formulation [3], and an appropriate damping condition [3,4,9,10] is applied to the spherical external surface Γ that surrounds the fluid domain. Then, the whole set of equations is :

$$\begin{bmatrix} [K_{uu}] - \omega^2[M] & K_{u\phi} & -[L] \\ K_{u\phi}^T & K_{\phi\phi} & 0^T \\ -\rho^2c^2\omega^2[L]^T & 0 & [H] - \omega^2[M_1] \end{bmatrix} \begin{bmatrix} U \\ \phi_o \\ P \end{bmatrix} = \begin{bmatrix} F \\ -\dfrac{I}{j\omega} \\ \rho c^2 \Psi \end{bmatrix} \quad (1)$$

where :

- U, F, P are vectors that contain the nodal values of the displacement field, the applied force and the pressure field,
- Ψ is a vector that represents the outgoing flow through Γ,
- ϕ_o is the applied voltage, while I is the entering current intensity,
- $[K_{uu}]$ and [H] are the solid and fluid rigidity matrices,
- [M] and $[M_1]$ are the solid and fluid mass matrices,
- $K_{u\phi}$ and $K_{\phi\phi}$ result from the assembling of the piezoelectric and dielectric matrices, followed by a static condensation of the internal potential degrees of freedom,
- [L] is the interface matrix,
- 0 is a null vector,
- ρ and c are the fluid density and sound speed,
- ω is the circular frequency,
- T means transposed.

Finally, using a dipolar damping condition [4,9,10,11,12] on the external boundary Γ :

$$\Psi = -\frac{1}{\rho c}\left(\frac{1}{R} + j\frac{\omega}{c}\right)[D]\,P + \frac{1}{\rho c}\,\frac{\dfrac{1}{R} - j\dfrac{\omega}{c}}{1 + \dfrac{\omega^2R^2}{c^2}}\,[D']\,P \quad (2)$$

where R is the radius of Γ. [D] and [D'] are obtained by assembling simple surface elements from Γ. The first term is the monopolar contribution associated with a spherical wave impedance and the second term is the dipolar contribution.

II. Description of the shell element

The equations of motion that have been retained to develop the axisymmetrical thin shell element of the ATILA code are deduced from the Flügge's theory [13]. In fact, though the effect of rotary inertia and shear deformation are neglected, this theory provides accurate results. The element is a doubly curved shell element with N nodes, where N is a given parameter equal to 2, 3 or 4. Following Webster [14], the displacement components v and w, that are respectively tangential and normal to the generating line, can be written :

$$v = \sum_{1}^{M} V_m \, s^{m-1}$$

$$w = \sum_{1}^{M'} W_m \, s^{m-1} \qquad (3)$$

where s is the curvilinear coordinate along the shell. According to the fact that each node possesses three degrees of freedom, two displacements and one rotation, minimal values of M and M' are related by : M + M' = 3N. The respective values of N and 2N have been retained. Thus, though the use of internal degrees of freedom can improve the element accuracy [15,16], it has not been developed. The following test has been analysed using three nodes per element.

To connect this element with a classical three dimensional elastic element, a transition element can be used following, for example, the work by Surana [17]. However, depending upon the geometrical shape of the transition region, simple assumptions can be retained, such as a clamping condition that is associated with a vanishing rotation degree of freedom and continuity of the displacement components. This condition is assumed in the following test example.

III. In air modal analysis of the test structure

In-air modal analysis of an elastic structure is performed by deleting the second and third lines of equation (1) and assuming that $\underline{F}$ is equal to zero. In-air modal analysis of a piezoelectric structure is performed by deleting the third line of equation (1) and assuming that $\underline{F}$ and ϕ_0 are equal to zero (resonance case) or $\underline{F}$ and I are equal to zero (antiresonance case). Table 1 gives the resonance frequencies of the shell part of the transducer, as obtained with a fine three dimensional mesh (case 1) or with shell elements (case 2), and compares them to experimental results, in the case of a symmetrical displacement field. These results are very accurate and gives confidence in the corresponding models. Table 2 gives the resonance frequencies of the whole structure, as obtained with a fine three dimensional mesh (case 1), a fine mesh with shell elements (case 2) and a coarse mesh with shell elements (case 3). It demonstrates that the coarse mesh is largely sufficient and can be retained for in-water analysis. Figure 3 presents the mode shapes for the first six symmetrical modes and shows that, though all the other modes exhibit a flexural motion of the shell, the first mode implies mainly a membrane behaviour. Table 2 also gives the resonance frequencies as obtained from a mixed plane wave-finite element analysis [6]. In this case, the piezoelectric part of the structure is described following a transfer matrix approach while the remaining part of the structure is described by a finite element model. Then, only the first line of equation (1) is kept and the applied force $\underset{\sim}{F}$ is related to the applied potential ϕ_0 by a simple relation deduced from the transfer matrix computation. Results also demonstrate a nice agreement. Finally, table (2) gives also the piezoelectric coupling coefficients [1,3] and shows that nearly all the modes are electrically active, particularly the first one.

IV. In-water harmonic analysis of the test structure

In-water harmonic analysis of the transducer has been performed using the mixed method and the mesh represented on figure 2. The radius of the spherical boundary Γ is equal to 0.6m, well inside the nearfield region, but another computation has been done with a radius equal to 1.2m, to test the damping condition. The retained condition is the dipolar damping, as given by equation (2). In the propagation direction, the mesh step is always less than one quarter of the smallest wavelength. The computed physical quantities are :

- the displacement and speed fields on the interface,
- the nearfield pressure,
- the transmitting voltage response S_V (in μPa/V at 1 meter from the acoustical center),
- the directivity patterns, that give the angular variations of the pressure amplitude in a meridian plane.

The two last quantities have to be determined in the farfield region, well beyond the Γ boundary. To do this, and extrapolation algorithm has been used that is based on the multipolar expansion of the pressure field [10]. The following results are displayed :

- the theoretical variations of the speed versus frequency, at the top of the headmass and in the symmetry plane (figure 4),
- the theoretical and experimental variations of S_V versus frequency (figure 5),
- the theoretical and experimental directivity patterns for six different frequencies (figure 6).

As mentioned above, theoretical results have also been obtained for a boundary radius of 1.2m but they are identical to the preceeding results and have not been reported.

As can be seen on figures 5 and 6, a nice agreement is obtained between numerical and experimental results, apart for a frequency shift of the upper region of the S_V curve that can be related to the structure model and will be analysed later in more details. Moreover, the various maxima of the S_V curve can be related to the resonance modes of the structure. The first peak (0.95) is clearly related to the first in-air mode, with a downward frequency shift of 0.05. The two following peaks (1.99 and 2.43) are probably related to the third and fourth in-air modes with large downward shifts of the frequency (more than 0.38). Finally, the fourth and fifth peaks (2.89 and 3.18) are related to the fifth and sixth modes, with small frequency shifts, the last peak being associated, as the first peak, to a large motion of the headmass. It has also to be pointed out that the large dip in the S_V curve (3.29) is related to a null speed everywhere, as shown on figure 4, due to a new nodal plane emergence at the top of the ceramic stack.

Conclusion

This paper has presented a detailed analysis of the acoustic behavior of a new flexural shell sonar transducer, using the finite element code ATILA, and has demonstrated the large interest of simple shell elements to describe the structure. New developments are carried on to model non-axisymmetrical flexural shell structure following the same approach.

References

1. D.A. Berlincourt, D.R. Curran, H. Jaffe, "Physical Acoustics: principles and methods", Vol. 1, Ed. by W.P. Mason, Academic Press, (1976).
2. G.A. Brigham, B. Glass, J. Acoust. Soc. Am., 68, 1046, (1980).
3. J.N. Decarpigny, "Application of the finite element method to the analysis of piezoelectric transducers", Thesis, Lille University, France, (1984).
4. R. Bossut, "Finite element modeling of flooded piezoelectric ring transducers", Thesis, Valenciennes University, France, (1985).
5. P. Tierce, "Finite element modeling of the Isabelle transducer", Thesis, Valenciennes University, France (1985).
6. J.N. Decarpigny, J.C. Debus, B. Tocquet, D. Boucher, J. Acoust. Soc. Am., 78, 1499, (1985).
7. J.N. Decarpigny, J.C. Debus, P. Tierce, B. Tocquet, D. Boucher, submitted to J. Acoust. Soc. Am.
8. The ATILA code has been developed for the "Groupe d'Etudes et de Recherche de Detection Sous-Marine", under contract with the "Direction des Constructions et Armes Navales" (DCAN Toulon, France).
9. R. Bossut, J.N. Decarpigny, Communication M4, 106th ASA Meeting, San Diego, (1983).
10. R. Bossut, J.N. Decarpigny, B. Tocquet, D. Boucher, Communication X8, 111th ASA Meeting, Cleveland, (1986).
11. A. Bayliss, M. Gunzburger, E. Turkel, ICASE Report N° 80/1, NASA

Langley, (1980).
12. O.C. Zienkiewicz, P. Bettess, T.C. Chiam, C. Emson, Proc. of the Winter Annual Meeting of the ASME, AMD Vol. 46, Ed. A.J. Kalinowski, (1981).
13. W. Flügge, "Stresses in shell", 2nd Edition, Springer Verlag Ed., (1973).
14. J.J. Webster, Int. J. Mech. Sci., 9, 559, (1967).
15. R. Cook, "Concepts and applications of finite element analysis", Wiley Ed., (1974).
16. S.K. Send, P.L. Gould, Journal of the Engineering Mechanics Division, ASCE, 100, EM2, (1974).
17. K.S. Surana, Int. J. Num. Meth. Eng. 15, 809, (1980).

Table 1	**Normalized resonance frequencies of the shell part**		
Mode	Case 1	Case 2	Experimental
1	1.44	1.43	1.40
2	3.01	2.97	2.98
3	3.18	3.14	3.15
4	3.36	3.31	-
5	3.56	3.52	-
6	3.69	3.66	-

Table 2	**In air resonance frequencies of the transducer**				
Mode	Case 1	Case 2	Case 3	Mixed Method	k*
1	1.	1.01	1.01	1.01	32
2	2.98	2.94	2.95	2.95	4
3	3.15	3.11	3.12	3.11	3
4	3.22	3.20	3.20	3.20	3
5	3.37	3.33	3.33	3.31.	0

* The coupling coefficient has been computed in case 1. It is given in %.

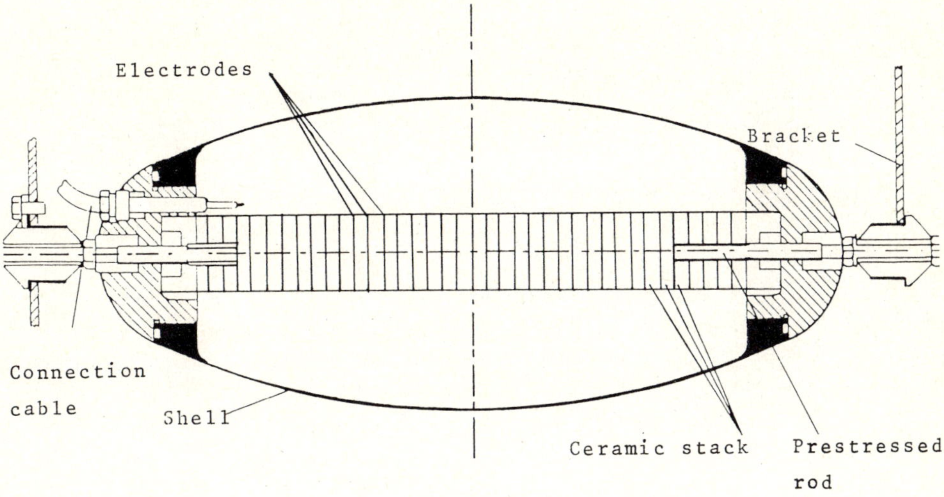

FIGURE 1

TRANSDUCER STRUCTURE

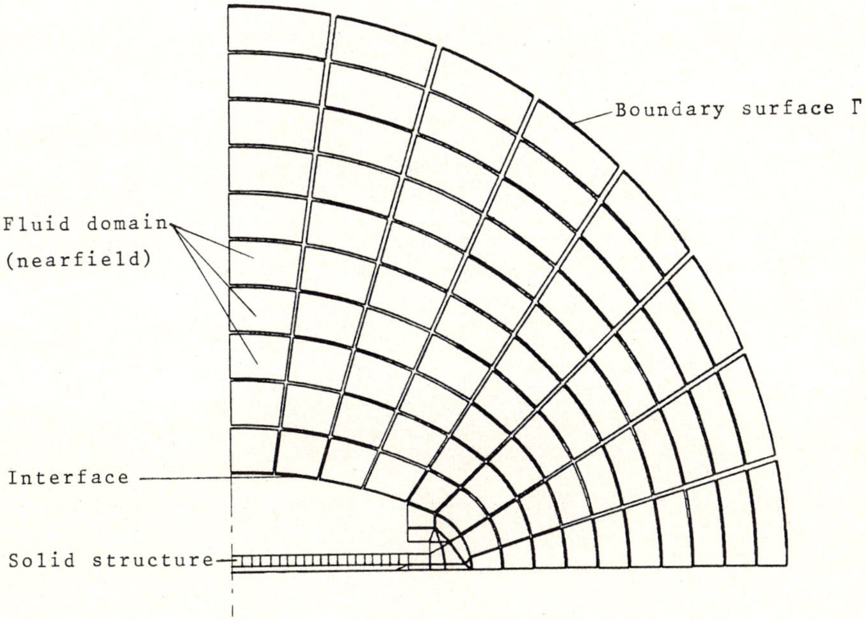

FIGURE 2

FINITE ELEMENT MESH

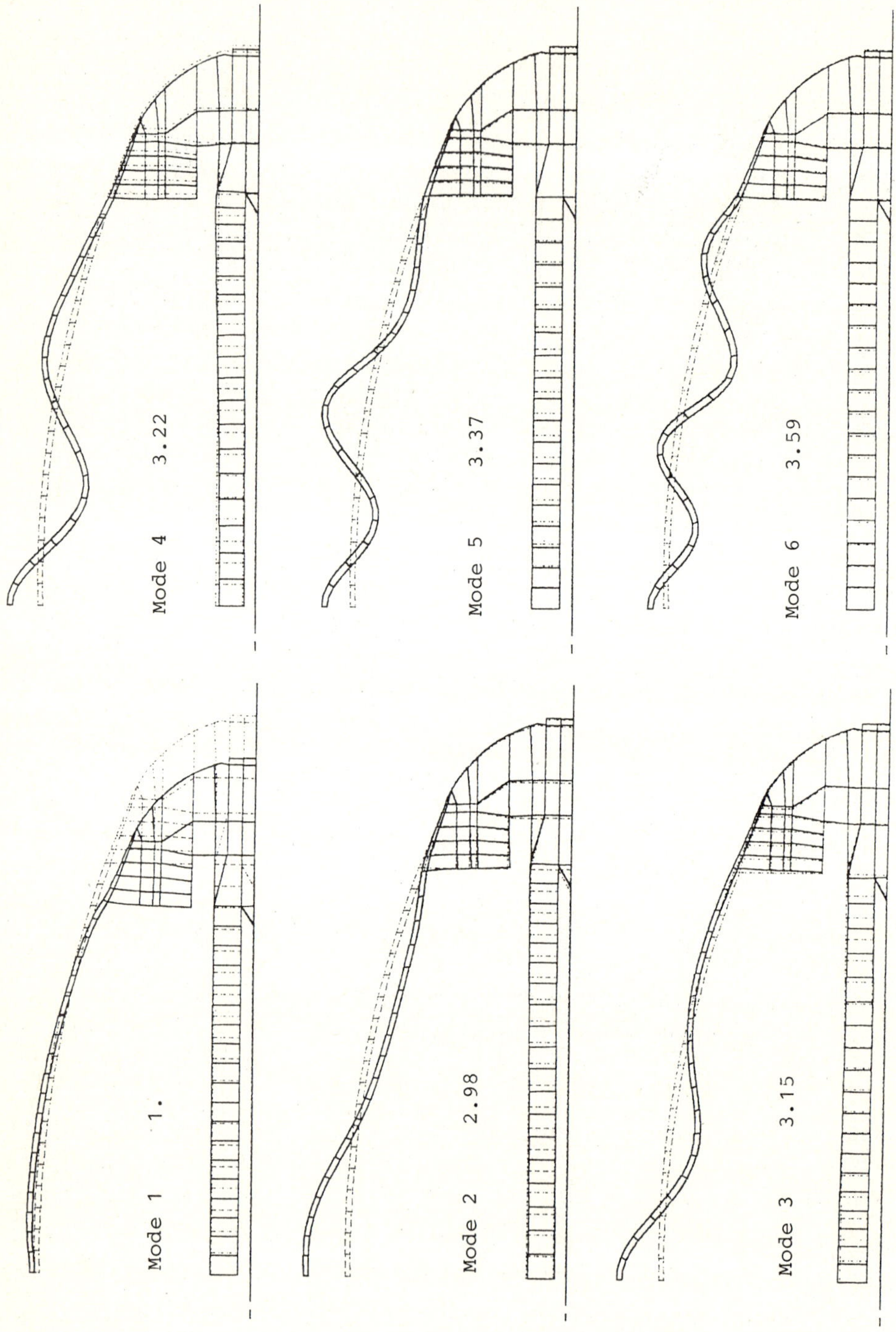

FIGURE 3 IN-AIR MODE SHAPES

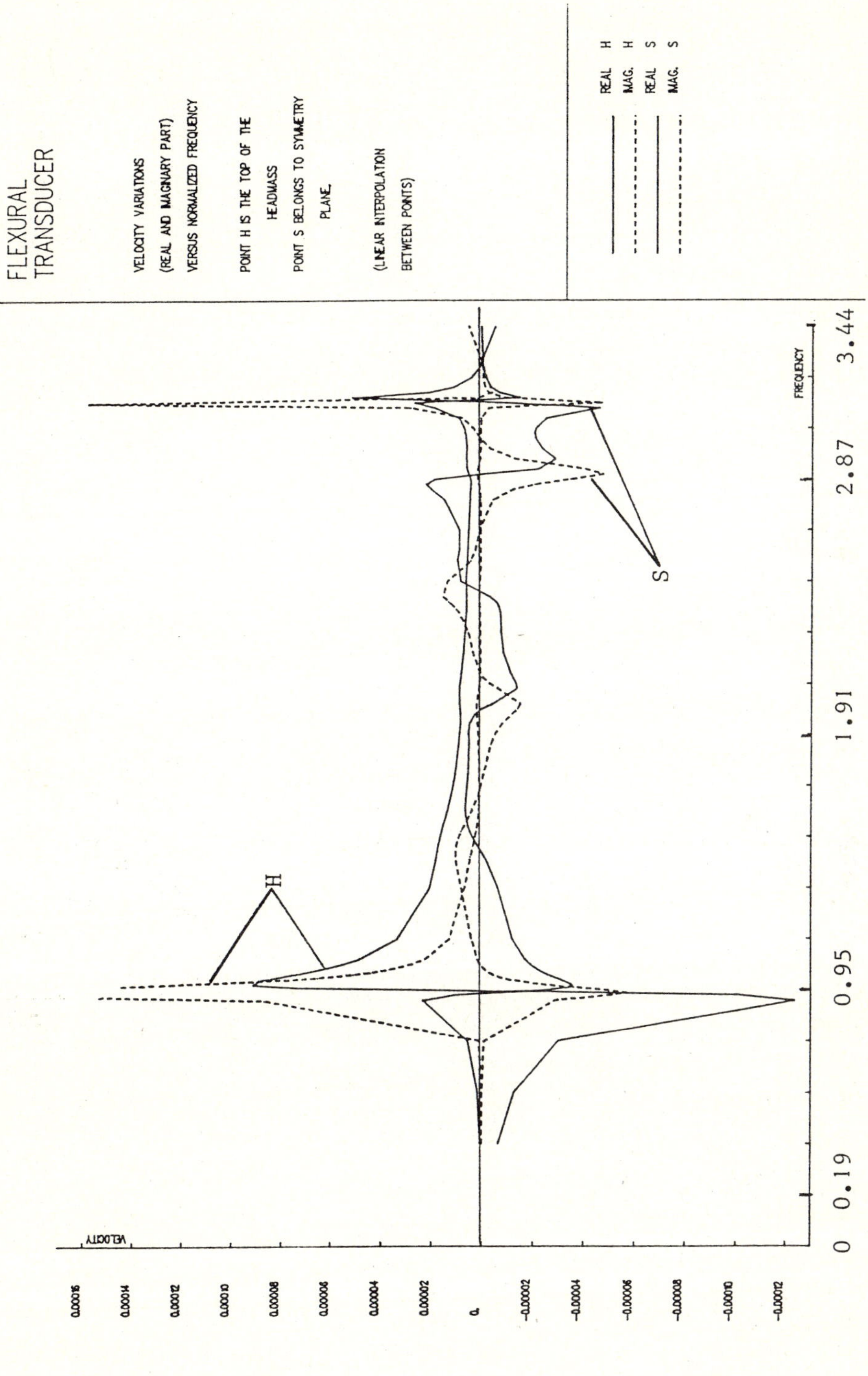
FLEXURAL
TRANSDUCER
VELOCITY VARIATIONS
(REAL AND IMAGINARY PART)
VERSUS NORMALIZED FREQUENCY
POINT H IS THE TOP OF THE
HEADMASS
POINT S BELONGS TO SYMMETRY
PLANE
(LINEAR INTERPOLATON
BETWEEN POINTS)
REAL H
IMAG. H
REAL S
IMAG. S
VELOCITY
FREQUENCY
0.00016
0.00014
0.00012
0.00010
0.00008
0.00006
0.00004
0.00002
0
-0.00002
-0.00004
-0.00006
-0.00008
-0.00010
-0.00012
0
0.19
0.95
1.91
2.87
3.44
H
S

FIGURE 4

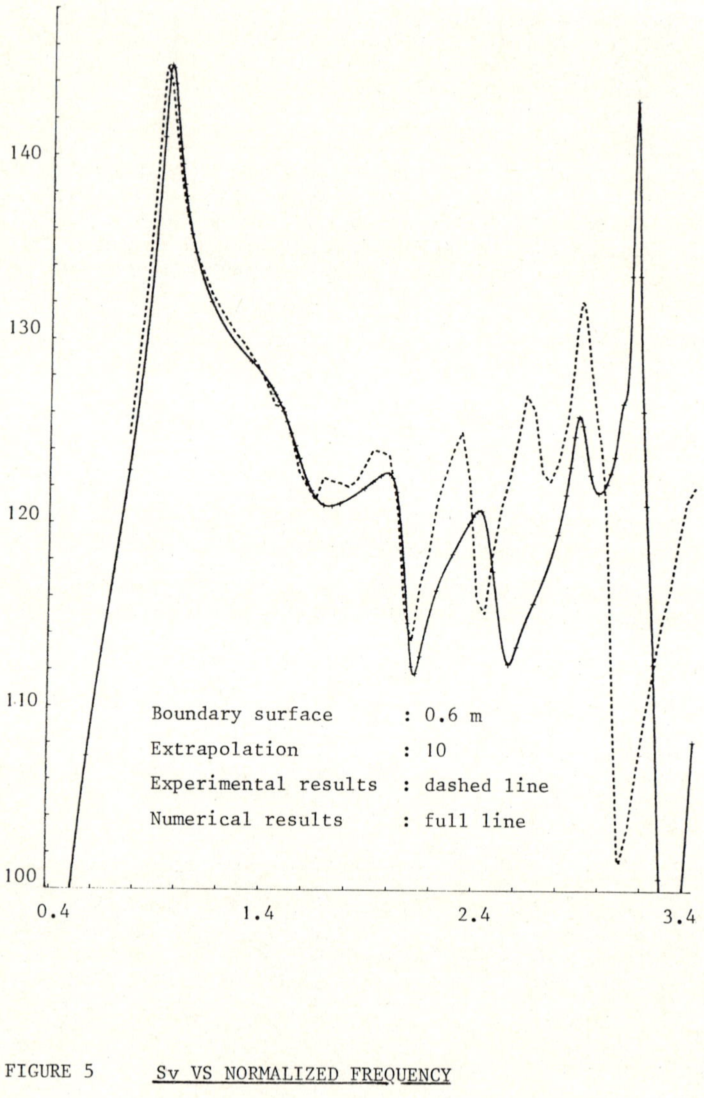

FIGURE 5 Sv VS NORMALIZED FREQUENCY

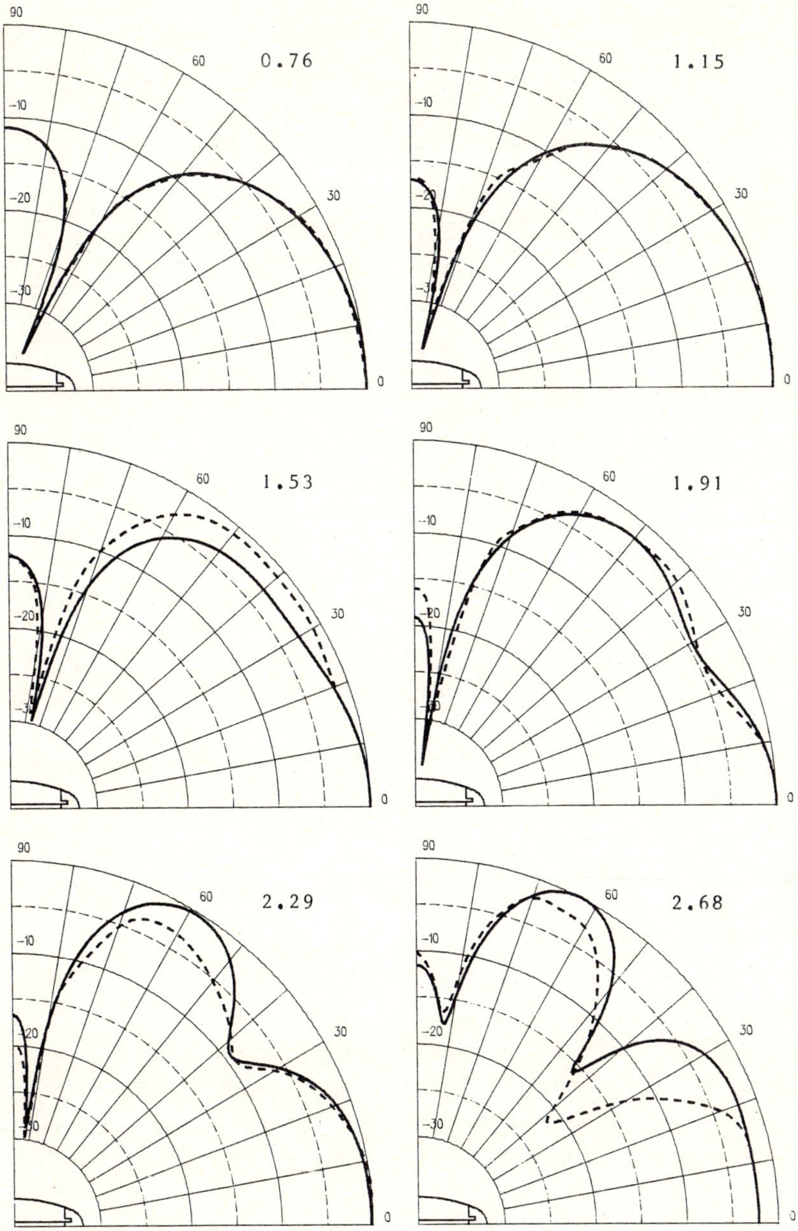

FIGURE 6

DIRECTIVITY PATTERNS

Comparison between experimental (dashed line) and numerical results (extrapolation 10. full line).

FINITE ELEMENT ANALYSIS OF CYLINDRICAL SHELLS UNDER CONCENTRATED LOADING

P. L. Gould, M. Hourani, J. S. Lin, and T. G. Harmon
Department of Civil Engineering
Washington University
St. Louis, MO 63130

K. J. Han
Department of Civil Engineering
University of Houston
Houston, TX 77004

ABSTRACT

The objective of the study was to determine the most appropriate support conditions for a reinforced concrete shell panel test specimen which will adequately reproduce internal forces and moments in a prototype cylindrical shell. The goal in the experiment is to investigate a punching shear failure in the vicinity of a concentrated load. For the analysis, four computer programs were employed, SHORE-III, two versions of RONALD, and ANSYS. The SHORE-III and RONALD (rotational) analyses considered a complete cantilevered cylinder, while the ANSYS (general) model treated a cylindrical panel free along the curved edges and constrained in various ways along the longitudinal edges. Comprehensive comparative results are provided for the linear elastic loading case, which indicate the magnitude and attenuation of concentrated loading effects on cylindrical shells and panels. Also, some selected results for the panel obtained from a high performance graphic package PATRAN/ANSYS are given.

COMPUTER PROGRAMS

This study utilized three programs for analyzing the complete cylindrical shell; (1) SHORE-III, based on rotational elements with the local loading represented by Fourier series [1]; (2) RONALD [2-4], version-II, using quadrilateral shell elements; and (3) RONALD, version-III, with quadrilateral solid elements in the vicinity of the loading. In both versions of RONALD, rotational elements are defined away from the loaded region.

For the stress analysis of the cylindrical panel, a popular commercial program ANSYS [5,6] was selected. Shell elements STIF63, STIF93 and STIF94 were employed in this study. Also, a related high performance graphics package ANSYS/PATRAN [7] was used to demonstrate advanced pre- and post-processing capabilities.

LOADING AND DISCRETIZATION

The shells shown in figure 1 were analyzed for the patch loading shown using the programs, SHORE-III and RONALD (version II and III) for the complete cylinder (figure 1a), and ANSYS for the panel (figure 1b). The cylinder represents a prototype concrete structure which is subjected to concentrated ice pressure loading of specified intensity and area, while the panel is a specimen to be tested in the laboratory. The patch loading shown was chosen to simulate the prototype ice load which might produce a punching shear failure. A desired outcome of the analysis was the determination of appropriate boundary supports for the test specimen which would adequately reproduce the internal forces and moments in the prototype structure in the vicinity of the load.

The sign convention for the stress resultants and stress couples is shown in figure 2 and the boundary conditions for the panel are indicated on figure 3. The patch loading was represented by 115 harmonics in the SHORE-III model and distributed directly on the quadrilateral elements in the RONALD and ANSYS models.

Figures 4 and 5 describe the meshs for the analyses. The mesh for SHORE-III is shown in figure 4(a). For RONALD, a refined mesh, detailed in figures 4(b) and 4(c), was defined in the region of the applied loading, while the mesh for the ANSYS analyses is described in figure 5.

RESULTS

First, the vertical displacement contours for case c of figure 3, obtained from PATRAN/ANSYS, are shown in figure 6. Following, results are compared between SHORE-III and RONALD, and also for various boundary conditions of the cylindrical panel (figure 3) using ANSYS. Corresponding results for the following quantities are presented:

1. The meridional moment resultant M_z along AB (figures 7 and 8).
2. The transverse shear force Q_z along AB (figures 9 and 10).
3. The circumferential stress resultant N_θ along CD (figures 11 and 12).
4. The circumferential moment resultant M_θ along CD (figures 13 and 14).
5. The transverse shear force Q_θ along CD (figures 15 and 16).
6. The circumferential moment resultant M_θ along AB (figures 17 and 18) and the corresponding deflected cross-section (figure 19).

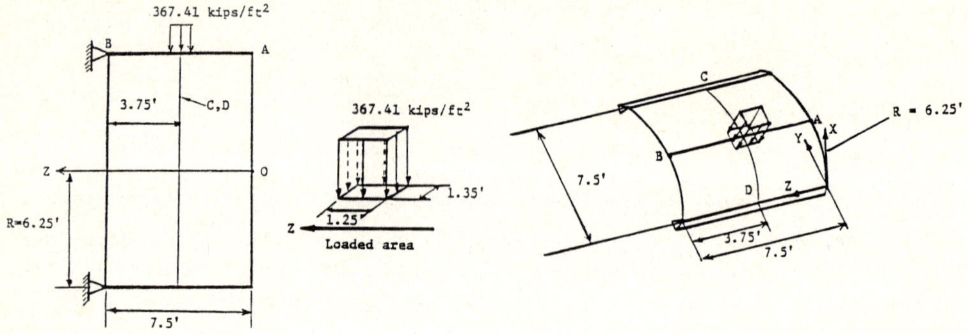

a) Complete cylindrical cantilever shell b) Cylindrical panel

Figure 1. Shells under Patch Loading

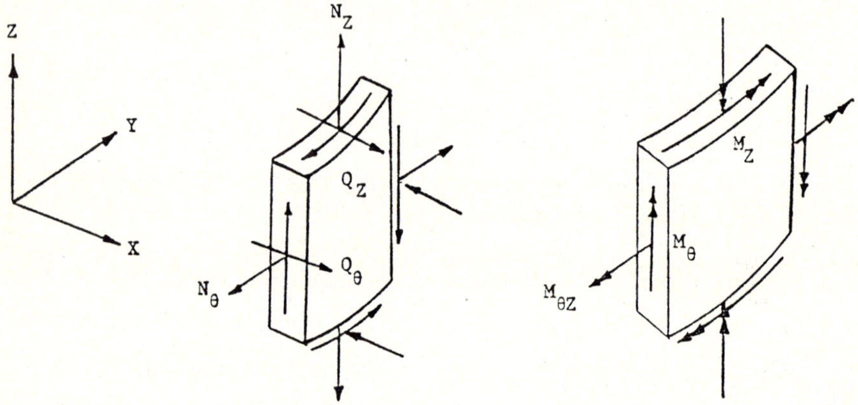

a) Stress resultants b) Stress couples

Figure 2. Sign Convention

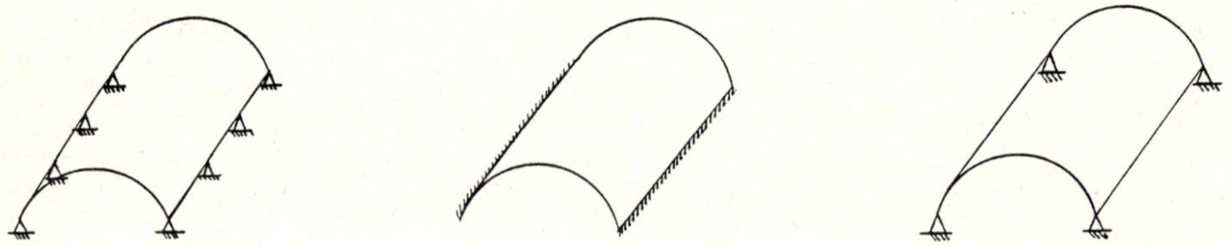

a) Simply supported b) Fixed c) Corner supported

Figure 3. Longitudinal Edge Boundary Conditions for ANSYS Models with Free Curved Edges

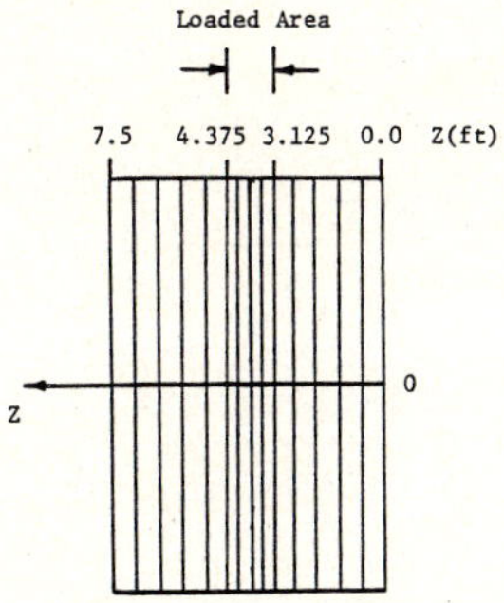

a) Mesh for SHORE-III

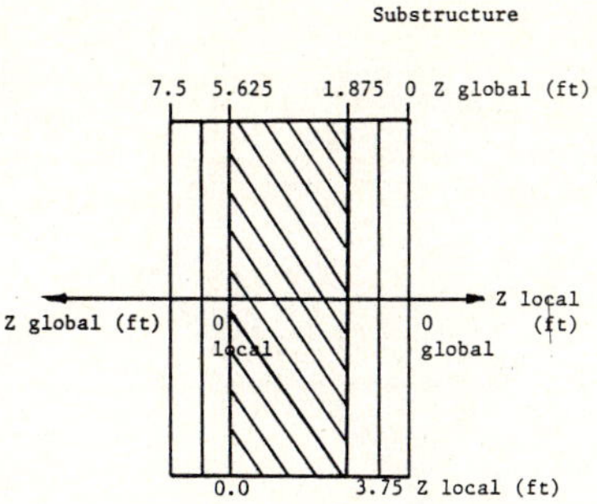

b) Mesh for RONALD

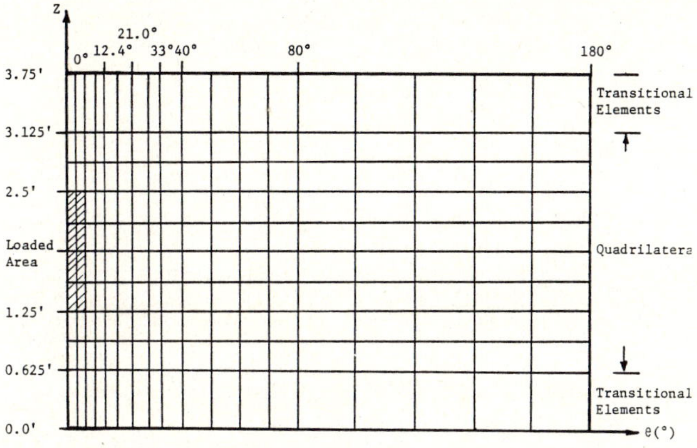

c) Mesh for substructure - 0° to 180°

Figure 4. Finite Element Mesh for SHORE-III and RONALD Analyses

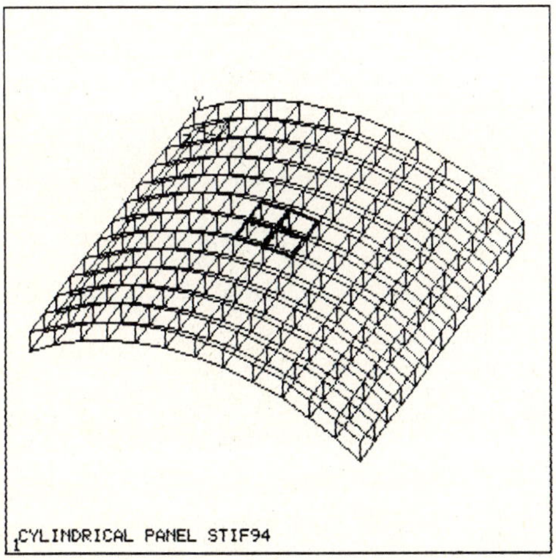

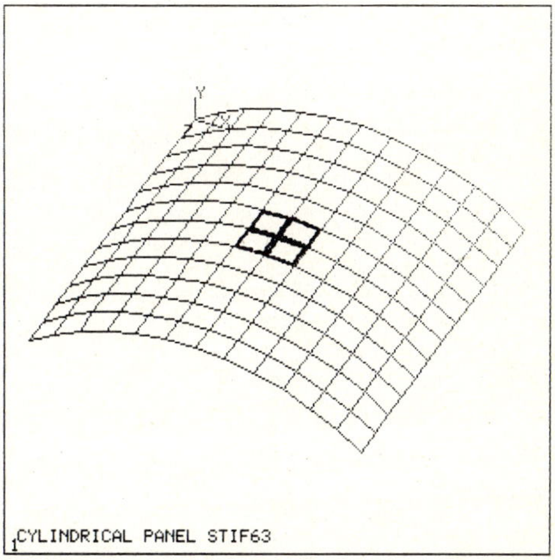

Figure 5. Finite Element Mesh for ANSYS Analysis

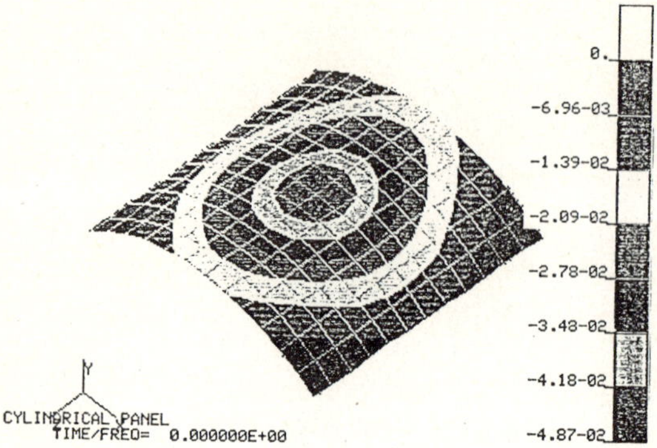

Figure 6. Vertical Displacement Contours (ft)

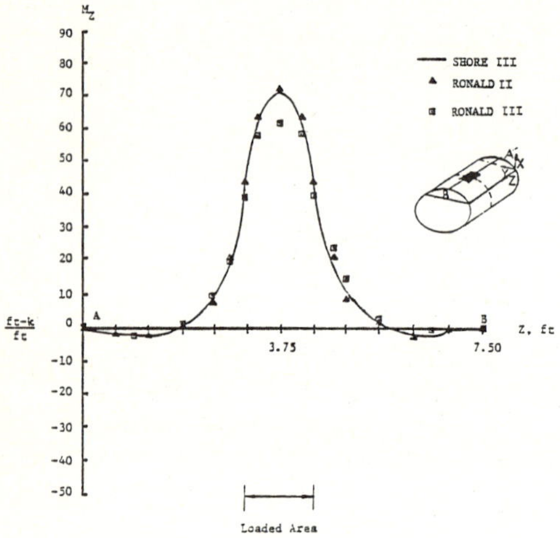

Figure 7. Moment Resultant M_Z vs. Z along AB

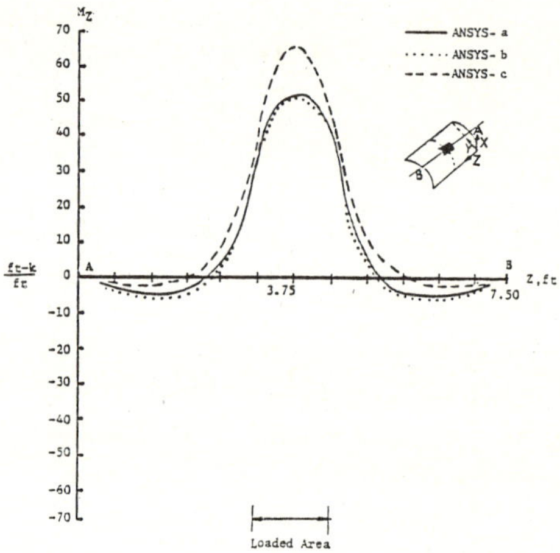

Figure 8. Moment Resultant M_Z vs. Z along AB

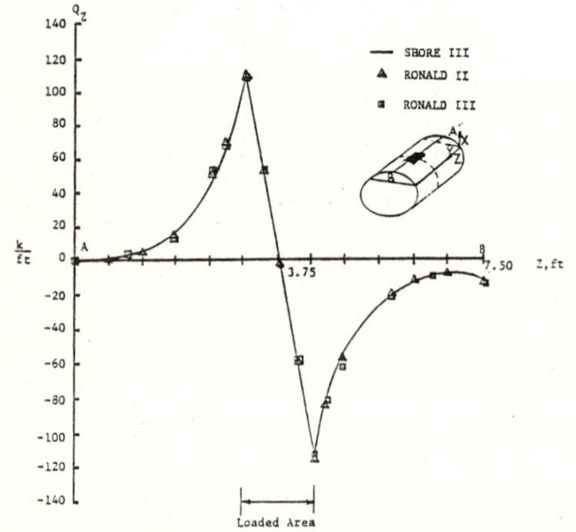

Figure 9. Shear Force Q_Z vs. Z along AB

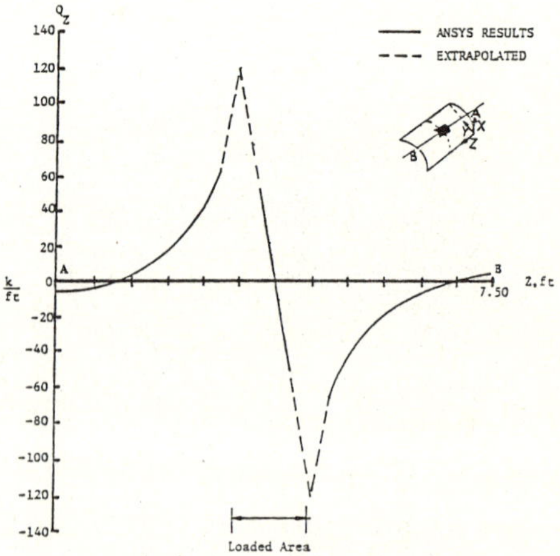

Figure 10. Shear Force Q_Z vs. Z along AB

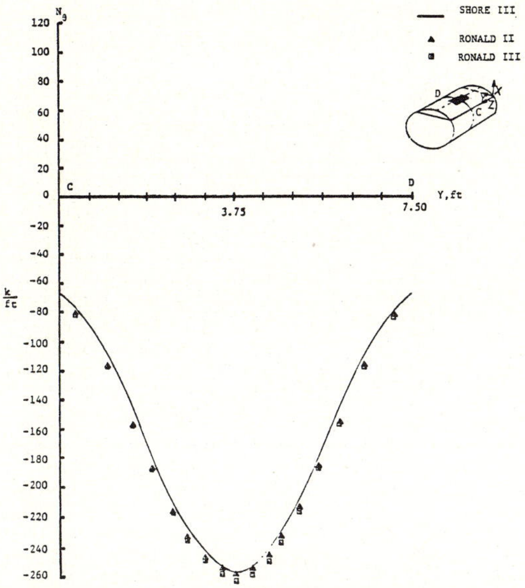

Figure 11. Stress Resultant N_θ vs. Y along CD

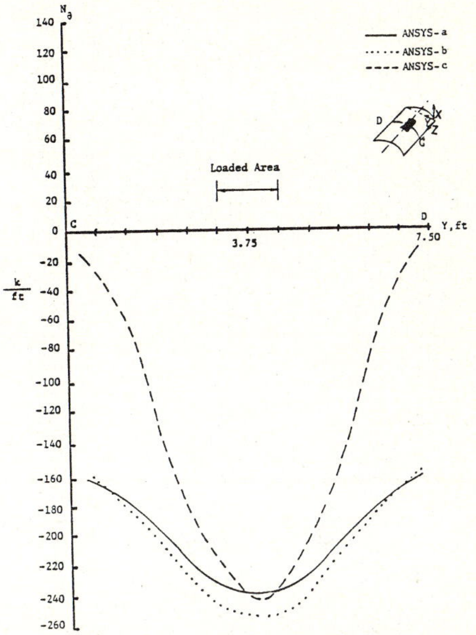

Figure 12. Stress Resultant N_θ vs. Y along CD

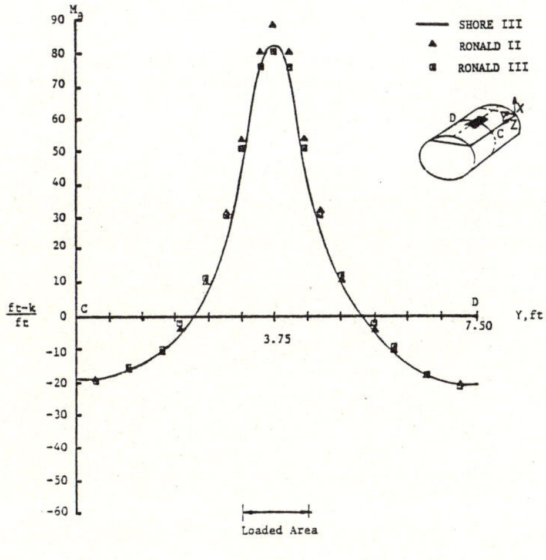

Figure 13. Moment Resultant M_θ vs. Y along CD

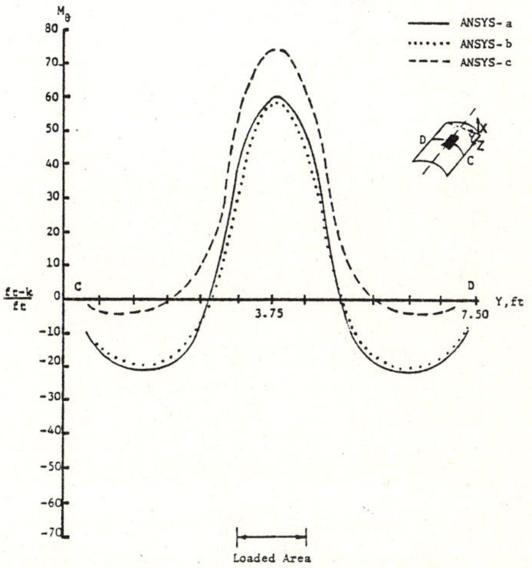

Figure 14. Moment Resultant M_θ vs. Y along CD

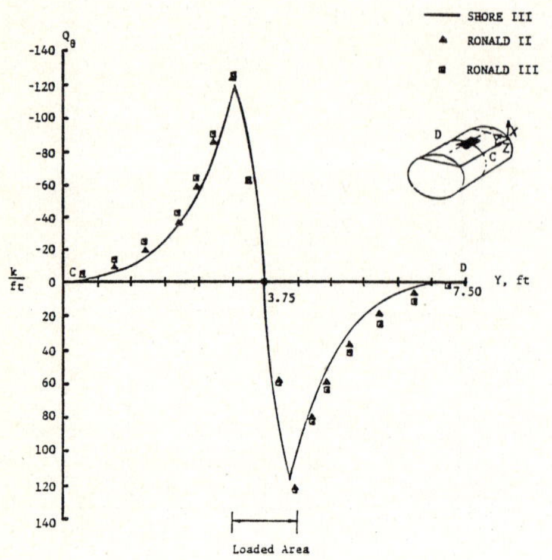

Figure 15. Shear Force Q_θ vs. Y along CD

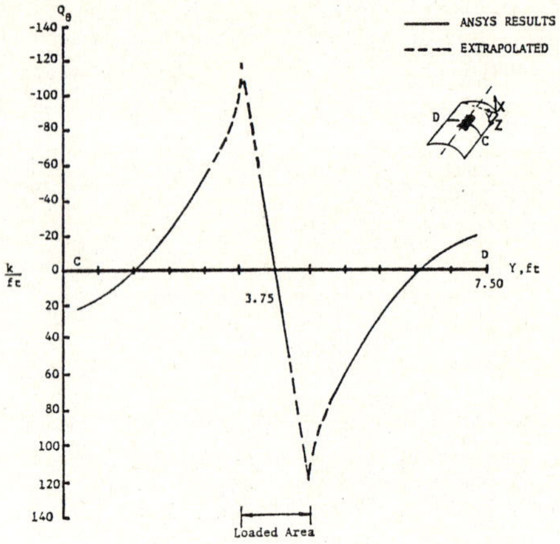

Figure 16. Shear Force Q_θ vs. Y along CD

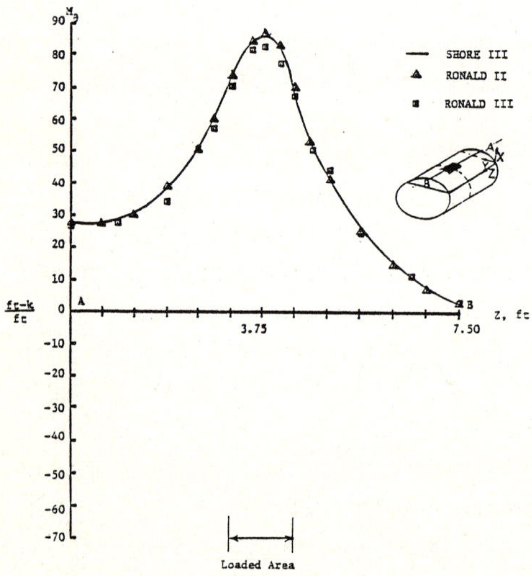

Figure 17. Moment Resultant M_θ vs. Z along AB

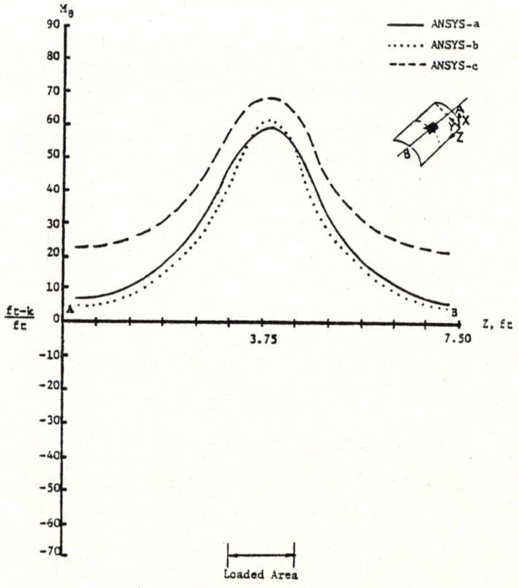

Figure 18. Moment Resultant M_θ vs. Z along AB

DISCUSSION

For the cylindrical shell, bending effects are especially important around the region of the applied load. Obviously, significant transverse shearing forces and bending moments can develop, as shown in figures 7-10 and 13-16. The shell exhibits large changes in the shearing

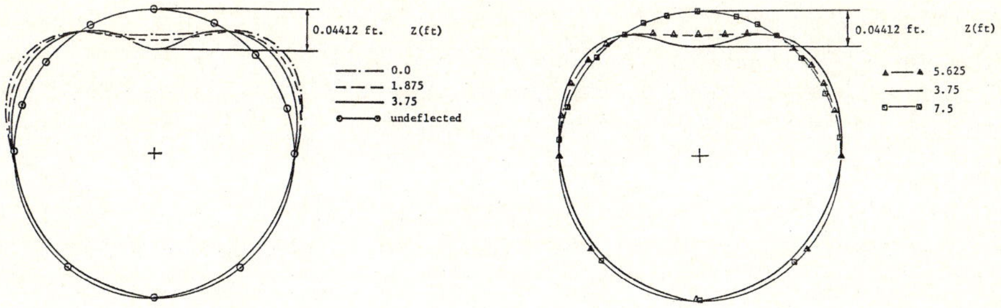

Figure 19. Deflected Cross-Section along AB

force at the edge of the load patch. The bending effects spread from the point of application by about 0.75λ along the meridian AB and 1.75λ along the circumference CD, where λ, the half-wavelength of bending, is given by

$$\lambda = \frac{\pi}{[3(1-\mu^2)]^{1/4}} \sqrt{Rt} = 4.283 \text{ ft} \tag{1}$$

R = the radius of the cylindrical shell = 6.25 ft
t = the thickness of the cylindrical shell = 0.5 ft
μ = Poisson's ratio = 0.24

For the complete cylinder, nodal point values of the transverse shear forces were obtained directly from RONALD. For the panel, STIF93 was used to obtain Q_z and Q_θ. In figures 10 and 16, the transverse shears obtained from STIF93 were graphically extrapolated to show the entire range within the panel, since only integration point results were available from ANSYS. Accurate values of these peak forces and the corresponding stresses are important in investigating the possibility of a punching shear failure around the perimeter of the patch load. In this case, a finer mesh in the vicinity of the load should give results closer to the nodal values. This will, of course, increase the cost of the analysis.

The applied load causes a marked peak value of all stress resultants, especially the hoop stress N_θ, as shown in figures 11 and 12. Therefore, the concentrated loading can be presumed to produce large negative values of the hoop stress, and the cylindrical shell may be susceptible to a local failure due to circumferential buckling. It should be noted from figure 12 that N_θ is affected appreciably by the longitudinal boundary conditions, even toward the center of the shell.

In observing the variation of the circumferential moment M_θ along the meridional axis, figures 17 and 18, it is apparent that the values are influenced by the boundary conditions. For the complete cylinder shown on Figure 17, the moment at the hinged boundary, point B, practically vanishes, while the value at the unrestrained longitudinal boundary, point A, remains significant. The circumferential curvature due to the inward concave deflection of the cylindrical shell is the major factor affecting M_θ.

The deflected cross-section along the axis of the cantilevered cylinder is shown in figure 19. The hinged boundary at end B forces the shell to maintain a circular shape, so that the M_θ value at this end is very small. The largest concave deflection, located the center of the loaded area, generates the peak value of M_θ. The free end, point A, still has significant concave deflection and circumferential curvature; therefore, it retains the M_θ bending effect, as shown on figure 17.

COMPARISON OF PROGRAMS

For peak values only, the computational costs of SHORE-III and RONALD-II are about the same. For accurate results away from the loaded area, the computer cost of RONALD-II is only about one-third of SHORE-III.

There are some small differences in the results between RONALD (version II and III) and SHORE-III. These programs are founded on different theories, i.e., SHORE-III is based on rotational thin shell theory; RONALD-II uses degenerated isoparametric shell elements locally; and RONALD-III employs full thickness isoparametric solid elements based on elasticity theory in the vicinity of the loading. Therefore, it is understandable that the results from the various programs may show some variations, and the small differences here appear to be reasonable and acceptable.

A solid element, STIF94, was employed to further examine the capabilities of ANSYS. The stress resultants and couples calculated using the latter element were very close to the values found with the other ANSYS elements, and also to the RONALD-III results.

CONCLUSION

It has been shown that the peak values of the stress resultants in a relatively thick cylindrical shell (R/t=12.5) under a concentrated loading can be predicted satisfactorily using either a rotational shell model or a general shell model, incorporating a shell theory which includes transverse shear. The rotational shell model can simulate the boundary conditions for a cylindrical panel and the general shell model can closely replicate the complete cylinder. Also, it is shown that various thick shell and solid elements can be satisfactorily employed in the vicinity of the loading, together with less involved thin shell elements over the remaining domain of the shell. This is advantageous from the standpoint of computational efficiency. The results are expected to be useful in planning, correlating and generalizing results obtained from a limited number of physical tests.

Currently, the research is focusing on the nonlinear behavior of reinforced concrete cylindrical shells, specifically, on the development of a three dimensional smeared non-orthogonal cracking model, including bond-slip and tension stiffening.

ACKNOWLEDGEMENTS

The material presented in this paper is partially based on research supported by the U.S. National Science Foundation under Grant No. CEE811797, the National Bureau of Standards under Grant No. 60NANB4D0014, and Swanson Analysis System, Inc. and PDA Engineering, under their Educational License Agreements. However, any opinions, findings, conclusions or recommendations expressed herein are those of the authors and do not necessarily reflect the views of the sponsors.

REFERENCES

1. Gould, P. L., Finite Element Analysis of Shells of Revolution, London, Pitman Publishing, 1985.
2. Han, K. J., "Local Deviations in Shells of Revolution", D.Sc. Dissertation, Washington University, St. Louis, MO, 1981.
3. Han, K. J. and Gould, P. L., "RONALD, Rotational Shell with Non-Axisymmetric Local Deviations - User's Manual", Research Report No. 64, Structural Division, Washington University, St. Louis, MO, Sep. 1982.
4. Han, K. J. and Gould, P. L., "Shell of Revolution with Local Deviations", International Journal for Numerical Methods in Engineering, Vol. 20, 1984.
5. ANSYS Theoretical Manual, Rev. 4.1, Swanson Analysis Systems, Inc., Houston, PA, 1983.
6. ANSYS User's Manual, Rev. 4.1, Swanson Analysis Systems, Inc., Houston, PA, 1983.
7. PDA/PATRAN User's Guide, Vol. I-II, (Rev. 1.6C), 1985, PDA Engineering, Santa Ana, CA.

A CURVED STIFFENER FINITE ELEMENT FOR SHELL STRUCTURES OF GENERAL GEOMETRY

San-Cheng Chang
Jhin-Hwei Lee
Department of Civil Engineering
National Taiwan University
Taipei, Taiwan, China

SUMMARY

A curved stiffener element for shell structures of general geometry is described. This element can be used for linear, geometrically nonlinear, and dynamic analyses in conjunction with an efficient triangular thin-shell finite element. The shell and the stiffener elements are mutually compatible with regard to the nodal degrees of freedom and are both based on the "natural strain" formulation. Derivation of the stiffener element is summarized and the performance for a number of benchmark problems is presented in this paper.

INTRODUCTION

Shell structures are characterized by their large area of coverage, long spans, and small thicknesses, and are frequently associated with structural stiffening members. It is therefore highly desirable that shell analysis tools be equipped with capabilities for both the shell and the stiffeners. To cope with structures of general geometry, triangular doubly-curved shell elements are generally preferred. However, there are only few thin-shell elements under this catagory with successful performance in both the linear and geometrically nonlinear range. A complete survey of shell elements was given by Wu [1].

In contrast to the case of doubly-curved triangular thin-shell elements, a number of good curved beam elements are available. Successful examples are those by Ferguson and Clark [2] and Venketash and Rao [3]. Their applications are limited however, due to the lack of an adequate shell element. Thomas and Webber [4] developed a "geodesic beam" element for use with Morley's shell element [5], the combination is nevertheless inapplicable to shells of general geometry.

SUMMARY OF THE SHELL ELEMENT

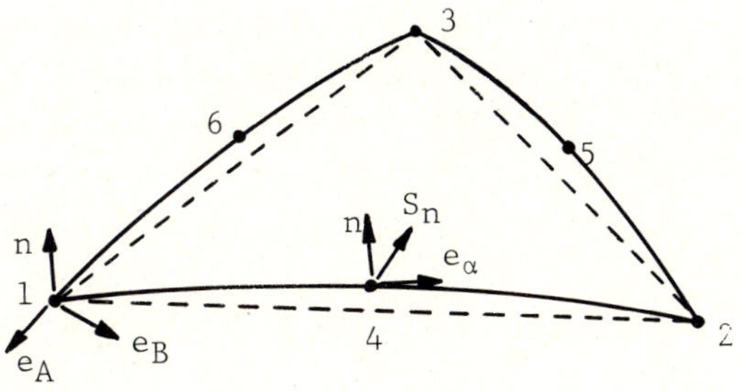

Figure 1. Triangular Thin-Shell Finite Element

The shell element compatible to the stiffener presented is a doubly-curved triangular thin-shell finite element of 36 degrees of freedom, as shown in Fig. 1. This element was jointly developed by Wu [2] and the first author [6]. It is adopted as the cornerstone for the stiffener as it is considered a good compromise between solution accuracy and computation efficiency. There are six nodes per element: three corner nodes and three mid-side nodes. Nine degrees of freedom are associated with each corner node and three degrees of freedom are with each mid-side node, as are identified below:

corner node d.o.f.: $u\ v\ w\ u_{,A}\ v_{,A}\ w_{,A}\ u_{,B}\ v_{,B}\ w_{,B}$

mid-side node d.o.f.: $u_{,Sn}\ v_{,Sn}\ w_{,Sn}$

e_A and e_B are the unit tangent vectors of the shell surface in the principal curvature directions, **n** is the unit surface normal vector, and u, v, and w are respectively the displacements in directions of e_A, e_B and n in the case of corner nodes, or e_α, S_n, n in the case of mid-side nodes. "A", "B" are the parameters in the principal directions, whereas $u_{,A}$ is the first derivative of displacement "u" in the first principal curvature direction. S_n is a surface tangent vector located at mid-side node normal to the element edge. Although the degrees of freedom are defined in the surface coordinate system of e_A-e_B-**n**, the formulation is actually carried out in the cartesian system. Cubic-order shape functions augmented with rational polynomials are used as the interpolants for both the geometry and the deformation. The rational polynomials are employed to provide first-order inter-element compatibility which is otherwisely achievable only through quintic-order interpolants, as of the case by Argyris et. al.[7]. Due to the singular second derivative of the rational terms in the shape functions, substitute shape function derivatives obtained through least-square procedure are employed for bending strain. The strain-displacement relationship is based upon a "natural strain" approach originally proposed by Argyris [7], and the rigid-body requirement is exactly satisfied. This shell element has given results of satisfactory accuracy in tested problems of linear, geometrically nonlinear, as well as free vibration analyses.

FORMULATION OF STIFFENER ELEMENT

The curved stiffener element derived is defined by three nodes and is associated with 21 degrees of freedom. The same set of nodal degrees of freedom used for the shell element is adopted, and the geometric data at nodes are also shared. The stiffener can be located arbitrarily on a free-form surface defined by a set of user-chosen curvilinear coordinate system, whereas coordinate transformations based on differential geometry is applied to convert them into that of the principal system. The configuration of the stiffening element is given in Fig. 2.

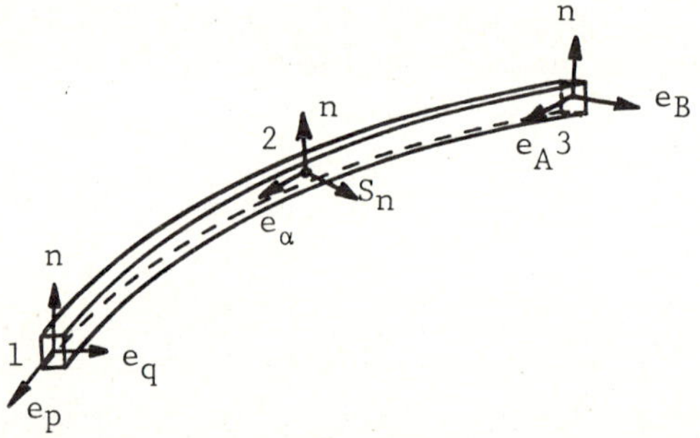

Figure 2. Curved Stiffener Element

To circumvent the need for using elaborated transcendental functions as the interpolant in satisfying the rigid-body mode requirement, the formulation is based on three-dimensional elasticity rather than the conventional curved beam theory. In addition, to facilitate elegant element formulation, the element properties are formed on the basis of the natural degrees of freedom. A series of transformation between the coordinate system upon which global degrees are defined and that which the element properties are formed is therefore required. In addition, transformations accounting for the effect of stiffener offset from the mid-surface of the shell where the nodes are located is necessary.

Natural Coordinate System

The natural coordinate system of the stiffener element is located in the neutral surface of the element. The neutral axis of the element is a parametric cubic curve of parameter "p" interpolated from the position vectors X_1, X_2 and the tangent vectors $X_{1,p}$, $X_{2,p}$ of the corner nodes through the well-known Hermite polynomials. The nodal position vectors may be taken from the definition of the shell surface directly, while the orientation of the neutral axis, $X_{1,p}$ and $X_{2,p}$, need to be found specifically for the stiffener. Let the unit direction vector between the corner nodes be j_p, and n be the surface unit normal vector, then the unit vectors at the corner nodes along the neutral axis may be expressed as

$$e_p=(j_p-(n\cdot j_p)n)/\sqrt{1-(n\cdot j_p)^2} \tag{1}$$

The lateral direction of the stiffener, given by vector e_q, can be obtained by normalizing the cross-product of n and e_p. With e_p and e_q, the definition for the neutral surface is complete (Fig. 2).

Once the natural directions are established, the differential operators in the principal system and the natural system on the corner nodes can be related easily, as was outlined in [1]. With these operators, it is possible to find the positional derivatives $X_{,p}$, $X_{,pq}$ and $X_{,q}$ in the natural directions at corner nodes. Based on these the natural directions at the mid-side node can be determined by interpolation through the Hermite polynomials:

$$X_{,p}=0.75(X_2-X_1)-0.25(X_{1,p}+X_{2,p}) \tag{2}$$

$$X_{,q}=0.50(X_{1,q}+X_{2,q})+0.25(X_{1,pq}-X_{2,pq}) \tag{3}$$

The derivative vectors of the nodes in the natural coordinate system can later be applied with the position vectors to interpolate the complete element geometry.

Interpolation

The element is based on isoparametric formulation where both the geometry and deformation, generically written as a cartesian vector function F(p,q), is interpolated within the neutural surface through the following equation:

$$
\begin{aligned}
F(p,q)=&N_1(p,q)F_1+N_2(p,q)F_2\\
&+N_3(p,q)F_{1,p}+N_4(p,q)F_{1,q}\\
&+N_5(p,q)F_{2,p}+N_6(p,q)F_{2,q}\\
&+N_7(p,q)F_{3,q}\\
&=\begin{bmatrix} \langle N\rangle & & \\ & \langle N\rangle & \\ & & \langle N\rangle \end{bmatrix}\{F\}\\
&=[D]\{F\}
\end{aligned}
\tag{4}
$$

In above, [D] is a supermatrix of rank 3-by-21 consisting of shape function vectors <N>, where <N> is simply the collection of entries $N_1(p,q)$ through $N_7(p,q)$. {F} is a vector collecting all the nodal parameters. The shape functions are concatenations of cubic or quadratic polynomials in "p" and linear functions in "q":

$$
\begin{aligned}
&N_1(p,q)=0.25(p^3-3p+2)\\
&N_2(p,q)=-0.25(p^3-3p-2)\\
&N_3(p,q)=0.25(p^3-p^2-p+1)\\
&N_4(p,q)=0.5q(p^2-p)\\
&N_5(p,q)=0.25(p^3+p^2-p-1)\\
&N_6(p,q)=0.5q(p^2+p)\\
&N_7(p,q)=q(1-p^2)
\end{aligned}
\tag{5}
$$

The derivatives of F(p,q) with respect to "p" and "q" for any point on the neutral surface can be found by differentiating the [D] matrix with respect to proper parameters. These relations are useful in interpolating strains within the element.

Strain-Displacement Relations

Strain-displacement transformation is not derived in detail, only the essential equations are outlined.

The axial strain along the neutral axis is derived based on the definition of Green's strain, and the expression of total axial strain is in the form of:

$$\varepsilon_A=e_p\cdot u_{,a}+0.5u_{,a}\cdot u_{,a} \tag{6}$$

The corresponding incremental form of axial strain is

$$\varepsilon_A=e_p\cdot \Delta u_{,a}+u_{,a}\cdot \Delta u_{,a} \tag{7}$$

The first term is the strain in proportion to the displacement, and the second term gives the strain due to the consideration of large deformation.

To find strain due to bending actions, the rotation vector for any point on the neutral axis is defined as:

$$\phi=(n\cdot u,_b)e_p-(n\cdot u,_a)e_q+0.5(e_q\cdot u,_a-e_p\cdot u,_b)n \tag{8}$$

The curvature in the p-direction is $\phi,_a$:

$$\begin{aligned}\phi,_a=&(n\cdot u,_{ab}+n,_a\cdot u,_b)e_p+(n\cdot u,_b)e_{p},_a\\ &-(n\cdot u,_{aa}+n,_a\cdot u,_a)e_q-(n\cdot u,a)e_{q},_a\\ &+0.5(e_q\cdot u,_{aa}+e_{q},_a\cdot u,_a-e_p\cdot u,_{ab}-e_{p},_a\cdot u,_b)n\\ &+0.5(e_q\cdot u,_a-e_p\cdot u,_b)n,_a\end{aligned} \tag{9}$$

It is assumed that any point on the same cross-section would experience the same rotation, hence the strain in the p-direction due to the rotation is the triple product:

$$\varepsilon_B=e_p\cdot(\phi,_a x r) \tag{10}$$

where r is given as the vector from the neutral axis to any point of the same section. Since the curvature vector in eqn(9) is a linear function of displacement, the incremental bending strain expression is straightforward, and is not elaborated here.

In addition to the bending actions in the directions of n and e_q, twisting about e_p also need to be considered. Shear strain caused by the twisting action is given by

$$\gamma_T=(r\cdot r)(e_p\cdot\phi,_a) \tag{11}$$

The incremental form of eqn(11) is also straightforward.

In conformity with the formulation of the shell element, tranverse shear strain in the n-e_q plane is not considered. However, shear strain in the e_p-e_q plane and the lateral strain in the e_q direction must be taken into account to avoid causing singularities in the element stiffness matrix in addition to those corresponding to the rigid-body modes. The incremental shear strain is taken as

$$\gamma_{pq}=e_q\cdot\Delta u,_a+e_p\cdot\Delta u,_b \tag{12}$$

and the incremental lateral strain is given as

$$\varepsilon_q=e_q\cdot\Delta u,_b \tag{13}$$

The contributions from γ_{pq} and ε_q are usually negligible in comparison to those from the shell element, but are nevertheless required for reason of numerical stability when the stiffener is being applied alone without the shell element.

Coordinate Transformations

Three levels of transformation are needed in the conversion of stiffener element properties between the natural coordinate system and the global surface system on the shell structure. The generalized nodal displacement in the natural coordinate system used in the element formulation is described in such a way that the displacement components are given in cartesian coordinate and derivatives in the e_p-e_q system. On the other hand, the global generalized displacements on the shell element is given in a system where both displacements and derivatives are expressed in the principal surface curvature system of e_A-e_B. Two sets of transformations, one on displacement components and one on directions of derivatives, are therefore required.

Secondly, the displacements on the shell surface must further be modified before being applied to the stiffener to account for the eccentric effect of the stiffener location. This is achieved througn a second-level transformation. Element matrices and vectors in the global system can be constructed through the above relations from their original natural counterparts.

EXAMPLES

Example 1 -- free vibration of stiffened shell

A closed cylinder is stiffened by 5 rows of intermediate circumferential ribs with equal spacing. Only the normal displacement of the ends are restrained, and the stiffening rings are located to the exterior of the shell surface. The fundamental vibration properties of this structure are sought. The shell and stiffeners are characterized by the followings:

E=29,600,000 psi (Young's modulus)

ν=0.29 (Poisson ratio)

R=4.25 inch (radius of cylinder)

L=18.0 inch (full length of cylinder)

t=0.152 inch (thickness of cylinder)

ρ=0.283 lb/in^3 (unit weight)

B=0.25 inch (width of ribs)

H=0.25 inch (depth of ribs)

ring spacing=3.0 inch

Both theoretical derivation and experiments on this the problem has been conducted by Al-Najafi and Warburton [8]. For the present study, a 180-degree portion with full length of the cylinder is discretized into a 12-by-18 (circumferential by longitudinal) and 16-by-24 mesh. The vibration modes are characterized by a number of full-waves ("n") in the circumferential direction and rippled half-waves in the axial direction ("m"). The "ripples" in the latter are due to the effect from the stiffening rings. The four modes with lowest frequencies are found, and the results are tabulated in table 1. The second mode, not found by [8], is one consisting mainly of rigid-body motions along the ciylinder except at locations close to the ends. This mode is attributed to the stiffness contribution from the rings. The other modes are associated with fairly consistent mode shapes to those of [8]. It is seen that the results from the 12-by-18 mesh of the current study have converged to stable values, and is of comparable accuracy to the results of Al-Najafi and Warburton which were obtained with 30-divisions along the axial direction of the cylinder with axisymmetric finite elements.

	current study		Ref[11]		wave number	
	12-18	16-24	theor.	exptl	m	n
mode-1	850.8	850.4	834.0	809.0	1	2
mode-2	869.8	873.4	-	-	1	1
mode-3	1068.6	1072.8	1065.0	1065.0	1	3
mode-4	1622.1	1624.0	1699.0	1658.0	2	3

Table 1 Vibration Frequencies (hertz) of Stiffened Hinged Cylinder

Example 2 -- snap-through of stiffened spherical cap

A point-loaded spherical cap, shown in Fig. 3, is used to study the geometrically nonlinear behavior of stiffened shell structure where snap-through phenomenon is involved. The spherical cap is hinged on all four boundaries. In particular, only a quarter of the cap need to be analyzed due to symmetry in geometry and loading. The following properties are used for analysis:

E=10,000 psi (Young's modulus)

v=0.3 (Poisson ratio)

R=100.0 inch (radius of sphere)

t=3.9154 inch (shell thickness)

B=H=0.0, 1.2, 2.4, 3.0 inches (beam width and depth)

The structure is initially assumed to be unstiffened, for which a load deflection curve consistent to that of a previous study by Thomas and Gallagher [9] was obtained for the tip of cap where the point load is applied. Stiffening members of various sizes are then added below the cap at locations defined in Fig. 3, and the load-displacement curves are traced again. It is seen from Fig. 3 that the snap-through is avoided only for stiffeners with sections above 3.0-inches square.

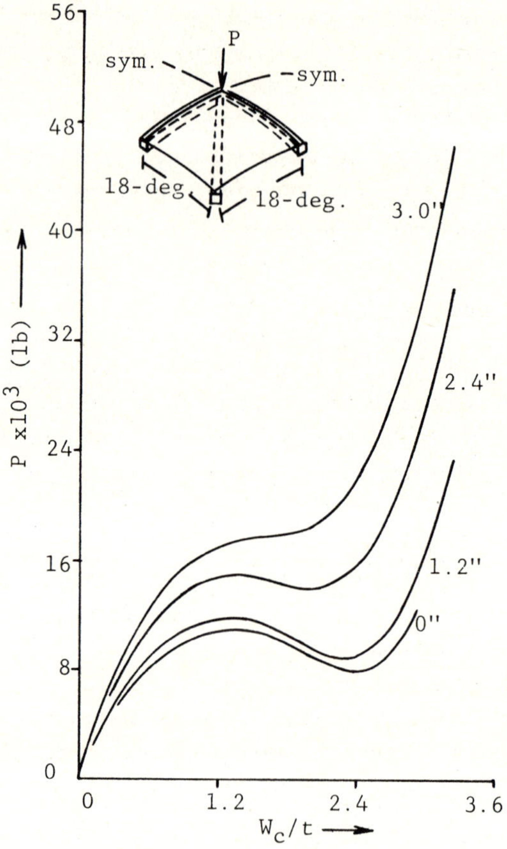

Figure 3. Spherical Cap

Example 3 -- parabolic shell
With the previous examples serving as benchmarks verifying the capability and accuracy of the derived element, an independent shell structure of practical application value is used as the last test case. The geometry is defined as a translational shell surface with two parabolas as the generator. The first generator lies in the XZ-plane of the cartesian coordinate system within range of x-coordinate 0-3000cm with the equation:

$$z=-x^2/3000+5x/6$$

The second generator lies in the cartesian YZ-plane and is given in range of y-coordinate 0-2000cm with the equation of

$$z=-7y^2/16000+39y/40$$

The structure is stiffened by curved beams of different cross-sections located below the shell surface. As shown in Fig. 4, four rows of stiffeners exist in the long direction and five rows in the short direction. This structure is analyzed for its

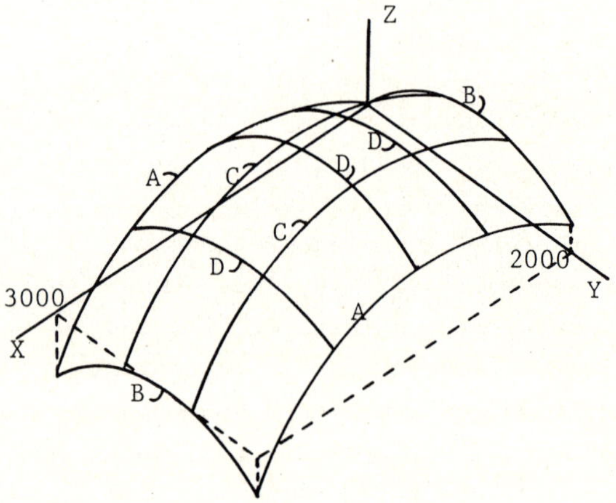

Figure 4. Parabolic Shell

linear static as well as free vibration behaviors. The shell surface is assigned the following properties:

$E=205{,}800\ N/cm^2$ (Young's modulus)

$\nu=0.29$ (Poisson ratio)

$t=20$ cm (thickness)

$\rho=0.0024\ kg/cm^3$ (mass density)

whereas the stiffeners are characterized by

$E=2{,}058{,}000\ N/cm^2$ (Young's modulus)

$\nu=0.3$ (Poisson ratio)

$\rho=0.0024\ kg/cm^3$ (mass density)

B=30 cm, H=80 cm (beam width and depth of type-A)

B=30 cm, H=60 cm (beam width and depth of type-B)

B=20 cm, H=60 cm (beam width and depth of type-C)

B=20 cm, H=40 cm (beam width and depth of type-D)

The stiffener offset values are obtained directly from the shell thickness and beam depth. The shell is supported only at the four corners with hinged condition, i.e., all three displacement components are restrained. A mesh size of 12-divisions along the first generator and 9-divisions along the second generator is used as the finite element discretization. A finer mesh of 16-by-12 was also adopted, but only negligible difference of less than 1% to those of the coarser mesh was found in the analysis result. The 12-by-9 mesh is therefore taken as the converged solution. For the case of linear static analysis, uniformly distributed normal pressure of magnitude $10\text{-}N/cm^2$ (per surface area) is applied. The overall deflected shape is tabulated in table 2. For the case of free vibration analysis, the lowest 3 fundamental modes are sought. The mode shapes are associated with corresponding angular velocities of 1.295 rad/sec, 1.599 rad/sec, 2.261 rad/sec respectively.

Location	X=0	X=750	X=1500	X=2250	X=3000
Y=0	0.0	10.1	2.34	-20.3	0.0
Y=667	-1.58	-5.24	-8.87	-13.8	-5.6
Y=1333	3.42	-3.75	-8.19	-14.5	-0.27
Y=2000	0.0	10.6	2.01	-18.2	0.0

Table 2 Normal Displacements of Selected Points on Shell

CONCLUSIONS

It is demonstatrated that through the combination of three-dimensional elasticity theory and differential geometry, the element formulation is made concise and elegant. The series of testing problems also verify the efficiency and accuracy of the derived element in the range of both small and large deformations and in stability and dynamic analyses. Coupled with the existing shell element, the current element is a valuable addition to the shell structure analysis capability.

REFERENCES

1. S.C. Wu, An integrated system for finite element analysis -- surface representation and curved shell element, Department of Structural Engineering Research Report 80-7, Cornell University, Ithaca, New York, 1980
2. G.H. Ferguson and R.D. Clark, "A variable thickness, curved beam and shell stiffening element with shear deformation", Int.J. Numer. Meth. Engng 14, 581-592, 1979
3. A. Venketash and K.P. Rao, "A Laminated anisotropic curved beam and shell stiffening finite element", Comp. Struc. 2, 197-201, 1982
4. I.T. Thomas and J.P.H. Webber, "A geodesic beam finite element for use in the static analysis of reinforced circular cylindrical shells", Int.J. Numer. Meth. Engng 18, 397-419, 1982
5. L.S.D. Morley, "Analysis of developable shells with special reference to the finite element method and circular cylinders", Phil.Rec. Soc. London. 281, 113-170, 1976
6. S.C. Chang, An integrated finite element nonlinear shell analysis system with interactive computer graphics, Department of Structural Engineering Research Report 81-4, Cornell University, Ithaca, New York, 1981
7. J.H. Argyris and D.W. Scharpf, "The SHEBA family of shell elements for the matrix displacement method", Aero.J. Roy. Aero. Soc. 72, 873-882, 1968
8. A.M.J. Al-Najafi and G.B. Warburton, "Free Vibration of Ring-Stiffened Cylindrical Shells", Journal of Sound and Vibration, Vol.13, pp.9-25, 1970
9. G.R. Thomas and R.H. Gallagher, "A triangular thin shell finite element: nonlinear analysis", NASA CR-2483, 1975

MEMBRANE STRESSES IN HYPERBOLIC PARABOLOIDAL SECTORIAL SHELL WITH CANTILEVERED FREE EDGE

I. SAJTOS
Institute of Design Development
and Technology
Budapest, Hungary

INTRODUCTION

The sectorial shells are beautiful shells (fig.1.A), but the sectorial shell with cantilevered free edge is one of the most interesting (fig. 1.B).

P.CSONKA presented the calculation of translational sectors by finite difference method in [1].
CANDELA was the first, who constructed that type of shell which is mentioned above. The membrane stresses were determined by solving the differential equation system of equilibrium with iteration [7].

In this paper, we are presenting the membrane stresses of hyperbolic paraboloidal, sectorial shell with cantilevered free edge by solving the PUCHER'S differential equation.

THE CONDITIONS

The conditions of membrane theory of shells are valid in this calculation. The ground-plane of the shell sector is borded by two straight lines and a parabola arc at the free edge (fig.2).
The edges along the radial straight lines are rigid, the edge along the parabola arc is free.
The middle surface of the shell is hyperbolic paraboloid.

$$z = Ax^2 - By^2 \qquad (1)$$

$$A = \frac{f_1}{a^2} \quad (1a) , \qquad B = \frac{f_2}{b^2} \quad (1b) .$$

The equation of the free edge's projection

$$y = D - Ex^2 \qquad (2)$$

where

$$D = b + c \quad (2a), \qquad E = \frac{c}{a^2} \quad (2b)$$

During the calculation we shall take into account only vertical loads (in the direction of axis "z").

THE SOLVING POSSIBILITIES

We can obtain the membrane stresses by solving the PUCHER's differential equation. The differential equilibrium equation in the orthogonal co-ordinate system is

$$\frac{\partial^2 z}{\partial y^2}\frac{\partial^2 F}{\partial x^2} - 2\frac{\partial^2 z}{\partial x\, \partial y}\frac{\partial^2 z}{\partial x\, \partial y} + \frac{\partial^2 z}{\partial x^2}\frac{\partial^2 F}{\partial y^2} = -g(x,y) \qquad (3)$$

where : $z = z(x,y)$ - function of the shell's middle surface,

$F = F(x,y)$ - stress function,

$g = g(x,y)$ - function of the vertical load intensity concerning to the ground-plane.

The differential equation can be solved with simple integral calculation, if (3) has the following shape :

$$\frac{\partial^2 F}{\partial u\, \partial v} = -Z(u,v) \qquad (4)$$

We can achieve (4) on two ways.

1. It is known from [8] [9], that a common differential equation can be ordered to the (3).

$$\frac{\partial^2 z}{\partial y^2}dy^2 - 2\frac{\partial^2 z}{\partial x\, \partial y}dx\, dy + \frac{\partial^2 z}{\partial x^2}dx^2 = 0 \qquad (5)$$

The characteristics are the solutions of the (5).
We can transform the partial differential equation into the canonical form, with the help of characteristics. It has the following shape :

$$\frac{\partial^2 F}{\partial u\, \partial v} = G(u,v) \qquad (7)$$

(in the hyperbolic case, if the middle surface of the shell is hyperbolic).
In this case we can get the canonical form using the following new co-ordinates :

$$\xi = y - \sqrt{\frac{A}{B}}\, x \qquad (8a)$$

$$\eta = y + \sqrt{\frac{A}{B}}\, x \qquad (8b)$$

2. We shall transform the equation of the middle surface of the shell into a new co-ordinate system, where

$$\frac{\partial^2 z}{\partial u^2} = 0 , \qquad \frac{\partial^2 z}{\partial v^2} = 0 \qquad (9a,b)$$

Then we can give the PUCHER's differential equilibrium equation in an oblique-angled co-ordinate system. That has the following shape (fig.3) :

$$\frac{\partial^2 z}{\partial u^2}\frac{\partial^2 F}{\partial v^2} - 2\frac{\partial^2 z}{\partial u\,\partial v}\frac{\partial^2 F}{\partial u\,\partial v} + \frac{\partial^2 z}{\partial v^2}\frac{\partial^2 F}{\partial u^2} = -g(u,v)\sin\omega \qquad (10)$$

where : $z = z(u,v)$ - function of the middle surface of the shell in the (uvz) co-ordinate system,

ω - angle between u and v axis,

$F = F(u,v)$ - the unknown stress function,

$g = g(u,v)$ - function of the vertical load intensity in the oblique-angled co-ordinate system, concerning to the ground-plane.

Because of the (9a,b) conditions, (10) will take the following shape :

$$-2\frac{\partial^2 z}{\partial u\,\partial v}\frac{\partial^2 F}{\partial u\,\partial v} = -g(u,v)\sin\omega \qquad (11)$$

We shall present this method in this paper.

THE TRANSFORMATION OF THE MIDDLE SURFACE OF THE SHELL

The new co-ordinate system is an oblique-angled co-ordinate system (fig.4). The relation between the system (uv) and (xy) is given by the following equations :

$$x = u'\cos\varphi - v'\sin\varphi \qquad (12a)$$

$$y = u'\sin\varphi + v'\cos\varphi \qquad (12b)$$

where :

$$u' = u + v\cos\omega \quad (13a), \qquad v' = v\sin\omega \quad (13b)$$

We substitute these in (1), and with the use of the conditions (9a,b) we can get the following :

$$tg\ \varphi = \sqrt{\frac{A}{B}} \quad (14), \qquad \omega = \pi - 2\varphi \quad (15)$$

(We considered just one of the possible values of the angle φ).
After that, substituting the equations (12), (13), (14), (15) into the equations (1), we obtain the following equation :

$$z = r.uv \qquad (16)$$

where :

$$r = -2\,[A\ \cos^2\varphi + B\ \sin^2\varphi] \qquad (17)$$

The equation (16) is the function of the middle surface of the shell in the oblique-angled co-ordinate system.
It is to be noted, that the used (12) transformation functions are also characteristics.
Substituting equations (12), (13), (14), (15) into equation (2) we obtain the equation of the free edge in the new co-ordinate system.

$$u_b = \frac{(2vE\ \cos^2\varphi - \sin\varphi) \pm \sqrt{(\sin\varphi - 2vE\ \cos^2\varphi)^2 - 4E\ \cos^2\varphi\ (v\ \sin\varphi + v^2\cos^2\varphi\ E - D)}}{2E\ \cos^2\varphi} \qquad (18a)$$

$$v_b = \frac{(2uE\ \cos^2\varphi - \sin\varphi) \pm \sqrt{(\sin\varphi - 2vE\ \cos^2\varphi)^2 - 4E\ \cos^2\varphi\ (u\ \sin\varphi + u^2\cos^2\varphi\ E - D)}}{2E\ \cos^2\varphi} \qquad (18b)$$

In the equations (18), only the positive sign has a meaning, if $\varphi > \alpha$ (fig.5). It is necessary to the existence and uniqueness of the membrane state of the shell, that $\varphi > \alpha$ [4] [8] .
The shell is in the membrane state if the free edge cuts all the characteristics only in one point (1), otherwise the shell isn't in the membrane state (2) (fig.6).

RELATION BETWEEN MEMBRANE STRESSES IN THE ORTHOGONAL CO-ORDINATE SYSTEM AND IN THE NEW CO-ORDINATE SYSTEM

The relations between membrane stresses in the orthogonal co-ordinate system and in the new oblique-angled co-ordinate system known [2] [6] [7] .
Using these in our case (fig.7), we get the following equations :

$$n_x = \frac{1}{2\ \mathrm{tg}\ \varphi}\ [n_u + n_v - 2\ n_{uv}] \qquad (19a)$$

$$n_y = \frac{\mathrm{tg}\ \varphi}{2}\ [n_j + n_v + 2\ n_{uv}] \qquad (19b)$$

$$n_{xy} = \frac{n_u - n_v}{2} \qquad (19c)$$

CALCULATION OF THE MEMBRANE STRESSES

Regarding (11) and (16) the PUCHER's differential equation of equilibrium in our case is the following :

$$-2r\ \frac{\partial^2 F}{\partial u\ \partial v} = -g(u,v)\ \sin\ \omega \qquad (20)$$

Stress function :

$$F = -\frac{1}{2r}\ \sin\ \omega \iint [-g(u,v)]\ du\ dv + C_1(u) + C_2(v) \qquad (21)$$

Reduced membrane stresses :

$$n_{uv} = \frac{1}{2r}\ \sin\ \omega\ [-g(u,v)] \qquad (22a)$$

$$n_u = \frac{\partial^2 F}{\partial v^2} \qquad (22b)$$

$$n_v = \frac{\partial^2 F}{\partial u^2} \qquad (22c)$$

Boundary conditions are along the free edge. We obtain the following conditions after the transformation :

$$\frac{\partial F}{\partial u} = 0 \qquad (\text{if } v = v_b(u)) \qquad (23a)$$

$$\frac{\partial F}{\partial v} = 0 \qquad (\text{if } u = u_b(v)) \qquad (23b)$$

The differential coefficients of the free edge which are necessary in the calculation :

$$u_b' = \frac{du_b}{dv} = \qquad (24a)$$

$$= 1 - \frac{2\ \sin\ \varphi}{\sqrt{(\sin\ \varphi - 2vE\ \cos^2\varphi)^2 - 4E\ \cos^2\varphi(v\ \sin\ \varphi + v^2\cos^2\varphi - D)}}$$

$$v'_b = \frac{dv_b}{du} = \tag{24b}$$

$$= 1 - \frac{2 \sin\varphi}{\sqrt{(\sin\varphi - 2uE \cos^2\varphi)^2 - 4E \cos^2\varphi(u \sin\varphi + u^2E \cos^2\varphi - D)}}$$

Relations between effective membrane stresses and reduced membrane stresses are known [1] [2] [3] [5] [6] .

The calculation of the stress function is very hard, but calculation of membrane stresses is easier.
The stress function (21) :

$$F = F_1(u,v) + C_1(u) + C_2(v) \tag{25}$$

From the (23) conditions :

$$0 = \frac{\partial F_1(u_b,v)}{\partial v} + \frac{\partial C_2}{\partial v} \tag{26b}$$

$$0 = \frac{\partial F_1(u,v_b)}{\partial u} + \frac{\partial C_1}{\partial u} \tag{26a}$$

We can compute $\frac{\partial^2 C_1}{\partial u^2}$ and $\frac{\partial^2 C_2}{\partial v^2}$, using them we can get the reduced membrane stresses :

$$n_u = \frac{\partial^2 F_{1(u,v)}}{\partial v^2} + \frac{\partial^2 C_2}{\partial v^2} \tag{27a}$$

$$n_v = \frac{\partial^2 F_{1(u,v)}}{\partial u^2} + \frac{\partial^2 C_1}{\partial u^2} \tag{27b}$$

and from the equation (22a) we can get the n_{uv}.

THE MEMBRANE STRESSES

1. IN THE CASE OF CONSTANT LOAD

$$g(x,y) = g(u,v) = p = \text{const.} \tag{28a}$$

$$n_{uv} = - \frac{\sin 2\varphi}{2r} p \tag{28b}$$

$$n_u = -\frac{\sin 2\varphi}{2r} p . u_b' \qquad (28c)$$

$$n_v = -\frac{\sin 2\varphi}{2r} p . v_b' \qquad (28d)$$

2. IN THE CASE OF OPTIONAL LOAD

We suppose, that the load function is a binary polynomial of x and y

$$q = \sum_{i=0}^{n} \sum_{j=0}^{m} p_{ij} x^i y^j$$

In the orthogonal co-ordinate system this polynomial is the following, for example :

$$q_{(x,y)} = p + p_2 x^2 + p_3 y^2 + p_6 x^4 + p_7 x^2 y^2 + p_8 y^4 + \ldots \qquad (29)$$

The load function and the membrane stresses :

$$\underline{g_{(x,y)} = p_2 x^2} , \; g(u,v) = p_2 \cos^2\varphi [u^2 - 2uv + v^2] , \qquad (30a)$$

$$n_{uv} = -\frac{\sin 2\varphi \cos^2\varphi}{2r} p_2 [u^2 - 2uv + v^2] , \qquad (30b)$$

$$n_u = \frac{\sin 2\varphi \cos^2\varphi}{2r} p_2 [-u^2 + 2uv - (3u_b^2 u_b' - (2u_b u_b' v + u_b^2) + u_b' v^2 + u_b 2v)], \qquad (30c)$$

$$n_v = \frac{\sin 2\varphi \cos^2\varphi}{2r} p_2 [-v^2 + 2uv - (2uv_b + u^2 v_b' - (v_b^2 + 2uv_b' v_b) + 3v_b^2 v_b')] . \qquad (30d)$$

$$\underline{g_{(x,y)} = p_3 y^2} , \; g(u,v) = p_3 \sin^2\varphi [u^2 + 2uv + v^2] , \qquad (31a)$$

$$n_{uv} = -\frac{\sin 2\varphi \sin^2\varphi}{2r} p_3 [u^2 + 2uv + v^2] , \qquad (31b)$$

$$n_u = \frac{\sin 2\varphi \sin^2\varphi}{2r} p_3 [u^2 + 2uv - (3u_b^2 u_b' + 2u_b u_b' v + u_b^2 + u_b' v^2 + u_b 2v)] , \qquad (31c)$$

$$n_v = \frac{\sin 2\varphi \sin^2\varphi}{2r} p_3 [v^2 + 2uv - (2uv_b + u^2 v_b' + v_b^2 + 2uv_b v_b' + 3v_b^2 v_b')] . \qquad (31d)$$

$$g_{(x,y)} = p_6 x^4 , \quad g(u,v) = p_6 (u-v)^4 \cos^4\varphi , \tag{32a}$$

$$n_{uv} = - \frac{\sin 2\varphi \cos^4\varphi}{2r} p_6 [u^4 - 4u^3 v + 6u^2 v^2 - 4uv^3 + v^4] , \tag{32b}$$

$$n_u = \frac{\sin 2\varphi \cos^4\varphi}{2r} p_6 [-u^4 + 4u^3 v - 6u^2 v^2 + 4uv^3 - (u_b^4 u_b' - 4u_b^3 u_b' v - u_b^4 +$$

$$+ 2(3u_b^2 u_b' v^2 + u_b^3 2v) - 2(2u_b u_b' v^3 + u_b^2 3v^2) + u_b' v^4 + u_b 4v^3)] , \tag{32c}$$

$$n_v = \frac{\sin 2\varphi \cos^4\varphi}{2r} p_6 [4u^3 v - 6u^2 v^2 + 4uv^3 - v^4 - (4u^3 v_b + u^4 v_b' -$$

$$- 2(3u^2 v_b^2 + u^3 2 v_b v_b') + 2(2uv_b^3 + u^2 3 v_b^2 v_b') - (v_b^4 + 4uv_b^3 v_b') + v_b^4 v_b')] . \tag{32d}$$

$$g_{(x,y)} = p_7 x^2 y^2 , \quad g(u,v) = p_7 (u-v)^2 (u+v)^2 \sin^2\varphi \cos^2\varphi , \tag{33a}$$

$$n_{uv} = - \frac{\sin 2\varphi \sin^2\varphi \cos^2\varphi}{2r} p_7 [u^4 - 2u^2 v^2 + v^4] , \tag{33b}$$

$$n_u = \frac{\sin 2\varphi \sin^2\varphi \cos^2\varphi}{2r} p_7 [- \tfrac{4}{3} u^3 v + 4v^3 u - (u_b^4 u_b' - \tfrac{2}{3} (3u_b^2 u_b' v^2 + u_b^3 2v) +$$

$$+ 4v^3 u_b + v^4 u_b')] , \tag{33c}$$

$$n_v = \frac{\sin 2\varphi \sin^2\varphi \cos^2\varphi}{2r} p_7 [4u^3 v - \tfrac{4}{3} uv^3 - (4u^3 v_b + u^4 v_b' -$$

$$- \tfrac{2}{3} (2uv_b^3 + 3u^2 v_b^2 v_b') + v_b^4 v_b')] . \tag{33d}$$

$$g_{(x,y)} = p_8 y^4 , \quad g(u,v) = p_8 (u+v)^4 \sin^4\varphi , \tag{34a}$$

$$n_{uv} = - \frac{\sin 2\varphi \sin^4\varphi}{2r} p_8 [u^4 + 4u^3 v + 6u^2 v^2 + 4uv^3 + v^4] , \tag{34b}$$

$$n_u = \frac{\sin 2\varphi \sin^4\varphi}{2r} p_8 [u^4 + 4u^3 v + 6u^2 v^2 + 4uv^3 - (u_b^4 u_b' + 4u_b^3 u_b' v + u_b^4 +$$

$$+ 2 (3u_b^2 u_b' v^2 + u_b^3 2v) + 2 (2u_b u_b' v^3 + u_b^2 3 v^2) + u_b' v^4 + u_b 4 v^3)] , \tag{34c}$$

$$n_v = \frac{\sin 2\varphi \sin^4\varphi}{2r} p_8 [4u^3v + 6u^2v^2 + 4uv^3 + v^4 - (4u^3v_b + u^4v_b') +$$

$$+ 2 (3u^2v_b^2 + u^3 2v_bv_b') + 2 (2uv_b^3 + u^2 3v_b^2v_b') + v_b^4 + u 4v_b^3v_b' + v_b^4v_b')] \quad . \tag{34d}$$

EXAMPLE

Let's take the following function for the middle surface of the shell (fig.8) :

$$z = 0{,}1015 \; x^2 - 4{,}3984.10^{-3} \; y^2$$

The function of the free edge :

$$y = 33{,}716 - 0{,}0455 \; x^2$$

The load intensity :

$$p = 1{,}0 \; kN/m^2$$

The constants :

$A = 0{,}1015$
$B = 4{,}3984.10^{-3}$
$\varphi = 1{,}3655$ rad
$r = -0{,}01686$
$D = 33{,}716$ m
$E = 0{,}0455 \; m^{-1}$

We compare the membrane stresses with P.CSONKA's solution [1].
The computed values from the proposed method corresponds well to the values of solution of finite difference method, as you can see in fig.9a,b,c.
Let's check a point of the free edge, if the boundary conditions are fulfiled?
Dates :

$x = 5{,}74$ m $\qquad n_x = -4{,}99$ kN/m
$y = 32{,}217$ m $\qquad n_y = -1{,}37$ kN/m
$\qquad n_{xy} = 2{,}6$ kN/m

Principal reduced membrane stresses :

$$n_{1,2} = \frac{n_x + n_y}{2} \pm \sqrt{\left(\frac{n_x - n_y}{2}\right)^2 + n_{xy}^2} \quad ; \quad tg\, 2\alpha_o = \frac{2\, n_{xy}}{n_x - n_y}$$

$$n_{1,2} = \frac{-4{,}99 - 1{,}37}{2} \pm \sqrt{\left(\frac{-4{,}99 + 1{,}37}{2}\right)^2 + 2{,}6^2} \quad ; \quad tg\, 2\alpha_o = \frac{2.2{,}6}{-4{,}99 + 1{,}38}$$

$$n_1 = 0 \qquad \alpha_o = 62{,}422°$$

$$n_2 = -6{,}34 \text{ kN/m}$$

Tangent of the free edge :

$$y' = -0{,}0910\, x = -0{,}0910.5{,}74$$

$$B = -27{,}578° \; ; \; B = \alpha_o - 90° = 62{,}422 - 90 = -27{,}578°$$

The boundary conditions are true, because $n_1 = 0$ and $n_{12} = 0$ and n_1 prependicular to the tangent of free edge.

CONCLUSION

The membrane stresses can be determined from the proposed formulaes. The proposed formulaes can be applied for design of hyperbolic paraboloidal, sectorial shells with cantilevered free edge.

REFERENCES

[1] CSONKA, P. : Shell Structures
Akadémiai Kiadó, Budapest, 1981

[2] FLÜGGE, W. : Stresses in Shells
Springer-Verlag, Göttingen, 1960

[3] BÖLCSKEI, E. - OROSZ, A. : Reinforced Concrete Structures Shells
Tankönyvkiadó, Budapest, 1983

[4] TARNAI, T. : On the Existence and Uniqueness Condition of Membrane State of Shell Structures
I. Hyperbolic Shells
Műszaki Tudomány 56 (1978), pp.19-47

[5] KOLLAR, L. : Beton Kalender 1984 Schalenkonstructionen
Berlin, 1984

[6] POLONYI, S. : Berechnung von Hyperbolischen Paraboloidschalen über beliebigen Viereck-Grundrissen
Beton und Stahlbetonbau 9 (1962), pp.213-220

[7] FABER, C. : Candela/The Shell Builder
New York, 1963

[8] KORN, G.A. - KORN, T.M. : Matematical Handbook for Scientist and Engineers
Műszaki Könyvkiadó, Budapest, 1975

[9] CZÁCH, L. : Partial Differential Equations
Magyar Geofizikusok Egyesülete
Budapest, 1969

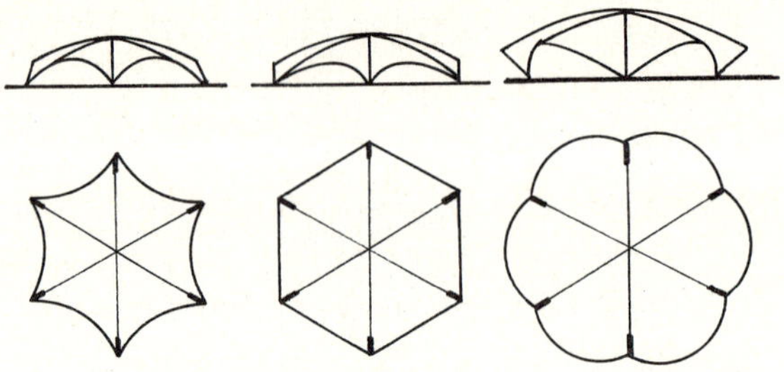

Fig. 1A.

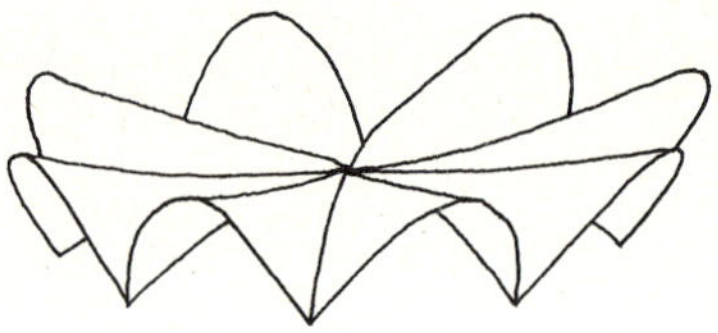

Fig. 1B.

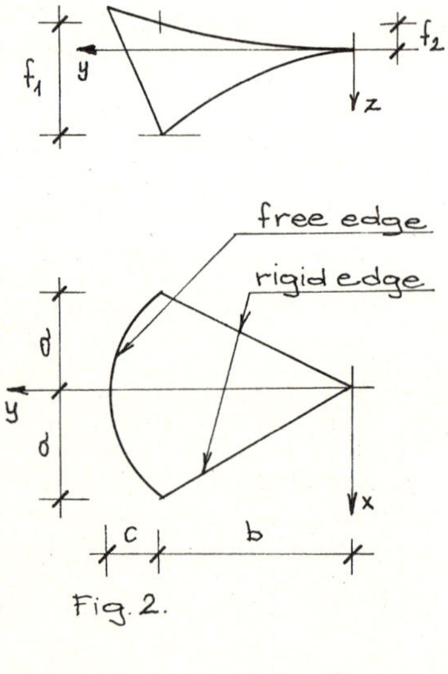

Fig. 2.

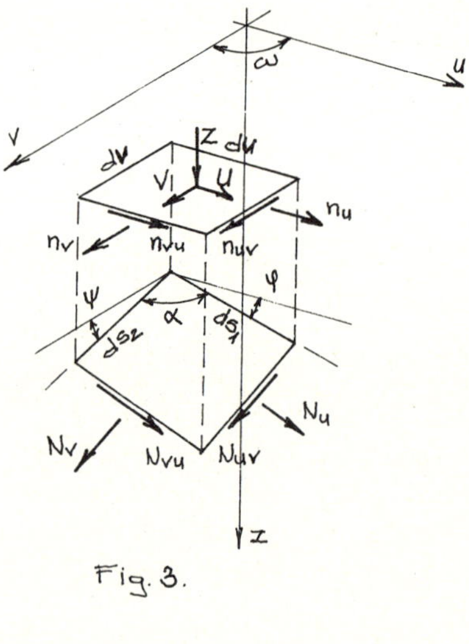

Fig. 3.

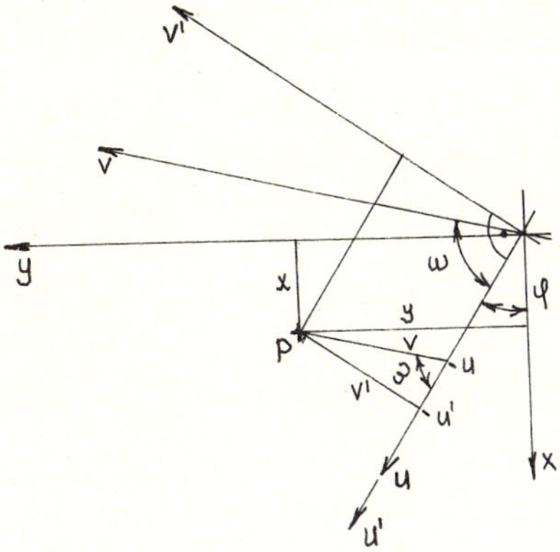

Fig. 4.

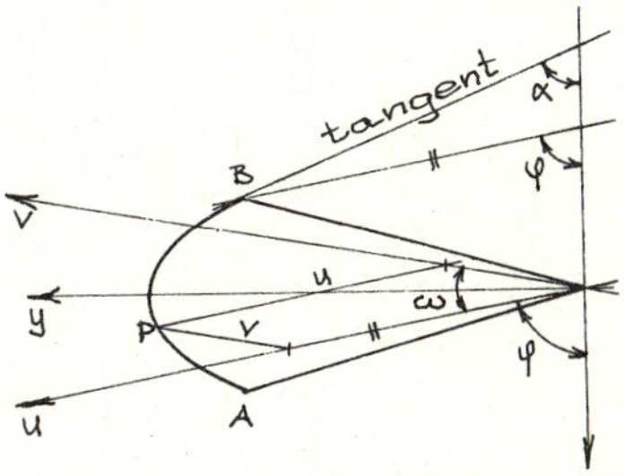

Fig. 5.

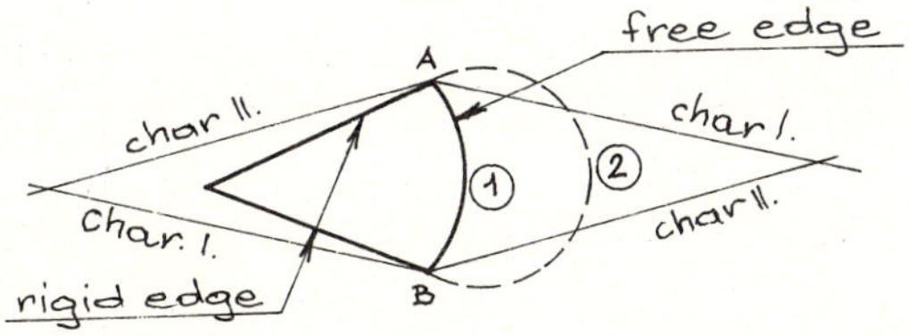

Fig. 6.

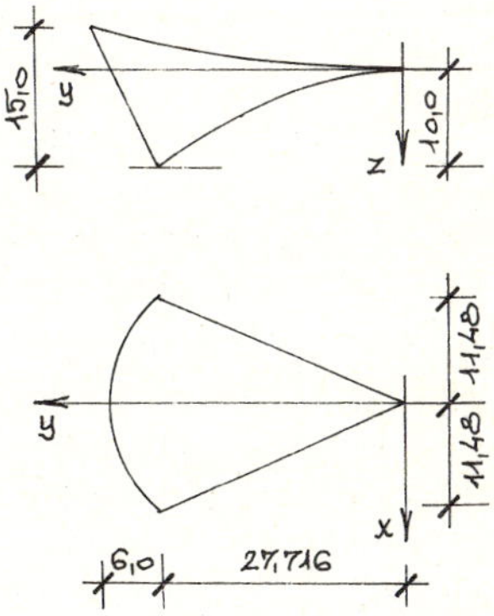

Fig. 8.

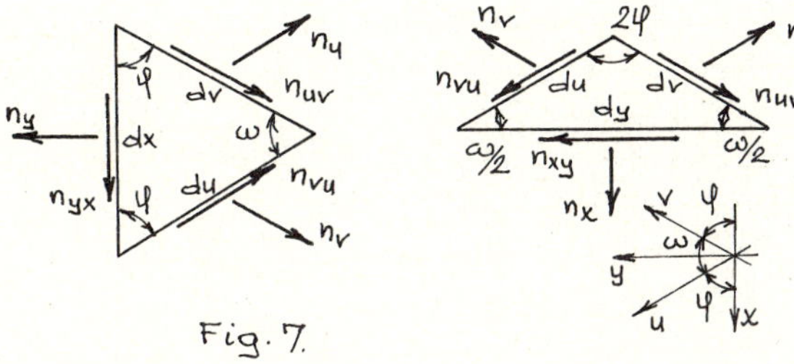

Fig. 7.

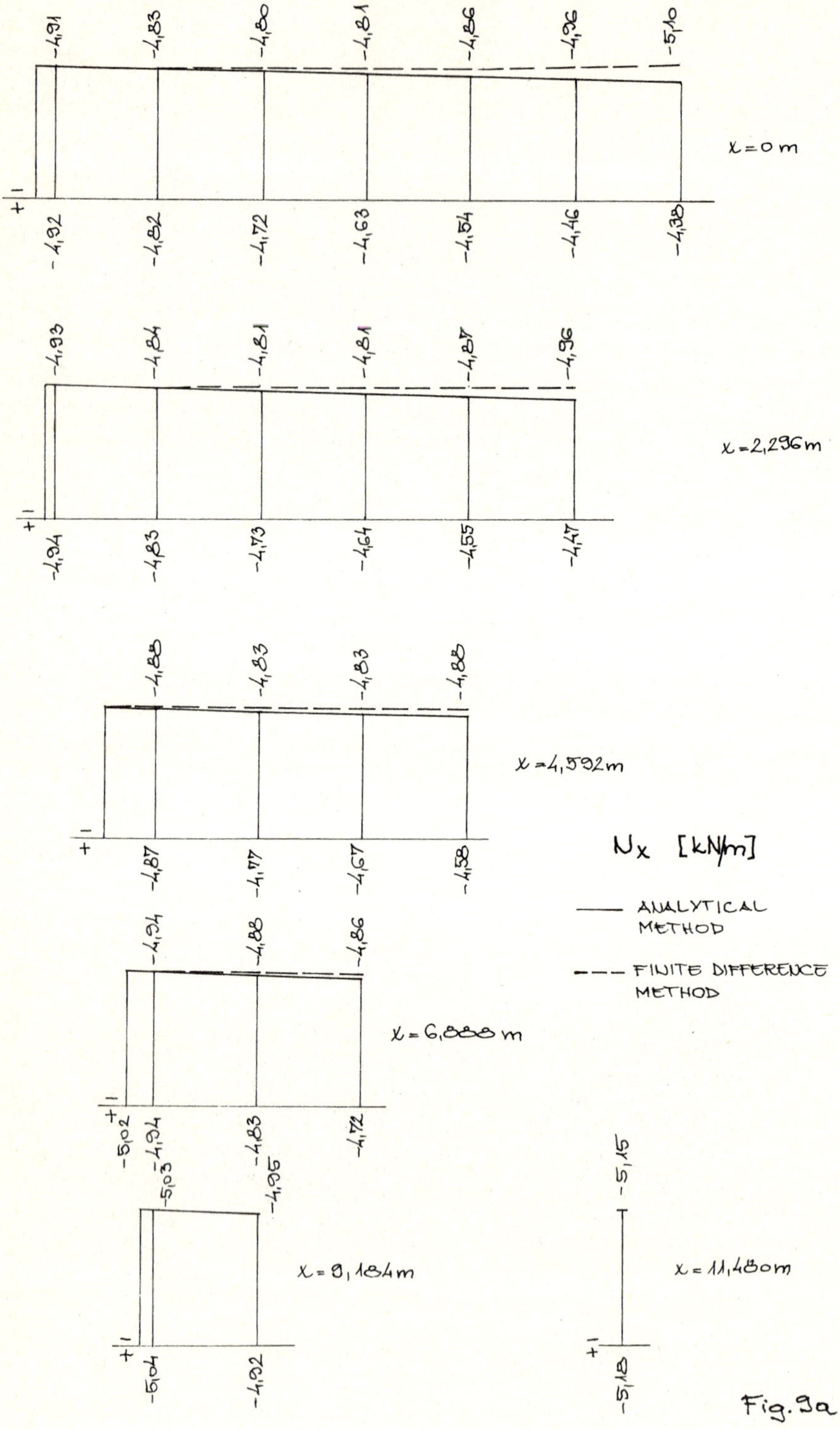
x=0 m
x=2,296m
x=4,592m
x=6,888 m
x=9,184m
x=11,480m
Nx [kN/m]
ANALYTICAL METHOD
FINITE DIFFERENCE METHOD

Fig. 9a

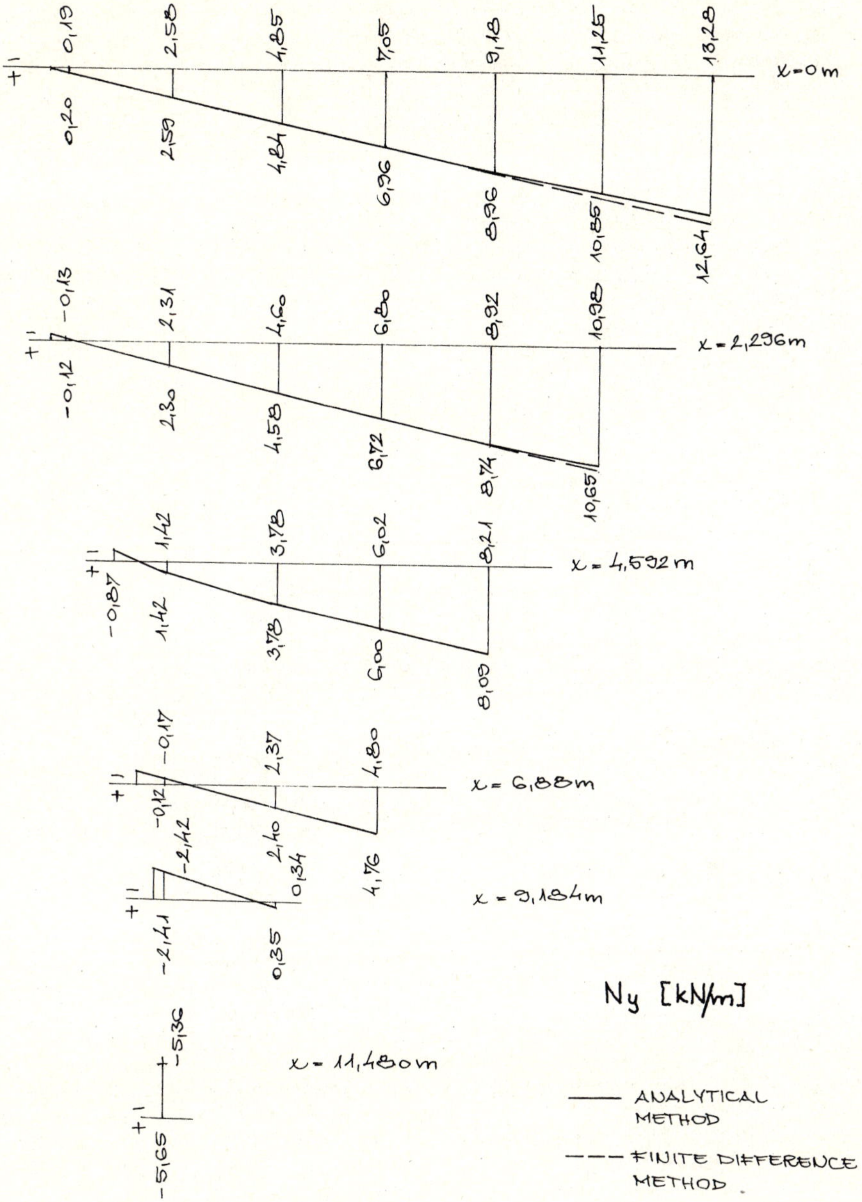
x = 0 m
x = 2,296 m
x = 4,592 m
x = 6,88 m
x = 9,184 m
x = 11,480 m
Ny [kN/m]
ANALYTICAL METHOD
FINITE DIFFERENCE METHOD

Fig. 9b

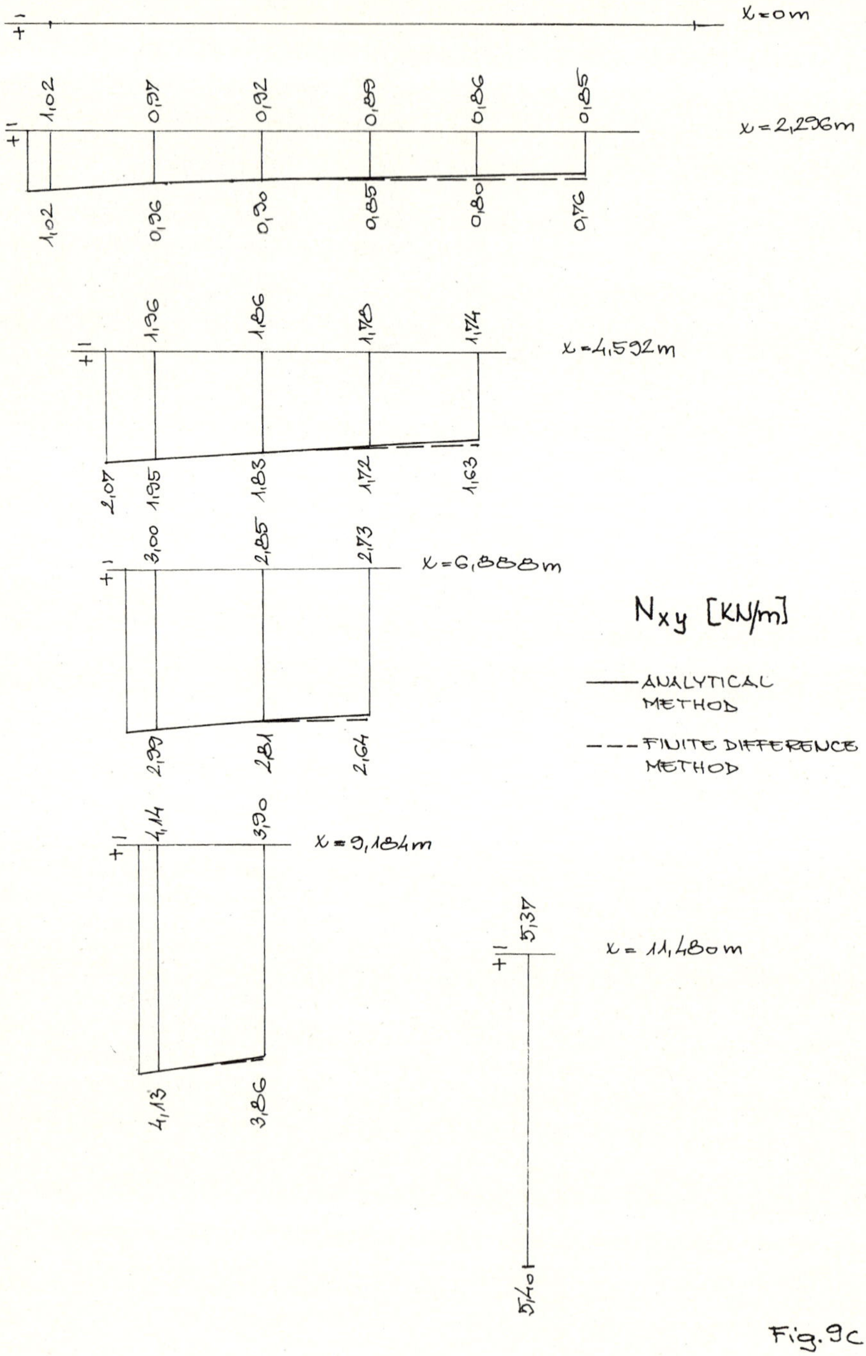
x=0m
x=2,296m
x=4,592m
x=6,888m
x=9,184m
x=11,480m
Nxy [KN/m]
ANALYTICAL METHOD
FINITE DIFFERENCE METHOD
1,02
0,97
0,92
0,89
0,86
0,85
1,02
0,96
0,90
0,85
0,80
0,76
1,96
1,86
1,78
1,74
2,07
1,95
1,83
1,72
1,63
3,00
2,85
2,73
2,99
2,81
2,64
4,14
3,90
4,13
3,86
5,37
5,40

Fig. 9c

APPROXIMATE ANALYSIS OF SADDLE CABLE NET ROOFS

DING SHEN-SHI
Northwestern Architectural
Engineering Institute
Xian, Shaanxi, China

SUMMARY

This paper presents a mixing method of finite integral and finite difference to analyse the structure of saddle cable net roofs. In the governing differential equations the vertical displacement W and its first derivative are changed into a function of the second derivative. The derivative of cable net joint's vertical coordinate Z can be expressed by finite difference. Thus the differential equations of cable net can be changed into algebraic simultaneous equations, which can be solved by iteration .

DIFFERENTIAL EQUATIONS OF CABLE NET

Fig.1a is a kind of saddle cable net roof structure. It can be analyzed according to membrane theory. We make the following computation assumptions : the cable is ideally flexible; the tension in the cable obeys Hooke's law; sagging cable and hogging cable are all the time tense and can only transfer vertical forces. Tensions in cable net are equal to the horizontal component.

From Fig.1a we take a microarea (Fig.1b). Under the acting of vertical load Δq and unit width tensions in cable net directions x and y, the microarea's vertical coordinate changes from primary state Z_o to loading state $Z = Z_o + W$, where W is vertical displacement of cable net under the acting of Δq. So the equilibrium equations of cable net under the action of loading increment Δq can be set up :

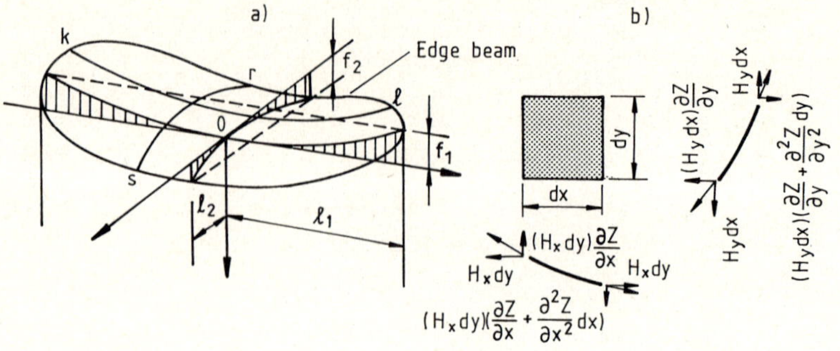

Fig.1

$$H_{xo} \frac{\partial^2 W}{\partial x^2} + H_{yo} \frac{\Omega^2 W}{\partial y^2} + \Delta H_x \left(\frac{\partial^2 Z_o}{\partial x^2} + \frac{\partial^2 W}{\partial x^2}\right) + \Delta H_y \left(\frac{\partial^2 Z_o}{\partial y^2} + \frac{\partial^2 W}{\partial y^2}\right) + \Delta q = 0 \quad [1] \qquad (1)$$

where : H_{xo}, H_{yo} - Cable net in primary state : pretension's horizontal component in the directions x and y (per unit width);

ΔH_x, ΔH_y - Cable net under the action of loading increment Δq : horizontal component of tension increment.

The extension of cable kl, rs (Fig.1a) caused by ΔH_x, ΔH_y and temperature difference Δt, can be expressed by the following deformation equations of cable net in the direction x and y :

$$\left.\begin{aligned} (u_l - u_k) + \int_0^{l_x} \left[\frac{\partial Z_o}{\partial x} \cdot \frac{\partial W}{\partial x} + \frac{1}{2}\left(\frac{\partial W}{\partial x}\right)^2\right] dx - l_x \alpha \Delta t &= \Delta H_x \frac{l_x}{EA_x} \\ (v_s - v_r) + \int_0^{l_y} \left[\frac{\partial Z_o}{\partial y} \cdot \frac{\partial W}{\partial y} + \frac{1}{2}\left(\frac{\partial W}{\partial y}\right)^2\right] dy - l_y \alpha \Delta t &= \Delta H_y \frac{l_y}{EA_y} \end{aligned}\right\} \quad [1] \qquad (2)$$

Where : u,v - Horizontal displacement of cable net in the direction x and y;

EA_x, EA_y, l_x, l_y - Unit width tensile rigidity and span of cable net in the direction x and y;

α - Linear expansion coefficient cable material.

The equations (1) and (2) are the differential equations of cable net roof structures to be solved.

SOLUTION OF DIFFERENTIAL EQUATIONS OF CABLE NET ROOF STRUCTURES

In equations (1) and (2) Z_o is known quantity, H_{xo}, H_{yo}, EA_x, EA_y can be first assumed, W, ΔH_x, ΔH_y are unknown quantities. The derivative of the known quantity Z_o can be expressed by finite difference. The unknown quantity W and its first order derivative $\frac{\partial W}{\partial x}$, $\frac{\partial W}{\partial y}$ can be expressed by function of its second order derivative $\frac{\partial^2 W}{\partial x^2}$, $\frac{\partial^2 W}{\partial y^2}$, i.e. expressed by finite integral. Thus differential equations can be converted into algebraic simultaneous equations which take its $\frac{\partial^2 W}{\partial x^2}$, $\frac{\partial^2 W}{\partial y^2}$ and ΔH_x, ΔH_y as unknown quantities. In order to simplify this method, we assume that cable net doesn't undergo horizontal displacements on boundary, i.e. $u_l - u_k = v_s - v_r = 0$ and we neglect influence of temperature difference. The influence of cable net horizontal displacements and temperature can be considered after preliminary computation. The cable net is divided into m, n sections in the direction x and y. We convert the vertical displacement W and its first order derivative W'_x, W'_y to functions of the second order derivative W''_x, W''_y by numerical integral :

$$\left.\begin{aligned}
W'_x &= \int_0^{ma} W''_x \, dx + W'_{x(0)} \\
W'_y &= \int_0^{nb} W''_y \, dy + W'_{y(0)} \\
W_x &= \int_0^{ma} W'_x \, dx + W_{x(0)} = \int_0^{ma}\int_0^{ma} W''_x \, dx\, dx + W'_{x(0)}x + W_{x(0)} \\
W_y &= \int_0^{nb} W'_y \, dy + W_{y(0)} = \int_0^{nb}\int_0^{nb} W''_y \, dy\, dy + W'_{y(0)}y + W_{y(0)}
\end{aligned}\right\} \quad (3)$$

[2]

Where : a,b - Size of cable net grid;

$W'_{x(0)}$, $W'_{y(0)}$, $W_{x(0)}$, $W_{y(0)}$ - Generalized displacement value of joint 0 on the cable net boundary.

The equation (3) can be approximately expressed by the matrix equations :

$$\left.\begin{aligned}
\{W'_x\} &= \frac{a}{12} [N] \{W''_x\} + W'_{x(0)}\{1\} \\
\{W'_y\} &= \frac{b}{12} [N] \{W''_y\} + W'_{y(0)}\{1\} \\
\{W_x\} &= (\frac{a}{12})^2 [N]^2 \{W''_x\} + W'_{x(0)}\{x\} + W_{x(0)}\{1\} \\
\{W_y\} &= (\frac{b}{12})^2 [N]^2 \{W''_y\} + W'_{y(0)}\{y\} + W_{y(0)}\{1\}
\end{aligned}\right\} \quad (4)$$

[2]

where : [N] - The matrix of integral operator,

$$[N] = \begin{bmatrix} 0 & 0 & 0 & 0 & 0 & 0 & 0 & 0 & 0 \\ 5 & 8 & -1 & 0 & 0 & 0 & 0 & 0 & 0 \\ 4 & 16 & 4 & 0 & 0 & 0 & 0 & 0 & 0 \\ 4 & 16 & 9 & 8 & -1 & 0 & 0 & 0 & 0 \\ 4 & 16 & 8 & 16 & 4 & 0 & 0 & 0 & 0 \\ 4 & 16 & 8 & 16 & 9 & 8 & -1 & 0 & 0 \\ \cdots & \cdots & \cdots & \cdots & \cdots & \cdots & \cdots & \cdots & \cdots \end{bmatrix}$$ [2]

When the grid length of cable kl or rs is ηa or ηb ($\eta < 1.0$) in Fig.1a.

$$[N] = \begin{bmatrix} 0 & 0 & 0 & 0 & 0 & 0 & 0 & 0 & 0 & 0 & 0 \\ c & d & & 0 & 0 & 0 & 0 & 0 & 0 & 0 & 0 \\ f & g & h & 0 & 0 & 0 & 0 & 0 & 0 & 0 & 0 \\ f & g & h+5 & 8 & -1 & 0 & 0 & 0 & 0 & 0 & 0 \\ f & g & h+4 & 16 & 4 & 0 & 0 & 0 & 0 & 0 & 0 \\ f & g & h+4 & 16 & 9 & 8 & -1 & 0 & 0 & 0 & 0 \\ \cdots & \cdots & \cdots & \cdots & \cdots & \cdots & \cdots & \cdots & \cdots & \cdots & \cdots \\ f & g & h+4 & 16 & 8 & 16 & 8 & 16 & 4+f & g & h \\ f & g & h+4 & 16 & 8 & 16 & 8 & 16 & 4+c & d & e \end{bmatrix}$$

$$c = \frac{1}{1+\eta}(4\eta^2 + 6\eta) \qquad d = 2\eta^2 + 6\eta \qquad e = -\frac{2\eta^3}{1+\eta}$$

$$f = \frac{1}{\eta}(4\eta^2 + 2\eta - 2) \qquad g = \frac{1}{\eta}(2\eta^3 + 6\eta^2 + 6\eta + 2) \qquad h = -2\eta^2 + 2\eta + 4 \;;$$

$[N]^2 = [N]\,[N]$

$\{x\}$, $\{y\}$ - Coordinate column matrix of cable net's joint in the direction x and y;

$\{1\}$ - Unit column matrix is composed of (m+1) or (n+1) elements;

$\{W_x\}$, $\{W_y\}$, $\{W'_x\}$, $\{W'_y\}$ - Generalized displacement column matrix of cable net's joint in the direction x and y.

Introducing boundary condition $W_{x(0)} = W_{x(ma)} = W_{y(0)} = W_{y(nb)} = 0$ and $W''_{x(0)} = W''_{x(ma)} = W''_{y(0)} = W''_{y(nb)} = 0$ into the integral edge value $W'_{x(0)}$, $W_{x(0)}$ and $W'_{y(0)}$, $W_{y(0)}$ of equation (4), gives :

$$W_{x(ma)} = \left(\frac{a}{12}\right)^2 [N]^2_{(m+1)} \{W''_x\} + W'_{x(0)}\, 1_x = 0$$

$$W_{y(nb)} = \left(\frac{b}{12}\right)^2 [N]^2_{(n+1)} \{W''_y\} + W'_{y(0)}\, 1_y = 0 \qquad (5)$$

where : $[N]^2_{(m+1)}$, $[N]^2_{(n+1)}$ - Row matrix of row (m+1) and row (n+1) in $[N]^2$.

From equation (5) :

$$\left.\begin{aligned} W'_{x(0)} &= -\frac{1}{1_x}\left(\frac{a}{12}\right)^2 [N]^2_{(m+1)} \{W''_x\} \\ W'_{y(0)} &= -\frac{1}{1_y}\left(\frac{b}{12}\right)^2 [N]^2_{(n+1)} \{W''_y\} \end{aligned}\right\} \qquad (6)$$

Combining the above equation with equation (4) :

$$\left.\begin{aligned} \{W'_x\} &= \left(\frac{a}{12}[N] - \left(\frac{a}{12}\right)^2 \frac{1}{1_x}\{1\}[N]^2_{(m+1)}\right)\{W''_x\} \\ \{W'_y\} &= \left(\frac{b}{12}[N] - \left(\frac{b}{12}\right)^2 \frac{1}{1_y}\{1\}[N]^2_{(n+1)}\right)\{W''_y\} \\ \{W_x\} &= \left(\left(\frac{a}{12}\right)^2 [N]^2 - \left(\frac{a}{12}\right)^2 \frac{1}{1_x}\{x\}[N]^2_{(m+1)}\right)\{W''_x\} \\ \{W_y\} &= \left(\left(\frac{b}{12}\right)^2 [N]^2 - \left(\frac{b}{12}\right)^2 \frac{1}{1_y}\{y\}[N]^2_{(n+1)}\right)\{W''_y\} \end{aligned}\right\} \qquad (7)$$

$\left(\frac{\partial W}{\partial x}\right)^2$, $\left(\frac{\partial W}{\partial y}\right)^2$ of equation (2) can be written as :

$$\left.\begin{aligned} \left(\frac{\partial W}{\partial x}\right)^2 &= \left(\frac{W_{x(i)} - W_{x(i-1)}}{a}\right)^2 = \frac{1}{a^2}\begin{Bmatrix} W_{x(i-1)} \\ W_{x(i)} \end{Bmatrix}^T \begin{bmatrix} 1 & -1 \\ -1 & 1 \end{bmatrix} \begin{Bmatrix} W_{x(i-1)} \\ W_{w(i)} \end{Bmatrix} \\ &= \frac{1}{a^2}\{W''_x\}^T [C_x]^T \begin{bmatrix} 1 & -1 \\ -1 & 1 \end{bmatrix} C_x \{W''_x\} \\ \left(\frac{\partial W}{\partial y}\right)^2 &= \frac{1}{b^2}\{W''_y\}^T [C_y]^T \begin{bmatrix} 1 & -1 \\ -1 & 1 \end{bmatrix} C_y \{W''_y\} \end{aligned}\right\} \qquad (8)$$

Where $[C_x] = \begin{bmatrix} d_{(i-1,1)} & d_{(i-1,2)} & \cdots & d_{(i-1,m+1)} \\ d_{(i,1)} & d_{(i,2)} & \cdots & d_{(i,m+1)} \end{bmatrix}$ is composed by (i-1) row and (i) row of square matrix $[D_x]$ defined by

$$\{W_x\} = \left(\left(\frac{a}{12}\right)^2 [N]^2 - \left(\frac{a}{12}\right)^2 \frac{1}{1_x}\{x\}[N]^2_{(m+1)}\right)\{W''\} = [D_x]\{W''_x\} .$$

$[C_y]$ can be obtained by the same method.
Equations (7) and (8) and a finite difference scheme can convert equations (1) and (2) into algebraic equations :

$$(H_{xo} + \Delta H_x)\{W''_x\} + (H_{yo} + \Delta H_y)\{W''_y\} + \Delta H_x \frac{1}{a^2} [1 \quad -2 \quad 1]\begin{Bmatrix} Z_{o(i-1,j)} \\ Z_{o(i,j)} \\ Z_{o(i+1,j)} \end{Bmatrix}\{1\} +$$

$$+ \Delta H_y \frac{1}{b^2} [1 \quad -2 \quad 1]\begin{Bmatrix} Z_{o(i,j-1)} \\ Z_{o(i,j)} \\ Z_{o(i,j+1)} \end{Bmatrix}\{1\} + \Delta q\{1\} = 0 \qquad (9)$$

$$\left.\begin{aligned} &\sum_0^{1_x} \left(\frac{Z_{o(i,j)} - Z_{o(i-1,j)}}{a} \left(\frac{a}{12} [N] - \left(\frac{a}{12}\right)^2 \frac{1}{1_x} \{1\} [N]^2_{(m+1)}\right)\{W''_x\}\right) a + \\ &\quad + \sum_0^{1_x} \left(\frac{1}{2} \frac{1}{a^2} \{W''_x\}^T [C_x]^T \begin{bmatrix} 1 & -1 \\ -1 & 1 \end{bmatrix} [C_x]\{W''_x\}\right) a = \Delta H_x \frac{1_x}{EA_x} \{1\} \\ &\sum_0^{1_x} \left(\frac{Z_{o(i,j)} - Z_{o(i,j-1)}}{b} \left(\frac{b}{12} [N] - \left(\frac{b}{12}\right)^2 \frac{1}{1_y} \{1\} [N]^2_{(n+1)}\right)\{W''_y\}\right) b + \\ &\quad + \sum_0^{1_x} \left(\frac{1}{2} \frac{1}{b^2} \{W''_y\}^T [C_y]^T \begin{bmatrix} 1 & -1 \\ -1 & 1 \end{bmatrix} [C_y]\{W''_y\}\right) b = \Delta H_y \frac{1_y}{EA_x} \{1\} \end{aligned}\right\} \qquad (10)$$

For joints near the boundary, with grid lengths ηa or ηb, the row matrix of third term and forth term in equation (9) is $\left[\frac{2}{1+\eta} \quad -\frac{2}{\eta} \quad \frac{2}{\eta(1+\eta)}\right]$.

When cable net roof plane has one or two symmetry axes, as its deflection curve is symmetric about midpoint of span, there are also boundary condition's $W'_{x(\frac{m}{2}a)} = W'_{y(\frac{n}{2}b)} = 0$. Taking $W'_{x(0)}$, $W'_{y(0)}$ into equation (4) after they are solved, gives :

$$\left.\begin{aligned} \{W'_x\} &= \left(\frac{a}{12}[N] - \frac{a}{12}\{1\}[N]_{(\frac{m}{2}+1)}\right)\{W''_x\} \\ \{W'_y\} &= \left(\frac{b}{12}[N] - \frac{b}{12}\{1\}[N]_{(\frac{n}{2}+1)}\right)\{W''_y\} \\ \{W_x\} &= \left(\left(\frac{a}{12}\right)^2[N]^2 - \frac{a}{12}\{x\}[N]_{(\frac{m}{2}+1)}\right)\{W''_x\} \\ \{W_y\} &= \left(\left(\frac{b}{12}\right)^2[N]^2 - \frac{b}{12}\{y\}[N]_{(\frac{n}{2}+1)}\right)\{W''_y\} \end{aligned}\right\} \qquad (11)$$

Where : $[N]_{(\frac{m}{2}+1)}$, $[N]_{(\frac{n}{2}+1)}$ - Row matrix of row $(\frac{m}{2}+1)$ and row $(\frac{n}{2}+1)$ in $[N]$.

Equation (2) can also be written by using equation (11) :

$$\left.\begin{aligned}
&\sum_{0}^{l_x}\left(\frac{Z_{o(i,j)} - Z_{o(i-1,j)}}{a}\left(\frac{a}{12}[N] - \frac{a}{12}\{1\}[N]_{(\frac{m}{2}+1)}\right)\{W''_x\}\right) a \; + \\
&+ \sum_{0}^{l_x}\left(\frac{1}{2}\frac{1}{a^2}\{W''_x\}^T [F_x]^T \begin{bmatrix} 1 & -1 \\ -1 & 1 \end{bmatrix} [F_x]\{W''_x\}\right) a = \Delta H_x \frac{l_x}{EA_x}\{1\} \\
&\sum_{0}^{l_x}\left(\frac{Z_{o(i,j)} - Z_{o(i,j-1)}}{b}\left(\frac{b}{12}[N] - \frac{b}{12}\{1\}[N]_{(\frac{n}{2}+1)}\right)\{W''_y\}\right) b \; = \\
&+ \sum_{0}^{l_x}\left(\frac{1}{2}\frac{1}{b^2}\{W''_y\}^T [F_y]^T \begin{bmatrix} 1 & -1 \\ -1 & 1 \end{bmatrix} [F_y]\{W''_y\}\right) b = \Delta H_y \frac{l_y}{EA_y}\{1\}
\end{aligned}\right\} \quad (12)$$

Where $[F_x]$ is composed by (i-1) row and (i) row of square matrix $[G_x]$ defined by $\{W_x\} = ((\frac{a}{12})^2 [N]^2 - \frac{a}{12}\{x\}[N]_{(\frac{m}{2}+1)}\{W''_x\} = [G_x]\{W''_x\}$ of equation (8). $[F_y]$ can be obtained by the same method.

We make use of iteration to solve equations (9) and (10) or equations (9) and (12).

EXAMPLE

The mixed method of finite integral and finite difference for analysing cable net roof structures, it suited for computer implementation. For simple instruction how to apply this method, take the saddle cable net roof structure of reference [3] for example. Geometrical size of roof is shown in Fig.2. Tensile rigidity $EA_x = 95000 \frac{KN}{m}$ and $EA_y = 47500 \frac{KN}{m}$; Intensity of the loading $\Delta q = 1 \frac{KN}{m^2}$; Pretensions of cable net in primary state $H_{xo} = 110 + 150 \frac{y^2}{l_2^2}$ (KN/m) and $H_{yo} = 70 \frac{KN}{m}$.

The cable net is divided into eight parts in both directions. As the cable net plane is symmetric about axis x and axis y, we only need to number joints of a fourth part of cable net plane. The joint numbers and coordinates are shown in Fig.2.

Column spacing of cable net rigid edge beams is small and cable net is simply supported at the edge beam. So joint's 14 ∿ 20 W = 0 and W" = 0. According to equation (9) we can write 13 equations for joints 1 ∿ 13. For example, equation of joint 1 is :

$$(110 + \Delta H_{14}) \begin{Bmatrix} 0 \\ W''_{14} \\ W''_{1} \end{Bmatrix} + (70 + \Delta H_{16y}) \begin{Bmatrix} W''_{16} \\ W''_{2} \\ W''_{1} \end{Bmatrix} =$$

$$= -1 - \Delta H_{14} \frac{-11.75 - 2(-9.925) - 9.169}{9.0^2} - \Delta H_{16y} \frac{2(-9,581) - 2(-9.925)}{9.0^2} \qquad (1)$$

According to equations (12) we can write 8 equations for the horizontal increments ΔH_{14}, ΔH_{15}, ΔH_{16x}, ΔH_{18}, ΔH_{16y}, ΔH_{17}, ΔH_{19}, ΔH_{20}. For example, equation of ΔH_{18} is :

$$2 \Big\{ \Big[\frac{-4.682 - (-5.14)}{6.75} \Big(\frac{6.75}{12} \begin{bmatrix} 0 & 0 & 0 \\ 3.857 & 5.625 & -0.482 \\ 2.333 & 14.292 & 4.375 \end{bmatrix} -$$

$$- \frac{6.75}{12} \begin{Bmatrix} 1 \\ 1 \\ 1 \end{Bmatrix} (2.333 \quad 14.292 \quad 4.375) \Big\} \begin{Bmatrix} W''_{18} \\ W''_{4} \\ W''_{5} \end{Bmatrix} \Big] 6.75 +$$

$$+ \Big[\frac{-4.44 - (-4.682)}{9.0} \Big(\frac{9.0}{12} \begin{bmatrix} 0 & 0 & 0 \\ 3.875 & 5.615 & -0.482 \\ 2.333 & 14.292 & 4.375 \end{bmatrix} -$$

$$- \frac{9.0}{12} \begin{Bmatrix} 1 \\ 1 \\ 1 \end{Bmatrix} (2.333 \quad 14.292 \quad 4.375) \Big) \begin{Bmatrix} W''_{18} \\ W''_{4} \\ W''_{5} \end{Bmatrix} \Big] 9.0 \Big\} +$$

$$+ 2 \Big\{ \frac{1}{2} \Big[\frac{1}{6.75^2} \begin{Bmatrix} W''_{18} \\ W''_{4} \\ W''_{5} \end{Bmatrix}^T \begin{bmatrix} 0 & 0 & 0 \\ -2.318 & -46.433 & -18.136 \end{bmatrix}^T \begin{bmatrix} 1 & -1 \\ -1 & 1 \end{bmatrix}$$

$$\begin{bmatrix} 0 & 0 & 0 \\ -2.318 & -46.433 & -18.136 \end{bmatrix} \begin{Bmatrix} W''_{18} \\ W''_{4} \\ W''_{5} \end{Bmatrix} \Big] 6.75 +$$

$$+ \frac{1}{2} \Big[\frac{1}{9.0^2} \begin{Bmatrix} W''_{18} \\ W''_{4} \\ W''_{5} \end{Bmatrix}^T \begin{bmatrix} -0.183 & -58.43 & -24.86 \\ 9.335 & -88.43 & -47.70 \end{bmatrix}^T \begin{bmatrix} 1 & -1 \\ -1 & 1 \end{bmatrix}$$

$$\begin{bmatrix} -0,183 & -58.43 & -24.86 \\ 9.335 & -88.43 & -44.79 \end{bmatrix} \begin{Bmatrix} W''_{18} \\ W''_{4} \\ W''_{5} \end{Bmatrix} \Big] 9.0 \Big\} = \Delta H_{18} \frac{15.75 \times 2}{95000} \qquad (17)$$

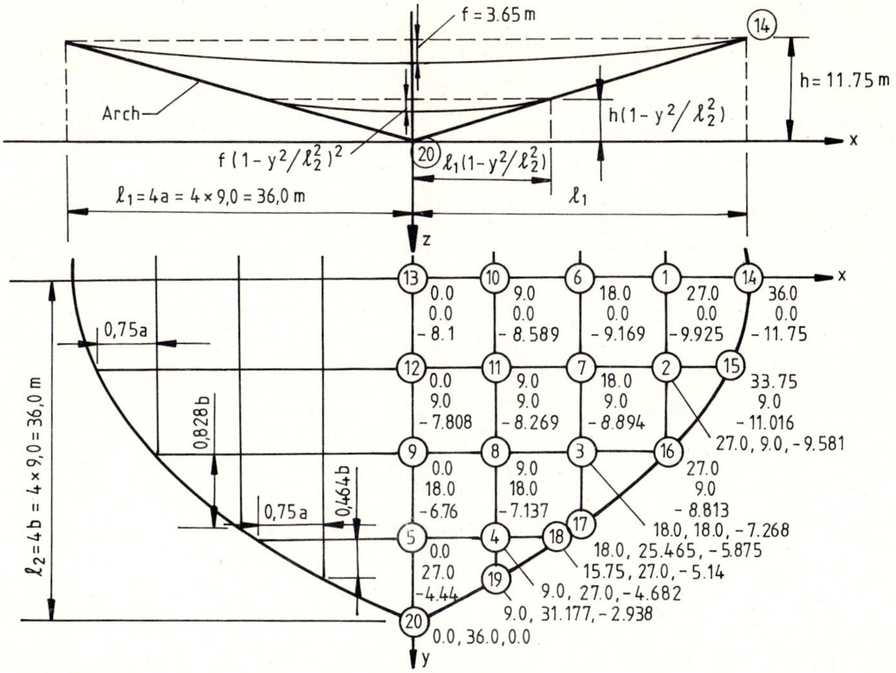

Fig.2

Solving the above 21 equations we can obtain solutions. The computation result agrees with reference [3] and is comparable to the result by stiffness method [4].

CONCLUSION

The accuracy of the mixed method of finite integral and finite difference for analysing cable-suspended roofs is superior to that of the finite difference method. So the number of unknown quantities can be reduced and application on microcomputer is straightforward.

REFERENCES

[1] Harbin Arch. Eng. Institute, Steel structure of large span building, China building Industry Press (1985).

[2] Huang Nong, Using Finite Integral Method Solve Problem of Lateral Torsional Buckling, Xian Inst. of Met. and Const. Eng., Science research Proceeding, China (1979).

[3] Shen Shi-Zhao and Xu Chong-Bao, Simplified Analysis of the acting of unsymmetric loading Cable-Net Roofs With Plane Parabolic Supporting Arches, Journal of Harbin Arch. Eng. Inst., China (1982.3).

[4] P. Krishna, Cable-Suspended Roofs. Mcgraw-Hill Book Company (1978).

SOLUTIONS OF THE EQUATIONS OF THE UNIFIED APPROACH IN THE ANALYSIS OF WALL BEAMS, PLATES, ARCHES AND SHELLS

Pimpo Hernández*, Enrique Ramírez**, José Soler*

SUMMARY

The Unified Approach[1] in the analysis of wall beams, plates, arches, and shells is an alternative elastic analysis method that surpass the limitations of classic methods: Methods of Elasticity Theory and Methods that come from the hypothesis of Navier, Love and Kirchoff.

Some of the advantages of this proposed method are:

1) It can be applied as well in thick elements than thin ones.

2) If the width of b = 1 is taken and some simplifications after, the theories of Navier, Love and Kirchoff are obtained as particular cases.

3) It can take in account the effect of the shear and transversal tensions in the deformations.

4) It permits calculate the transversal tensions.

In this work are shown the general equations of Unified Approach for wall beams in any cross section and under different border conditions and load regimes. Different solutions alternatives of these general equations as the sliddings, transference and mixed methods are presented.

1.- INTRODUCTION.

The steps to follow in the application of this method in concrete cases are:

a) The equations of the Elasticity are established.

Only the equations of the equilibrium, the boundary conditions and the equation of the strain energy-in stress- are necessary.

b) The body forces, displacements and tensions are developed in infinite series. The series of tensions must fulfil the boundary conditions in the intrados and extrados.

c) The equations of internal equilibrium specific to this method are

* Professor.- Facultad de Construcciones. ISPJAE, La Habana, Cuba.
** Professor.- Escuela de Ingeniería Civil. U.A.P. Puebla, México.

obtained, introducing the series of tensions and body forces in the equations of equilibrium of Elasticity.

d) After, if one applies the principle of virtual works, the geometrical equations specifics of the method as well the boundary conditions are obtained.

e) The geometrical equations are transformed separating the bending effect and the shear one.

f) Deriving the density of the strain energy, the physical equations are obtained.

g) Finally, the system of equations are solved.

2.- EQUATIONS OF ELASTICITY.

a) Equations of Equilibrium.

The equations of equilibrium are:

$$\frac{\partial \sigma_x}{\partial x} + \frac{\partial(\tau_{xy} b)}{b \partial y} + X = 0 \qquad [1]$$

$$\frac{\partial \tau_{xy}}{\partial x} + \frac{\partial(\sigma_y b)}{b \partial y} + Y = 0 \qquad [2]$$

$$\tau_{xy} = \tau_{yx} \qquad [3]$$

These equations can be directly obtained from the equilibrium of the element of figure 1:

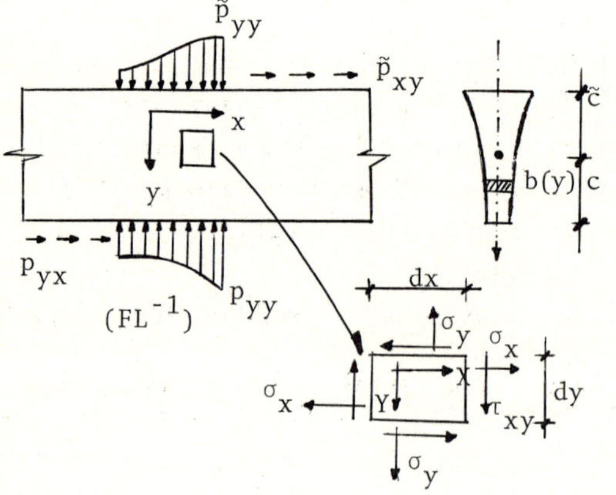

Figure 1

b) Geometrical equations

The geometrical equations are the same that the equations of the plane stress state:

$$\varepsilon_x = \frac{\partial u}{\partial x} \quad [4] \qquad \varepsilon_y = \frac{\partial v}{\partial y} \quad [5] \qquad \gamma_{xy} = \frac{\partial u}{\partial y} + \frac{\partial v}{\partial x} \quad [6]$$

c) Physical equations

The physical equations are the well known of the Theory of Elasticity (Hooke's and Lamé's equations).

d) Strain Energy

The density of the strain energy is stated in the form:

$$U = \tfrac{1}{2}(\sigma_x \varepsilon_x + \sigma_y \varepsilon_y + \tau_{xy}\gamma_{xy}) \qquad [7]$$

The equation [7] can be established in tensions:

$$U = \frac{1}{2\varepsilon}(\sigma_x^2 + \sigma_y^2 - 2\nu\sigma_x\sigma_y) + \frac{1}{2G}\tau_{xy}^2 \qquad [8]$$

or, in deformations:

$$U = \frac{E}{2(1-\nu^2)}(\varepsilon_x^2 + \varepsilon_y^2 + 2\nu\varepsilon_x\varepsilon_y) + \frac{G}{2}\gamma_{xy}^2 \qquad [9]$$

3.- RESUME OF GENERAL EQUATIONS FOR THE WALL BEAM

After, the indicated steps in paragraph 1 are applied, we obtain the following equations:

3.1 Equations of Equilibrium

$$\frac{dN}{dx} + p_o = 0 \qquad \text{(one equation)} \quad [10]$$

$$\frac{dMi}{dx} - Q_i + p_i = 0 \qquad \forall i \geq 1 \text{ (i equations)} \quad [11]$$

$$\frac{dQ1}{dx} + q_1 = 0 \qquad \text{(one equation)} \quad [12]$$

$$\frac{dQi}{dx} - T_i + q_i = 0 \qquad \forall i \geq 2 \text{ (i-1 equations)} \quad [13]$$

We can observe there is a system of (2i+1) equations with the following unknowns:

Ni; Mi (i unknowns); Ti (i-1 unknowns); Qi (i unknowns); it means 3i unknowns or in other words there are (i-1) unknowns more than equations, which are equal to the number of transversal tensions Ti.

In the general case there are not enough equations to obtain the unknowns and is necessary to have recourse to the geometrical and physical equations, whatever the boundary conditions.

If i=1, there are three equations and three unknowns that are the cu-rrent ones of the Streght of Materials.

3.2 Geometrical Equations

$$\kappa_i = \frac{du_i}{dx} \quad (i+1 \text{ equations}) \qquad [14]$$
$i \geq 0$

$$\gamma_i = u_i + \frac{dv_i}{dx} \quad (i \text{ equations}) \qquad [15]$$
$i \geq 1$

$$\zeta_i = v_i \quad (i-1 \text{ equations}) \qquad [16]$$
$i \geq 2$

That is a system of 3i equations with (2i + 1) unknowns: u_i (i + 1 unk-nowns) and v_i (i unknowns).

3.3 Boundary Conditions

The boundary conditions stated in forces or in sliddings are:

$$\begin{aligned} p_{xxi} &= M_{iA} \text{ —— } u_{iA} = 0, \ \forall i \geq 0 \quad (i+1 \text{ conditions}) \\ \tilde{p}_{xxi} &= -M_{iB} \text{ —— } u_{iB} = 0, \ \forall i \geq 0 \quad (i+1 \text{ conditions}) \\ p_{xyi} &= Q_{iA} \text{ —— } v_{iA} = 0, \ \forall i \geq 1 \quad (i \text{ conditions}) \\ \tilde{p}_{xyi} &= -Q_{iB} \text{ —— } v_{iB} = 0, \ \forall i \geq 1 \quad (i \text{ conditions}) \end{aligned} \qquad [17]$$

3.4 Physical Equations

We will stop the physical equations in the term number i:

$$M_i = E'[I_i\kappa_i + \mu_c \nu \, (\sum_{j=2}^{\infty} \rho''_{ij} \; \zeta_i) \quad (i+1 \text{ equations}) \qquad [18]$$
$\forall i \geq 0$

$$T_i = E'[\mu_c \nu \, (\sum_{j=0}^{\infty} \rho''_{ij} \; X_j) + \sum_{j=2}^{\infty} I''_{ij}\zeta_j] \quad (i-1 \text{ equations}) \qquad [19]$$
$\forall i \geq 2$

$$Q_i = G\sum_{j=1}^{\infty} I'_{ij}\gamma_j \quad (i \text{ equations}) \qquad [20]$$
$\forall i \geq 1$

In general, the system of equations can be solved in sliddings[2] (Method of Displacements) or in sliddings and forces (Mixed Method). In the first case we obtain a system of (2i+1) differential equations of se-cond order with (4i+2) boundary conditions.

In the second case we obtain a system of (4i+2) differential equations of first order with (4i+2) boundary conditions.

4.- SOLUTION IN SLIDDINGS AND FORCES.

The equations of equilibrium, the geometrical ones and the physical ones can be written in matrix form as shown:

a) Equilibrium equations:

$$\underset{\sim}{S}_1' - B_1^t \underset{\sim}{S}_1 - B_2^t \underset{\sim}{S}_2 + \underset{\sim}{q} = \underset{\sim}{\Theta} \qquad [21]$$

Where:

$$\underset{(2i+1)(1)}{\underset{\sim}{S}_1} = \begin{Bmatrix} \underset{\sim}{M}_i \\ \underset{\sim}{Q}_i \end{Bmatrix} \quad [22]; \qquad \underset{\sim}{S}_2 = \begin{Bmatrix} T_i \end{Bmatrix} \qquad [23]$$

$$B_1^t = \left[\begin{array}{c|c} \underset{(i+1)(i+1)}{\Theta} & \underset{(i+1)i}{J_1^t} \\ \hline \underset{i(i+1)}{\Theta} & \underset{(ixi)}{\Theta} \end{array}\right] \qquad [24]$$

$$\underset{(i+1)i}{J_1^t} = \left[\begin{array}{c} \underset{(1xi)}{\Theta} \\ \hline \underset{(ixi)}{I} \end{array}\right] \quad [25] \qquad B_2^t = \left[\begin{array}{c} \underset{(i+1)(i-1)}{\Theta} \\ \hline \underset{i(i-1)}{J_2^t} \end{array}\right] \qquad [26]$$

$$\underset{i(i-1)}{J_2^t} = \left[\begin{array}{c|c} \underset{(i-1)1}{\Theta} & \underset{(i-1)(i-1)}{I} \end{array}\right] \quad [27] \qquad \underset{\sim}{q} = \begin{Bmatrix} \underset{\sim}{p}_i \\ \underset{\sim}{q}_i \end{Bmatrix} \qquad [28]$$

Note that the generalized internal actions $\underset{\sim}{S}_1$ are those that figure in the boundary conditions.

b) Geometrical equations:

$$\underset{\sim}{\xi}_1 = \underset{\sim}{u}' + B_1 \underset{\sim}{u} \quad [29] \qquad \underset{\sim}{\xi}_2 = B_2 \underset{\sim}{u} \qquad [30]$$

Where

$$\underset{(2i+1)1}{\underset{\sim}{\xi}_1} = \begin{Bmatrix} \kappa_i \\ \gamma_i \end{Bmatrix} \quad [31] \qquad \underset{(i-1)1}{\underset{\sim}{\xi}_2} = \begin{Bmatrix} \zeta_i \end{Bmatrix} \quad [32] \qquad \underset{\sim}{u} = \begin{Bmatrix} \underset{\sim}{u}_i \\ \underset{\sim}{v}_i \end{Bmatrix} \qquad [33]$$

c) Physical equations:

$$\begin{Bmatrix} \underset{\sim}{S}_1 \\ \underset{\sim}{S}_2 \end{Bmatrix} = \begin{bmatrix} K_{11} & K_{12} \\ K_{12}^t & K_{22} \end{bmatrix} \begin{Bmatrix} \underset{\sim}{\xi}_1 \\ \underset{\sim}{\xi}_2 \end{Bmatrix} \quad [34] \qquad K_{11} = \begin{bmatrix} \lceil K_M \rfloor & \Theta \\ \Theta & K_Q \end{bmatrix} \qquad [35]$$

$$K_{12} = \begin{bmatrix} K_M \\ \Theta \end{bmatrix} \quad [36] \qquad K_{22} = \begin{bmatrix} K_{TT} \end{bmatrix} \qquad [37]$$

We can define the state vector $\underset{\sim}{Z}$ that contains all the unknowns which can appear in the boundary conditions:

$$\underset{\sim}{Z} = \begin{Bmatrix} \underset{\sim}{u} \\ \underset{\sim}{S}_1 \end{Bmatrix} \qquad [38]$$

Basically the mixed Method consists in reducing the system of equations of equilibrium, the geometrical ones and the physical ones to a system as:

$$\underset{\sim}{Z}' = A\underset{\sim}{Z} + \underset{\sim}{a} \qquad [39]$$

Following we develop the obtainment of A and $\underset{\sim}{a}$.

First:

S_2 is stated as a function of z, which is necessary because the S_2 are not part of the state vector.

From the second equation [34]

$$S_2 = K_{12}^t \xi_1 + K_{22} \xi_2 \qquad [40]$$

but from the [30]

$$\xi_2 = C_2 \; Z \qquad [41]$$

Where:

$$\underset{(i-1)(4i+2)}{C_2} = \underset{(i-1)(2i+1)}{\left[B_2 \;\middle|\; \Theta \right]} \qquad [42]$$

From the first equation [34]:

$$K_{11}\xi_1 = S_1 - K_{12}\xi_2 = S_1 - K_{12}C_2Z \qquad [43]$$

and

$$\xi_1 = K_{11}^{-1}S_1 - K_{11}^{-1}K_{12}C_2Z \qquad [44]$$

or $\xi_1 = C_1 \; Z$ Where: $C_1 = \left[\underset{(2i+1)(2i+1)}{K_{11}^{-1} K_{12}B_2} \;\middle|\; \underset{(2i+1)(2i+1)}{K_{11}^{-1}} \right]$ [45]

By introducing the equations [41] and [44] in [40] we obtain

$$S_2 = D_2 \; Z \qquad [46]$$

in which:

$$D_2 = \left[K_{22}^{*}B_2 \;\middle|\; K_{12}^{t}K_{11}^{-1} \right] \qquad [47]$$

Where:

$$K_{22}^{*} = K_{22} - K_{12}^{t}K_{11}^{-1}K_{12} \qquad [48]$$

As a verification, introducing the equations [41] and [44] in the first of [34] one obtains:

$$S_1 = \left[\Theta \;\middle|\; I \right] Z$$

and the vector of S_1 is stated in function of the Z

Second:

The equation [29] is stated substituting ξ_1 according to the function [44]

$$u' = \left[-B_1 - K_{11}^{-1}K_{12}B_2 \;\middle|\; K_{11}^{-1} \right] Z \qquad [49]$$

Third:

The equations of equilibrium [21] are stated introducing S_2 according to [46]

$$S_1' = \left[B_2^tK_{22}^{*}B_2 \;\middle|\; B_1^t + B_2^tK_{12}^tK_{11}^{-1} \right] Z - q \qquad [50]$$

Fourth:

Joining the equations [49] and [50] one obtains:

$$\begin{Bmatrix} \underset{\sim}{u} \\ \underset{\sim}{S}_1 \end{Bmatrix}' = \left[\begin{array}{c|c} A_{11} & A_{12} \\ \hline A_{21} & A_{22} \end{array}\right] \begin{Bmatrix} \underset{\sim}{u} \\ \underset{\sim}{S}_1 \end{Bmatrix} + \begin{Bmatrix} \underset{\sim}{a}_1 \\ \underset{\sim}{a}_2 \end{Bmatrix} \qquad [51]$$

being

$$A_{11} = -(B_1+K_{11}^{-1}K_{12}B_2);\; A_{12} = K_{11}^{-1};\; A_{21} = B_2^t K_{22}^* B_2;\; A_{22} = -A_{11}^t \qquad [52]$$

$$\underset{\sim}{a}_1 = \underset{\sim}{\Theta} \qquad \underset{\sim}{a}_2 = -\underset{\sim}{q} \qquad [53]$$

The system [51] and the boundary conditions [17] can be solved using classic methods.[2,3] One of them, the method of the transfer matrix is particularly useful because it permits the analysis of the wall beams with discontinuities of form and load.[3]

5.- SOLUTION IN SLIDDINGS.

It could be convenient to solve the problem through a solution in sliddings. For example, in the case of a wall beam fixed in both ends, in which the boundary conditions are merely: $\underset{\sim}{u}_A = \underset{\sim}{u}_B = \Theta$

From the equation [51] one obtains:

$$\underset{\sim}{u}' = A_{11}\underset{\sim}{u}+A_{12}\underset{\sim}{S}_1 \quad [54] \qquad \underset{\sim}{S}_1' = A_{21}\underset{\sim}{u}+A_{22}\underset{\sim}{S}_1-\underset{\sim}{q} \quad [55]$$

Solving for $\underset{\sim}{S}_1$ in the equation [54] and bearing in mind the values of A_{11} and A_{12} given by [52] one obtains:

$$\underset{\sim}{S}_1 = K_{11}\underset{\sim}{u}'+(K_{11}B_1+K_{12}B_2)\underset{\sim}{u} \qquad [56]$$

By substituting [56] in [55] and after some simplifications one obtains:

$$K_{11}\underset{\sim}{u}''+H\underset{\sim}{u}'-M\underset{\sim}{u}+\underset{\sim}{q}=\underset{\sim}{\Theta} \qquad [57]$$

Where:

$$H=(K_{11}B_1+K_{12}B_2)-(K_{11}B_1+K_{12}B_2^t) \qquad [58]$$

and

$$M = \langle B_1^t \; B_2^t \rangle \left[\begin{array}{c|c} K_{11} & K_{12} \\ \hline K_{12}^t & K_{22} \end{array}\right] \begin{Bmatrix} B_1 \\ \hline B_2 \end{Bmatrix} \qquad [59]$$

If in the equations [58] and [59] one introduces B_{11}; B_{21}; K_{12}; K_{22}; etc., according to the equations [24], [26], [35], [36] and [37] one obtains:

$$H = \begin{bmatrix} \Theta & K_{MT}J_2 \\ K_QJ_1 & \Theta \end{bmatrix} - \begin{bmatrix} \Theta & K_{MT}J_2 \\ K_QJ_1 & \Theta \end{bmatrix}^t \qquad [60]$$

being:

$$\underset{(i+1)i}{K_{MT}J_2} = \begin{bmatrix} \underset{(i+1)(1)}{\Theta} & \underset{(i+1)(i-1)}{K_{MT}} \end{bmatrix} \quad ([61] \qquad \underset{(i)(i+1)}{K_QJ_1} = \begin{bmatrix} \underset{(i+1)}{\Theta} & \underset{(i+1)}{K_Q} \end{bmatrix} \quad [62]$$

$$M = \begin{bmatrix} J_1^tK_QJ_1 & \Theta \\ \Theta & J_2^tK_TJ_2 \end{bmatrix} \qquad [63]$$

Where:

$$J_1^tK_QJ_1 = \begin{bmatrix} \underset{(1x1)}{0} & \underset{(1xi)}{\Theta} \\ \underset{(ix1)}{\Theta} & \underset{(ixi)}{K_Q} \end{bmatrix} \quad [64] \qquad J_2^tK_QJ_2 = \begin{bmatrix} \underset{(1x1)}{0} & \underset{1(i-1)}{\Theta} \\ \underset{(i-1)1}{\Theta} & \underset{(i-1)(i-1)}{K_T} \end{bmatrix} \quad [65]$$

The system [57] is solved by the classic methods of solution of the systems of differential equations.[4]

6.- STIFFNESS MATRIX OF THE WALL BEAM.

When it is necessary to analize the system of wall beams (for example a continuous wall beam) it can be convenient to obtain the stiffness matrix of the beam of lenght L and analize it by the Method of Displacements. One of the possible methods is to obtain the stiffness matrix from the transfer matrix[3] and this can be obtained from the matrix A defined by the equations [51] and [52].
If the transfer matrix is named T, for a complete beam we have (Fig.2).

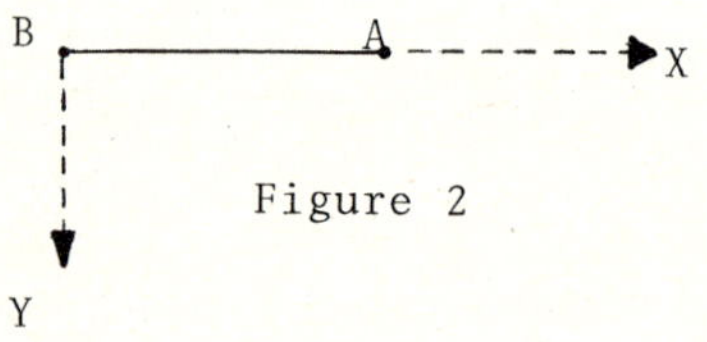

Figure 2

From the figure 2:

$$\underset{\sim}{Z}_A + T\,\underset{\sim}{Z}_B \qquad [66]$$

or, by partitioning:

$$\begin{matrix} \underset{\sim}{u}_A \\ \underset{\sim}{S}_{1A} \end{matrix} = \begin{matrix} T_{11} & T_{12} \\ T_{21} & T_{22} \end{matrix} \; \begin{matrix} \underset{\sim}{u}_B \\ \underset{\sim}{S}_{1B} \end{matrix} \qquad [67]$$

but, the displacements $\underset{\sim}{d}_A$ and $\underset{\sim}{d}_B$, the forces $\underset{\sim}{p}_A$ and $\underset{\sim}{p}_B$ are related with the sliddings $\underset{\sim}{u}_A$ and $\underset{\sim}{u}_B$ and with the generalized internal actions $\underset{\sim}{S}_{1A}$ and $\underset{\sim}{S}_{1B}$ by the boundary conditions of the equation [17] which is written in a vectorial form:

$$\underset{\sim}{u}_A = \underset{\sim}{d}_A \; ; \; \underset{\sim}{u}_B = \underset{\sim}{d}_B \; ; \; \underset{\sim}{S}_{1A} = \underset{\sim}{p}_A \; ; \; \underset{\sim}{S}_{1B} = - \underset{\sim}{p}_B \qquad [68]$$

By substituting:

$$\underset{\sim}{d}_A = T_{11}\underset{\sim}{d}_B - T_{12}\underset{\sim}{p}_B \quad [69] \qquad \underset{\sim}{p}_A = T_{21}\underset{\sim}{d}_B - T_{22}\,\underset{\sim}{p}_B \qquad [70]$$

Solving by p_B in the equation [69] and substituting in the equatio [70] one obtains:

$$\begin{Bmatrix} \underset{\sim}{p}_B \\ \underset{\sim}{p}_A \end{Bmatrix} = \left[\begin{array}{c|c} T_{12}^{-1}\,T_{11} & T_{12}^{-1} \\ \hline T_{21}-T_{22}T_{12}^{-1}T_{11} & T_{22}T_{12}^{-1} \end{array}\right] \begin{Bmatrix} \underset{\sim}{d}_A \\ \underset{\sim}{d}_B \end{Bmatrix} \qquad [71]$$

It is obvious that:

$$K_v = \left[\begin{array}{c|c} T_{12}^{-1}\quad T_{11} & T_{12}^{-1} \\ \hline T_{21}\;\; -T_{22}\;\; T_{12}^{-1}\;\; T_{11} & T_{22}\quad T_{12}^{-1} \end{array}\right] \qquad [72]$$

being K_v the complete stiffness of the wall beam.

REFERENCES.

1.- HERNANDEZ, P., RAMIREZ, E., and SOLER, J. : "Vigas pared, Ecuaciones generales. Primera parte", en Ingeniería Estructural, Vol. IV, No. 2, pp. 131 - 166, ISPJAE, Ciudad de la Habana, mayo de 1983.

2.- FILONENKO - BORODICH, M.: Teoría de la Elasticidad, Ed. Platina, Buenos Aires, Argentina, 1963.

3.- PESTEL, E., and LECKIE, F.: Matrix in Elastomechanics, Mc Graw Hill, New York, San Francisco, Toronto, London, 1963.

4.- COURANT, R. and HILBERT, D.: Methods of Mathematical Physics, Ed. Interscience Publishers, 1965.

APPENDIX

Following, we point out some of the terms used in the former paragraphs.

Mi = Generalized moment, about of the center of section, of stress σ_x:

$$M_i = \int_{\bar{c}}^{c} \sigma_x b\, p_i d_y$$

Q_i = Generalized shear force:

$$Q_i = \int_{\tilde{c}}^{c} \tau_{xy} b p_i' \, dy$$

Ti = Generalized transversal force: $T_i = \int_{\tilde{c}}^{c} \sigma_y b p_i'' \, dy$

Pi, orthogonal polynomials in the interval $(c,\tilde{c})$ which can be obtained by orthogonalization of the power of the variable.

p_i', p_i'', first and second derivative from p_i.

κ_i, general term of the series development of ε_x

$$\varepsilon_x = \varepsilon_o p_o + \kappa_1 p_1 + \kappa_2 p_2 + \ldots + \kappa_i p_i + \ldots$$

ζ_i, general term of the series development of ε_y

$$\varepsilon_y = \zeta_2 p_2'' + \zeta_3 p_3'' + \ldots + \zeta_i p_i'' + \ldots$$

γ_i, general term of the series development of γ_{xy}

$$\gamma_{xy} = \gamma_i p_i' + \gamma_2 p_2' + \ldots + \gamma_1 p_1' + \ldots$$

Alternatively, one can obtain from the strain energy:

$$\kappa_i = \frac{1}{E} \int_{\tilde{c}}^{c} (\sigma_x - \mu_c \nu \sigma_y) \frac{\partial \sigma_x}{\partial M_i} b dy$$

$$\zeta_i = \frac{1}{E} \int_{\tilde{c}}^{c} (\mu_T \sigma_y - \mu_c \nu \sigma_x) \frac{\partial \sigma_y}{\partial T_i} b dy$$

$$\gamma_i = \frac{\mu_s}{G} \int_{\tilde{c}}^{c} \tau_{xy} \frac{\partial \tau_{xy}}{\partial Q_i} b dy$$

Where μ_c, μ_t and μ_s are coefficients that can be equal to 1 or 0, according to if we take in account or not the correspondent term in the expression of the strain energy.

$$E' = \frac{E}{1-\mu^2\nu^2} \qquad \rho_{ij}'' = \int_{\tilde{c}}^{c} p_i p_j'' b dy \qquad \forall i \geq 0, \ \forall j \geq 2$$

$$I_{ij}'' = \int_{\tilde{c}}^{c} p_i'' p_j'' \, b dy \quad \forall i \geq 2, \ \forall j \geq 2; \qquad I_{ij}' = \int_{\tilde{c}}^{c} p_i' p_j' \, b dy \quad \forall i \geq 1, \ \forall j \geq 1$$

$$I_i = \int_{\tilde{c}}^{c} p_i^2 b dy \quad = \text{Generalized moment of inertia}$$

$$K_{M} = E' \begin{bmatrix} A & & & & \\ & I_1 & & & \\ & & \ddots & & \\ & & & I_i & \\ & & & & \ddots \\ & & & & & I_n \end{bmatrix} = E' \lceil I_n \rfloor ; \qquad K_{MT} = K_{TM}^t = \mu_c \nu \, E' [\rho_{ij}''] \quad \forall i \geq 0, \ j \geq 2$$

$$K_{TT} = E'[I''_{ij}]; \quad \forall i \geq 2, j \geq 2 \qquad\qquad K_Q = G \, [I'_{ij}] \quad \forall i \geq 1, j \geq 1$$

BOND SLIP EQUATIONS OF REINFORCED CONCRETE SHELLS

M. Kurata, Dr. Eng.
Department of Architecture
College of Engineering, Nihon University
Koriyama, Fukushima-ken, 963 Japan

INTRODUCTION

The bond slip behavior at the interface between a steel bar and concrete is of importance, in paticular, in the deformation stage of initial cracked, cracked and crashed concrete. In order to predict this behavior, the finite element method has been used, frequently, by many researchers. In this approach, this interface is modeled either by springs or by interface elements, which is based on the relation between the local bond stress and the relative slip of the bar. But the reports on the geometrical nonlinearities of the bond slip in the reinforced concrete plate and shells are not found.

In this paper, we shall try to derive the nonlinear basic equations of the reinforced concrete shells, which include the effect of the bond slip. Using the equations of equilibrium and the relation between the bond shear stress and the relative slip of the bar after deformation, the principle of virtual work is derived and this is applied to the reinforced concrete shells, Finally, the principle of incremental virtual work is presented and the application to a finite element formulation will be discussed briefly.

PRINCIPLE OF VIRTUAL WORK OF REINFORCED CONCRETE BODY

We use the curvilinear coordinate Z^K ($K=1,2,3$) as a coordinate system of the reinforced concrete body as shown in Fig.1.

By introducing the base vector $\mathbb{A}_K$ associated with the Z^K axes after deformation, we can write the stress vector $\mathbb{T}^K$ and the body force $\mathbb{B}$ as follow:

$$\mathbb{T}^K = \sigma^{KL}\mathbb{A}_L \qquad \mathbb{B} = B^K\mathbb{A}_K \tag{1}$$

where σ^{KL} are the Euler stress tensors. Then, we have the following equations of equilibium after deformation.

$$\begin{aligned} {}_c\sigma^{KL}{}_{\shortparallel L} + {}_cB^K &= 0 \qquad \text{in } V_c \\ {}_s\sigma^{KL}{}_{\shortparallel L} &= 0 \qquad \text{in } V_s \end{aligned} \tag{2}$$

where V_c , V_s and ($\shortparallel$) are the concrete volume, the steel bar volume and the covariant derivative after deformation respectively.

we assume the relations between the bond shear stress ${}_b\sigma^{\Lambda 3}$ ($\Lambda=1,2$) and the bond slip ${}_bU_\Lambda$ on the boundary surface S as follow:

$${}_b\sigma^{\Lambda 3} + \mathcal{k}\, {}_bU^\Lambda = 0 \qquad \text{on } S \tag{3}$$

where $\mathcal{k}$ and ${}_bU^\Lambda$ are the bond slip stiffness and the components of the bond slip vector ${}_b\mathbb{U}$ after deformation.

The bond slip vector ${}_b\mathbb{U}$ is defined by

$${}_b\mathbb{U} = {}_s\mathbb{U} - {}_c\mathbb{U} \tag{4}$$

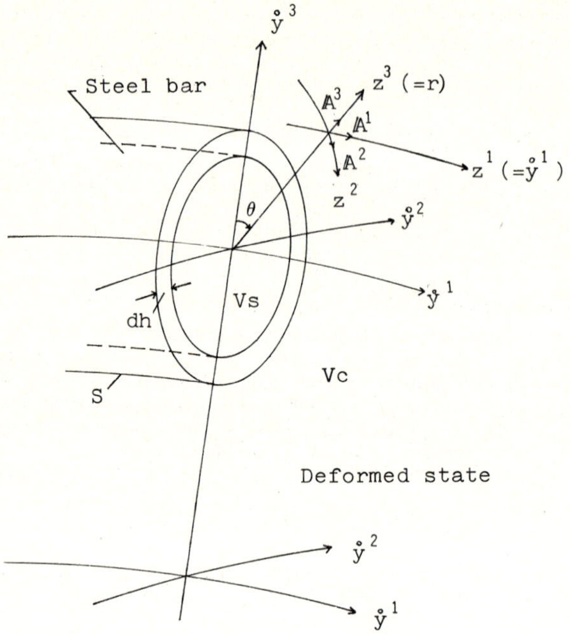

Fig.1 Coodinate system of reinforced concrete body.

where $_s\mathbb{U}$, $_c\mathbb{U}$ are the displacement vectors on the steel bars and the concrete respectively. By resolving $_b\mathbb{U}$, $_s\mathbb{U}$, $_c\mathbb{U}$ in the directions of $\mathbb{A}^k$, we have

$$_b\mathbb{U} = {}_bU_\Lambda \mathbb{A}^\Lambda \quad , \quad {}_s\mathbb{U} = {}_sU_K \mathbb{A}^K \qquad {}_c\mathbb{U} = {}_cU_K \mathbb{A}^K \tag{5}$$

$$_bU_\Lambda = {}_sU_\Lambda - {}_cU_\Lambda \quad , \quad {}_bU_3 = {}_sU_3 - {}_cU_3 = 0 \tag{6}$$

where $\mathbb{A}^\Lambda$ is the contravariant base vectors.

When the body is assumed to execute an arbitary set of the virtual displacements $\delta_S U_K$, $\delta_c U_K$, $\delta_b U_\Lambda$ from the equilibrating configuration of Eqs.(2) and (3), we have the principle of virtual work in deformed state as follow:

$$\int_{V_C}[{}_c\sigma^{KL}\delta_c\varepsilon_{KL}]dV_c + \Sigma\Big[\int_{V_S}[{}_s\sigma^{KL}\delta_S\varepsilon_{KL}]dV_s + \int_S[\mathcal{k}\,{}_bU^\Lambda\delta_bU_\Lambda]ds\Big]$$

$$= \int_{V_C}[{}_cB^K\delta_cU_K]dV_c + \int_{S_\sigma}[{}_c\bar{\sigma}^K\delta_cU_K]ds_\sigma + \Sigma\int_{S_{s\sigma}}[{}_s\bar{\sigma}^K\delta_SU_K]ds_{s\sigma} \tag{7}$$

where $_c\bar{\sigma}^K$, $_s\bar{\sigma}^K$ are the external forces on the boundary surfaces S_σ , $S_{s\sigma}$ of the concrete body and the steel bars respectively, the notation Σ means summation over all the steel bars and $_c\varepsilon_{KL}$, $_s\varepsilon_{KL}$ are defined by

$$_c\varepsilon_{KL} = \frac{1}{2}({}_cU_{K\|L} + {}_cU_{L\|K}) \quad , \quad {}_s\varepsilon_{KL} = \frac{1}{2}({}_sU_{K\|L} + {}_sU_{L\|K}) \tag{8}$$

substituting Eq.(6) into Eq.(8), we have

$$_s\varepsilon_{KL} = {}_c\varepsilon_{KL} + {}_b\varepsilon_{KL} \tag{9}$$

where

$$_b\varepsilon_{KL} = \frac{1}{2}({}_bU_{K\parallel L} + {}_bU_{L\parallel K}) \qquad (10)$$

The stress-strain relations of steel bars are given by

$$_s\sigma^{KL} = {}_sE^{KLMN}{}_s\varepsilon_{MN} \qquad (11)$$

where ${}_sE^{KLMN}$ are the fourth-order tensors of steel bars, substituting Eq.(9) into Eq.(11), we have

$$_s\sigma^{KL} = {}_{sc}\sigma^{KL} + {}_{sb}\sigma^{KL} \qquad (12)$$

$$_{sc}\sigma^{KL} = {}_sE^{KLMN}{}_c\varepsilon_{MK} \qquad {}_{sb}\sigma^{KL} = {}_sE^{KLMN}{}_b\varepsilon_{MN} \qquad (13)$$

Concequently, the substitution of Eqs.(9) and (12) into Eq.(7) and the use of the Kirchhoff-Stress tensors S^{KL} and the Cauchy-Green strain tensors e_{KL} transform Eq.(7) into the Principle virtual work in V_c , V_s , S , S_σ , $S_{s\sigma}$ before deformation as

$$\int_V[{}_cS^{KL}\delta_c e_{KL}]dV + \Sigma\int_{VS}[({}_{sc}S^{KL} - {}_cS^{KL})\delta_c e_{KL} + {}_{sb}S^{KL}\delta_c e_{KL}]dV_s$$
$$+\Sigma[\int_{VS}[{}_sS^{KL}\delta_b e_{KL}]dV_s + \int_S[K\,{}_bU^\wedge\,\delta_b U_\wedge]d_s]$$
$$= \int_{VC}[{}_cb^K\delta_c U_K]dV + \int_{S\sigma}[{}_c\bar{S}^K\delta_c U_K]ds_\sigma + \Sigma\int_{S_{S\sigma}}[{}_s\bar{S}^K\delta\,\mathring{\hat{U}}_K]ds_{S\sigma} \qquad (14)$$

where is the bond stiffness before deformation and ${}_cU_K$, $\mathring{\hat{U}}_K$ are the displacement vector components defined later by Eqs.(19),(24) and ${}_cb^K$, ${}_c\bar{S}^K$, ${}_s\bar{S}^K$ are the body forces, the external forces which do work on ${}_cU_K$, $\hat{U}_K$ respectively.

If only the axial slip and its strain of the steel bar are assumed to occur, we have

$$\int_V[{}_cS^{KL}\delta_c e_{KL}]dV + \Sigma\int_{Cs}A_s[({}_{sc}S^{KL} - {}_cS^{KL})\delta_c e_{KL} + {}_{sb}S^{11}\delta_c e_{11}]_{Co}\,dc_s$$
$$+\Sigma\int_{Cs}[A_s[{}_sS^{11}\delta_b e_{11}]_{z^3=0} + \psi[\tau^1\delta_b U_1]_{z^3=r_s}]dc_s$$
$$= \int_{VC}[{}_cb^K\delta_c U_K]dV + \int_{S\sigma}[{}_c\bar{S}^K\delta_c U_K]ds_\sigma + \Sigma A_{S\sigma}[{}_s\bar{S}^K\delta\,\mathring{\hat{U}}_K]_{S_{S\sigma},z^3=0} \qquad (15)$$

where

$$\tau^1 = K\,{}_bU^1 \qquad (16)$$

and r_s , ψ , A_s and $A_{S\sigma}$ are the radius, the perimeter, the cross sectional area and the area sectioned by the boundary surface $S_{S\sigma}$ of a steel bar before deformation respectively.

BOND SLIP EQUATIONS

When we use normal coordinates (y^α , y^3)(α =1,2) as shown in Fig. 2 and 3, the position vector $\mathbb{r}$ (y^α , y^3) of an arbitrary point on V_c before deformation is given by

$$\mathbb{r}(y^\alpha, y^3) = \mathbb{r}(y^\alpha, 0) + y^3\mathbb{g}_3 \qquad (17)$$

The assumption of this thin structure allows us to express the displacement vector $\mathbb{u}$ as follow:

$$\mathbb{u}(y^\alpha, y^3) = \mathcal{V}(y^\alpha) + y^3\phi(y^\alpha) \qquad (18)$$

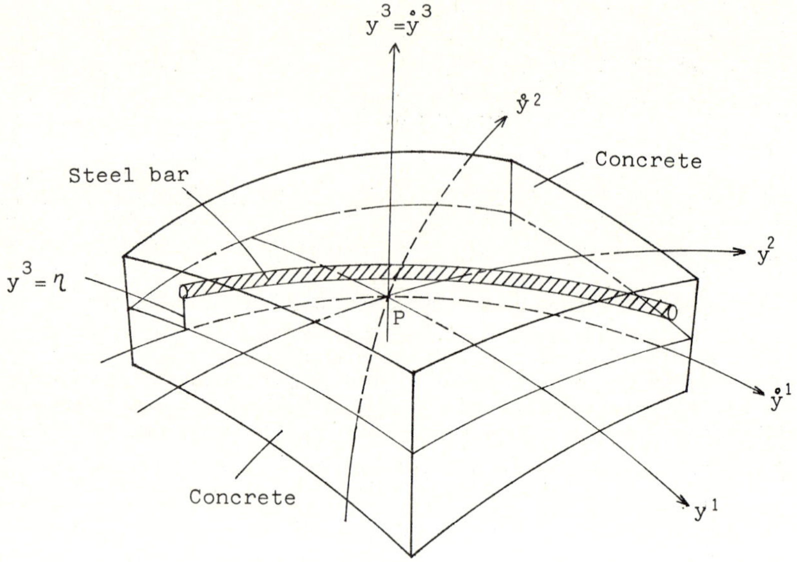

Fig.2 Normal coodinate system of R.C. shell.

If we express $\mathbb{u}$, $\mathscr{v}$, ϕ by the base vectors (g_α , g_3),(g^α , g^3) and ($\bar{g}_\alpha$, $\bar{g}_3$),($\bar{g}^\alpha$, $\bar{g}^3$), we obtain

$$\mathbb{u} = {}_cu^i g_i = {}_cu_i g^i = {}_c\bar{u}^i \bar{g}_i = {}_c\bar{u}_i \bar{g}^i \qquad (i=1,2,3)$$

$$\mathscr{v} = {}_cv^\alpha \bar{g}_\alpha + {}_cw\bar{g}_3 = {}_cv_\alpha \bar{g}^\alpha + {}_cw\bar{g}^3$$

$$\phi = {}_c\phi_\alpha \bar{g}^\alpha \tag{19}$$

Then the displacement vector components u_i and the rotation vector components are approximated as follow:

$$_cu_i = {}_c\bar{u}_i \qquad {}_c\phi_\alpha = -w_{/\alpha} \tag{20}$$

$$_cu_\alpha = {}_cv_\alpha + y^3{}_c\phi_\alpha \quad , \quad {}_cu_3 = {}_cw \tag{21}$$

where (/) is the covariant derivative with respect to $\bar{g}_{\alpha\beta}$ (= $\bar{g}_\alpha \cdot \bar{g}_\beta$). Under the above equations and Kirchhoff-Love assumptions, we have the following strain tensors ${}_ce_{\alpha\beta}$ on V_c of the shell:

$$2{}_ce_{\alpha\beta} = ({}_c\phi_{\alpha\beta} + {}_c\phi_{\alpha\beta}) + y^3(2{}_c\phi_{\alpha/\beta}) + {}_c\phi_\alpha\, {}_c\phi_\beta \ , \ e_{\alpha 3} = e_{33} = 0 \tag{22}$$

where

$$_c\phi_{\alpha\beta} = {}_cv_{\alpha/\beta} - b_{\alpha\beta}\, {}_cw \ , \ {}_c\phi_{\alpha/\beta} = -\, {}_cw_{/\alpha\beta} \tag{23}$$

and where $b_{\alpha\beta}$ are the second foundamental tensors.

Next, we shall assume that the curved axes of steel bars may specified by $\mathring{y}^1$ and $\mathring{y}^3 = y^3 = \eta$ of the curvilinear normal coordinate ($\mathring{y}^\wedge$, y^3)($\wedge$ =1,2) and the bond slip vector $_b\mathbb{U}$ is approximated as shown in Fig.3, namely

$$_b\mathbb{U} = {}_bU^1 \mathbb{A}_1 \doteqdot {}_b\mathbb{u} = \mathring{U}^1 \mathring{G}_1 \qquad \mathbb{A}_1 \doteqdot \mathring{G}_1$$

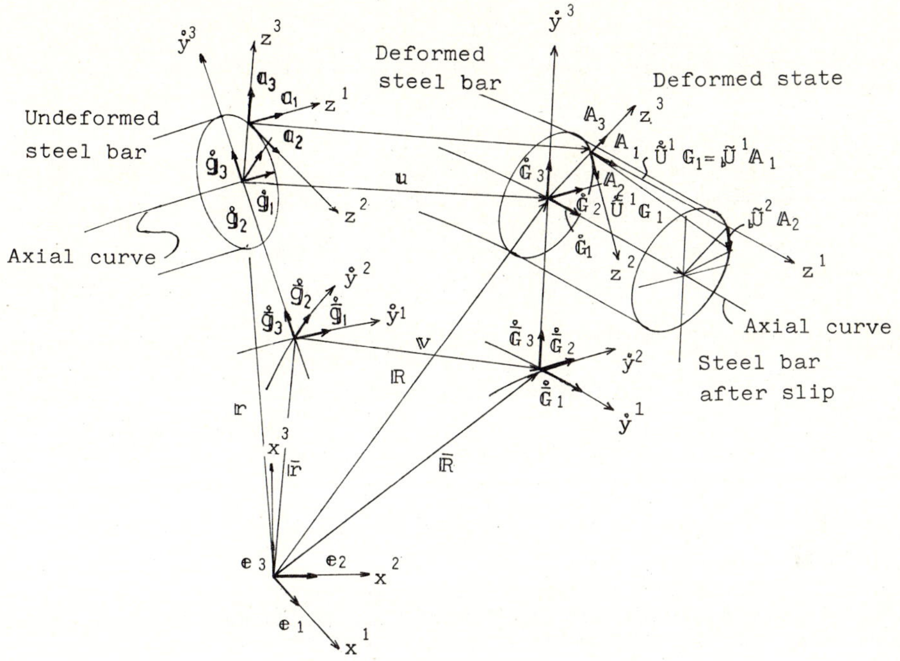

Fig.3 Deformation of a steel bar.

$$ {}_b\mathbb{U} = \mathring{\tilde{U}}_\Lambda g^\Lambda + \mathring{\tilde{U}}_3 \mathring{g}^3 = \mathring{\tilde{v}}_\Lambda \mathring{\bar{g}}^\Lambda + \mathring{\tilde{w}} \mathring{\bar{g}}^3 \quad , \quad \mathring{g}^\Lambda = \mathring{\bar{g}}^\Lambda \ , \ \mathring{g}^3 = \mathring{\bar{g}}^3 $$

$$ {}_bU^1 \doteqdot \mathring{\tilde{U}}^1 \quad , \quad \mathring{\tilde{U}}_\Lambda \doteqdot \mathring{\tilde{v}}_\Lambda \quad , \quad \mathring{\tilde{U}}_3 \doteqdot \mathring{\tilde{w}} \tag{24} $$

where ($\mathring{g}^\Lambda$, $\mathring{g}^3$),($\mathring{\bar{g}}^\Lambda$, $\mathring{\bar{g}}^3$) are contravariant base vectors of ($\mathring{y}^\Lambda$, $\mathring{y}^3$) and ($\mathring{y}^\Lambda$, 0) respectively.

Therefore, the displacement vector components ${}_s\mathring{U}_\Lambda$, ${}_s\mathring{U}_3$ on the axes of steel bars are given in the simmilar forms as

$$ {}_s\mathbb{U} = \mathbb{U} + {}_b\mathbb{U} $$

$$ {}_s\mathbb{U} = {}_s\mathring{U}_\Lambda \mathring{g}^\Lambda + {}_s\mathring{U}_3 \mathring{g}^3 \qquad \mathbb{U} = {}_c\mathring{U}_\Lambda \mathring{g}^\Lambda + {}_c\mathring{U}^3 \mathring{g}^3 $$

$$ {}_s\mathring{U}_\Lambda = {}_c\mathring{U}_\Lambda + \mathring{\tilde{U}}_\Lambda \qquad {}_s\mathring{U}_3 = {}_c\mathring{U}_3 + \mathring{\tilde{U}}_3 $$

and

$$ \mathbb{U} = \mathcal{V} + y^3 \phi \qquad \mathcal{V} = {}_c\mathring{v}_\Lambda \mathring{\bar{g}}^\Lambda + {}_c\mathring{w} \mathring{\bar{g}}^3 \qquad \phi = {}_c\mathring{\phi}_\Lambda \mathring{\bar{g}}^\Lambda $$

$$ {}_s\mathring{U}_\Lambda = ({}_c\mathring{v}_\Lambda + y^3_c \mathring{\phi}_\Lambda) + \mathring{\tilde{v}}_\Lambda \quad , \quad {}_s\mathring{U}_3 = {}_c\mathring{w} + \mathring{\tilde{w}} \tag{25} $$

By the use of the above equations, we obtain the strain tensors ${}_s\mathring{e}_{\Lambda\Gamma}$ on the axes of steel bars as follows:

$$ {}_s\mathring{e}_{\Lambda\Gamma} = {}_c\mathring{e}_{\Lambda\Gamma} + {}_b\mathring{e}_{\Lambda\Gamma} \quad , \quad {}_s\mathring{e}_{\Lambda 3} = {}_s\mathring{e}_{33} = 0 $$

$$ 2{}_c\mathring{e}_{\Lambda\Gamma} = ({}_c\mathring{\phi}_{\Lambda\Gamma} + {}_c\mathring{\phi}_{\Gamma\Lambda}) + y^3 (2{}_c\mathring{\phi}_{\Lambda/\Gamma}) + {}_c\mathring{\phi}_\Lambda {}_c\mathring{\phi}_\Gamma $$

$$ {}_b\mathring{e}_{11} = \mathring{\tilde{\phi}}_{11} + {}_c\mathring{\phi}_1 \mathring{\tilde{\phi}}_1 \quad , \quad {}_b\mathring{e}_{12} = {}_b\mathring{e}_{21} = {}_b\mathring{e}_{22} = 0 \tag{26} $$

and

$$ {}_c\mathring{\phi}_{\Lambda\Gamma} = {}_c\mathring{v}_{\Lambda/\Gamma} - \mathring{b}_{\Lambda\Gamma\, c}\mathring{w} \ , \quad {}_c\mathring{\phi}_{\Lambda} = -{}_c\mathring{w}_{/\Lambda} \ , \quad {}_c\mathring{\phi}_{\Lambda/\Gamma} = -{}_c\mathring{w}_{/\Lambda\Gamma} $$

$$ \mathring{\tilde{\phi}}_{11} = \mathring{\tilde{v}}_{1/1} - \mathring{b}_{11}\mathring{\tilde{w}} \ , \quad \mathring{\tilde{\phi}}_{1} = \mathring{\tilde{w}}_{/1} + \mathring{b}_{11}\mathring{\tilde{v}}^1 \tag{27} $$

where (/) is the covariant derivative with respect to $\bar{g}_{\Lambda\Gamma}$ $(= \bar{g}_\Lambda \cdot \bar{g}_\Gamma)$ and $b_{\Lambda\Gamma}$ are the second foundamental tensors on ($\mathring{y}^\Lambda$, 0).

Next, we shall derive the relations between the bond slip vector components $\mathring{\tilde{U}}_1$, $\mathring{\tilde{U}}^1$ and $\mathring{\tilde{U}}_k$, $\mathring{\tilde{U}}^k$ (k =1,2,3). From Eq.(24), we have

$$ \mathring{\tilde{U}}_1 = \mathring{\tilde{U}}^i \mathring{g}_i \, \mathring{G}_1 \ , \quad \mathring{\tilde{U}}^1 = \mathring{\tilde{U}}^i \mathring{g}_i \, \mathring{G}^1 \qquad (i=1,2,3) \tag{28} $$

The base vectors $\mathring{G}^k$ are given by

$$ \mathring{G}_i = \frac{\partial \mathbb{R}}{\partial \mathring{y}^i} = \frac{\partial}{\partial \mathring{y}^i}(\mathbb{r} + \mathbb{U}) = \mathring{g}_i + \mathbb{U}_{,i} = (\delta_i^k + {}_c\mathring{U}^k{}_{\perp i})\,\mathring{g}_k \quad (i=1,2,3) \tag{29} $$

as shown in Fig.3. The notation ($\perp$) denotes the covariant derivative with respect to $\mathring{g}_{ij}$ $(= \mathring{g}_i \cdot \mathring{g}_j)$.

Neglecting the terms higher than the displacement terms of the first order, we obtain the contravariant base vector G^k as follows:

$$ \mathring{G}^k = (\delta_i^k - {}_c\mathring{U}^k{}_{\perp i})\,\mathring{g}^i $$

Substituting Eqs.(29),(30) and (24) into Eq.(28), we have the following approximate relationship equations as

$$ \mathring{\tilde{U}}_1 = (\delta_1^k + {}_c\mathring{U}^k{}_{\perp 1})\mathring{\tilde{U}}_k = \mathring{\tilde{v}}_1 + {}_c\mathring{\phi}_1 \mathring{\tilde{w}} $$

$$ \mathring{\tilde{U}}^1 = (\delta_k^1 - {}_c\mathring{U}^1{}_{\perp k})\mathring{\tilde{U}}^k = \mathring{\tilde{v}}^1 - {}_c\mathring{\phi}^1 \mathring{\tilde{w}} \tag{31} $$

where

$$ \mathring{\tilde{v}}^1 = \mathring{g}^{1\Lambda}\mathring{\tilde{v}}_\Lambda \ , \quad {}_c\mathring{\phi}^1 = \mathring{\bar{g}}^{1\Lambda}{}_c\mathring{\phi}_\Lambda \ , \quad \mathring{\bar{g}}^{1\Lambda} = \mathring{\bar{g}}^1 \, \mathring{\bar{g}}^\Lambda \tag{32} $$

Substituting Eqs.(22),(26) and (31) into Eq.(15) and using the formulae as follows:

$$ dV = \lambda dy^3 ds \qquad n_\alpha ds_\sigma = \nu_\alpha \lambda dy^3 dc_\sigma \qquad \lambda = 1 - 2y^3 H + (y^3)^2 K \tag{33} $$

where H and K denote, respectively, the mean and Gaussian curvatures. ds and ds_σ are the element of areas on the middle surface S and on the boundary surface s_σ . dc is the line element of the bounding cure c_σ . n_α and ν_α are components of the outward unit normal vectors, respectively, on s_σ and on c_σ .

We obtain finally,

$$ \int_S [{}_cN^{\alpha\beta}\delta_c\phi_{\alpha\beta} + {}_cM^{\alpha\beta}\delta({}_c\phi_{\alpha/\beta}) + {}_cN^{\alpha\beta}{}_c\phi_\alpha \delta_c\phi_\beta]\, ds $$

$$ + \sum \int_{C_s} \begin{bmatrix} \overset{*}{N}{}^{\Lambda\Gamma}\delta_c\mathring{\phi}_{\Lambda\Gamma} + \overset{*}{M}{}^{\Lambda\Gamma}\delta({}_c\mathring{\phi}_{\Lambda/\Gamma}) + \overset{*}{N}{}^{\Lambda\Gamma}{}_c\mathring{\phi}_\Lambda \delta_c\mathring{\phi}_\Gamma \\ + N^{11}\delta_c\mathring{\phi}_{11} + M^{11}\delta({}_c\mathring{\phi}_{1/1}) + N^{11}{}_c\mathring{\phi}_1 \delta_c\mathring{\phi}_1 \end{bmatrix} dcs $$

$$ + \sum \int_{C_s} \begin{bmatrix} \tilde{N}^{11}\delta\mathring{\tilde{\phi}}_{11} + \tilde{N}^{11}{}_c\mathring{\phi}_1 \delta\mathring{\tilde{\phi}}_1 \\ + T^1\delta\mathring{\tilde{v}}_1 + T^1{}_c\mathring{\phi}_1 \delta\mathring{\tilde{w}} \end{bmatrix} dcs $$

$$= \int_S [P^\alpha \delta_c v_\alpha + P \delta_c w] ds + \int_{C_\sigma} [\bar{N}^\alpha \delta_c v_\alpha + \bar{N} \delta_c w + \bar{M}^\alpha \delta_c \phi_\alpha] dc_\sigma$$

$$+ \sum [\bar{\overset{*}{N}}{}^1 \delta \mathring{\bar{v}}_1 + \bar{\overset{*}{N}} \delta \mathring{\bar{w}}]_{c_{s\sigma}} \tag{34}$$

$${}_cN^{\alpha\beta} = \int_{y^3} \lambda_c S^{\alpha\beta} dy^3 , \quad {}_cM^{\alpha\beta} = \int_{y^3} \lambda_c S^{\alpha\beta} y^3 dy^3$$

$$\overset{*}{N}{}^{\Lambda\Gamma} = A_s ({}_{sc}\mathring{S}^{\Lambda\Gamma} - {}_c\mathring{S}^{\Lambda\Gamma}) , \quad \overset{*}{M}{}^{\Lambda\Gamma} = A_s ({}_{sc}\mathring{S}^{\Lambda\Gamma} - {}_c\mathring{S}^{\Lambda\Gamma}) \eta$$

$$N^{11} = A_s \, {}_{sb}\mathring{S}^{11} , \quad M^{11} = A_s \, {}_{sb}\mathring{S}^{11} \eta$$

$$\tilde{N}^{11} = A_s \, {}_sS^{11} , \quad T^1 = \psi \mathring{\tau}^1 \tag{35}$$

and p^α , p are the external forces. $\bar{N}^\alpha$, $\bar{N}$, $\bar{M}^\alpha$ and $\bar{\overset{*}{N}}{}^\alpha$, $\bar{\overset{*}{N}}$ are the boundary forces, respectively, for the displacements of the shell and for the bond slip. Equation (34) is the principle virtual work of thin reinforced concrete shells including the effect of the bond slip.

By the use of Eqs.(19) and (25), we have the transformation law between (y^α , y^3) and ($\mathring{y}^\Lambda$, $\mathring{y}^3$) for the displacement components as

$${}_c\mathring{v}_\Lambda = A^\alpha_\Lambda \, {}_cv_\alpha , \quad {}_c\mathring{w} = {}_cw , \quad {}_c\mathring{\phi}_\Lambda = A^\alpha_\Lambda \, {}_c\phi_\alpha \tag{36}$$

where $A^\alpha_\Lambda = \bar{g}^\alpha \mathring{g}_\Lambda$.

Substituting Eqs.(23) and (27) into Eq.(34) and using Eq.(36), concequently, we obtain the nonlinear equations of equilibrium and the mechanical boundary conditions as follows:

$${}_cN^{\alpha\beta}{}_{/\beta} + \sum A_s [A^\alpha_\Lambda ({}_{sc}\mathring{S}^{\Lambda\Gamma} - {}_c\mathring{S}^{\Lambda\Gamma})_{/\Gamma} + A^\alpha_1 \, {}_{sb}\mathring{S}^{11}{}_{/1}] + P^\alpha = 0$$

$$({}_cM^{\alpha\beta}{}_{/\alpha} - {}_cN^{\alpha\beta} \, {}_c\phi_\alpha)_{/\beta} + ({}_cN^{\alpha\beta} b_{\alpha\beta}) + P$$

$$+ \sum A_s [[({}_{sc}\mathring{S}^{\Gamma\Lambda} - {}_c\mathring{S}^{\Gamma\Lambda})_{/\Lambda} \eta - ({}_{sc}\mathring{S}^{\Lambda\Gamma} - {}_c\mathring{S}^{\Lambda\Gamma}) {}_c\mathring{\phi}_\Lambda]_{/\Gamma}$$

$$+ ({}_{sc}\mathring{S}^{\Lambda\Gamma} - {}_c\mathring{S}^{\Lambda\Gamma}) \mathring{b}_{\Lambda\Gamma} + ({}_{sb}\mathring{S}^{11}{}_{/1} \eta - {}_{sb}\mathring{S}^{11} \, {}_c\mathring{\phi}_1)_{/1} + {}_{sb}\mathring{S}^{11} \mathring{b}_{11}]] = 0$$

$$A_s [({}_sS^{11} \, {}_c\mathring{\phi}_1)_{/1} + {}_s\mathring{S}^{11} b_{11}] - \psi \mathring{\tau}^1 \, {}_c\mathring{\phi}_1 = 0$$

$$A_s ({}_s\mathring{S}^{11}{}_{/1} - {}_s\mathring{S}^{11} b^1_1 \, {}_c\mathring{\phi}_1) - \psi \mathring{\tau}^1 = 0 \tag{37}$$

and

$$\bar{N}^\alpha - [{}_cN^{\alpha\beta} \nu_\beta + A_s \{A^\alpha_\Lambda ({}_{sc}\mathring{S}^{\Lambda\Gamma} - {}_c\mathring{S}^{\Lambda\Gamma}) \mathring{\nu}_\Gamma + A^\alpha_1 \, {}_{sb}\mathring{S}^{11} \mathring{\nu}_1\}] = 0$$

$$\bar{N} - [{}_cM^{\beta\alpha}{}_{/\alpha} - {}_cN^{\alpha\beta} \, {}_c\phi_\alpha + A_s [\{({}_{sc}\mathring{S}^{\Gamma\Lambda} - {}_c\mathring{S}^{\Gamma\Lambda})_{/\Lambda} \eta - ({}_{sc}\mathring{S}^{\Lambda\Gamma} - {}_c\mathring{S}^{\Lambda\Gamma}) {}_c\phi_\Lambda\} \mathring{\nu}_\Gamma$$

$$+ \{{}_{sb}\mathring{S}^{11}{}_{/1} \eta - {}_{sb}\mathring{S}^{11} \, {}_c\mathring{\phi}_1\} \mathring{\nu}_1]] = 0$$

$$\bar{M}^\alpha - [{}_cM^{\alpha\beta} \nu_\beta + A_s [A^\alpha_\Lambda ({}_{sc}\mathring{S}^{\Lambda\Gamma} - {}_c\mathring{S}^{\Lambda\Gamma}) \eta \mathring{\nu}_\Gamma + A^\alpha_1 \, {}_{sb}\mathring{S}^{11} \eta \mathring{\nu}_1]] = 0$$

$$\bar{\overset{*}{N}}{}^1 - A_s \, {}_s\mathring{S}^{11} n_1 = 0$$

$$\bar{\overset{*}{N}} - A_s \, {}_s\mathring{S}^{11} \, {}_c\mathring{\phi}_{11} n_1 = 0 \tag{38}$$

PRINCIPLE OF INCREMENTAL VIRTUAL WORK

Using Eq.(34) and neglecting the terms of higher order product of the incremental displacements, we obtain after some manipulations

$$\Delta U - \Delta W - \Delta E = 0 \tag{39}$$

where

$$\Delta U = \Delta \overset{(0)}{U} + \Delta \overset{(1)}{U} + \Delta \overset{(2)}{U}$$

$$\Delta \overset{(0)}{U} = \int_S [\Delta_C N^{\alpha\beta}\,\delta(\Delta_C \Phi_{\alpha\beta}) + \Delta_C M^{\alpha\beta}\,\delta(\Delta_C \Phi_{\alpha/\beta})]\,ds$$
$$+ \sum\int_{CS} [\Delta \overset{*}{N}{}^{\Lambda\Gamma}\,\delta(\Delta_C \overset{\circ}{\Phi}_{\Lambda\Gamma}) + \Delta \overset{*}{M}{}^{\Lambda\Gamma}\,\delta(\Delta_C \overset{\circ}{\Phi}_{\Lambda/\Gamma})$$
$$+ \Delta N^{11}\,\delta(\Delta_C \overset{\circ}{\Phi}_{11}) + \Delta M^{11}\,\delta(\Delta_C \overset{\circ}{\Phi}_{1/1})]\,dc_S$$
$$+ \sum\int_{CS} [\Delta \widetilde{N}^{11}\,\delta(\Delta \overset{\circ}{\widetilde{\Phi}}_{11}) + \Delta T^{1}\,\delta(\Delta \overset{\circ}{\widetilde{v}}_1)]\,dc_S$$

$$\Delta \overset{(1)}{U} = \int_S [{}_C\Phi_\alpha\,\Delta_C N^{\alpha\beta}\,\delta(\Delta_C \Phi_\beta)]\,ds$$
$$+ \sum\int_{CS} [{}_C\overset{\circ}{\Phi}_\Lambda\,\Delta \overset{*}{N}{}^{\Lambda\Gamma}\,\delta(\Delta_C \Phi_\Gamma) + {}_C\overset{\circ}{\Phi}_1\,\Delta N^{11}\,\delta(\Delta_C \overset{\circ}{\Phi}_1)]\,dc_S$$
$$+ \sum\int_{CS} [{}_C\overset{\circ}{\Phi}_1\,\Delta \widetilde{N}^{11}\,\delta(\Delta \overset{\circ}{\widetilde{\Phi}}_1) + {}_C\overset{\circ}{\Phi}_1\,\Delta T^{1}\,\delta(\Delta \overset{\circ}{\widetilde{w}})]\,dc_S$$

$$\Delta \overset{(2)}{U} = \int_S [{}_C N^{\alpha\beta}\,\Delta_C \Phi_\alpha\,\delta(\Delta_C \Phi_\beta)]\,ds$$
$$+ \sum\int_{CS} [\overset{*}{N}{}^{\Lambda\Gamma}\,\Delta_C \overset{\circ}{\Phi}_\Lambda\,\delta(\Delta_C \overset{\circ}{\Phi}_\Gamma) + N^{11}\,\Delta_C \overset{\circ}{\Phi}_1\,\delta(\Delta_C \overset{\circ}{\Phi}_1)]\,dc_S$$
$$+ \sum\int_{CS} [\widetilde{N}^{11}\,\Delta_C \overset{\circ}{\Phi}_1\,\delta(\Delta \overset{\circ}{\widetilde{\Phi}}_1) + T^{1}\,\Delta_C \overset{\circ}{\Phi}_1\,\delta(\Delta \overset{\circ}{\widetilde{w}})]\,dc_S$$

$$\Delta W = \int_S [\Delta P^\alpha\,\delta(\Delta_C v_\alpha) + \Delta P\,\delta(\Delta_C w)]\,ds$$
$$+ \int_{C\sigma} [\Delta \bar{N}^\alpha\,\delta(\Delta_C v_\alpha) + \Delta \bar{N}\,\delta(\delta_{\Delta C} w) + \Delta \bar{M}^\alpha\,\delta(\Delta_C \Phi_\alpha)\,dc_\sigma$$
$$+ \sum [\Delta \overset{*}{\bar{N}}{}^{1}\,\delta(\Delta \overset{\circ}{\widetilde{v}}_1) + \Delta \overset{*}{\bar{N}}\,\delta(\Delta \overset{\circ}{\widetilde{w}})]\,c_{S\sigma}$$

$$\Delta E = \Delta \overset{(1)}{\mathcal{E}} + \Delta \overset{(2)}{\mathcal{E}} + \Delta_{\mathcal{E}} W$$

$$\Delta \overset{(1)}{\mathcal{E}} = \int_S [{}_C N^{\alpha\beta}\,\delta(\Delta_C \Phi_{\alpha\beta}) + {}_C M^{\alpha\beta}\,\delta(\Delta_C \Phi_{\alpha/\beta})]\,ds$$
$$+ \sum\int_{CS} [\overset{*}{N}{}^{\Lambda\Gamma}\,\delta(\Delta_C \overset{\circ}{\Phi}_{\Lambda\Gamma}) + \overset{*}{M}{}^{\Lambda\Gamma}\,\delta(\Delta_C \overset{\circ}{\Phi}_{\Lambda/\Gamma})$$
$$+ N^{11}\,\delta(\Delta_C \overset{\circ}{\Phi}_{11}) + M^{11}\,\delta(\Delta_C \Phi_{1/1})]\,dc_S$$
$$+ \sum\int_{CS} [\widetilde{N}^{11}\,\delta(\Delta \overset{\circ}{\widetilde{\Phi}}_{11}) + T^{1}\,\delta(\Delta \overset{\circ}{\widetilde{v}})]\,dc_S$$

$$\overset{(2)}{\Delta E} = \int_S [{}_cN^{\alpha\beta}\, {}_c\phi_\alpha\, \delta(\Delta_c\phi_\beta)]\,ds$$
$$+\sum \int_{CS} [\overset{*}{N}{}^{\Lambda\Gamma}\, {}_c\mathring{\phi}_\Lambda\, \delta(\Delta_c\mathring{\phi}_\Gamma) + N''_c\, \mathring{\phi}_1\, \delta(\Delta_c\mathring{\phi}_1)]\,dc_s$$
$$+\sum \int_{CS} [\tilde{N}''_c\, \mathring{\phi}_1\, \delta(\Delta\mathring{\tilde{\phi}}_1) + T'_c\, \mathring{\phi}_1\, \delta(\Delta\mathring{\tilde{w}})]\,dc_s$$
$$\Delta_E W = \int_S [P^\alpha\, \delta(\Delta_c v_\alpha) + P\, \delta(\Delta_c w)]\,ds$$
$$+\int_{C\sigma} [\bar{N}^\alpha\, \delta(\Delta_c v_\alpha) + \bar{N}\, \delta(\Delta_c w) + \bar{M}^\alpha\, \delta(\Delta_c\phi_\alpha)]\,dc_\sigma$$
$$+\sum [\overset{*}{\bar{N}}{}'\delta(\Delta\mathring{\tilde{v}}_1) + \overset{*}{\bar{N}}\, \delta(\Delta\mathring{\tilde{w}})]_{Cs\sigma} \tag{40}$$

and where $\overset{(0)}{\Delta U}$, $\overset{(1)}{\Delta U}$ and $\overset{(2)}{\Delta U}$ are the incremental terms, the initial displacement terms and the initial stress terms respectively. ΔW is the external force term and ΔE is the residial term.

We can use Eq.(39) to a finite element formulation. If the displacements within each finite element is approximated by the shape functions and incremental nodal point desplacements $\{\Delta u\}$, we can expressed Eq.(39) in a matrix form as

$$\delta\{\Delta u\}^T\cdot[\,([\overset{(0)}{K}] + [\overset{(1)}{K}] + [\overset{(2)}{K}])\{\Delta u\} - \{\Delta P\} - \{\Delta E\}\,] = 0 \tag{41}$$

on conditions that

$${}_cN^{\alpha\beta} = {}_cB^{\alpha\beta\gamma\delta}({}_c\phi_{\gamma\delta} + \frac{1}{2}{}_c\phi_\gamma\, {}_c\phi_\delta) + {}_cC^{\alpha\beta\gamma\delta}\, {}_c\phi_{\gamma/\delta}$$
$${}_cM^{\alpha\beta} = {}_cC^{\alpha\beta\gamma\delta}({}_c\phi_{\gamma\delta} + \frac{1}{2}{}_c\phi_\gamma\, {}_c\phi_\delta) + {}_cD^{\alpha\beta\gamma\delta}\, {}_c\phi_{\gamma/\delta}$$
$${}_cB^{\alpha\beta\gamma\delta} = \int_y {}_ca^{\alpha\beta\gamma\delta}\,dy^3 \ , \quad C^{\alpha\beta\gamma\delta} = \int_{y^3} {}_ca^{\alpha\beta\gamma\delta}\, y^3\,dy^3$$
$${}_cD^{\alpha\beta\gamma\delta} = \int_y {}_ca^{\alpha\beta\gamma\delta}\,(y^3)^2\,dy^3$$

$$\overset{*}{N}{}^{\Lambda\Gamma} = A_s({}_s\mathring{a}^{\Lambda\Gamma\Sigma\Omega} - {}_c\mathring{a}^{\Lambda\Gamma\Sigma\Omega})\{({}_c\mathring{\phi}_{\Sigma\Omega} + \frac{1}{2}{}_c\mathring{\phi}_\Sigma\, {}_c\mathring{\phi}_\Omega) + \eta\, {}_c\mathring{\phi}_{\Sigma/\Omega}\}$$
$$\overset{*}{M}{}^{\Lambda\Gamma} = \overset{*}{N}{}^{\Lambda\Gamma}\cdot\eta$$
$$N'' = A_s\, {}_s\mathring{a}^{1111}(\mathring{\tilde{\phi}}_{11} + {}_c\mathring{\phi}_1\, \mathring{\tilde{\phi}}_1)$$
$$M' = N''\,\eta$$
$$\tilde{N}'' = A_s \begin{bmatrix} {}_s\mathring{a}^{11\Lambda\Gamma}({}_c\mathring{\phi}_{\Lambda\Gamma} + \frac{1}{2}{}_c\mathring{\phi}_\Lambda\, {}_c\mathring{\phi}_\Gamma + y^3{}_c\mathring{\phi}_{\Lambda/\Gamma}) \\ + {}_s\mathring{a}^{1111}(\mathring{\tilde{\phi}}_{11} + {}_c\mathring{\phi}_1\, \mathring{\tilde{\phi}}_1) \end{bmatrix}$$
$$T' = \psi K(\mathring{\tilde{v}}{}' - {}_c\mathring{\phi}{}'\, \mathring{\tilde{w}}) \tag{42}$$

where $[\overset{(0)}{K}]$, $[\overset{(1)}{K}]$ and $[\overset{(2)}{K}]$ are the incremental stiffness matrix, the initial displacement stiffness matrix and the initial stress stiffness matrix respectively. ${}_{c}a^{\alpha\beta\gamma\delta}$ and ${}_{c}a^{\Lambda\Gamma\Sigma\Omega}$ are the elasticity tensors of concrete with respect to $g_{\alpha\beta}$ and $g_{\Lambda\Gamma}$ respectively and ${}_{s}a^{\Lambda\Gamma\Sigma\Omega}$ are the elasticity tensors of a steel bar.

CONCLUSIONS

The Principle of virtual work for the reinforced concrete shells is presented in the form after deformation and the bond slip equations are derived. In addition, the principle of incremental virtual work for a finite element formulation is presented.

Though we cannot estimate the order among the displacements of Eq.(31), it seems that the geometrical nonlinearities of bond slip have an influence on the elastic-plastic behavior of reinforced concrete shells.

REFERENCES

1) F.R. Hand, D.A. Pecknold and W.C. Schnobrich., "Nonlinear Layered Analysis of RC Plates and Shells", ASCE, Vol.99, ST7, July, 1973, pp.1491-1505.

2) M.K. Wanchoo and G.W. May., "Cracking Analysis of Reinforced Concrete Plates", ASCE, Vol.101, ST1, January 1975, pp.201-215.

3) C.S. lin and A.C. Scordelis., "Nonlinear Analysis of RC Shells of General Form", ASCE, Vol.101, ST3, March, 1975, pp.523-538.

4) P. Desayi and A.B. Kulkarni., "Load-Deflection Behavior of Restrained R/C Slabs", ASCE, Vol.103, ST2, February, 1977, pp.405-419.

5) F.K. Bashur and D. Darwin., "Nonlinear Model for Reinforced Concrete Slabs", ASCE, Vol. 104, ST1, January, 1978, pp.157-170.

6) T. Ichinose., "Analysis of Reinforced Concrete Frames with Bond-Slip", Trans. of AIJ, No.327, May, 1983, pp.61-70.

7) D.Z. Yankelevsky., "New Finite Element for Bond-Slip Analysis", Journal of Structural Engineering, ASCE, Vol.111, No.7, July 1985, pp.1533-1542.

8) K. Washizu., "Variational Methods in Elasticity and Plasticity", Pergamon Press, 1982.

9) Y. Minagawa., "Geometrical Nonlinearities Effects on the Elastic-Plastic Behavior of Reinforced Concrete Slabs", Trans. of AIJ, No.361, March 1986, pp.132-139.

ANALYSIS METHODS FOR SPATIAL STRUCTURES

M. Papadrakakis
Institute of Structural Analysis
National Technical University, Athens, Greece

Abstract

A state-of-the-art survey of all trends and applications developed for the static analysis of spatial structures is presented. Emphasis is given to the numerical matrix structural analysis approach for treating geometric and material nonlinearities. The behavior of members considered either pin-jointed or end-restrained in the inelastic post-buckling range is described and the response of a 3-D beam element in a large displacement environment is discussed. Numerical techniques for the solution of the resulting linear and nonlinear systems of equations as well as for tracing the nonlinear response near limit points are also discussed.

1. Introduction

During the last twenty years and especially during the last decade spatial structures have influenced the architectural scene all over the world. The reason for this interest is due to several factors; the most important is the introduction and widespread use of high-speed electronic computers, the development of highly efficient standardized connections and the remarkable amount of recent scientific research into the elastic and non-elastic behavior of spatial structures with the determination of modes of failure under excessive loading and the development of effective numerical solution procedures. The introduction of digital computers completely changed the approach used in the analysis of complex structures. Modern spatial structures consist, as a rule, of so many members that analysis by "traditional" techniques becomes extremely time-consuming and in most practical cases virtually impossible. One of the problems involved in the analysis of complex spatial structures is the sheer amount of data preparation for the computer analysis. Techniques based on the concept of formex algebra are considered as a breakthrough in the algebraic representation and processing of highly sophisticated structural configurations.

It has long been recognized that in practice the response of spatial structures can be significantly affected by both geometric and material nonlinearities. Accordingly, the necessity of providing effective tools for the analysis of simultaneous plastic collapse and buckling behavior of spatial structures in a large displacement environment is self-evident. This creates a challenging engineering problem, a solution of which can only be expected through the use of sophisticated numerical methods of nonlinear structural mechanics.

1.1 The analysis problem

Methods which have been used for the investigation of the structural behavior of spatial structures may be divided into four broad categories: discrete field analysis, equivalent continuum analysis, matrix analysis and experimental analysis. The term discrete field analysis denotes the body of concepts used to obtain field or functional solutions for structures most accurately represented as a lattice or pattern of elements. The approach is closely analogous to that employed in the classical plate or shell theory for continuous surfaces although the mathematical model now consists of difference equations as opposed to differential equations which are associated with the continua. The procedures for a discrete field analysis are distinguished into micro and macro approaches by Dean [1,2].

In the equivalent continuum analysis the structural state of a regular spatial structure is approximately represented as functions of continuous variables instead of discrete variables even when a discrete field analysis is not tractable. Two approaches have been used to determine the appropriate relations. The first involves the conversion of the governing difference equations of the system into approximating differential equations by means of Taylor expansions [3,4]. The second method is to relate force or deformation characteristics, or both, of a small segment of the structure to those of a small element of a continuum [5-8]. From these relationships equivalent properties such as modulus of elasticity, Poisson's ratio, and plate or shell thickness can be found. A detailed study of the merits and limitations of the equivalent shell analysis when applied to the investigation of the elastic buckling response of reticulated shell structures has been carried out by Forman and Hutchinson [9].

Matrix analysis methods express the fundamental equations in matrix form and proceed to the solution by utilizing suitable numerical methods via computer implementation. Early difficulties existing in the application of this approach, when large structures with complex topology are encountered, have been almost completely alleviated through the astounding expansion of computer capabilities and the immense development of numerical methods in recent years. The application of the finite element concept to the analysis of spatial structures resulted in an almost unlimited capacity for treating material and geometric nonlinearities which both previous methods of analysis are unable to predict. For these reasons the subsequent discussion will be focused on matrix analysis methods.

Experimental results can always provide a useful means of verification of theoretical analyses and can sometimes lead to the discovery of new phenomena which existing theoretical approaches or numerical procedures are unable to foresee. This approach, however, is extremely time-consuming and expensive and for this reason discriptions in the literature of experimental work have almost invariably been related to specific design project and, as such, it is difficult to make general statements concerning methodology.

1.2 The collapse analysis problem

In spatial structure design, in order to obtain a realistic evaluation of the safety of the structure, it is very important to predict accurately ultimate strength and post-ultimate strength response until final collapse of the structure. We can classify the various approaches currently used in the collapse analysis of spatial structures into the plastic mechanism approach, elastic stability approach, and numerical approach. The first approach involves finding a plastic mechanism within a structure so that the structure is in equilibrium and no member carries a load greater than its squash load. Modern methods of structural analysis can be used by utilizing mathematical programming and optimization techniques [10-12]. Plastic mechanism analysis, however, can only accomodate stability effects in a very approximate manner. Also, the solution of the mathematical programming problem is not trivial and can be very expensive.

The elastic stability analysis concerns the calculation of certain critical states at which the stiffness of a structure, with respect to a small disturbing force, becomes zero or negative. Typical elastic stability problems may be identified as bifurcation buckling, snap-through buckling, and submission buckling. The "initial"or "linearized" stability problem, which is valid under small displacement theory, is much more simplified towards the "nonlinear" stability problem with the assumption of large displacement theory. Engineers are generally most familiar with the initial stability problem which can be posed as a linear eigenvalue problem. An initial stability analysis will not necessarily produce the correct critical loads since in general the equilibrium and kinematic equations must refer to the deformed geometry of the structure immediately prior to buckling. In conclusion we may say that the stability of spatial structures is affected by several factors such as geometry, boundary conditions, type of loading, and material properties. Plastic yielding can only be considered in a stability analysis very approximately by reducing the modulus of elasticity.

Finally, the numerical approach can fully take into consideration all kinds of nonlinearities that are present in the structure as a result of the change in geometry or of the material properties of the structure. Therefore, it is the only approach which can predict the inelastic post-buckling response to any desirable degree of accuracy and to any loading or displacement stage. For most problems, what is required is the limit or bifurcation load without the need to trace the post-buckling behavior. However, in some cases, for example in studying the effects of a concentrated load on a restricted part of the structure, it is important to obtain information of the nature of the load shedding after the occurrence of a local instability, in order to assess the response of the whole structure. It is this potentiality of the numerical approach, of being able to solve all kind of problems in a general and systematic way, together with the immense development of high-speed electronic computer that led to its widespread popularity. In what follows an investigation will be attempted of all current trends of numerical methods as applied to linear and nonlinear analysis of spatial structures.

2. Behavior of members

Member behavior of spatial structures has a paramount effect on the response of the structure particularly when large displacement post-buckling analysis is considered. Member modelling is primarily affected by the design of the connections and secondarily by the form and arrangement of the structure. In many connection designs the bending and torsional stresses could be treated as secondary stresses which do not substantially affect the behavior of the structure. Then, the connections can be approximated by pin-joints. When the connections are designed so that they can transfer bending and twisting moments in addition to axial forces, then their effects should be taken into account. When elastic or inelastic post-buckling behavior is considered, then irrespective of the end conditions of the member, the flexural characteristics should be considered.

Several types of models have been developed for treating materially nonlinear bean-type problems. In the "section" type of model, the inelastic behavior is defined for the cross-section as a whole, and not for individual points. The bending and twisting moments and the resultant axial force at a cross-section interact with each other to produce yielding of the cross-section. The yielding may be treated either as being distributed over the length of the element or lumped at generalized plastic hinges with the beam element between the hinges remaining linearly elastic. In the "fibre" type of model, the member cross-section is divided into a number of areas (layers). Each area is assumed to be uniaxially stressed and the element stiffness is obtained by numerical integration throughout the volume of the beam. The "fibre" type of model provides a better simulation of spread of plasticity in an element, whereas the "section" type of model is computationally more efficient.

2.1 Pin-jointed members

In many patented space truss systems, the method of connection of members at the joints approaches the pin-ended connections so that there is little moment interaction between adjacent members. The form of the real load versus axial deformation relationship for the compressive member will therefore be known ab initio, whether by experiment or detailed analysis. Extensive studies regarding the cyclic inelastic behavior of columns have been reported, mainly in the USA and Japan, and have been directed toward both thoretical and experimental investigations.

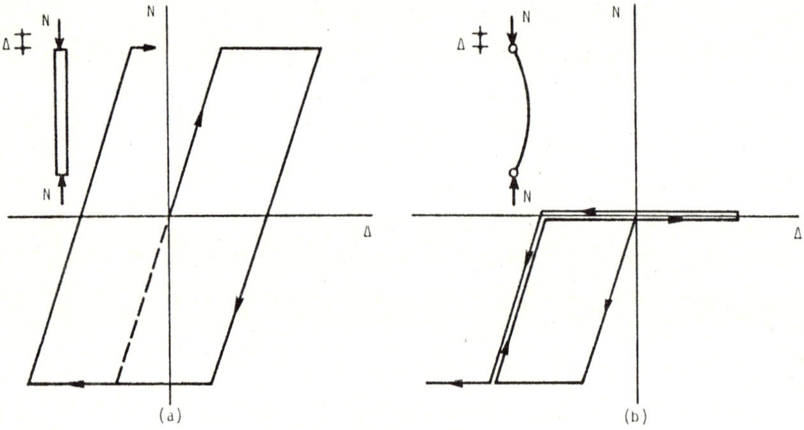

Fig. 1 N-Δ Relationship for Stocky and Slender Struts

Fig. 1 illustrates the axial force N versus axial deflection Δ relationship of two simply supported members with a very small and very large slenderness ratio under reversed loading. Fig. 2 indicates a more complex hysteretic behavior which is characteristic of struts with intermediate slenderness ratios, often used in practical designs. Range OA represents the uniaxial compression, AB represents the elastic buckling, BC corresponds to the plastic hinge formation in compression, CD is the elastic recovery stage, DE corresponds to the plastic hinge formation in tension and EF represents the yielding in tension under constant stress. Theoretical studies performed on pin-jointed members followed three main approaches: the plastic hinge concept [13-19], the direct analysis by means of the finite element method [20-26] and intermediate approaches without referring to either the plastic hinge concept or to finite element formulation [27-29] .

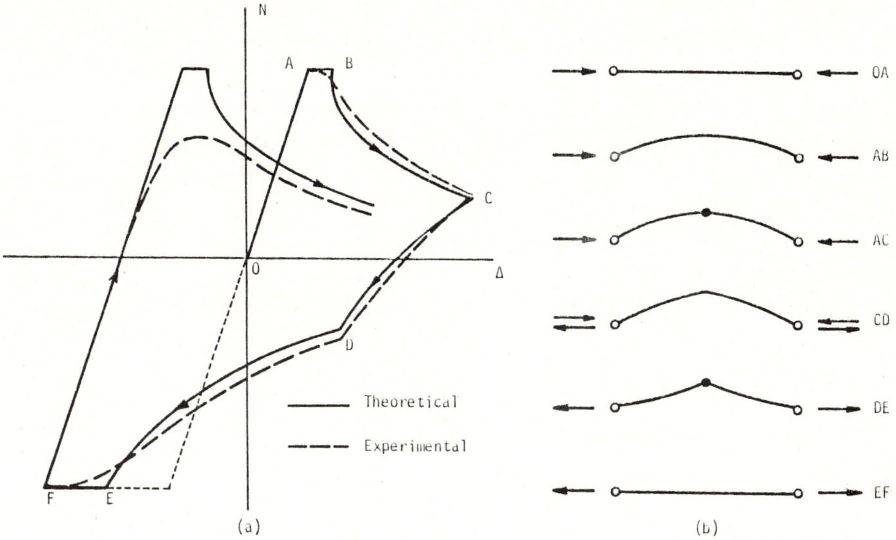

Fig. 2 N-Δ Relationship for a Strut with Intermediate Slenderness Ratio; (a) Hysteresis Loops; (b) Shapes on Various Stages

The plastic hinge concept, which was originally developed by Nonaka [14], has led, under certain assumptions, to elegant closed form solutions for any history of axial loading and any form of cross-section in the range of small deformations. However, change of geometry due to lateral deflection in the evaluation of the axial displacement is considered. Also to predict the axial growth at the end of a complete cycle, plastic axial deformation of the yield hinge based on the plastic flow theory is also considered [14,15] . Large deformation studies have been performed in [30] and it has been confirmed that in practice the influence of the additional terms due to large deformation theory could be neglected. The inelastic response of imperfect columns is of particular interest because imperfect column behavior could significantly influence the load displacement path of the entire structure. Theoretical and experimental results have been reported in [15,16,19, 31-33].

The main deficiency of the hinge method remains its inability to anticipate the drop of the critical load after the first cycle which has been observed during experimental studies (Fig.2). This phenomenon is attributed to the strain hardening and the Bauschinger effects which prevent the bar from regaining its initial straightness after the tensile strength has been reached. Also, closed form solutions need some additional numerical calculations when the member is incorporated in the structural response. What an analyst needs is a formula to provide the axial and/or the flexural stiffness of the member at each displacement level. As this is an almost impossible task when theoretical expressions are followed, researches have established analytical simplified expressions based on theoretical and experimental results [34-40]. Fig. 3 gives some analytical idealizations that have been used.

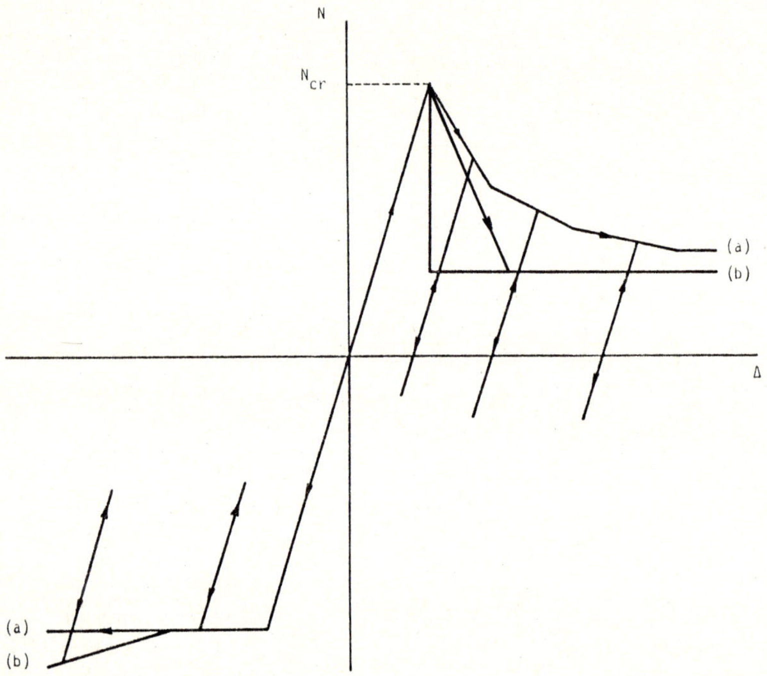

Fig. 3 N-Δ Simplified Curves

2.2 End-restrained member

The nonlinear inelastic analysis of a rigid jointed space frame is a relatively complicated and expensive procedure. The most accurate way of treating this problem is to consider all sorts of nonlinearities that are inherent in this phenomenon. The effect of change of geometry may take into account, the change of member lateral stiffness due to the effect of axial forces, the change of member length due to bowing, finite deflection of joints, rigid body rotations and finite rotation of joints. To model the effects of material yield a "section" or a "fibre" type of models may be considered. The general 3-D nonlinear beam formulation is not a simple extension of a 2-D formulation, because in 3-D analysis large rotations have to be accounted for that are not vector quantities. Thus, the development of general nonlinear beam elements is probably one of the most challenging and unresolved research task of F.E. structural analysis and receives intense attention from researchers all over the world. The best known 3-D finite elements are the Hermitian and isoparametric beam elements [41,42] which are consistently treated on the basis of the equations of continuum mechanics, the natural element [20,43] which is based on the natural mode technique and the beam element based on the beam-column approach [44] .

The Hermitian and isoparametric elements are introduced by Bathe and Bolourchi [42] and Bathe [41] , respectively, through a consistent development of a total and an updated Lagrangian formulations.[84] In treating beams of constant cross-sections that can undergo large displacements including large rotations but small strains, the Hermitian element employs the Hermitian functions to interpolate the transverse bending displacements, and linear functions to interpolate the torsional and longitudinal displacements. The beam element is assumed straight and by using convected coordinate axes the usual beam kinematic assumptions are referred to the current beam geometry. The isoparametric beam element is based on the isoparametric F.E. formulation in which the interpolation of the element coordinates and element displacements are using the same interpolation functions and are defined in a natural coordinate system. In comparing the two elements we may say that the isoparametric beam element has twice the number of degrees of freedom of the Hermitian element for cubic interpolation of transverse bending displacements and needs almost five times more computational time. For linear analyses the Hermitian element is recommended mainly because its stiffness matrix is obtained explicitly, while for the isoparametric element a numerical evaluation is required. The isoparametric element, however, has the advantage of being compatible with isoparametric plate and shell elements and therefore can be combined easily in the analysis of plate and shell structures with line elements. It can also better model curved line elements. Material nonlinearities are taken into account by employing a "section" type of model with distributed yielding over the length of the element and considering appropriate constitutive descriptions [41,45]. The use of the isoparametric beam element and the total Lagrangian formulation are recommended when considering material nonlinearities.

The natural 3-D beam element is based on the natural mode technique developed by Argyris and his co-workers [20,43]. On the simplest level of explanation the terminology "natural method"refers to the use of separate rigid body displacements and natural deformations for the description of the current state of a finite element. Individual deformations of elements are considered through the subelement concept, which constitutes the backbone of the natural mode technique. The natural modes are then defined for the simplest possible finite element types, such as straight beam in space, to cope with the analysis of a curved beam in space. This formulation handles the noncommutative nature of rotations by the introduction of a consistent set of commutative rotations called semitangential torques. The extension of the geometric natural mode nonlinear analysis to include physical nonlinearities is materialized by considering all plastic effects to be concentrated at critical cross-sections at the ends of the members. In [20] several plastic interaction surfaces are explicitly quoted which have proved to be effective in the numerical analysis of the problems considered. Fig. 4 shows comparative results for a beam clamped at both ends using two elements and 10 and 50 "layered" beam elements. In comparing the computation time it was found that the natural element needs only a tenth of the CPU time required for the computation with 50 layered elements.

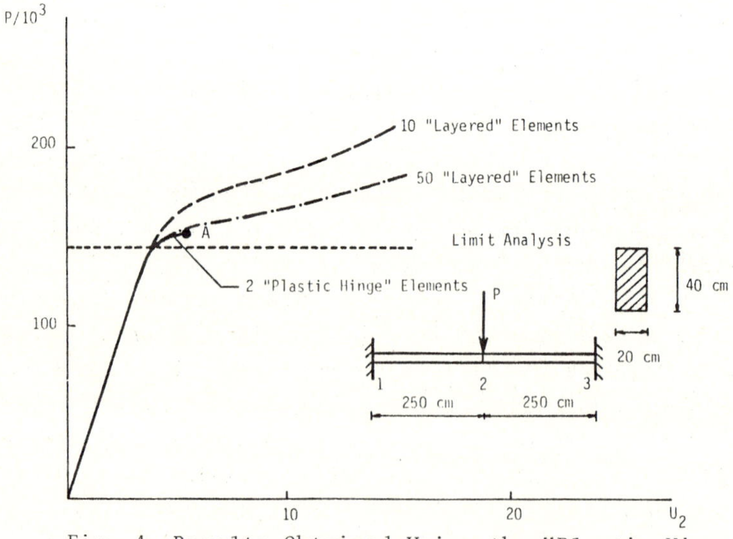

Fig. 4. Results Obtained Using the "Plastic-Hinge" Element and "Layered" Elements; Argyris et al. [20]

Other formulations for 3-D geometric nonlinear analysis may be found in [21, 46-49] and for geometric and material nonlinearities in [50,51]. When the relative transverse displacements of the two ends of a beam may be assumed negligible closed form solutions for the cyclic inelastic response based on the plastic hinge concept have been reported [15,19,29,52].

3. Solution procedures

The modelling and the mathematical formulation constitutes an important stage in the analysis of a structure, but of equal importance is the use of efficient procedures for the solution of the resulting system of equations. The ideal choice among solution procedures for the same purpose is not only case-dependent and their efficiency, but in some cases hardware restrictions such as available core size, may enter into consideration. This means that the cost of an analysis and, in fact, its practical feasibility depend to a considerable degree on the algorithms available for the solution of the equations.

3.1 Linear solution techniques

The solution of the equations that arise in the linear analysis problems

$$KU = R \tag{1}$$

where K = the stiffness matrix, U = the displacement vector, and R = the load vector, is of great importance not only in the case of linear analysis but in the case of nonlinear and dynamic analyses too where the solution of linear systems of equations is frequently required at intermediate stages. Essentially, there are two different families of methods for the solution of eq. (1): direct and iterative solution techniques.

The most effective direct solution techniques are basically applications of Gauss elimination which can take advantage of the specific properties of the F.E. stiffness matrix. Static condensation, substructuring, the frontal solution method and compact storage techniques have been used for handling excessive storage requirements. In contrast to the elimination methods, iterative methods do not require the formation and storage of the global stiffness matrix and operations could be performed on the "element" level of the structure. The most commonly used iterative methods are propably the successive overrelaxation and the conjugate gradient method. The latter, moreover, is found to possess a remarkable property of behaving like a direct method in exact arithmetic. A renewed interest in iterative methods has been observed in recent years because of the widespread application of minicomputers and the special features of the new generation of parallel computers in which iterative methods render their advantages of performing the operations on the "element" level. Thus, a number of techniques have been proposed aiming at accelerating the convergence properties of iterative methods. Preconditioning and block elimination have considerably improved the speed of convergence at the expense of using extra computer storage.

3.2 Nonlinear solution techniques

The classification of nonlinear solution techniques is not an easy task since all may be considered iterative or can be used in an incremental way and sometimes there is a combination of iterative and incremental. The basic equation to be solved may be written as

$$f = R - F = 0 \tag{2}$$

where f =the force unbalance and F = the nodal point forces that are equivalent to the element stresses. The method of successive substitutions (Fig. 5a) is characterized by the recursion relation

$$KU^{(i+1)} = R + F^{*(i)} \qquad i = 0,1,2,... \tag{3}$$

where F* = pseudo force due to nonlinear effects. The Newton-Raphson (NR) method (Fig. 5b) can be written as

$$K_T^{(i)} \Delta U^{(i+1)} = f^{(i)}$$

and

$$U^{(i+1)} = U^{(i)} + \Delta U^{(i+1)} \tag{4}$$

where $K_T^{(i)}$ = the tangent stiffness matrix in iteration i. For geometric nonlinear analysis the NR method has proved itself to be one of the best methods of solution available, mainly in a mixed formulation where incremental load steps and iterations are combined. However, in material nonlinear analysis or dynamic nonlinear response the load steps that can be employed are restricted in size by other stability and accuracy considerations. For these reasons most analyses now use the modified NR (mNR) wherein K_T is held constant for several iterations and is updated only when the rate of convergence begins to deteriorate (Fig. 5c). The mNR is particularly appealing when the complete load deflection path of a complicated structure is required.

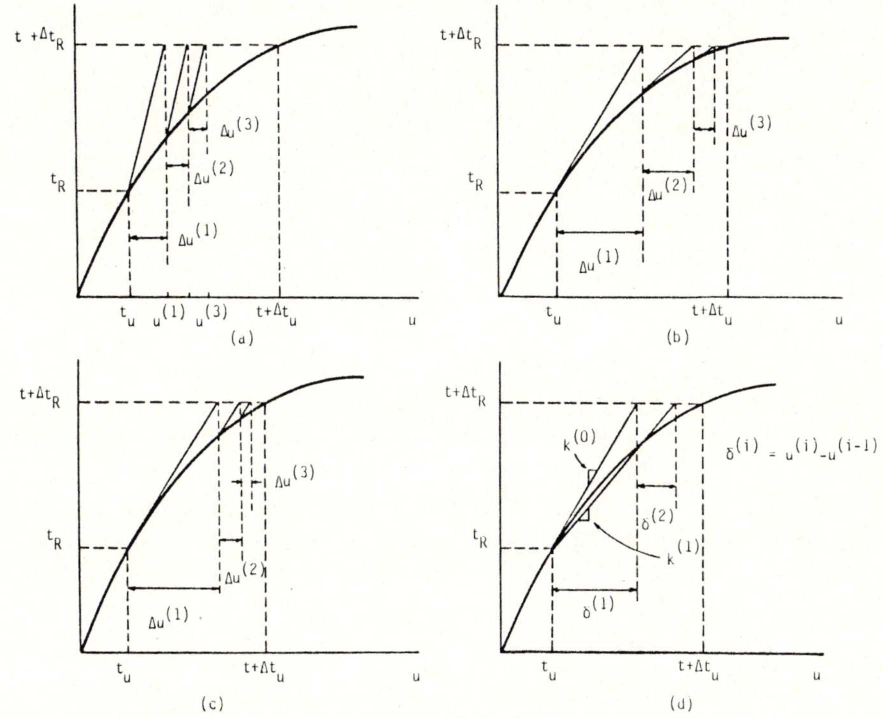

Fig. 5. Solution Techniques for One-Dimensional Problem; (a) Successive Approximation; (b) Newton-Raphson; (c) Modified Newton Raphson; (d) BFGS

A second family of nonlinear methods, called vector iteration methods [53], are based on iterative methods for the solution of linear systems and may be expressed as the three-term recursion formula of Rutishauser [54]

$$U^{(i+1)} = -\frac{f^{(i)}}{q_i} + \left(\frac{e_{i-1}}{q_i} + 1\right) U^{(i)} - \frac{e_{i-1}}{q_i} U^{(i-1)} \tag{5}$$

where q_i, e_{i-1} = iteration coefficients and they are characteristic of the method. The basic advantage of these methods over elimination methods is that only four vectors need be stored since $f^{(i)}$ is the unbalance force vector in iteration i.

As an alternative to forms of Newton iteration, a class of methods have been developed known as matrix update methods or quasi-Newton methods. These methods provide a compromise between the full reformulation of the K_T and the use of a stiffness matrix from a previous configuration as is done in mNR. Among the quasi-Newton methods available the BFGS (Broyden-Fletcher-Goldfarb-Shanno) method appears to be the most effective. In the BFGS the following procedure is employed to evaluate $U^{(i+1)}$ and $K^{(i+1)}$ (Fig. 5d):

$$\Delta\bar{U} = (K^{-1})^{(i)} f^{(i)}$$

$$U^{(i+1)} = U^{(i)} + \beta \Delta\bar{U} \tag{6}$$

$$K^{(i+1)} = \left[A^{(i+1)}\right]^T (K^{-1})^{(i)} A^{(i+1)}$$

where matrix $A^{(i+1)}$ is an n x n matrix of a simple form. Matthies and Strang [55] developed a modification to the original algorithm that reduces the computer storage requirements.

Since the need for nonlinear analysis of structures has significantly increased during recent years and will continue to do so, and since ever more complex problems are being tackled, much emphasis is currently being placed on improving the convergence properties of the methods and developing more general and automatic solution techniques. A secant Newton method based on the variable metric advancement toward the solution [56,57] , the conjugate - Newton method [58] and modified conjugate - Newton [59] , and the preconditioned CG methods [60-62] have been proposed recently in an effort to combine the adantages of direct and iterative numerical techniques. Also the coupling between Newton and quasi-Newton methods [63,64] and the use of line searches [64,65] to "damp" or scale-down the solution vector are found effective for highly nonlinear problems.

3.3 Tracing the nonlinear response near limit points

For most practical problems, it is quite unnecessary to trace such a convoluted load-deflection path with snap-through and snap-back buckling phenomena, as shown in Fig. 6, which possess some of the most difficult problems in nonlinear structural analysis. For some problems, all that may appear to be required is the load level at the first limit point. However, without analysis techniques that allow the limit points to be passed, even this information may be unavailable or unreliable. "Collapse loads" are often associated with a failure to achieve convergence with the iterative solution procedure. However, it way only be the terative solution procedure that has collapsed (possibly as a concequence of round-off errors). For other problems, the analysis may be

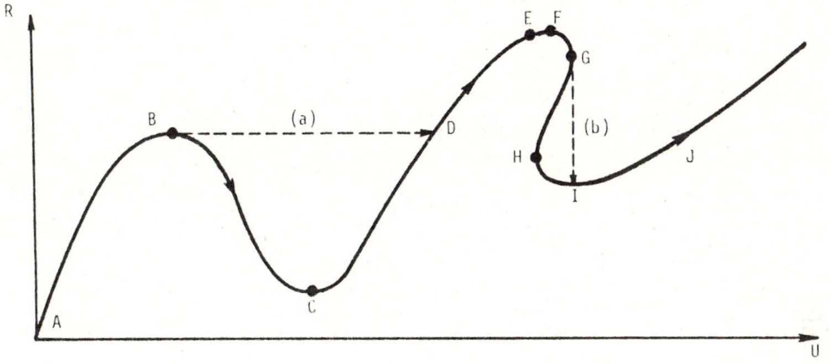

Fig. 6. "Snap-Buckling"; (a) Dynamic Snap under Load Control; (b) Dynamic Snap Under Displacement Control.

performed on a local part of a structure. In such a case, it may be important to obtain information on the nature of the load shedding, following the limit point, in order to assess the performance of the complete structure. Consequently, nonlinear finite element computer programs should be provided with solution procedures that will handle such "snapping phenomena" automatically without resorting to many interventions.

Amongst the earliest techniques to circumvent limit points are the artificial-spring method [66], Bergan's "current stiffness parameter" [67] and the displacement incrementation introduced first by Argyris [68]. The interchange of dependent and independent variables is based on an incremental - iterative solution procedure depicted in Fig. 7. For proportional loading the loads may be expressed by one load factor ${}^{t}\lambda$: ${}^{t}R = {}^{t}\lambda R$ within one increment from configuration t to t+Δt. Following the nomenclature of Fig. 7 the incremental equilibrium equations may be expressed by the linearized expressions

$$K^{(i)}\ \delta U^{(i+1)} = \delta\lambda^{(i)}\ R + f^{(i)} \tag{7}$$

If we assume the prescribed displacement component $\delta U_2^{(i+1)} = \hat{U}_2$ is the last one in the displacement vector $\Delta U^{(i+1)}$, then eq. (7) may be written after interchanging variables

$$\begin{bmatrix} K_{11} & -R_1 \\ K_{21} & -R_2 \end{bmatrix} \begin{bmatrix} \delta U_1 \\ \delta\lambda \end{bmatrix}^{(i+1)} = \begin{bmatrix} f_1 \\ f_2 \end{bmatrix}^{(i)} - \begin{bmatrix} K_{12} \\ K_{22} \end{bmatrix}^{(i)} \hat{U}_2 \qquad (8)$$

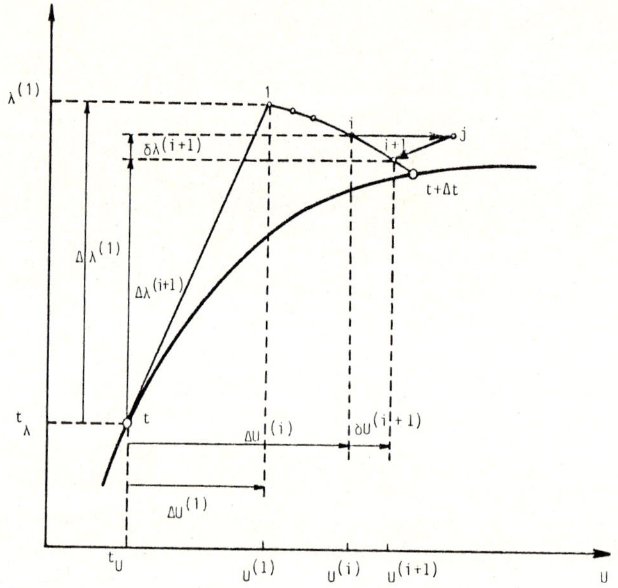

Fig. 7. Tracing the Equilibrium Path-Notation

To avoid the loss of the symmetrical and banded structure of the resulting matrix, the first line of eq. (8) is decomposed in two parts.

$$K_{11}^{(i)} \left[\delta U_1^{(i+1)}\right]^{I} = R_1 \qquad (9)$$

$$K_{11} \left[\delta U_1^{(i+1)}\right]^{II} = f_1^{(i)} - K_{12}\, \hat{U}_2 \qquad (10)$$

$$\delta U_1^{(i+1)} = \delta\lambda^{(i+1)} \left[\delta U_1^{(i+1)}\right]^{I} + \left[\delta U_1^{(i+1)}\right]^{II} \qquad (11)$$

The displacement $\delta U^{(i+1)}$ of eq. (11) is now introduced into the second part of eq.(8). This allows the determination of the load parameter $\delta\lambda^{(i+1)}$.

The displacement control method obviously fails whenever the structure snap-backs from one load level to a lower one. Therefore, an obvious modification is to relate the procedure to a measure including all displacements rather than the one single component. Based on this idea Riks [69] and Wempner [70] independently introduced a constrained equation to limit the load step $\delta\lambda^{(1)}$

$$\left[\delta U^{(1)}\right]^T \delta U^{(1)} + \delta\lambda^{(1)} = dS^2 \qquad (12)$$

where dS is a prescribed value. Then the iteration path follows a "plane" normal to the tangent (Fig. 8a) so that

$$\left[\delta U^{(1)}\right]^T \delta U^{(i+1)} + \delta\lambda^{(1)} \delta\lambda^{(i+1)} = 0 \qquad (13)$$

A modification to the Riks-Wempner method was reported by Ramm [85] in which the normal "plane" was updated in every iteration cycle and by Crisfield [71] (Fig. 8b) in which a "spherical" path was used and eq. (12) is replaced by

$$\left[\Delta U^{(i+1)}\right]^T \Delta U^{(i+1)} = dS^2 \qquad (14)$$

It was found that except for very large load steps the differences resulted from these two formulations are minor. Park [72] used the generalized constrained equation

$$\left[\Delta U^{(i+1)}\right]^T \Delta U^{(i+1)} + b\,(\Delta\lambda^{(i+1)})^2\, R^T R = dS^2 \qquad (15)$$

where b is a scalar parameter which is related to Bergan's [73] current stiffness parameter. A scaling of the parameter dS which automatically adjusts the size of the increment in regions with increased nonlinear behavior is suggested in [83]. Finally other techniques for tracing equilibrium paths may be found in [63, 74-78].

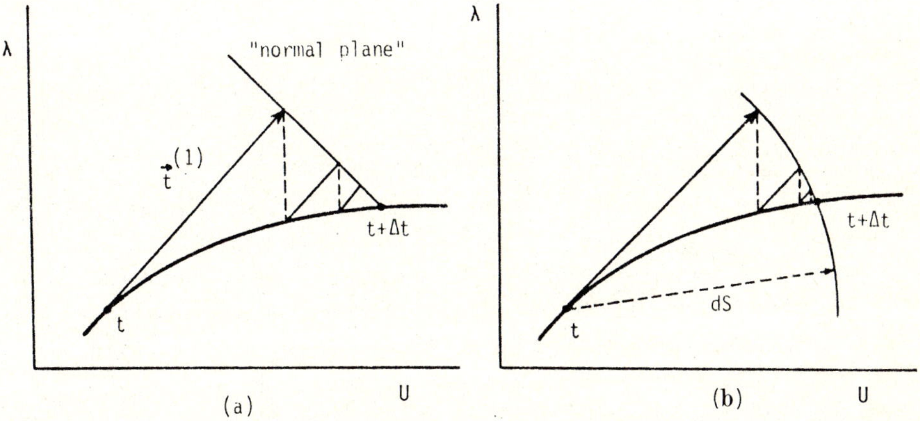

Fig. 8. (a) Riks-Wempner Method
(b) Crisfield Method

3.4 Truss analysis techniques

The aforementioned numerical solution procedures are general and can handle any set of nonlinear equations which describe the response of any structure and naturally any spatial structure. For pin-jointed truss problems, however, several simplified but effective numerical procedures have been reported which can handle the limit state analysis and predict the behavior of the structure in the post-buckling range. Schmidt and Gregg [36] have reported a dual load method that allows the inelastic nonlinear member behavior to be followed. The method assumes that the inelastic strut response is known and the piecewise linearization of the behavior is acceptable. Depending on the axial deformation of each member, an appropriate stiffness and residual force is known and a first order linear analysis for two separate loading conditions is then carried out on the structure whose members possess these newly determined stiffnesses. The two loading conditions are the unit external load and the residual forces for each member. A similar approach called the "fictitious force technique" is presented by Klimke and Posch [79] to the analysis of the large deflection problem combined with the ultimate load analysis. Papadrakakis [80] used the bending moment - curvature relationship to model the cyclic response of the members together with the dynamic relaxation method for the solution of the geometrically and materially nonlinear equations. Smith [81] developed a nonlinear stepwise linearization analysis which does not require the repeated updating of the stiffness matrix. The method can handle the chordal snap-through phenomenon and member displacement sense reversal. Other specified solutions may be found in [82,83].

4. Numerical and Experimental Tests

Example 1.- The simple plane arch structure shown in Fig.9 was used by many investigators to test the mathematical formulation and the nymerical solution procedure. In Fig. 9 only geometric nonlinearities are considered. Hangai and Kawamata [86]used the static petrurbation technique with various degrees of approximations in the nonlinear force-displacement relation. Curves 1 HK up to 5 HK are obtained by taking into account up to the fifth order term in the strain-displacement relations respectively. Bergan and Søreide [87] obtained curves 1 BS and 2 BS with an automatic load incrementation and a constant load incrementation respectively. Kani et al. [51] employed a 2 and 3 node isoparametric beam elements with an updated Lagragian formulation coupled with Bergan's current stiffness parameter to circumvent the limit point. Papadrakakis [88] used two "vector iteration" methods, namely the conjugate gradient method and the dynamic relaxation method, for the solution of equations. A displacement incrementation technique is used to trace the equilibrium path. Fig 10 shows the response of the structure when material nonlinearities are considered in addition to the geometric ones. The results depicted are obtained in [80] by considering linearly elastic, perfectly plastic behavior (curve b), elastic buckling (curve c), elastic-perfectly plastic buckling (curve d), and elastoplastic buckling (curve e). Kani et al. [51] also reported results identical with curve b using the previously mentioned formulation.

Example 2.- (24 - member hexagonal star-shaped shallow dome). For this space truss it becomes clear (Fig. 11) that Hangai-Kawamata's static petrurbation technique, with retention of up to the third order terms of displacements, is inadequate to predict the post-buckling behavior of such structures. Jagannathan et al. [89] used a total Lagrangian formulation with a displacement incrementation of the NR method, and Rothert et al. [90] a modification to the Riks-Wempner method. They too failed to predict the accurate behavior of the structure because of the large member rotations that occur. It is clear from Fig.[11] that their formulation failed to satisfy the geometric constraints of the problem at U=2 cm. The results obtainet by Papadrakakis [87] appeared to be in complete agreement with those obtained by Paradiso and Tempesta [91], where a secant-tangent approach is used by successive linear approximation. Watson and Holzer [92] used a TL formulation and Crisfield's method, Meek and Tan [93] used a beam-column large rotation formation and Crisfield's method and finally Cichon [21] employed a TL formulation and a numerical procedure based on Waszsyszyn [78] . Fig. 12 shows the response of the structure when material nonlinearities are considered [80] .

Example 3.- (6 x 6 bay double layer offset grid space truss). The truss shown in Fig. 13 is simply supported to each perimeter joint and loaded symmetrically at the four points at the corners of the central 2ft square of the bottom chord. The truss was fabricated from aluminium tube, coined at the ends to fit patented "Triodetic"hubs. Schmidt et al. [34 , 36 , 94] in a number of tests reported the load deflection responses, external load-member responses, and compression and tension test responses on the members. The authors point out the reduction in the moment of inertia of the members at the joint face and because of the rigidity of the joints they used the joint face to joint face distance for the effective length of the members. Schmidt's truss tests show that the compression members did reach force values greater tham the pin-ended buckling load but did not achieve the values obtained in the individual member tests before system failure. In a study performed by Smith [81] for the same truss, the effective length was taken between the end joint faces multiplied by an effective length factor, taken to be 1.0 and 0.9 in two different computer runs. The smaller effective length factor, 0.9, qualitatively accounts for the small but non-zero moment of inertia and plastic moment of resistance at the joint face and gives an effective length similar to that obtained by a classical elastic stability analysis. In the structural analysis, the joints were assumed to be rigid and thus the stiffness of the members was modified accordingly.

Example 4.- (12-member hexagonal frame). This space frame (Fig. 14) has been experimentally studied by Griggs [95] . The pre-buckling behavior has been solved by [96, 97] , while in [87, 93, 98] the solution was traced into the elastic post-buckling range. Brendel and Ramm [98] used finite element large rotation Langangian formulation with four elements per member. Connor et al. [97] used a small rotation derivation of the equilibrium equations and the NR method, while Chu and Rempetsreiter [96] employed a combined secant tangent stiffnes approach with the mNR method and neglected the effect of length shortening due to bowing. Papadrakakis [87] used the beam column approach with the effect of axial force in the flexural stiffness, of flexural shorteming of members, and of large deflections with moderate rotations have been taken into acount. A displacement incrementation of the vector iteration methods was employed to trace the equilibrium path. Meek and Tan [93] used a beam column geometric formulation with arbitrary large rotations which are accommodated by making use of an updated joint orientated matrix [44] . The NR method and Crisfield constant arc-length technique are employed.

Example 5.- (18-member space framed dome). As a final example the space frame shown in Fig. 15 was considered under the prescribed downward displacement at the top joint. The material and geometric data were taken from [43] . Papadrakakis and Ghionis [99] used the same methodology discussed for example 5, while Argyris et al. used the 3-D natural element discussed previously, taking into consideration both geometric and material nonlinearities.

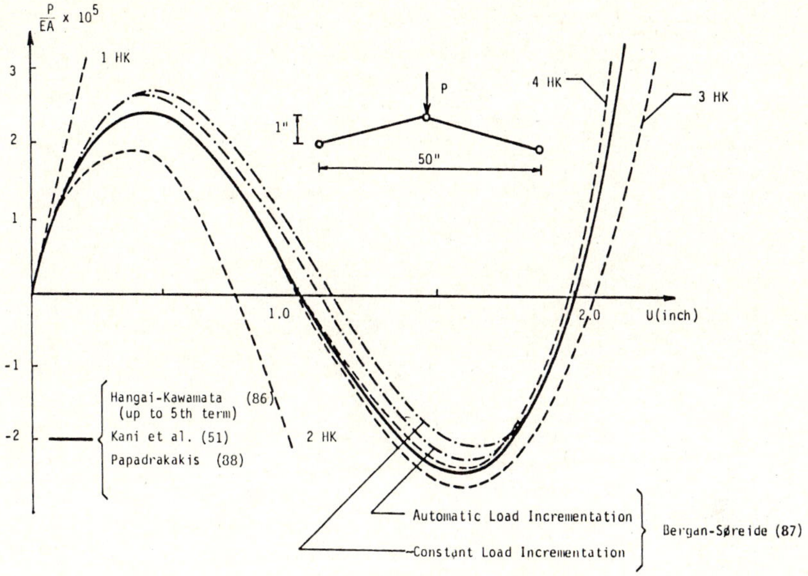
P/EA x 10^5
3
2
1
-1
-2
1 HK
P
1"
50"
4 HK
3 HK
1.0
2.0
U(inch)
Hangai-Kawamata (86)
(up to 5th term)
Kani et al. (51)
Papadrakakis (88)
2 HK
Automatic Load Incrementation
Constant Load Incrementation
Bergan-Søreide (87)

Fig. 9

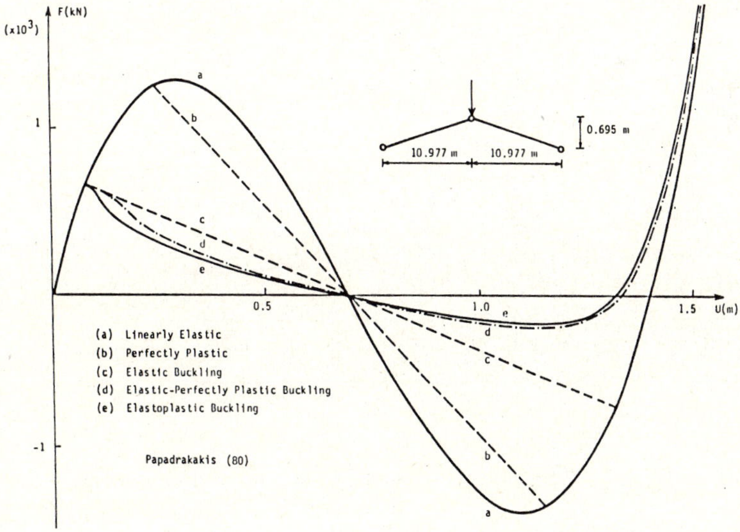
F(kN)
(x10^3)
1
-1
a
b
c
d
e
10.977 m
10.977 m
0.695 m
0.5
1.0
1.5
U(m)
(a) Linearly Elastic
(b) Perfectly Plastic
(c) Elastic Buckling
(d) Elastic-Perfectly Plastic Buckling
(e) Elastoplastic Buckling
Papadrakakis (80)

Fig. 10

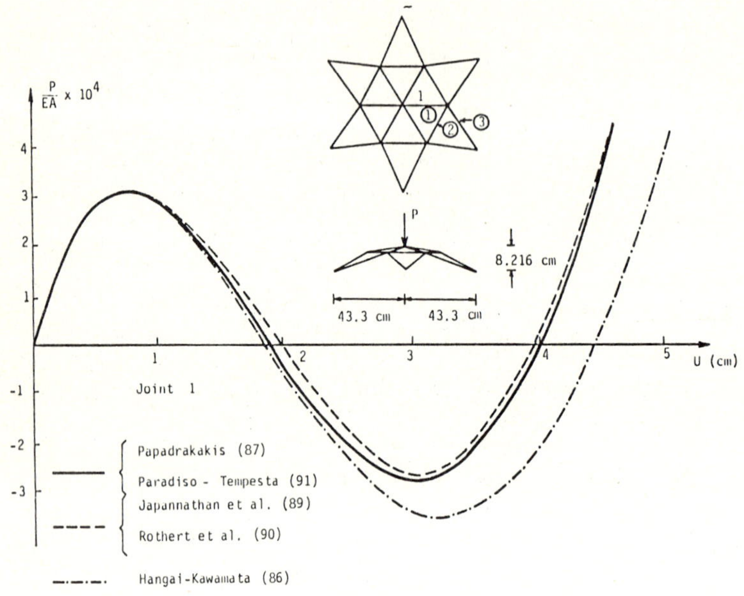
P/EA x 10^4
4
3
2
1
-1
-2
-3
1
2
3
4
5
U (cm)
Joint 1
1
①
②
③
P
8.216 cm
43.3 cm
43.3 cm
Papadrakakis (87)
Paradiso - Tempesta (91)
Japannathan et al. (89)
Rothert et al. (90)
Hangai-Kawamata (86)

Fig. 11

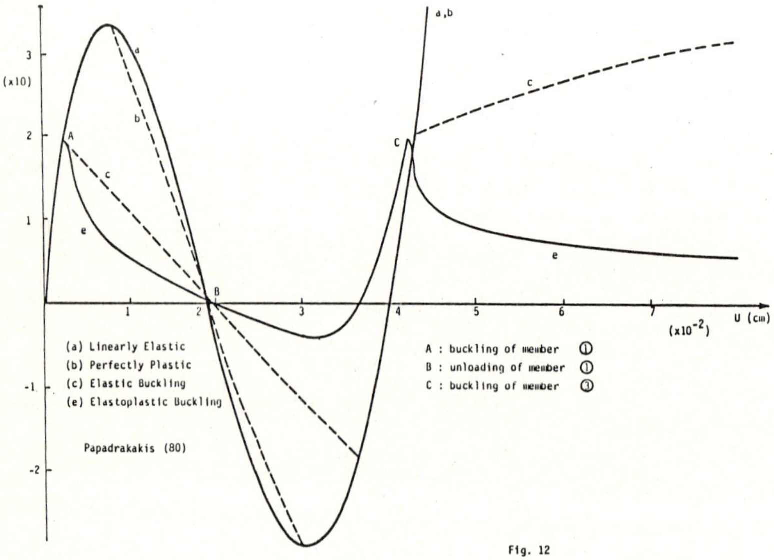
3
(x10)
2
1
-1
-2
1
2
3
4
5
6
7
U (cm)
(x10^-2)
a,b
c
d
b
A
B
C
c
e
e
(a) Linearly Elastic
(b) Perfectly Plastic
(c) Elastic Buckling
(e) Elastoplastic Buckling
Papadrakakis (80)
A : buckling of member ①
B : unloading of member ①
C : buckling of member ③

Fig. 12

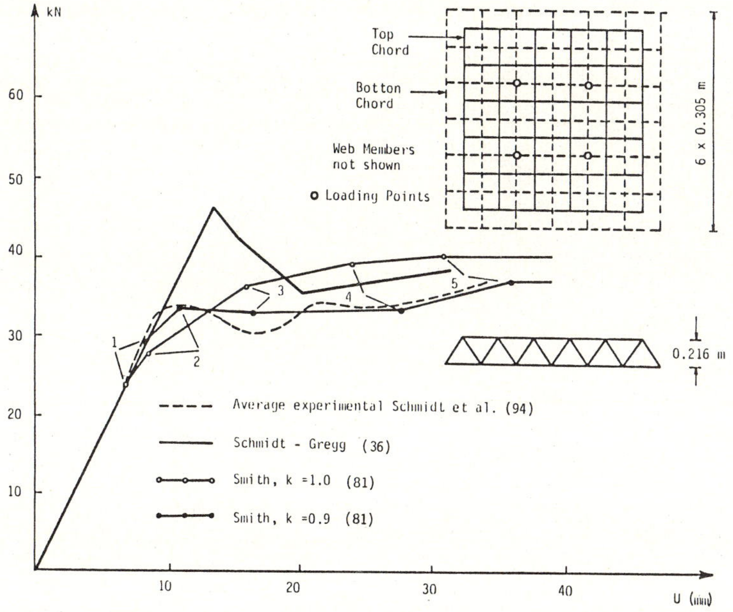
kN
60
50
40
30
20
10
10
20
30
40
U (mm)
Top Chord
Botton Chord
Web Members not shown
o Loading Points
6 x 0.305 m
0.216 m
1
2
3
4
5
Average experimental Schmidt et al. (94)
Schmidt - Gregg (36)
Smith, k =1.0 (81)
Smith, k =0.9 (81)

Fig. 13

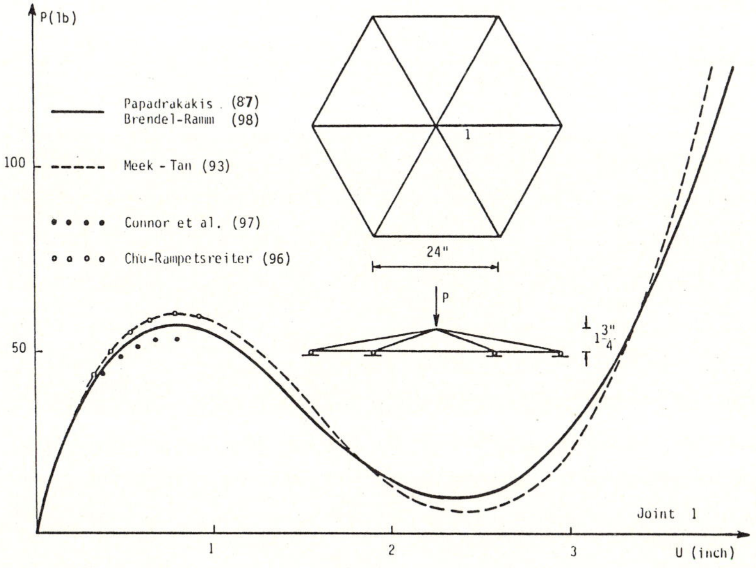
P(lb)
100
50
1
2
3
U (inch)
Papadrakakis (87)
Brendel-Ramm (98)
Meek - Tan (93)
Connor et al. (97)
Chu-Rampetsreiter (96)
24"
P
Joint 1

Fig. 14

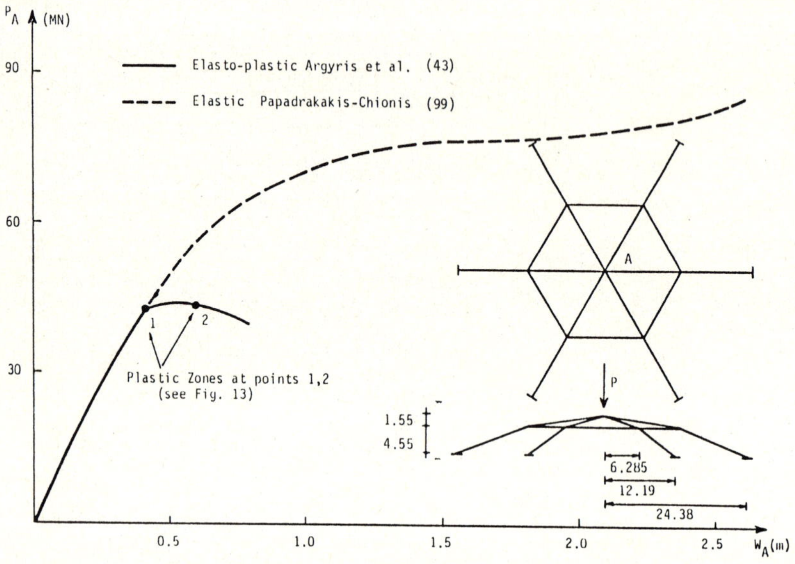

P_A (MN)
90
60
30
Elasto-plastic Argyris et al. (43)
Elastic Papadrakakis-Chionis (99)
1
2
Plastic Zones at points 1,2
(see Fig. 13)
A
P
1.55
4.55
6.285
12.19
24.38
0.5
1.0
1.5
2.0
2.5
W_A(m)

Fig. 15

References

1. Dean, D.L., "Discrete Field Analysis of Structural System" CISM Courses and Lectures, No 203, Udine, Italy, Springer-Verlag, Wien, 1976.
2. Dean, D.L., and Avent, R.R., "State of the Art Discrete Field Analysis of Space Structures, "Proc. of the 2nd Int. Conf. on Space Structures, Guilford, England, 1975, pp. 7-16.
3. Heki, K., and Fujitani, Y., "The Stress Analysis of Grids under the Action of Bending and Shear", Space Structures, Blackwell Publ., Oxford, 1967, pp. 33-43.
4. Renton, J.D., "On the Gridword Analogy for Plates", J. of the Mech. and Phys. of Solids, Vol. 13, 1965, pp. 413-420.
5. Fluge, W., "Stresses in Shells", Springer-Verlag, Berlin, 1962, pp. 293-307.
6. Heki, K., "The Effect of Shear Deformation on Double Layer Lattice plates and Shells, Proc. of the 2nd Int. Conf. on Space Structures, Guilford, England, 1975, pp. 189-198.
7. Saka, T., and Heki, K., "The Effect of Joint on the Strength of Space Trusses", Third Int. Conf. on Space Structures, Guilford, England, H. Nooshin ed., Applied Science Publishers Ltd., London, 1984, pp. 417-422.
8. Wright, D.T., "Membrane Forces and Buckling in Reticulated Shells"J. of the Str. Div., ASCE, Vol. 21, No. ST1, 1965, pp. 173-201.
9. Forman, S.E., and Hutchinson, J.W., "Buckling of Reticulated Shell Structures", Int.J.of Solids and Str. Vol. 6, 1970, pp. 909-932.
10. Cohn, M.Z., Ghosh, S.K., and Parimi, S.R., "Unified Approach to Theory of Plastic Structures", J. of the Eng. Mech. Div. ASCE, Vol. 98, No. EM5, 1972, pp. 1133-1158.
11. Grierson. D.E., and Abdel-Baset, S.B., "Plastic Analysis under Combined Stresses", J. of the Eng. Mech. Div., ASCE, Vol.103, No.EM5 1977,pp.837-854.
12. Watwood, V.B., "Mechanism Generation for Limit Analysis of Frames", J. of the Str. Div., ASCE, Vol. 109, ST1, 1978, pp. 1-15.
13. Higginbotham, A.B., and Hanson, R.D., "Axial Hysteretic Behavior of Steel Members", J. of the Str. Div. ASCE, Vol. 102, No. ST7, 1975, pp. 1365-1381.
14. Nonaka, T., "An Elastic-Plastic Analysis of a Bar under Repeated Axial Loading", Int. J. of Solids and Str., Vol. 9, 1973, pp. 569-580.
15. Papadrakakis, M., and Loukakis, K., "Inelastic Cyclic Analysis of Imperfet Columns with End Restraints", Submitted for Publication.
16. Prathuangsit, D., Goel, S.C., and Hanson, R.D., "Axial Hysteresis Behavior with End Restraints", J. of the Str. Div., ASCE, Vol. 104, No. ST6, 1978, pp.883-898.
17. Supple, W.J., and Collins, I., "Post-Critical Behaviour of Tubular Struts", Engineering Structures, Vol. 2, Oct. 1980, pp. 225-229.
18. Toma, S., and Chen, W.F., "Inelastic Cyclic Analysis of Pin-Ended Tubes", J. of Str. Div. ASCE, Vol. 108, No. ST10, 1982, pp. 2279-2293.
19. Wakabayashi, M., Matsui, C., and Mitani, I., "Cyclic Behavior of a Restrained Steel Brace under Axial Loading", Proc. of the 6th, World Conference on Earthquake Enging. New Delhi, 1977, pp. 3181-3187.
20. Argyris, J.H., et al., "Finite Element Analysis to Two-and Three-Dimensional Elastro-Plastic Frames". - The Natural Approach", Comp. Meth.in Appl. Mech. and Enging. Vol. 35, 1982, pp. 221-248.
21. Cichon, C., "Large Displacements in-Plane Analysis of Elastic-Plastic Frames", Comp. and Str., Vol. 19, No. 5/6, 1984, pp. 737-745.
22. El-Zanaty, M.H., and Murray, D.W., "Nonlinear Finite Element Analysis of Steel Frames", Journal of Structural Engineering, ASCE, Vol. 109, No. 2 1983, pp. 353-368.
23. Fujimoto, et al., "Nonlinear Analysis for K-Type Braced Steel Frames", Transactions, Architectural Institute of Japan, No. 209, July, 1973, pp. 41-51, (in Japanese).
24. Kassimali, A., "Large Deformation Analysis of Elastic Plastic Frames", J. of Str., ASCE, Vol. 109, No. 8, 1983, pp. 1869-1886.
25. Madi, U.R., and Smith, D.L., "A Finite Element Model for Determining the Constitutive Relation of a Compression Member", Third Int. Conf. on Space Structures, Guilford, England, H. Nooshin ed., Applied Science Publishers Ltd.,

London, 1984, pp. 625-629.
26. Wakabayashi, M., and Shibata, M., "Studies on the Post-Buckling Behavior of Braces", Part 4, Abstracts Annual Meeting Kinki Branch, Architectural Institute of Japan, 1976, p. 201.
27. Chen, W.F., and Sugimoto, H., "Inelastic Cyclic Behavior of Tubular Members in Offshore Structures", Proc. of the 8th World Conf. on Earthquake Engng., San Francisco, 1984.
28. Shibata, M., "Analysis of Elastic-Plastic Behavior of a Steel Brace subjected to Repeated Axial Force", Int. J. of Solids and Str. Vol. 18, No. 3, 1982, pp. 217-228.
29. Toma, S., and Chem, W.F., "Cyclic Analysis of Fixed-Ended Steel Beam-Columns", J. of Str. Div. ASCE, Vol. 108, No. ST6, 1982, pp. 1385-1399.
30. Nonaka, T., "An Analysis of Large Deformation of an Elastic-Plastic Bar under Repeated Axial Loading", Int.J.of Mech. Sci.,Vol. 19, 1977, pp. 619-638.
31. Papadrakakis, M., and Chrysos, L., "Inelastic Cyclic Analysis of Imperfect Columns", J. of Str. Engng., ASCE, Vol. 111, No. 6, 1985.
32. Smith, E.A., and Epstein, H.I., "Hartford Coliseum Roof Collapse: Structural Collapse Sequence and Lessons Learned", Civil Engng. ASCE, April. 1980,pp.59-62.
33. Toma, S., and Chem. W.F., "Post-Buckling Behavior of Tubular Beam-Columns", J. of Str. Engng., Vol. 109, No. 8, 1983, pp. 1918-1932.
34. Schmidt, L.C., Morgan, P.R., and Clarkson, J.A., "Space Trusses with Brittle-Type Strut Buckling", J. of Str. Engng. ASCE, Vol. 102, No.ST7, 1976,pp.1479-1492.
35. Suzuki, T., Kubodera, I., and Ogawa, T., "An Experimental Study on Load-Bearing Capacity", Third Int. Conf. on Space Structures, Guilford, England, H. Nooshin, ed., Applied Science Publishers, Ltd., London, 1984, pp. 571-576.
36. Schmidt L.C., Gregg, B.M., "A Method for Space Truss Analysis in the Post-Buckling Range", Int. J. for Num. Methods in Engng. Vol. 15, 1980, pp. 237-247.
37. Singh, P., and Goel, S.C., "Hysteresis Model of Bracing Members for Earthquake Response of Braced Frames", Proc. of the 6th World Conf. on Earthquake Engng., New Delhi, Vol. 11, Jan. 1977, pp. 43-48.
38. Wakabayashi, M., et al., "Hysteretic Behavior of Steel Braces Subjected to Horizontal Load due to Earthquake", Proc. of the 6th, World Conf. on Earthquake Engng., New Delhi, 1977, pp. 3188-3194.
39. Maison, B.F., and Popov, E., "Cyclic Response Prediction for Braced Steel Frames", J. of the Str. Div., ASCE, Vol. 106, No. ST7, 1980, pp. 1401-1416.
40. Roeder, C., and Popov, E., "Inelastic Behavior of Eccentrically Braced Steel Frames under Cyclic Loadings", Report No. UCB/EERC-77/18, Univ. of California, Berkeley, Aug. 1977.
41. Bathe, K.-J., "Finite Element Procedures in Engineering Analysis", Prentice-Hall, Inc., New Jersey, 1982.
42. Bathe, K.-J., and Bolourchi, S., "Large Displacement Analysis of Three-Dimensional Beam Structures", Int. J. for Num. Methods in Engng., Vol. 14, 1979, pp.961-986.
43. Argyris, J.H., et al. "Finite Element Method - The Natural Approach" Comp. Meth. in Appl. Mech. and Engng., Vol. 17/18, 1979, pp. 1-106.
44. Oran, C., "Tangent Stiffness in Space Frames", J. of the Str. Div., ASCE, Vol. 99, No. ST6, 1973, pp. 987-1001.
45. Bathe, K.-J., and Ozdemir, H., "Elastic-Plastic Large Deformation Static and Dynamic Analysis", Comp. and Str., Vol. 6, 1976, pp. 81-92.
46. Wen, R.K., and Rahimzadeh, J., "Nonlinear Elastic Frame Analysis by Finite Element", J. of Str. Engng., ASCE, Vol. 109, No. 8, 1983, pp. 1952-1971.
47. Wood, R.D., and Zienkiewicz O.C., "Geometrically Nonlinear Finite Element Analysis of Beams, Frames, Arches and Axisymmetric Shells", Comp. and Str. Vol. 7, 1977, pp. 725-735.
48. Remseth, S.N., "Nonlinear Static and Dynamic Analysis of Framed Structures", Comp. and Str., Vol. 10, 1979, pp. 879-897.
49. Boswell, L.F., "A Small Strain Large Rotation Theory and Finite Element Formulation of Thin Curved Lattice Members", Third Int. Conf. on Space Structures, Guilford, England, H. Nooshin, ed., Applied Science Publishers Ltd., London, 1984, pp. 375-380.
50. Tang, S.C., Yeung, K.S., and Chon, C.T., "On the Tangent Stiffness Matrix in a Convected Coordinate System", Comp. and Str.,Vol. 12, 1980, pp. 849-856.
51. Kani, I.M. McConnel, R.E., and See, T., "The Analysis and Testing of a Single

Layer, Shallow Braced Dome", Third Int. Conf. on Space Structures, Guilford, England, H. Nooshin ed., Applied Science Publishers, London, 1984, pp. 613-618.
52. Jain, A.K., Goel, S.C., and Hanson, R.D., "Inelastic Response of Restrained Steel Tubes", J. of the Str. Div., ASCE, Vol. 104, No. ST6, 1978, pp. 897-910.
53. Papadrakakis, M., "A Family of Methods with Three-Term Recursion Formulae", Int. J. for Num. Methods in Engng., Vol. 18, 1982, pp. 1785-1799.
54. Engeli, M., et al., "Refined Iterative Methods for Computation of the Solution and the Eigenvalues of Self-Adjoint Boundary Value Problems", Birkhauser Verlag, Basel/Stuttgart, 1959.
55. Matthies, H., and Strang, G., "The Solution of Nonlinear Finite Element Equations", Int. J. for Num. Methods in Engng. Vol. 14, 1979, pp. 1613-1626.
56. Crisfield, M.A., "A Faster Modified Newton-Raphson Iteration" Comp. Meth. in Appl. Mech. and Engng., Vol. 20, 1979, pp. 267-278.
57. Crisfield, M.A., "Incremental/Iterative Solution Procedures for Non-Linear Structural Analysis", Numerical Methods for Nonlinear Problems, C. Taylor, E. Hinton, and D.R.J. Owen, eds., Pineridge Press, Swansea U.K., 1980, pp.261-290.
58. Irons, B., and Elsawaf, A., The Conjugate-Newton Algorithm for Solving Finite Element Equations", Proc. U.S. - German Symposium on Formulations and Algorithms in Finite Element Analysis, K.-J. Bathe, J.T. Oden, and W. Wunderlich, eds., MIT press, 1977, pp. 656-672.
59. Gambolati, G., "Fast Solution to Finite Element Flow Equations by Newton Iteration and Modified Conjugate Gradient Method", Int. J. for Num. Meth. in Engng., Vol. 15, 1980. pp. 661-675.
60. Buckley, A.G., "A Combined Conjugate-Gradient Quasi-Newton Minimization Algorithm", Math. Progr., Vol. 15, 1978, pp. 200-210.
61. Jennings, A., and Malik, G.M., "The Solution of Sparse Linear Equations by the Conjugate Gradient Method", Int. J. for Num. Meth. in Engng., Vol. 12, 1978, pp. 141-158.
62. Papadrakakis, M., "Accelerating Vector Iteration Methods", J. of Appl. Mech., ASME, Vol. 53, 1986, pp. 291-297.
63. Bathe, K.-J., and Dvorkin, E., "On the Automatic Solution of Nonlinear Finite Element Equations", Comp. and Str., Vol. 17, No. 5-6, 1983, pp. 871-879.
64. Geradin, M., Idelsohn, S., and Hogge, M., "Computational Strategies for the Solution of Large Nonlinear Problems via Quasi-Newton Methods", Computers and Structures", Vol. 13, 1981, pp. 73-81.
65. Crisfield, M.A., "An Arc-Length Method Including Line Searches and Accelerations", Int. J. for Num. Meth.in Engng., Vol. 19, 1983, pp. 1269-1289.
66. Wright, E.W., Gaylord, E.H., "Analysis of Unbraced Multi-Story Steel Rigid Frames", J. of the Str. Div., ASCE, Vol. 94, 1968, pp. 1143-1163.
67. Bergan, P.G., "Solution Algorithms for Nonlinear Structural Problems", Proc. of the Conf. on Engng. Applics.of the Finite Element Method", Høvik, Norway 1979, published by A.S. Computas.
68. Argyris, J.H., "Continua and Discontinua, Matrix Methods in Structural Mechanics", Proc. of the Conf. on Matrix Methods, Wright-Patterson Air Force Base, Ohio 1965.
69. Riks, E., "The Application of Newton's Method to the Problem of Elastic Stability", J. of Appl. Mech., Vol. 39, 1972, pp. 1060-1066.
70. Wempner, G.A., "Discrete Approximations Related to Nonlinear Theories of Solids", Int. J. of Solids and Str., Vol. 7, 1971, pp. 1581-1599.
71. Crisfield, M.A. "A Fast Increamental/Iterative Solution Procedure that handles "Snap-Through", Comp. and Str., Vol. 13, 1981, pp. 55-62.
72. Park, K.C., "A Family of Solution Algorithms for Nonlinear Structural Analysis Based on Relaxation Equations", Int. J. for Num. Methods in Engng., Vol. 18, 1982, pp. 1337-1347.
73. Bergan, P.G., "Solution Algorithms for Nonlinear Structural Problems", Comp. and Str., Vol. 12, 1980, pp. 497-509.
74. Fried, I., Orthogonal Trajectory Accession to the Nonlinear Equilibrium Curve", "Comp. Meth. in Appl. Mech. and Engng., Vol. 47, 1984, pp. 283-297.
75. Noor, A.K., and Peters, J.M., "Tracing Post-Limit-Point Paths with Reduced Basis Technique", Comp. Meth. in Appl. Mech. and Engng., Vol. 28, 1981, pp.217-240.
76. Powell, G., and Simons, J., "Improved Iteration Strategy for Nonlinear Structures", Int. J. for Num. Meth. in Engng., Vol. 17, 1981, pp. 1455-1467.
77. Riks, E., "An Incremental Approach to the Solution of Snapping and Buckling

Problems", Computers and Structures", Vol. 15, 1979, pp. 259-551.
78. Waszczyszyn, Z., "Numerical Problems of Nonlinear Stability Analysis of Elastic Structures", Comp. and Str., Vol. 17, No. 1, 1983, pp. 13-24.
79. Klimke, H., and Posch, J., "A Modified Fictitius Force Method for the Ultimate Load Analysis of Space Trusses", Third Int. Conf. on Space Structures, Guilford, England, H. Nooshin, ed., Applied Science Publishers, London, 1984, pp. 589-593.
80. Papadrakakis, M., "Inelastic Post-Buckling Analysis of Trusses", Journal of Structural Engineering, ASCE, Vol. 109, No. 9, 1983, pp. 2129-2147.
81. Smith, E.A., "Space Truss Nonlinear Analysis", J. of Str. Engng., ASCE, Vol. 110, No. 4, 1984, pp. 688-705.
82. Cichon, C., "Stability Analysis of Elastic Space Trusses", Third Int. Conf. on Space Structures, Guilford, England, H. Nooshin ed., Applied Science Publishers Ltd, London, 1984, pp. 567-570.
83. McConnel, R.E., and Klimke, H., "Geometrically Nonlinear Pin-Jointed Space Frames", Numerical Methods for Nonlinear Problems, Vol. 1, C. Taylor, E. Hinton, and D.R.J. Owen, eds., Pineridge Press, Swansea, 1980, pp. 333-342.
84. Bathe, K.-J., Ramm, E., and Wilson, E.L., "Finite Element Formulations for Large Deformation Dynamic Analysis", Int. J. for Num. Meth. in Enging., Vol. 9, pp. 353-386, 1975.
85. Ramm, E., "Strategies for tracing nonlinear responses near limit points" in Nonlinear Finite Element Analysis in Structural Mechanics, Wunderlich, Stein, Bathe, eds, Springer - Verlag, 1981.
86. Hangai, Y., and Kawamata, S., Nonlinear Analysis of Space Frames and Snap - Through Buckling of Reticulated Shell Structures", Proceedings, 1971 IASS Pacific Symposium on Tension Structures and Space Frames, Architectural Institute of Japan, Tokyo, 1972, pp. 803-816.
87. Bergan, P.G., and Søreide, T., "A Comparative Study of Different Numerical Solution Techniques as Applied to a Nonlinear Structural Problem", Computer Methods in Applied Mechanics and Engineering, Vol. 2, 1973, pp. 185-201.
88. Papadrakakis, M., "Post-Buckling Analysis of Spatial Structures by Vector Iteration Methods" Computers and Structures, Vol. 14, No. 5-6, 1981, pp. 393-402.
89. Jagannathan, D., Epstein, H.I., and Christiano, P., "Nonlinear Analysis of Reticulated Space Trusses", Journal of the Structural Division, ASCE, Vol. 101, No. ST12, pp. 2641-2658.
90. Rothert, H., Dickel, T., and Renner D., "Snap-Through Buckling of Reticulated Space Trusses", Journal of the Structural Division, ASCE, Vol. 107, 1981, pp. 129-143.
91. Paradiso, M., and Tempesta, G., "Member Buckling Effects in Nonlinear Analysis of Space Trusses", Numerical Methods for Nonlinear Problems, Vol. 1, C. Taylor, E. Hilton, and D.R.J. Owen eds., Pineridge Press, Swansea, 1980, pp. 395-405.
92. Watson, L.T., and Holzer, S.M., "Quadratic Convergence of Crisfield's Method", Computers and Structures, Vol. 17, No. 1, 1983, pp. 69-72.
93. Meek, J.L., and Tan, H.S., "Geometrically Nonlinear Analysis of Space Frames by an Incremental Iterative Technique", Computer Methods in Applied Mechanics and Engineering", Vol. 47, 1984, pp. 261-282.
94. Schmidt, L.C., Morgan P.R., and Hanaor, A., "Ultimate Load Testing on Space Trusses", Journal of the Structural Division, ASCE, Vol. 108, No. ST6, 1982, 1324-1335.
95. Griggs, H.P., "Experimental Study of Instability in Elements of Shallow Space Frames", Research Report, Department of Civil Engineering, MIT, Cambridge, MA, 1966.
96. Chu, K.H., and Rampetsreiter, R.H., "Large Deflection Buckling of Space Frames", Journal of the Structural Division, ASCE, Vol. 98, No. ST12, 1972, pp.2701-2722.
97. Connor, J.J., Logcher, D., and Chan, S.C., "Nonlinear Analysis of Elastic Framed Structures", Journal of the Structural Division, ASCE, Vol. 94, No. ST6, 1968, pp. 1525-1547.
98. Brendel, B., and Ramm, E., "Stabilitätsuntersuchungen Weitgespannter Tragwerke mit der Methode der Finiten Elemente", Proceedings of the International Symposium on Wide Span Surface Structures, Universität Stuttgart, 1976.
99. Papadrakakis, M., and Ghionis, P., "Conjugate Gradient Algorithms in Nonlinear Structural Analysis Problems", Computer Methods in Applied Mechanics and Engineering, to be Published.

ANALYSIS OF LARGE SCALE SPACE FRAMES BY MINIMIZATION OF TOTAL POTENTIAL ENERGY

H Tabar-heydar
Marchant Filer Dixon, Consulting Engineers and Designers
Bell House, 32 Bell Street, Romsey SO51 8GW, United Kingdom

1. ABSTRACT

An approach suitable for the analysis of large scale space frames is presented which offers a number of advantages over conventional methods. Chief amongst these is the ability to monitor the analysis in progress to obtain estimates of results or detect errors early on in the process. The method is also more economical than those based on the direct solution of the equations of equilibrium both in terms of the number of operations and primary storage requirements.

2. INTRODUCTION

The principal of minimum potential energy for a conservative system states that:- amongst all admissible deflected forms, only those that are stationary points of the total potential energy produce an equilibrium state and that such an equlibrium is stable only if it corresponds to a minimum of the total potential energy. It is, therefore, evident that, if the total potential energy of a structure could be formulated as a function of deflections, then the problem of calculating the response of the structure to the action of a given loading regime can be reduced to one of minimizing the energy functional.

The method described in this paper is based on the approach above. The total potential energy is formulated as a function of nodal displacements and the Conjugate Gradients algorithm is used to carry out the minimization.

3. CONJUGATE GRADIENTS METHOD

The method of conjugate gradients was developed independently by E. Stiefel and M.R. Hestenes with the first reports appearing in 1951-52 (1). The use of the method has been further popularized by R.Fletcher and C.M.Reeves (2).

The algorithm can be summarized thus:

a. Given routines for calculating the values of the function $f(x)$ and its derivative $g(x)$ for any point and a scaling matrix H, select an estimate x_0 of the solution $\bar{x}$.

b. Generate the conjugate direction v_k from:

$$v_k = -Hg_k \qquad ,k=0$$

$$v_k = -Hg_k + R_k v_{k-1} \qquad ,k>0$$

where

$$R_k = g_k^T Hg_k \Big/ g_{k-1}^T Hg_{k-1}$$

c. Carry out a line search along v to calculate s such that:

$$\partial f\left(x+sv\right) \Big/ \partial s = 0$$

d. Update the approximation to $\bar{x}$ from:

$$x_{k+1} = x_k + s_k v_k$$

e. Continue the iterative process until x_k is sufficiently close to $\bar{x}$.

The scaling matrix H can be constructed in a number of ways to modify the behaviour of the algorithm. In this paper, H is selected as a diagonal matrix with the elements calculated from:

$$h_{i,i} = 1/\sqrt{k_{i,i}}$$

where K is the structural stiffness matrix.

The scalar s is often calculate by quadratic or cubic approximation (3,4). However, as the function is known in the present application, an analytical expression for s is developed.

4. TOTAL POTENTIAL ENERGY OF A SPACE FRAME

Given a structural system deflecting elastically under the action of static loads, the total potential energy at a given deflected form is given by:

$$W = U + V$$

Where U is the elastic strain energy stored in the system and V is the potential energy of the loading. Considering a space frame with M elements and N degrees of freedom and assuming that the loading regime is defined by the vector f then:

$$U = \sum_{m=1}^{M} U_m$$

$$V = -f^T x$$

The contribution to the strain energy from the mth element, U_m, is given by:

$$U_m = \frac{1}{2} \frac{EA}{L} e^2$$

In order to calculate this contribution, some relationship between the elongation e and the vector of displacements x needs to be developed. Consider an element of length L with its projections onto the global axes denoted by X_1, X_2 and X_3. Now if this element is displaced and extended such that its projections onto the global axes are (X_1+x_1), (X_2+x_2) and (X_3+x_3) then it is clear that the new length is given by:

$$\left(L+e\right)^2=\sum_{i=1}^{3}\left(X_i+x_i\right)^2$$

$$L^2+2Le+e^2=\sum_{i=1}^{3}X_i^2+2\sum_{i=1}^{3}X_i x_i+\sum_{i=1}^{3}x_i^2$$

If we assume small elongations and small displacements, terms in e^2 and x^2 will disappear to yield the familiar strain/displacement relationship:

$$e=\frac{1}{L}\sum_{i=1}^{3}X_i x_i$$

This linear relationship is adequate for most space frame structures, however, the formulation can be extended to deal with geometric nonlinearities. Assuming small elongations so that e^2 can be ignored and preserving the terms in x^2

$$e=\frac{1}{L}\left(\sum_{i=1}^{3}X_i x_i+\frac{1}{2}\sum_{i=1}^{3}x_i^2\right)$$

The derivative of the total potential energy, $g=dW/dx$, can be identified as the vector of out of balance forces and calculated from contributions from elements and external loads. The element m, will contribute to the terms corresponding to the degrees of freedom of nodes at its either end. Denoting by T the direct force in the element, the contribution, say in the Y direction, is found by projecting T onto that direction:

$$\frac{X_2}{L}T$$

Where

$$T=\frac{EA}{L}e$$

5. CALCULATION OF SCALAR s

As mentioned earlier, s is calculated from the condition:

$$\partial W_{k+1}/\partial S_k=0$$

Now

$$W_k = \sum_{m=1}^{M} \left(\frac{EA}{2L^3} \left(\sum_{i=1}^{3} x_i x_i \right)^2 \right)_m - \sum_{n=1}^{N} f_n x_n$$

From which W_{k+1} can be calculated by substituting x_{k+1} for x_k :

$$W_{k+1} = s_k^2 \sum_{m=1}^{M} \left(\frac{EA}{2L^3} \left(\sum_{i=1}^{3} x_i v_i \right)^2 \right)_m +$$
$$s_k \left(\sum_{m=1}^{M} \left(\frac{EA}{L^3} \left(\sum_{i=1}^{3} x_i x_i \right) \left(\sum_{i=1}^{3} x_i v_i \right) \right)_m - \sum_{n=1}^{N} f_n v_n \right) +$$
$$\sum_{m=1}^{M} \left(\frac{EA}{2L^3} \left(\sum_{i=1}^{3} x_i x_i \right)^2 \right)_m - \sum_{n=1}^{N} f_n x_n$$

Which Leads to the expression for s :

$$s_k = - \left(\sum_{m=1}^{M} \left(\frac{EA}{L^3} \left(\sum_{i=1}^{3} x_i x_i \right) \left(\sum_{i=1}^{3} x_i v_i \right) \right)_m - \sum_{n=1}^{N} f_n v_n \right) \Big/ \sum_{m=1}^{M} \left(\frac{EA}{L^3} \left(\sum_{i=1}^{3} x_i v_i \right)^2 \right)_m$$

6. TERMINATION CRITERION

A number of termination criteria have been proposed (5). The choice would depend on the nature of the iterative process and the relative numerical values of the quantities involved. In the present application where the stiffness coefficients in the various degrees of freedom are of the same order, a simple termination criterion would suffice. The criterion here proposed would require the norm of the vector of out of balance forces to fall below a given percentage of the initial value. In practical problems, it is sometimes difficult to achieve values below 1% due to rounding errors. Values above 10% are also not recommended as they can lead to misleading results in certain cases.

7. SUMMARY OF THE ALGORITHM

The algorithm described above can be summarised in the following steps:

a. Calculate the derivative g

a.1 Loop through the elements calculating the elongation and hence the direct force and its contribution to g .

a.2 Loop through the vector of external loads and add their contribution to g .

a.3 Loop through the restraints, setting to zero the corresponding elements of g .

b. Check for convergence.

c. Calculate the descent vector v .

d. Calculate the scalar s .

d.1 Loop through the elements to calculate their contributions.

d.2 Loop through the vector of external loads to calculate their contribution.

e. Update the displacement vector x .

f. Return to step a.

8. PERFORMANCE OF THE ALGORITHMS

It can be shown that, barring the effects of rounding off errors, the algorithm converges in at most n iterations where n is the number of degrees of freedom. Furthermore, it has been found that for a range of practical space frames, the time taken to perform n iterations is roughly equivalent to solution by conventional finite element programes. However, practical structures would rarely require more than a fraction of n iterations to converge.

The rate of convergence is difficult to predict and is dependent on a number of factors including the pattern of eigen values of the stiffness matrix and the initial values of x and g .

The table below would give an indication of the number of iterations usually required to reduce the norm of the out of balance force vector to 1% its initial value.

Description of the model	no of degrees of freedom	no of elements	no of iterations	no of iterations as a % of ndf
13m diameter, 4 frequency geodesic dome	213	250	24	11.3%
20 x 20m double layer flat frame simply supported on the edges	602	800	77	12.8%
Bahrain Grand Mosque Dome 23m diameter, partially double layer	4,136	3,960	146	3.5%
Selangor State Mosque Dome 50.7m diameter double layer	12,195	13,864	536	4.4%
Proposed Shah Alam Sports Stadium Roof Triple layer crescent shape roof with span of 280m covering 16,300m^2.	16,692	24,309	771	4.6%

It can be seen that the algorithm is substantially more economical than direct solution of the quilibrium equations specially in the case of larger problems. Furthermore, the primary storage requirements are limited to a few vectors making the algorithm particularly attractive for implementation on micro or small mini computers.

9. MONITORING THE PROGRESS OF THE ALGORITHM

It is recommended to output the scalar quantities: total potential energy, norm of the out of balance force, and the step length at each iteration. Figures 1 to 3 show the plots of these quantities against the iteration numbers for the Bahrain Grand Mosque Dome.

It can be seen that the value of the total potential

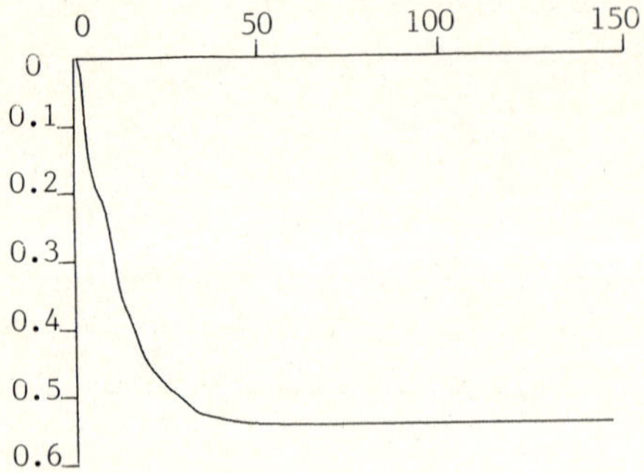

Fig 1 Total Potential Eiergy/iteration

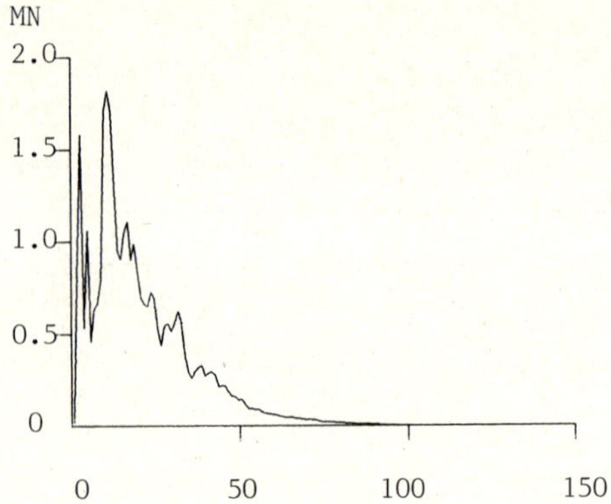

Fig 2 Norm of OBF/iteration

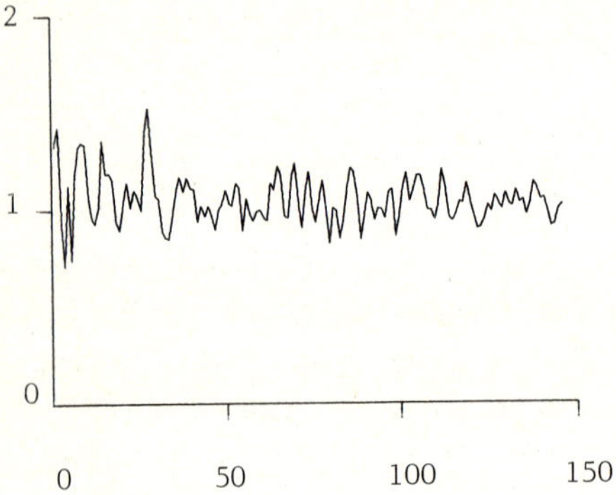

Fig 3 Step length/iteration

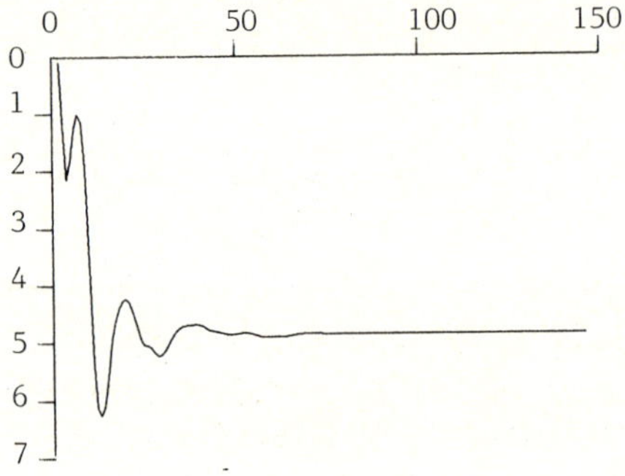

Fig 4 Displacement/iteration

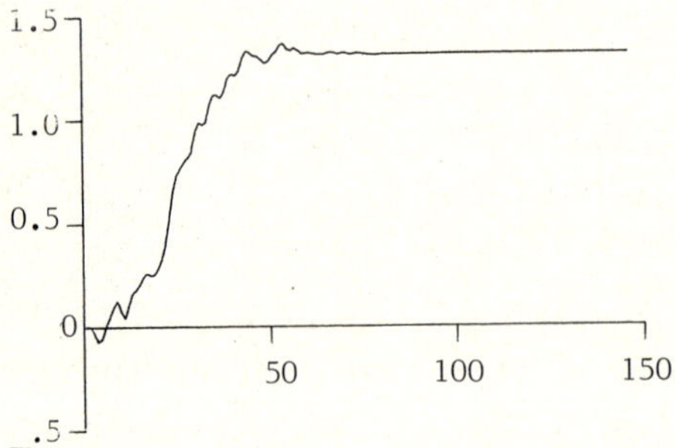

Fig 5 Tension/iteration

energy reduces rapidly at the start of the process and becomes very gradual at the end. Any deviation from this pattern would indicate an error.

Figure 2 shows the pattern of the norm of the out of balance force. Usually, this value shows a marked increase during the first few iterations after which it starts to decrease with considerable variation. It is difficult to judge the progress from this quantity alone. However, a steady high value can normally be interpreted as a sign of a difficult problem.

Figure 3 shows the step length. The application of scaling should cause this quantity to remain in the order of unity throughout the iterations. Very small quantities coupled with high values for the norm of the out of balance force would normally indicate free body displacement.

Figures 4 and 5 illustrate the displacements at a node and the direct force in an element selected arbitrarily from the structure. It is seen that the values stabilize after a relatively few iterations, in this case roughly 50 or 1/3 of iterations necessary for convergence. Clearly this will not be the case if the structure was not behaving appropriately and printing out of the largest displacement would quickly identify mechanisms or free body motions. Furthermore, when used with care, the characteristic would enable examination of the behaviour of the structure early in the process and is an invaluable facility specially in the case of large problems.

10. RESTARTING THE PROCESS

As an examination of the algorithm would show, only the values of the vector of derivatives , the descent vector and the displacement vector are carried over from one iteration to the next. This characteristic would indicate that if these vectors were occasionally written to back up storage then the process could be restarted from that point providing security in the case of interruptions. In fact only the vector of displacements is necessary as occasional restarts with the steepest descent direction would not significantly affect the performance.

The characteristic can also be used in re-analysis or re-design situations. If the mathematical model is varied while the displacement pattern remains more or less unchanged then starting with the results of the previous analysis would normally yield considerable economies.

11. CONCLUSIONS AND COMMENTS

An approach to solution of space frames is outlined which offers the following advantages over the more conventional methods:

1. Substantial economy of analysis for all but the rare pathological models.

2. The facility to monitor the progress of the analysis and obtain valuable information long before the analysis is completed.

The approach can also be extended to cater for nonlinearities where it offers even more substantial economics.

It is also possible to extend the algorithm to accept any type of element, although the application of the conjugate gradients method as a general solver to conventional finite element formulations has not proved advantageous (6). In such cases, most of the effort is expended in calculating the vector of derivatives. However, if the formulation is varied to make this task more affordable, the balance of economics may well change.

REFERENCES

1. HESTENES M.R. and STIEFEL E., Methods of conjugate gradients for solving linear systems, Journal of Research of the National Bureau of Standards, 49 (1952) 409-436.
2. FLETCHER R. and REEVES C.M., Function minimisation by conjugate gradients, Computer Journal, 7 (1964) 149-154.
3. POWELL M.J.D., An efficient method for finding the minimum of a function of several variables without calculating derivatives, The Computer Journal, 7 (1964) 155-162.
4. Davidon W.C., Variable metric method for minimisation, A.E.C.R. & D. Report ANL-5990 (Rev.) Argonne National Laboratory (1959).
5. BERGAN P.G. and CLOUGH R.W., Convergence criteria for iterative process, AIAA Journal, 10 8 (1972) 1107-1108.
6. IRON B and AHMAD S, Techniques of finite elements, Ellis Press, (1981).

Influence of Boundary Conditions Upon Stress Distribution in Single-Layer Diagonal Grids

Professor Z S Makowski, Mr D Maranzano and Mr G A R Parke
Space Structures Research Centre
University of Surrey
Guildford
UK

Introduction

There are various types of single-layer grid structures used in civil engineering practice.
The progressive structural designers are now paying special attention to the diagonal grid configuration consisting of beams arranged in two parallel sets, equally spaced and usually interconnected at 90° and placed obliquely (usually at 45° with respect to the boundary lines [1].
The advantages offered by such a system of construction are such that it is being used intensively by an ever-increasing number of architects and structural engineers especially for long-span floors and roofs in steel or reinforced concrete.
Extensive research work, [2] and successful application of tests on models, as well as on actual structures, highlights the great rigidity of diagonal grids leading to:
1) reduction of the structural depth and
2) increase in clear span.
The experience shows that heavy concentrated loads can be easily and economically accommodated without the provision of excessive deep and costly beams. Diagonal grids can be applied with advantage in a large variety of arrangements to all types of floors and roofs, whether flat, curved or folded and whether regular or irregular in plan. Diagonal grids have been successfully used in the construction of large-span industrial buildings, garages, assembly halls and also in some cases for floors in multi-storey buildings. The exceptional strength of diagonal grids has been realised in the construction of hangars, [3]. The typical examples are the prefabricated steel roofs over the

Heathrow Airport hangars (with clear openings of 137m.) and recently at the Changi Airport in Singapore (with a record clear span of 217m.). Electric arc welding is frequently used in the fabrication of single-layer steel diagonal grids giving full continuity of the interconnected beams. This leads to a co-operative action of all the grid members, an essential feature of the diagonal grids. The node connections are neat and unobtrusive providing a structure of exceptional architectural appeal.
The diagonal grid has been used by one of the authors in a number of practical applications including steel flat roofs over petrol stations.
During the feasibility study for structures of this type several different boundary conditions have been considered during the design and the detailed resulting stresses and deflections compared.
The paper contains a comparison of the study of 12 different boundaries of a diagonal grid, covering a square area of 4 subdivisions along each side, as shown in Figures 1 and 2.

Analysis

Single-layer diagonal grids have been considered, all of them consisting of members of identical cross-section. All structures have been analysed under uniformly distributed loading of intensity $1kN/m^2$. Figure 1 illustrates twelve different boundary conditions;

case 1 - refers to a grid simply supported along all four sides
case 2 - grid is encastre along all four sides,
case 3 - grid is supported by a single column at the centre,
case 4 - grid is simply supported at four corners,
case 5 - grid is simply supported at four middle points,
case 6 - refers to a grid supported at four inner points,
case 7 - the structure is supported at 5 points,
case 8 - the structure is encastre along three sides,
case 9 - the structure is encastre along two sides,
case 10 - the structure is cantilevered, being encastre at one side only,
case 11 - fully encastre at four corners,
case 12 - fully encastre at four middle points.

In order to determine the nodal deflections and detailed stress distribution within the grid, the structure was analysed using the standard stiffness method.
In this technique the grid is considered to be an assembly of discrete members rigidly interconnected. By considering the force-displacement

relationship for each individual member, and then by satisfying both the compatibility and equilibrium requirements at each node, a set of linear simultaneous equations is formed for the complete structure. This system of linear simultaneous equations is expressed in the usual matrix notation as

$$[K][d] = [w], \qquad \text{eq.(1)}$$

where [K] is the primary stiffness matrix of the structure, [d] is a vector of unknown nodal displacements and [w] is the appended load vector of the structure. Modification of the above equation to allow for the various support constraints (used in the cases 1-12) will result in a consistent set of equations. These equations may then be solved to give a unique set of nodal displacements, which, in turn, will be used to obtain individual member forces.

The single-layer flat grids are analysed by allowing for only three degrees of freedom at each node; it is assumed that in flat grids being under the action of lateral loads only, the in-plane axial forces and moments are negligible.

To generate the input data required for the stiffness analysis computer program, extensive use was made of Formex Algebra developed at the Space Structures Research Centre by H. Nooshin, [4]. Using the diagonal grid configuration shown in Figure 2 the Formex formulation required to generate the grid topology is given by Formex F3 as follows:

E = [1,1;3,1]
F = [1,1;2,2]
F1 = RINID (4,4,2,2)¦ROSID(2,2)¦F
F2 = ROSID (5,5)¦RIN(1,4,2)¦E
F3 = F1#F2

Normally it is usual to make use of any symmetry occurring within the structure as this will simplify the data generation and reduce computer storage requirements.

In our case making full use of symmetry for the single-layer diagonal grid would cause the formation of several non-conformable joint displacements occurring at some of the boundary nodal points. To prevent this happening and consequently avoiding modification to the basic program, the complete structure was analysed. In an attempt to minimise the computer storage requirements the overall stiffness matrix was stored in a banded form. The system was then solved to yield the nodal displacement using the direct method of Gaussian elimination. To simplify the comparison of displacement and stress distributions obtained for each separate analysis one of the post-processing techniques developed in conjunction with Formex Algebra was used.

Comparison of results

The enclosed diagrams show deflection, bending and twisting moment and shears. Figure 3 illustrates the arrangement. Space limitation only allows the inclusion of the full diagrams for the following six cases:
case 1 (simply supported)
case 2 (encastre)
case 3 (supported by a single column in the middle)
case 4 (simply supported at four corners)
case 5 (simply supported at four middle points along the boundary)
case 6 (supported at four inner points)
Table I gives the comparison for all twelve boundary conditions. Cases 3, 4, 5, and 6 are of particular interest. As expected, a structure supported only at one central point will experience the greatest deflections and maximum stresses. Comparison of cases 4 and 5 shows clearly the great improvement in the stress distribution through the movement of four corner supports to the middle points. The reduction of resulting deformation in case 5 will probably come as a surprise to many designers.

Case 6 should also receive special attention. Columns supporting the grid at four inner points produce a very uniform stress distribution mainly through the effect of cantilevered portions of the diagonal grid along all four sides. This particular arrangement has been used in several practical design cases of petrol stations. The central portion between four columns has been occupied by an office, where the four outside portions allowed free access to cars.

The other boundary conditions, whether simply supported or encastre will be of interest to designers dealing with construction of diagonal grids in buildings.

References

1. Z S Makowski -"Modern grid structures" Architectural Science Review, 3 (1960), pp. 52-65.

2. H Nooshin and Z S Makowski -"Some aspects of the behaviour of square diagonal grids". International Journal of Building Science, December (1974), pp. 269-75.

3. G M Crook and Z S Makowski - "Analysis, design and construction of the space frame roofs for the BOAC 747 hangars at London Airport". Proceedings of the IASS Symposium on Tension Structures and Space Frames, Tokyo and Kyoto, Architectural Institute of Japan, 1972, pp. 487-501.

4. H Nooshin - "Formex configuration processing in structural engineering", Elsevier Applied Science Publishers, London 1984.

BOUNDARY CONDITIONS	CASE	MAX DEFLECTION mm	MAX SHEAR FORCE kN	MAX. POSITIVE B.M. kNm	MAX. NEGATIVE B.M. kNm	MAX. TWISTING MOMENT kNm
Simply supported along four sides	1	2.7042	2.073	2.185	1.709	0.272
Fully encastre along four sides	2	0.8598	1.490	0.654	1.131	0.385
Supported at centre	3	9.008	edge 0.484 8.000	edge 1.814 0.683	edge - 7.878	edge 0.266 0.925
Simply supported at four corners	4	11.975	edge 2.536 2.677	edge 3.603 2.910	edge 0.503 -	edge 0.546 1.057
Simply supported at four middle points	5	2.319	edge 1.674 2.201	edge 1.101 1.521	edge 1.267 0.810	edge 0.471 0.196
Supported at four inner points	6	0.839	edge 0.330 2.370	edge 0.1 0.466	edge 0.422 1.997	edge 0.057 0.281
Supported at four corners and at centre	7	0.5843	edge 1.09 2.084	edge 0.5 0.922	edge 1.146 2.248	edge 0.093 0.237
Fully encastre along three sides	8	1.3085	edge 1.391 1.770	edge 1.05 0.808	edge 1.672 1.473	edge 0.099 0.555
Fully encastre along two opposite sides	9	1.5818	edge 1.381 1.723	edge 1.094 0.792	edge 1.688 1.611	edge 0.128 0.650
Fully encastre along one side	10	66.123	edge 3.147 3.749	edge - 0.309	edge 12.35 7.866	edge 0.554 4.758
Fully encastre at four corners	11	3.86	edge 2.003 3.744	edge 1.477 1.386	edge 2.473 4.266	edge 0.803 0.571
Fully encastre at four middle points	12	0.888	edge 1.075 2.800	edge 0.382 0.979	edge 1.139 2.039	edge 0.227 0.158

TABLE 1

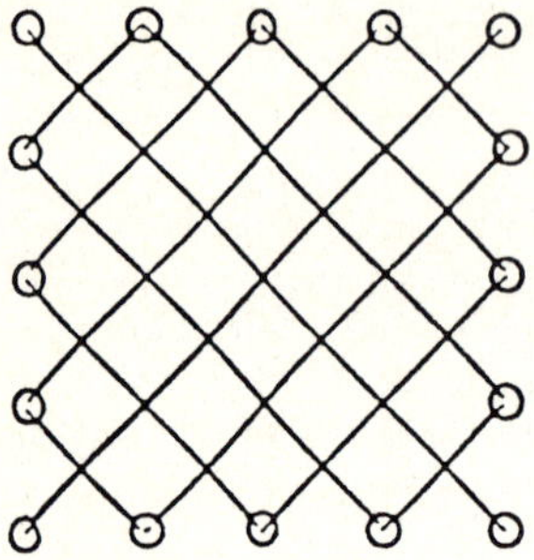

Case 1
simply supported along all four sides

Case 2
encastre along all four sides

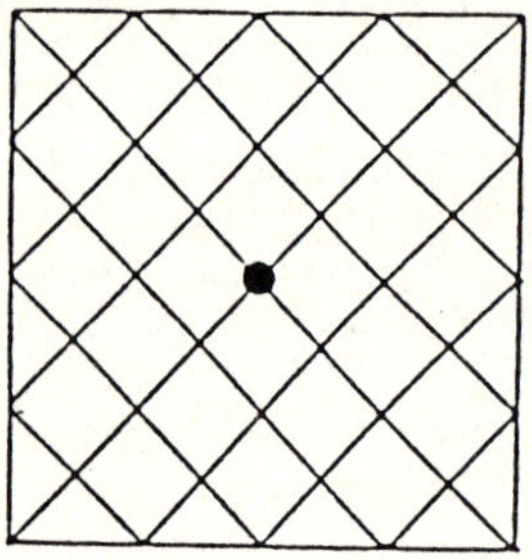

Case 3
supported at the centre

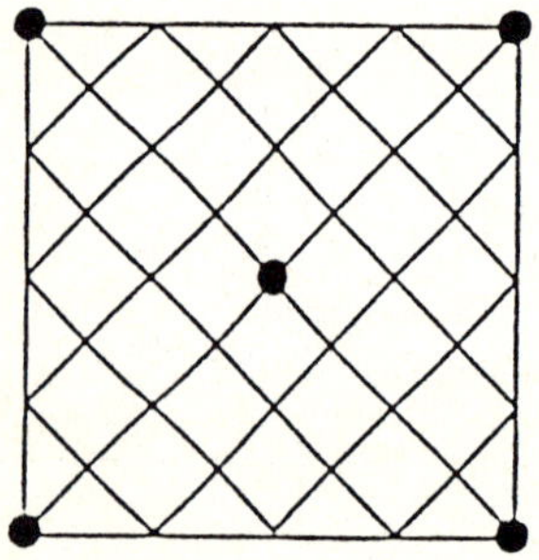

Case 7
supported at five points

Case 8
encastré along three sides

Case 9
encastré along two sides

Case 4

simply supported at four corners

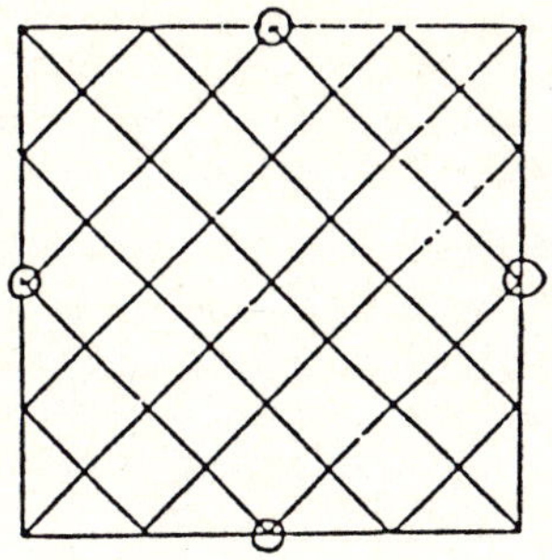

Case 5

simply supported at four middle points

Case 6

supported at four inner points

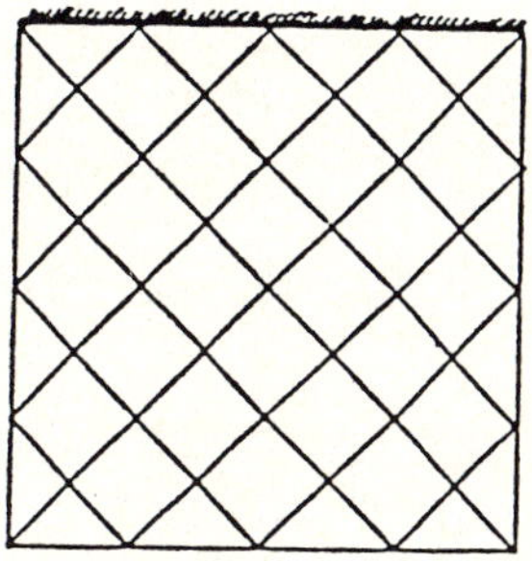

Case 10

encastré at one side cantilevered

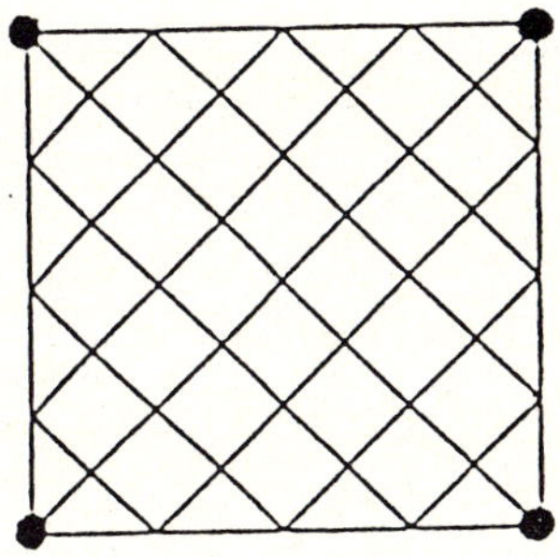

Case 11

supported at four corners

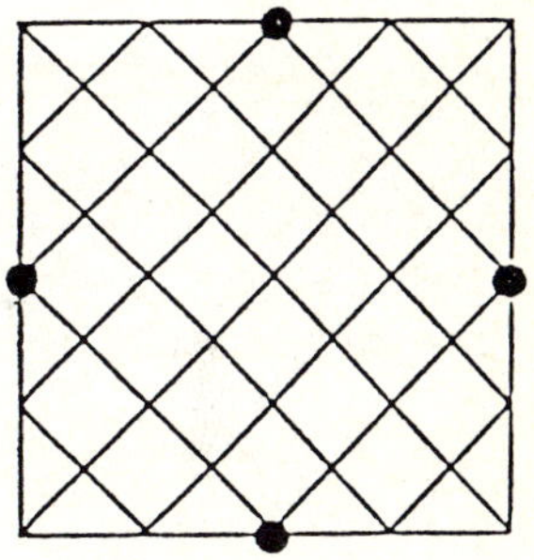

Case 12

supported at four middle points along the boundary

FIGURE 1

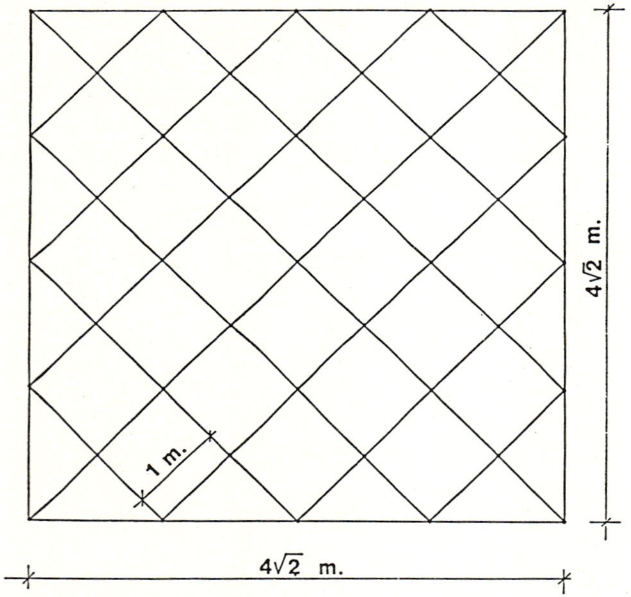

Single-layer diagonal grid

members size 100·100 mm.

load 1 kN./m.^{2} (concetrated on purlins)

modulus of elasticity 200 kN./mm.^{2}

FIGURE 2

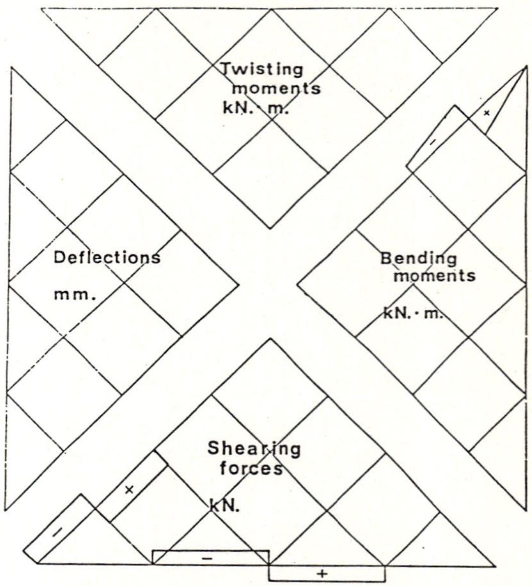

Single-layer diagonal grid

support conditions

- ● encastré
- o simply support
- encastré

FIGURE 3

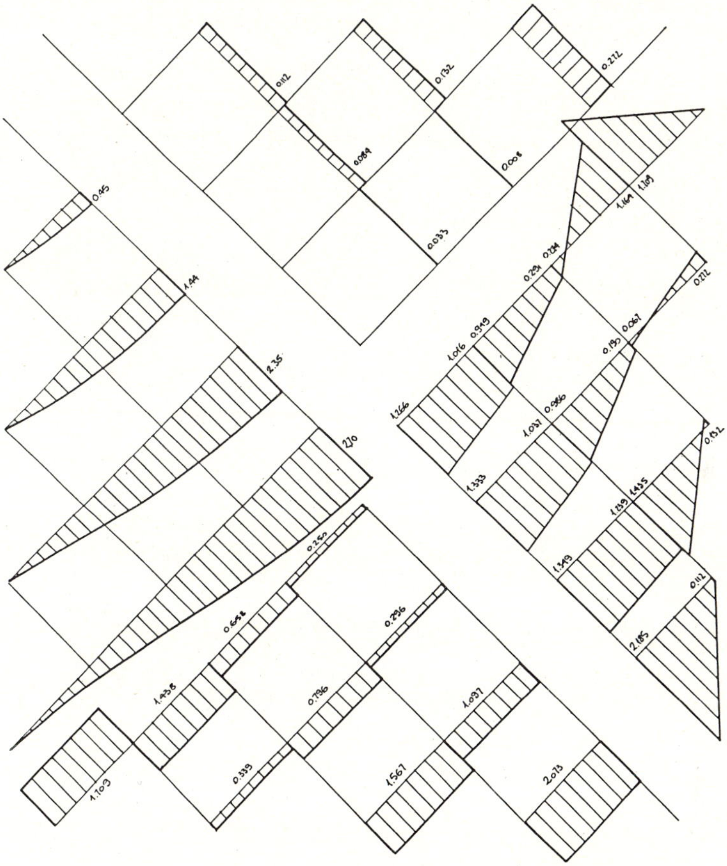
0.112
0.132
0.272
0.084
0.008
0.45
1.164
1.169
0.033
1.44
0.291
0.234
0.272
2.35
1.016
0.949
0.130
0.067
1.266
1.037
0.966
2.70
0.132
1.333
1.239
1.435
0.250
1.349
0.658
0.296
0.112
2.185
1.438
0.796
1.097
0.339
1.567
2.073
1.109

CASE 1

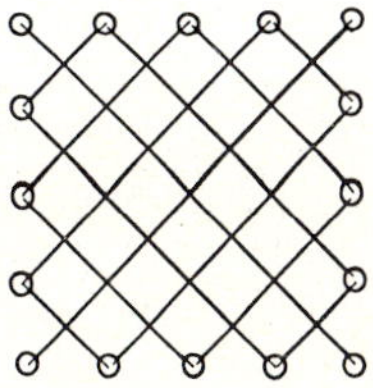

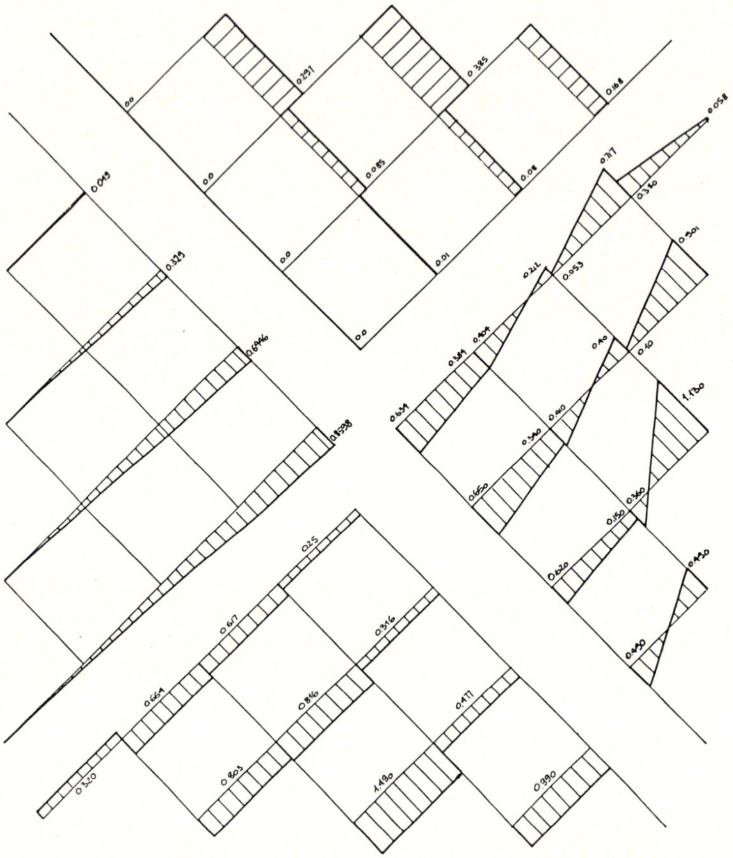

CASE 2

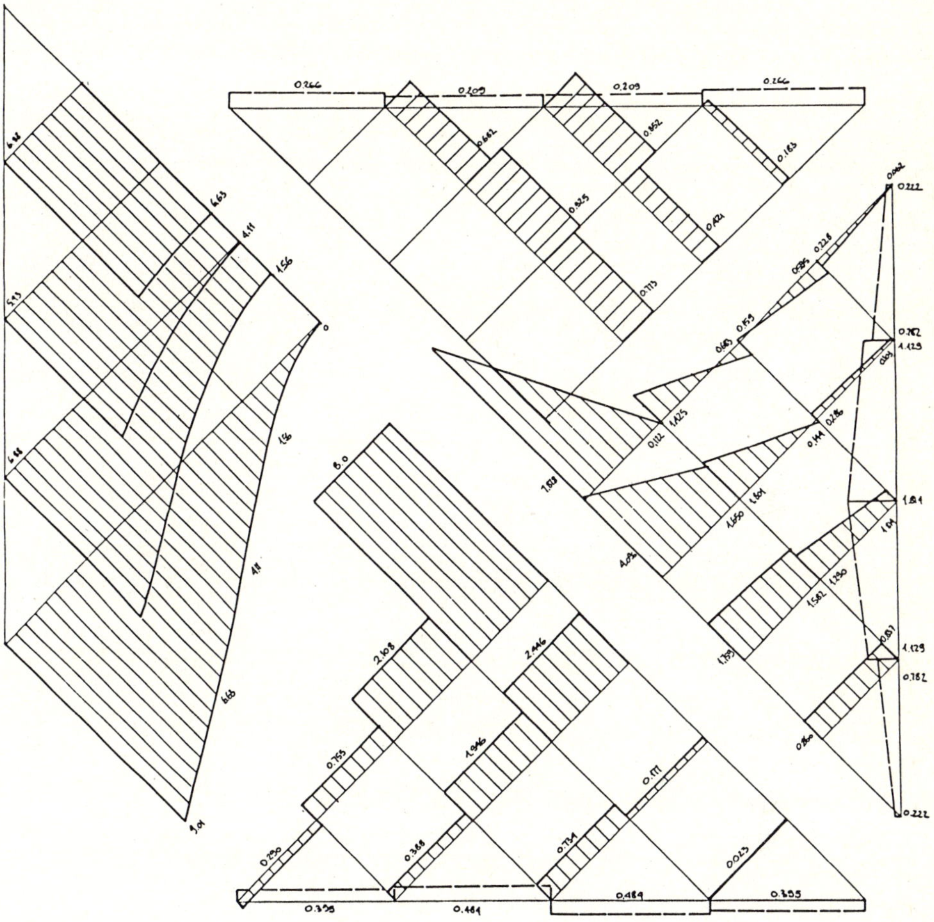

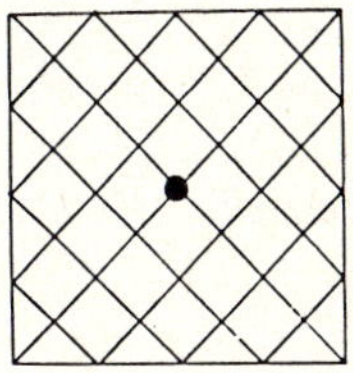

CASE 3

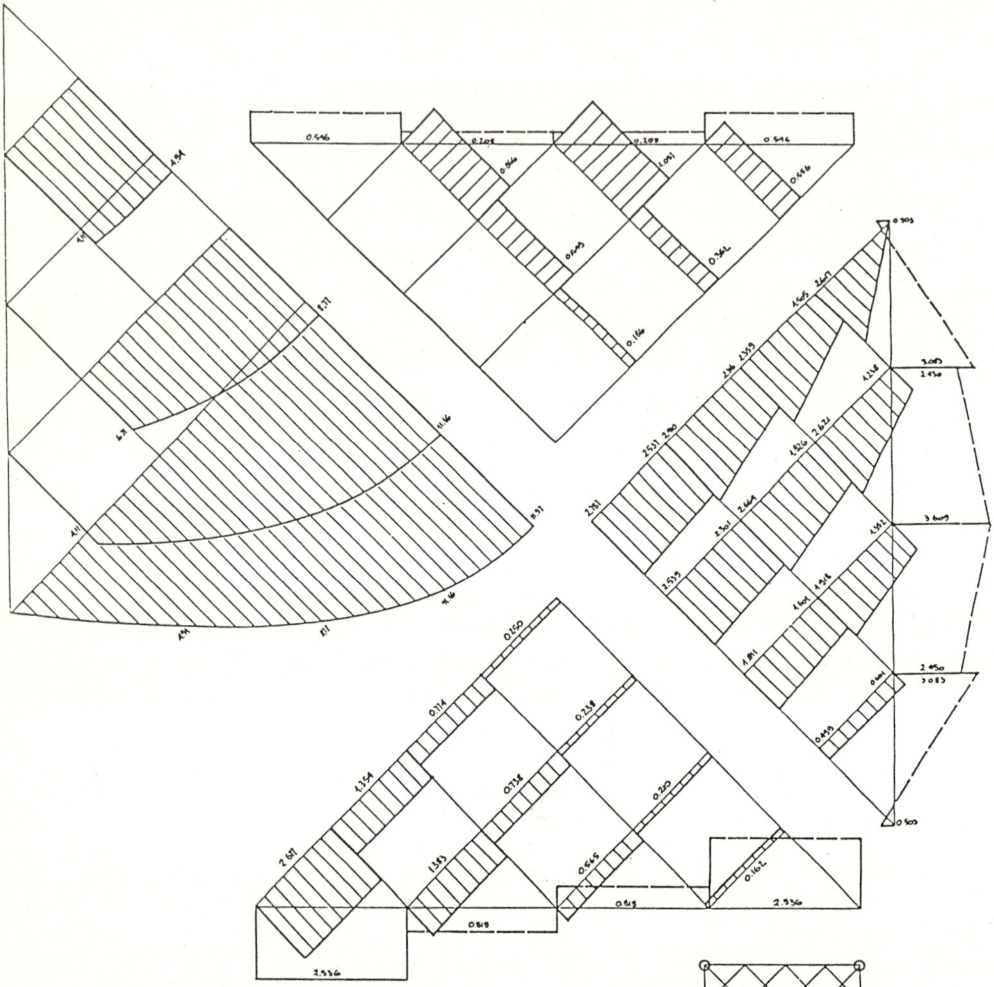

CASE 4

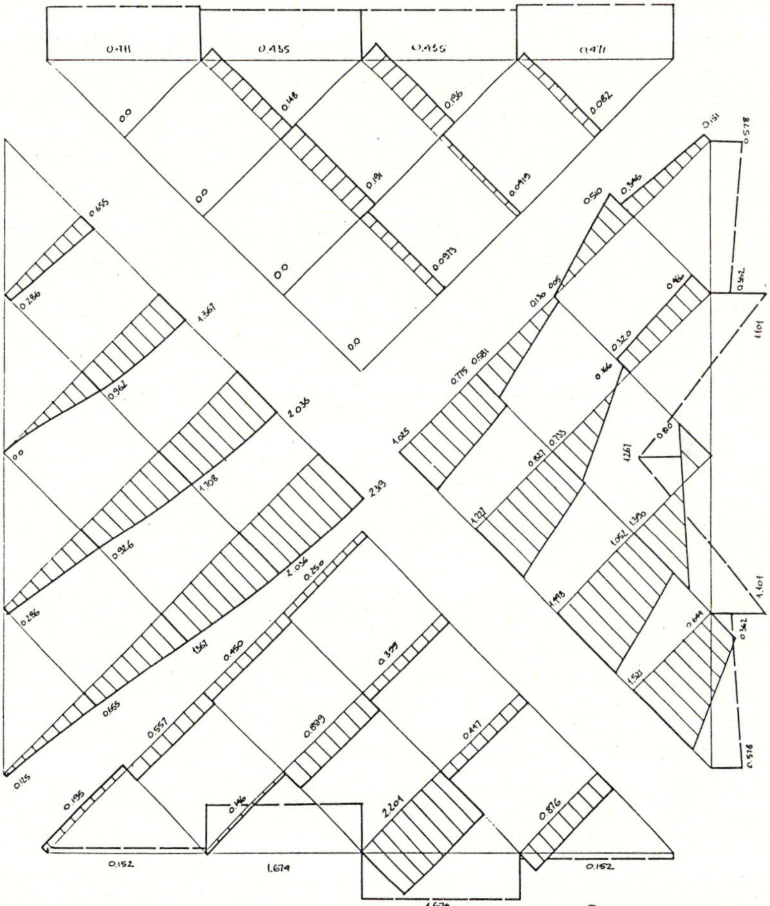

CASE 5

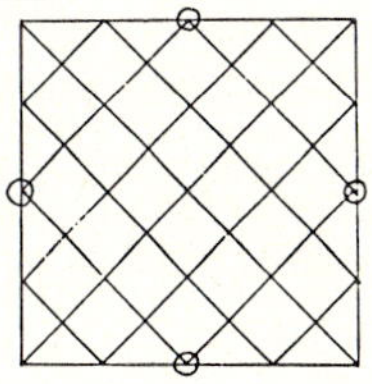

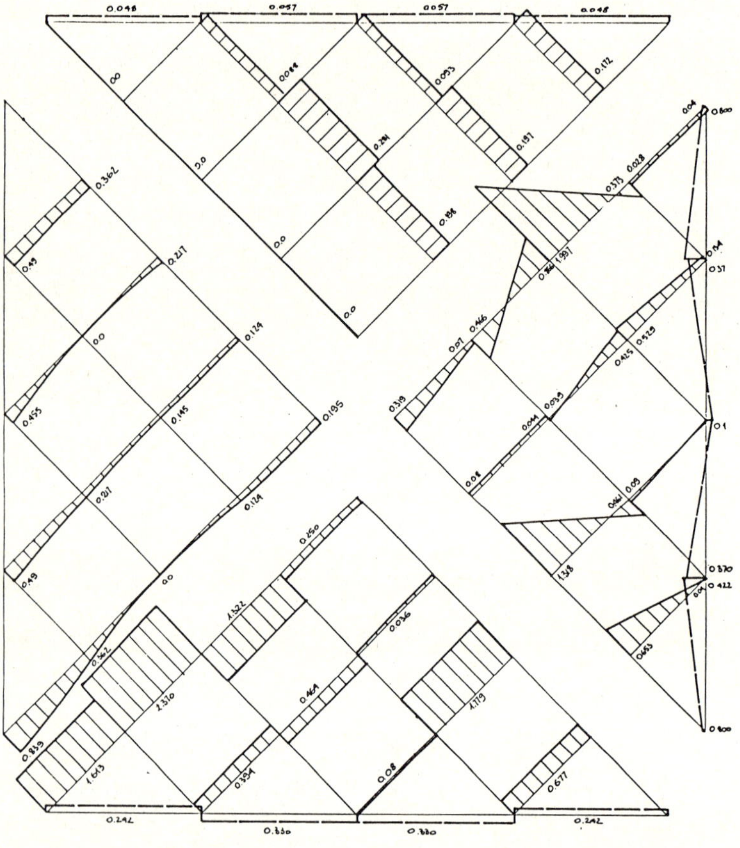

CASE 6

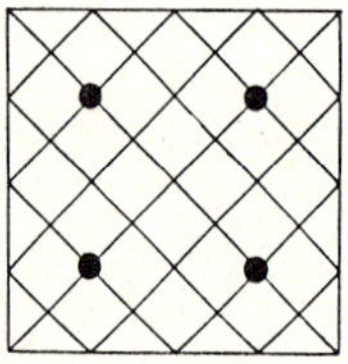

EXTENSIBLE CABLE NETWORK ANALYSIS
THROUGH A NONLINEAR OPTIMIZATION CODE

J.P. COYETTE and P. GUISSET
Université de Louvain
Civil Engineering Department
Place du Levant 1
B-1348 Louvain-la-Neuve
Belgium

ABSTRACT

Cable networks are geometrically highly nonlinear structures. This paper presents an elegant way to find the equilibrium shape of extensible cable networks by minimization of total potential energy. Both theoritical and practical aspects of the method are described below. Some applications are presented, including cases with special boundary conditions such as unilateral contact without friction. Finally, it is shown how a dynamic analysis can be performed and some information about the developed software is given.

INTRODUCTION

Because of their flexibility, cables offer interesting possibilities for shaping attractive three-dimensional forms. The main difficulty encountered in cable structure analysis is related to the lack of flexural rigidity which leads to large displacements. The geometric nonlinearities arise because the deformation of the structure must cause sufficient strain in the cables to produce the required changes of tension in order to satisfy the static equilibrium.

Due to unavailability of analytical solutions (except for simple cases), structural analysis of cable structures implies generally the use of numerical methods. A classical approach in this field is the (stiffness) displacement method which leads to set up equilibrium equations in terms of displacements. Such an approach is quite satisfactory but often get to troubles especially when the initial shape is far from the solution to be obtained.

A better solution can be achieved in this case by reporting to the basic energy concepts underlying the stiffness approach. The theorem of minimum of total potential energy is very attractive in this context, especially when its use is based on a powerful nonlinear optimization code.

A finite element approach supports the evaluation of this energy function (the objective function of the minimization process). Various constraints can be considered, enabling for example the treatment of unilateral contact without friction.

After determination of the static solution, a dynamic analysis can be performed easily by direct integration of the motion equation in the time domain, using the Newmark scheme.

BASIC CONCEPTS

In order to simplify the analysis, the following hypotheses are taken into account :

(i) cables are perfectly flexible and possess only axial stiffness;
(ii) material is linear elastic;
(iii) only small strains are involved.

The determination of the equilibrium position is based on the theorem of stationary potential energy (Washizu, 1982) :

$$\delta\pi = 0 \qquad (1)$$

where

$$\pi = U + V \qquad (2)$$

is the total potential energy, sum of the strain energy U and the potential energy of the external loads V. In order to equilibrate the cable system under the action of static loading, the total potential energy must be minimum :

$$\begin{matrix}\text{Stable} \\ \text{Equilibrium}\end{matrix} \iff \pi \text{ minimum} \qquad (3)$$

Strain energy. Strain energy can be evaluated from

$$U = \frac{1}{2} \int_S EA\, \varepsilon^2\, ds \qquad (4)$$

where EA is the axial rigidity of a cable, ε is the axial strain, ds is the infinitesimal element along a cable and S denotes all the cables of the structure.

With reference to Fig. 1, axial strain is calculated from

$$\varepsilon = \frac{dp - ds}{ds} = \frac{dp}{ds} - 1 \qquad (5)$$

where dp is the actual (strained) length of an initial (unstrained) element of length ds.

$$dp = \sqrt{(dx)^2 + (dy)^2 + (dz)^2} \qquad (6)$$

So, (5) becomes

$$\varepsilon = \sqrt{\left(\frac{dx}{ds}\right)^2 + \left(\frac{dy}{ds}\right)^2 + \left(\frac{dz}{ds}\right)^2} - 1 \qquad (7)$$

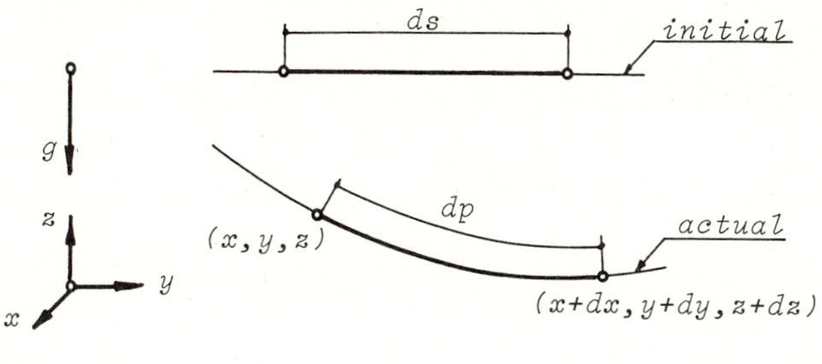

Fig. 1

External potential. The external potential is defined according to

$$V = V_g + V_F \qquad (8)$$

where V_g is the potential related to dead loads and V_F the potential of concentrated loads. Gravity forces are assumed to act in the opposite direction to z-axis. So, V_g is given by :

$$V_g = \int_S \rho g A z\, ds \qquad (9)$$

where ρ is the mass density and g the gravity acceleration. Potential V_F can be computed from :

$$V_F = - \Sigma \, (F_{x,i} \cdot x_i + F_{y,i} \cdot y_i + F_{z,i} \cdot z_i) \tag{10}$$

where the components of the ith force and coordinates of its application point are denoted respectively $(F_{x,i}, F_{y,i}, F_{z,i})$ and (x_i, y_i, z_i).

The above relations are the basic expressions on which will be developed the finite element model.

FINITE ELEMENT DISCRETIZATION

The purpose of FE discretization is double : firstly to compute numerically the potential energy function and then to achieve the minimization of that function, nodal coordinates being the decision variables of the problem.

Energy function evaluation. Substituting the continuous cable by connected finite elements, we are able to write :

$$\pi = \sum_e \pi_e + V_F \tag{11}$$

π_e is the element contribution to the objective function :

$$\pi_e = U_e + V_{g,e} \tag{12}$$

where U_e is the element strain energy and $V_{g,e}$ the element gravity loads potential. Noting ξ the normalized local coordinate, we can write for a three-node cable element (Fig. 2) :

$$ds = h.d\xi \tag{13}$$

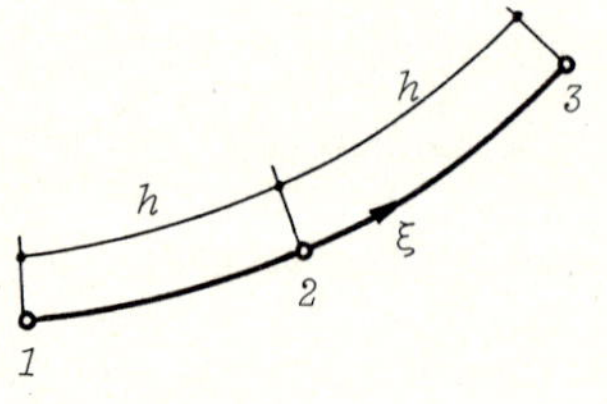

node		ξ
1	(x_1, y_1, z_1)	-1
2	(x_2, y_2, z_2)	0
3	(x_3, y_3, z_3)	+1

Fig. 2

Assuming that the cross-sectional area is constant on the length of each element, relations (4), (7) and (13) lead us to :

$$U_e = \frac{1}{2} EAh \int_{-1}^{+1} \left[\frac{1}{h} \sqrt{(\frac{dx}{d\xi})^2 + (\frac{dy}{d\xi})^2 + (\frac{dz}{d\xi})^2} - 1 \right]^2 d\xi \qquad (14)$$

and from (9) and (13), we get :

$$V_{g,e} = \rho g A h \int_{-1}^{+1} z \, d\xi \qquad (15)$$

Continuous cartesian cable coordinates must be expressed in terms of nodal coordinates, using the Lagrange interpolation functions :

$$x = \sum_i N_i(\xi) x_i \qquad y = \sum_i N_i(\xi) y_i \qquad z = \sum_i N_i(\xi) z_i \qquad (16)$$

where $N = \left(\frac{1}{2}\xi(\xi-1) \; , \; 1-\xi^2 \; , \; \frac{1}{2}\xi(\xi+1) \right)$ (17)

Therefore, relations (14) and (15) can be rewritten as :

$$U_e = \frac{1}{2} EAh \int_{-1}^{+1} \left[\frac{1}{h} \sqrt{(\Sigma \frac{dN_i}{d\xi} x_i)^2 + (\Sigma \frac{dN_i}{d\xi} y_i)^2 + (\Sigma \frac{dN_i}{d\xi} z_i)^2} - 1 \right]^2 d\xi \qquad (18)$$

$$V_{g,e} = \rho g A h \int_{-1}^{+1} (\Sigma N_i z_i) \, d\xi \qquad (19)$$

These two integrals have to be computed numerically with the Gauss-Legendre integration scheme.

Energy gradient vector evaluation. Optimization processes are more efficient if objective gradient vector is known. This vector will also be computed numerically. To establish its expression, consider node j, with coordinates (x_j, y_j, z_j). The first derivative of energy function with respect to x_j is given by :

$$\frac{\partial \pi}{\partial x_j} = \frac{\partial U}{\partial x_j} + \frac{\partial V_g}{\partial x_j} + \frac{\partial V_F}{\partial x_j} \qquad (20)$$

and similar expressions can be computed for y_j and z_j derivatives.

If we note

$$P_\xi = \sqrt{(\Sigma \frac{dN_i}{d\xi} x_i)^2 + (\Sigma \frac{dN_i}{d\xi} y_i)^2 + (\Sigma \frac{dN_i}{d\xi} z_i)^2} \qquad (21)$$

we get from (11) and (18) the following result :

$$\frac{\partial U}{\partial x_j} = \sum_e \frac{\partial U_e}{\partial x_j} \quad , \quad \frac{\partial V_g}{\partial x_j} = \sum_e \frac{\partial V_{g,e}}{\partial x_j} \tag{22}$$

where

$$\frac{\partial U_e}{\partial x_j} = EAh \int_{-1}^{+1} \frac{P_\xi/h - 1}{P_\xi h} \left(\Sigma \frac{dN_i}{d\xi} x_i\right) \frac{dN_j}{d\xi} \, d\xi \tag{23}$$

Derivation of the strain energy according to y_j and z_j gives similar results.

Morever, we have from (19)

$$\frac{\partial V_{g,e}}{\partial x_j} = 0 \quad , \quad \frac{\partial V_{g,e}}{\partial y_j} = 0 \quad , \quad \frac{\partial V_{g,e}}{\partial z_j} = \rho g A h \int_{-1}^{+1} N_j \, d\xi \tag{24}$$

Finally, we get directly from (10)

$$\frac{\partial V_F}{\partial x_j} = -F_{x,j} \quad , \quad \frac{\partial V_F}{\partial y_j} = -F_{y,j} \quad , \quad \frac{\partial V_F}{\partial z_j} = -F_{z,j} \tag{25}$$

Mass matrix evaluation. Further dynamic analysis implies mass matrix evaluation. This mass matrix M is computed from the relations

$$M = \sum_e m_e \tag{26}$$

$$m_e = \int N^T \rho \, N \, A ds \tag{27}$$

Numerical integration and diagonal condensation lead to the lumped mass matrix

$$m_e = \frac{\rho A h}{3} \operatorname{diag}(1,1,1,4,4,4,1,1,1) \tag{28}$$

MINIMIZATION OF ENERGY FUNCTION

In the previous section, we developed the tool to compute the total potential energy of a cable network. We just need now to find the cable shape which will minimize this energy function called the objective function.

The problem is then expressed in the following form :

$$\underset{p}{\text{Minimize}} \ \pi(p) \tag{29}$$

subject to boundary conditions.

In this expression, p is the nodal coordinates vector (the decision variable vector).

Usually, boundary conditions are related to fixed points; it means that some components of vector p are known and invariable and do not anymore participate in the optimization process. In this case, the problem is an uncontrained nonlinear problem. In other cases, it is possible to include some particular boundary conditions in the form of a constraint which limits the feasible region. For example, it is really easy to treat so some problems with rolling-ends or contact without friction problems.

This nonlinear optimization problem (29) is processed by an efficient optimization code called MINOS 5.0 (Murtagh, 1983). MINOS solves such problems using a reduced-gradient algorithm in conjunction with a quasi-Newton algorithm. These algorithms and their implementations are described elsewhere (Murtagh, 1978).

CABLE DYNAMICS

If dynamic effects are present, the equation of motion for the displaced position of the cable is

$$M \ddot{p} + K p = F \qquad (30)$$

where M is the mass matrix, K is the stiffness matrix, F is the nodal load vector and p the nodal coordinates vector.

Equation (30) can be rewritten

$$M \ddot{p} + G = 0 \qquad (31)$$

where

$$G = K p - F \qquad (32)$$

is the gradient of total potential energy at the time considered.

Usually, initial conditions are provided for displacements and velocities at time t=0 :

$$p = p_0 \ , \ \dot{p} = \dot{p}_0 \tag{33}$$

So initial acceleration $\ddot{p}_0$ can be calculated from :

$$\ddot{p}_0 = - M^{-1} G_0 \tag{34}$$

Direct integration of equation (32) at time t

$$M \ddot{p}_t + G_t = 0 \tag{35}$$

can be performed with the explicit central difference method :

$$\ddot{p}_t = \frac{1}{\Delta t^2} (p_{t-\Delta t} -2 p_t + p_{t+\Delta t}) \tag{36}$$

$$\dot{p}_t = \frac{1}{2\Delta t} (- p_{t-\Delta t} + p_{t+\Delta t}) \tag{37}$$

These relations lead to the following recurrence formula :

$$p_{t+\Delta t} = p_t + \Delta t \ \dot{p}_t - \frac{\Delta t^2}{2} M^{-1} G_t \tag{38}$$

$$\dot{p}_{t+\Delta t} = \dot{p}_t - \frac{\Delta t}{2} M^{-1} (G_t + G_{t+\Delta t}) \tag{39}$$

Conditional stability is the main attribute of this method. Small time steps have to be used (see Lewis (1984) for a conservative estimation of critical time step). Direct integration is further facilitated by the use of a lumped mass matrix.

Alternatively, the Newmark's scheme can be used

$$p_{t+\Delta t} = p_t + \Delta t \ \dot{p}_t + \Delta t^2 \ [(\tfrac{1}{2} - \alpha) \ \ddot{p}_t + \alpha \ \ddot{p}_{t+\Delta t}] \tag{40}$$

$$\dot{p}_{t+\Delta t} = \dot{p}_t + \Delta t \ [(1 - \delta) \ \ddot{p}_t + \delta \ \ddot{p}_{t+\Delta t}] \tag{41}$$

A predictor-corrector formulation (Hughes, 1978) is formulated from the modified equilibrium equations at time $t+\Delta t$:

$$M \ddot{p}_{t+\Delta t} + \tilde{G}_{t+\Delta t} = 0 \tag{42}$$

where

$$\tilde{G}_{t+\Delta t} = G(\tilde{p}_{t+\Delta t}) \tag{43}$$

is a predictor value of $G_{t+\Delta t}$.

Predictor values of displacements and velocities result from :

$$\tilde{p}_{t+\Delta t} = p_t + \Delta t\, \dot{p}_t + \Delta t^2 \left(\frac{1}{2} - \alpha \right) \ddot{p}_t \qquad (44)$$

$$\tilde{\dot{p}}_{t+\Delta t} = \dot{p}_t + \Delta t \, (1 - \delta) \, \ddot{p}_t \qquad (45)$$

Once accelerations are calculated from (42), updated corrector values can be computed from :

$$p_{t+\Delta t} = \tilde{p}_{t+\Delta t} + \Delta t^2 \, \alpha \, \ddot{p}_{t+\Delta t} \qquad (46)$$

$$\dot{p}_{t+\Delta t} = \tilde{\dot{p}}_{t+\Delta t} + \Delta t \; \delta \, \ddot{p}_{t+\Delta t} \qquad (47)$$

DEVELOPED SOFTWARE

Four FORTRAN computer programs have been written to form a complete package for static and dynamic cable analysis. The set of data to be given by the user is reduced to its minimum by pre-processing by program CABGEN which furnishes an initial shape and all the data files needed by the static analysis. This initial shape is computed by a procedure given by Peyrot and Goulois (1979). The static analysis program is called CABSTA and uses the MINOS 5.0 subroutines for the optimization process. On basis of the static solution, a dynamic analysis can be executed by program CABDYN and results of the dynamic analysis are graphically represented by the animation program CABANI.

At each step of the analysis, files compatible with the MOVIE.BYU (Christiansen, 1983) finite element post-processor are generated, making possible a graphical check of the cable shape.

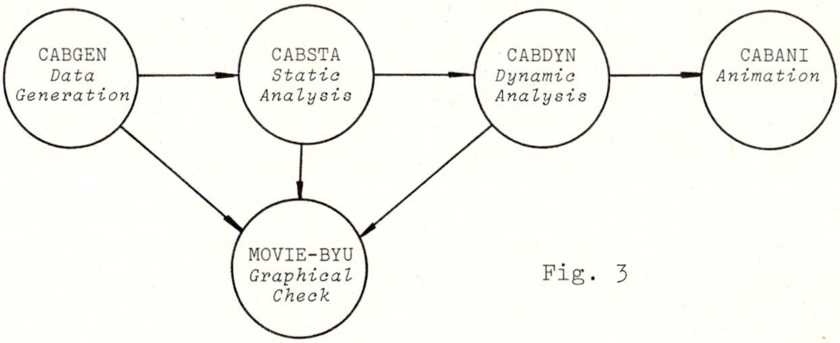

Fig. 3

APPLICATIONS

To illustrate the energy minimization method for cable analysis, three adimensional applications are presented below. Firstly, let us analyze a simple network of four cables (Peyrot, 1979), subject to gravity loads and a horizontal ponctual force. Data and cable equilibrium shape are given at Fig. 4. Results are quite satisfactory.

DATA

EA = 290000.
L = length

Coordinates :

B	*(-400, 0, -400)*
C	*(400, 300, 0)*
D	*(400, -300, 0)*
E	*(- , - , 290)*

RESULTS : Point A coordinates :

CABSTA	*(26.11, 40.42, -2.89)*
PEYROT (1979)	*(26.47, 41.14, -2.87)*

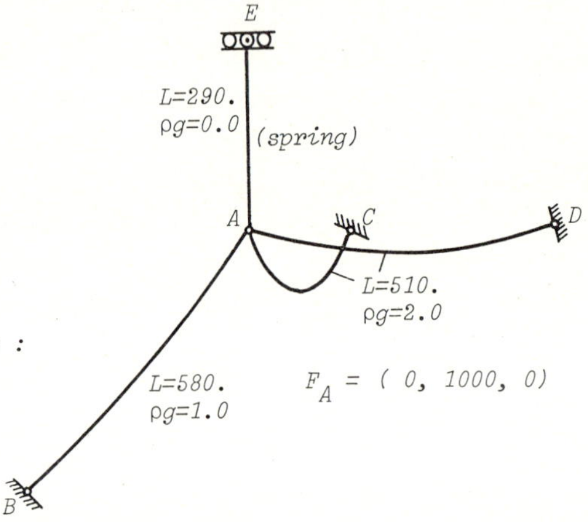

Fig. 4

The following example highlights the ability to consider special boundary conditions. A cable lying on the ground, fixed at one end, is rised at the other (Fig. 5). The cable is discretized in 15 three-node elements. Unilateral contact without friction is easily taken into account by constaints of non-negativity of z-coordinates.

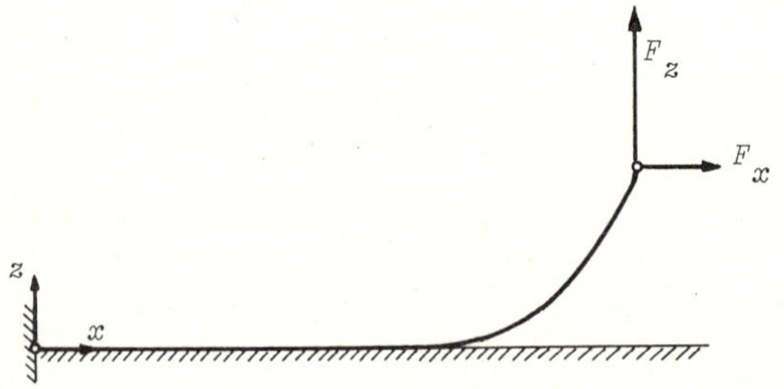

EA = 500.
Length = 1.0
ρg = 1.0
$F_x = 0.2$
$F_z = 0.4$

Fig. 5

The last application shows a dynamic analysis of a single cable, released at its right end (Fried, 1982). Time integration follows the explicit central difference method with a time step of 0.001 seconds. Cable shape has been drawn at Fig. 6 every 200 time steps.

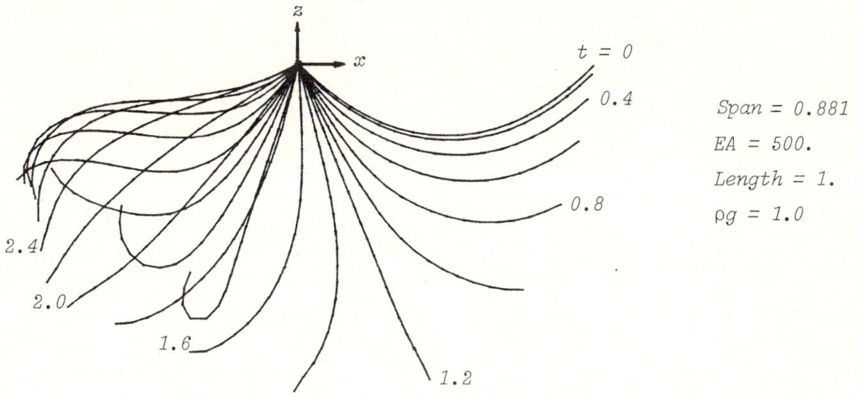

Fig. 6

CONCLUSIONS

The energy approach described in this paper for static analysis of cable structures presents many advantages :

- Complex networks are processed in exactly the same manner than single cables, because the method manages only a scalar value, the total potential energy.
- It is very well suited for distended cables which involve very large displacements.
- There is no absolute need to provide a "good" initial shape.
- Complex boundary conditions are easily taken into account.

REFERENCES

Christiansen, H., and M. Stephenson (1983). MOVIE-BYU A general purpo-computer graphics system, Brigham Young University.

Fried, I. (1982). Large deformation static and dynamic finite element analysis of extensible cables. Comput. Structures, 15, 315-319.

Hughes, T.J.R., and W.K. Liu (1978). Implicit-explicit finite elements in transient analysis : implementation and numerical examples. J. Appl. Mech., 45, 375-378.

Lewis, W.J., and al. (1984). Dynamic relaxation analysis of the non-linear static response of pretensioned cable roofs. Comput. Structures, 18, 989-997.

Murtagh, B.A., and M.A. Saunders (1978). Large scale linearly constrained optimization. Math. Prog., 14, 41-72.

Murtagh, B.A., and M.A. Saunders (1983). MINOS 5.0 User's Guide, Technical report SOL 83-20, Standford University.

Peyrot, A.H., and A.M. Goulois (1979). Analysis of cable structures. Comput. Structures, 10, 805-813.

Washizu, K. (1982). Variational Methods in Elasticity and Plasticity. Pergamon Press, London.

PRACTICAL FORMEX ANALYSIS OF SPACE STRUCTURES

H Nooshin   Space Structures Research Centre
Department of Civil Engineering
University of Surrey
GUILDFORD
Surrey GU2 5XH
United Kingdom

P G King   Space Decks Limited
CHARD
Somerset TA20 2AA
United Kingdom

ABSTRACT

Formex algebra has proved to be a suitable mathematical tool for automated data generation in computer aided analysis and design of space structures.

The objective of the present paper is to demonstrate the process of data generation for the analysis of space structures using the ideas of formex algebra (Ref 1) together with the programming language Formian (Ref 2) and the concept of plenices (Ref 3). The material is presented in terms of a simple practical example.

A small subset from the ideas of formex algebra together with even smaller subsets from the concepts of Formian and plenices are employed in the paper. The reader is not required to have any previous familiarity with the topics.

AN EXAMPLE

Consider the double layer grid whose plan view is shown in Fig 1, where the top layer elements are drawn in full line, the bottom layer elements are drawn in dashed line and the diagonal elements (that is, elements interconnecting the top and bottom layers) are drawn in dotted line. The overall plan dimensions of the grid are 43.35m and 28.05m as shown and the depth of the grid is 2.05m. This implies that all the top and bottom layer elements are of length 2.55m and the diagonal elements are of length 2.73m, where the member lengths are measured from joint centre to joint centre. The grid has 321 top layer elements, 247 bottom layer elements and 568 diagonal elements.

The structure is supported at 14 bottom layer joints. These joints are indicated by little circles in Fig 1, where a hollow circle implies a single vertical constraint and a solid circle implies complete translational constraint in all directions.

It is required to analyse the grid as a pin-connected system and the program which is to be used for the analysis is assumed to employ the standard stiffness method of structural analysis using a band-solution routine for the simultaneous equations.

Two loading cases are to be considered. The first loading case consists of equal vertical loads of magnitude 8.5kN applied at all the top layer joints and the second loading case consists of 17 equal vertical loads of magnitude 14.8kN applied at the bottom layer joints situated along the line indicated by two arrows in Fig 1.

The grid is to be designed involving no more than three different member cross-sections with all the top layer elements having the same cross-sectional area, say A1, all the bottom layer elements having the same cross-sectional area, say A2, and all the diagonal elements having the same cross-sectional area, say A3. This implies that A1 should be chosen with respect to the worst combination of effects that may occur for a member in the top layer of the grid and A2 and A3 should be chosen on a similar basis with relation to the bottom layer and diagonal elements, respectively.

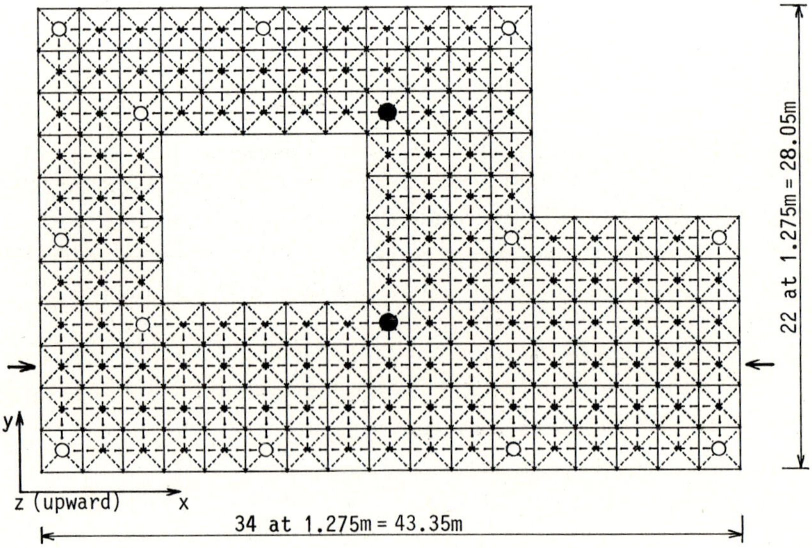

Fig 1

DATA PREPARATION

The data for the analysis consists of information about

(1) interconnection pattern of the grid,

(2) joint coordinates,

(3) support positions,

(4) load positions and

(5) cross-sectional areas, modulus of elasticity of the material, load magnitudes and constraint particulars.

Information relating to items 1 to 4 above may be conveniently generated using the concepts of formex algebra and the manner in which this may be achieved is described in the sequel.

The interconnection pattern of the grid may be specified in terms of a simple reference system which is referred to as a 'normat'. A suitable normat for the problem under consideration is a Cartesian type integer coordinate system which is shown together with a view of the lower left corner of the grid in Fig 2.

In terms of this normat, an element such as R1 in Fig 2 may be represented by

[1,1,2; 3,1,2].

This construct, which is a simple example of a formex, indicates a connection between joint J1 (represented by the normat coordinates 1,1,2) and joint J2 (represented by the normat coordinates 3,1,2). One may also use a variable, say F1, to denote the above formex and write

F1 = [1,1,2; 3,1,2].

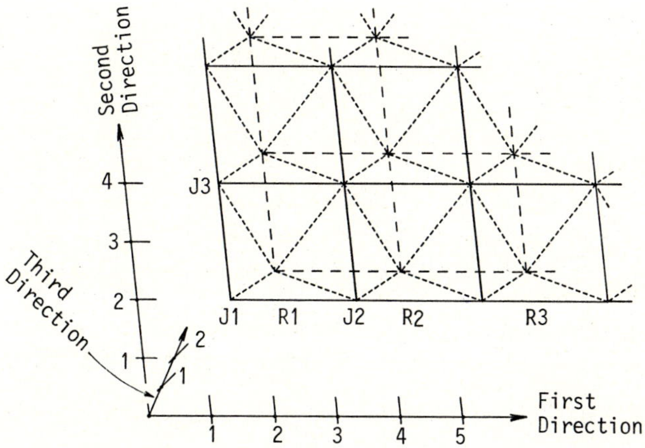

Fig 2

Similarly, element R2 may be represented by

$F2 = [3,1,2;\ 5,1,2]$

and element R3 may be represented by

$F3 = [5,1,2;\ 7,1,2]$.

Also, one may represent the combination of elements R1, R2 and R3 with a larger formex written as

$F = \{[1,1,2;\ 3,1,2],\ [3,1,2;\ 5,1,2],\ [5,1,2;\ 7,1,2]\}$

where each one of the parts that is enclosed in square brackets represents a component of the structure and the curly brackets enclose a sequence of such parts.

FORMEX COMPOSITION

The above formex F may alternatively be written as

F = F1 # F2 # F3.

Here, one has used the operation of 'formex composition' to create a formex that is the composition of those given on the right-hand side of the equation. The operator for formex composition is the symbol # which is read as 'duplus'.

SOME FORMEX FUNCTIONS

The above formex F may also be obtained using a transformation called a 'rindle function'. The term rindle is an old English word meaning a 'watercourse' and in formex algebra it is used to imply a serial composition of successively translated objects.

Thus, one may write

F = rin(1,3,2)¦F1

where the significance of different parts in the above equation are as shown in Fig 3.

Using the rindle function, one may now represent all the top layer elements, that are along the lower edge of the grid by

Eo = rin(1,17,2)¦F1.

Also, the formex

E1 = rin(2,12,2)¦Eo

will represent the array of 204 elements shown in Fig 4. The formex E1 represents all the top layer elments of the grid that are parallel to J1-J2. However, the elements represented by E1 also include the ones that extend over the cut-out portions of the grid. One may dispose of these unwanted parts in formex E1 in a number of ways. However, before this operation is attempted, it is useful to mention that the formex E1 may in fact be generated in a simpler manner by writing

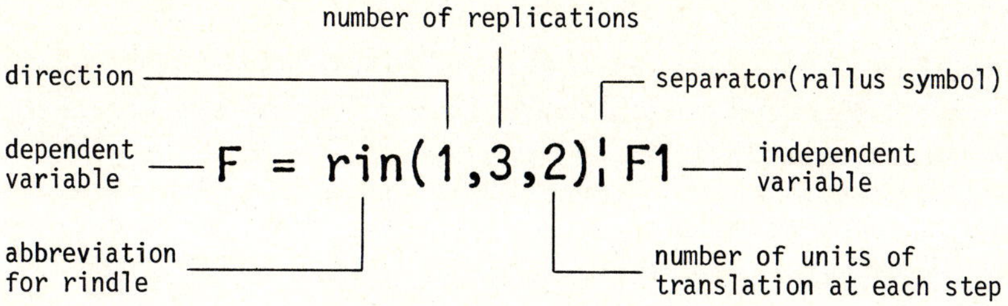

Fig 3

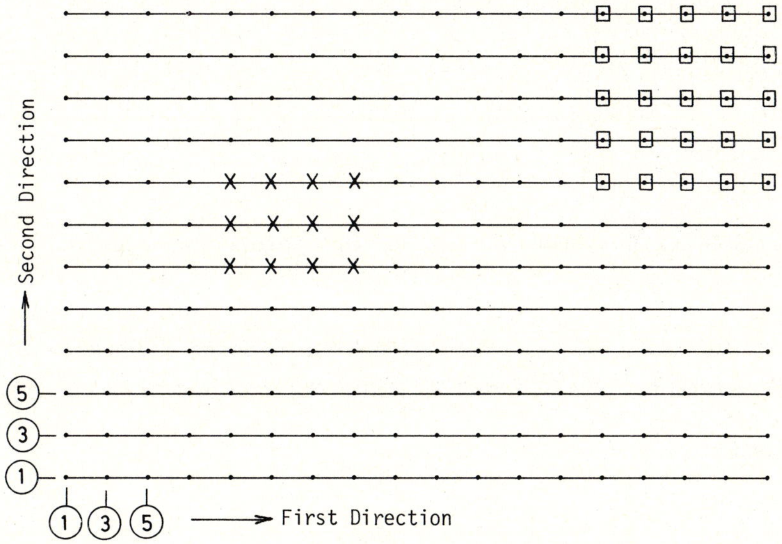

Fig 4

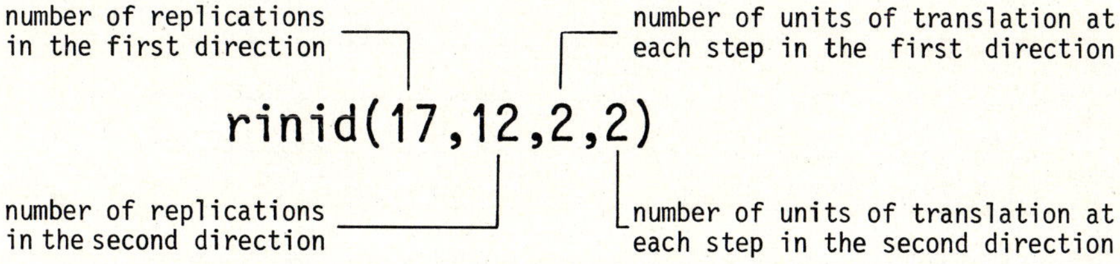

Fig 5

E1 = rinid(17,12,2,2)¦F1

where 'rinid(17,12,2,2)' is a compound rindle function implying operation in the first and second directions with the significance of the terms in parentheses indicated in Fig 5.

Returning to the problem of removal of the unwanted parts of E1, one may begin by specifying the joint positions that lie in the empty regions of the top layer of the grid. This may be done by writing

G1 = rinid(4,3,2,2)¦[9,11,2],

G2 = rinid(5,5,2,2)¦[27,15,2]

and G = G1 # G2.

Here G1 represents the array of 12 joints that are indicated by crosses in Fig 4 and [9,11,2] represents the lower left joint in this array. Also, G2 represents the array of 25 joints that are indicated by small squares in Fig 4 and [27,15,2] represents the lower left joint in this array. Finally, G represents the combination of both of the indicated groups of joints. The required formex may now be obtained by writing

T1 = lux(G)¦E1

where T1 is a formex obtained from E1 by removing every part of it that represents a member connected to one of the joints represented in G. The construct

lux(G)

is referred to as a 'luxum function' with 'lux' being an abbreviation for the word 'luxum'. The elements that are represented by T1 are shown in Fig 6.

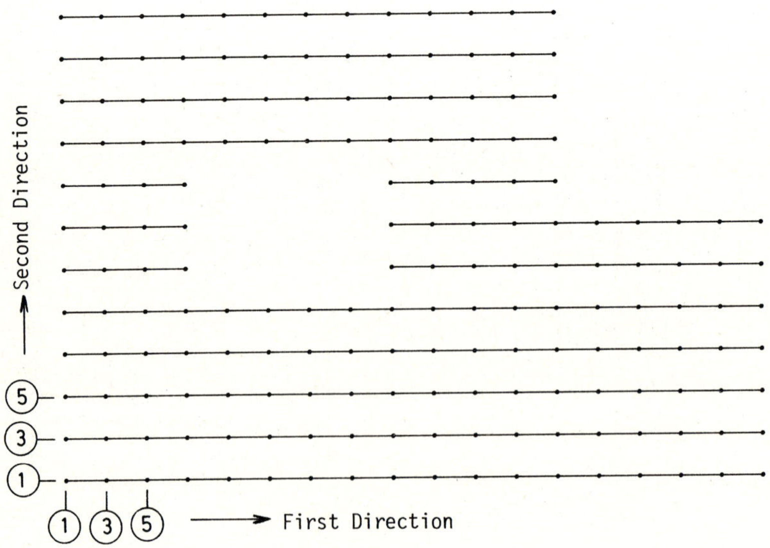

Fig 6

The elements in the top layer of the grid that are parallel to J1-J3 in Fig 2 may be similarly represented by

T2 = lux(G)¦E2

where

E2 = rinid(18,11,2,2)¦[1,1,2; 1,3,2].

Finally, all the top layer elements of the grid may be represented by

T = T1 # T2.

The elements that are represented by T are shown in Fig 7.

Using the same procedure, one may obtain a formex B representing all the bottom layer elements by writing

Bo = rinid(16,11,2,2)¦[2,2,1; 4,2,1] #
rinid(17,10,2,2)¦[2,2,1; 2,4,1],

H = rinid(5,4,2,2)¦[8,10,1] #
rinid(5,5,2,2)¦[26,14,1]

and B = lux(H)¦Bo.

The elements represented by Bo are shown in Fig 8. Also, the joints represented by H are indicated by crosses and small squares in Fig 8 and the elements represented by B are shown in Fig 9.

A formex D representing all the diagonal elements may be created in a similar fashion by writing

Do = rinid(17,11,2,2)¦{[1,1,2; 2,2,1], [3,1,2; 2,2,1],
[1,3,2; 2,2,1], [3,3,2; 2,2,1]}

and D = lux(H)¦Do

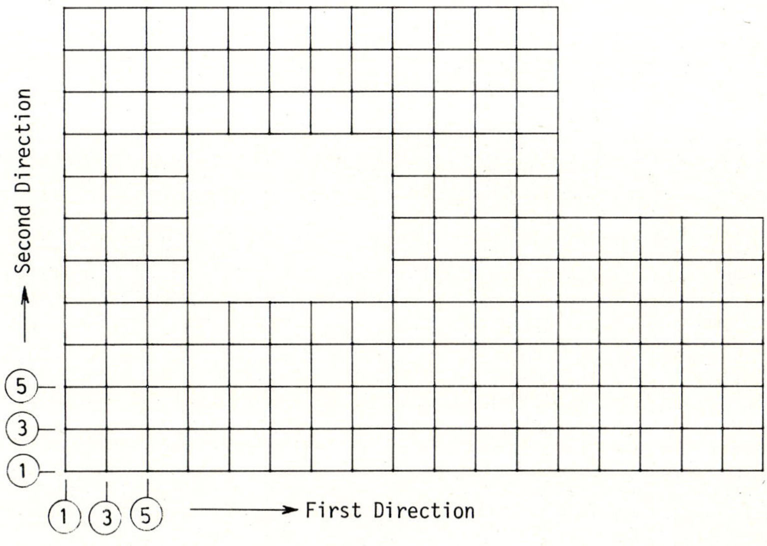

Fig 7

where H is the formex representing the joint positions in the empty regions of the bottom layer of the grid, obtained earlier.

Finally, the interconnection pattern of the whole grid may be represented by

GRID = T # D # B.

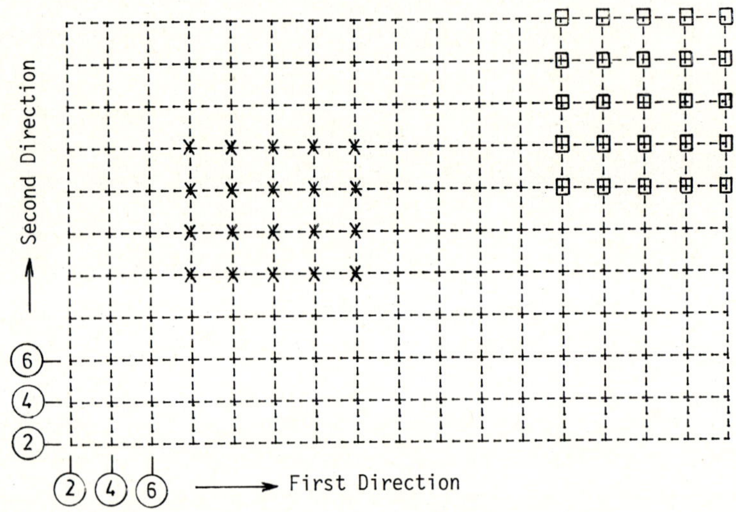

Fig 8

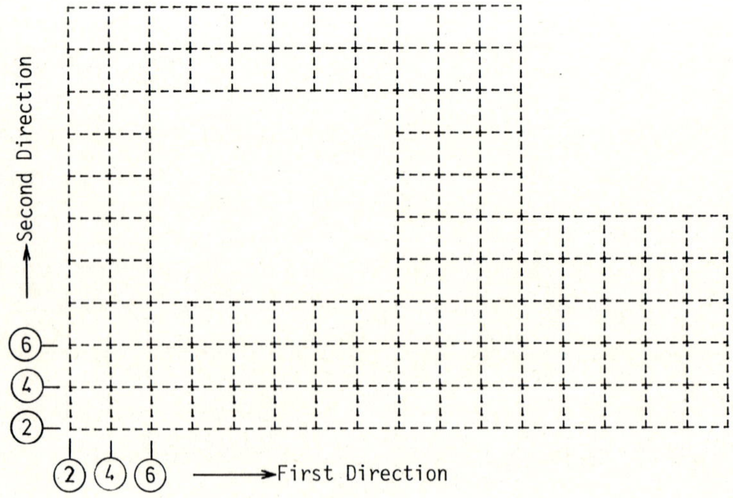

Fig 9

THE ROLE OF FORMIAN

Formian is the name of an interactive programming language which may be employed for data generation using the concepts of formex algebra as explained in Ref 2. Formian is operated through a program which is installed at a number of places including the Space Structures Research Centre of the University of Surrey and Space Decks Limited. Using Formian, it would be possible to type in equations identical to the ones discussed so far through a computer terminal and create the required formices. In addition to dealing with formex formulations, Formian has facilities that allow efficient handling of graphics. For instance, having created the formex that represents the configuration of the grid, as discussed above, and after specification of a view point and certain other particulars, the Formian statement

```
DRAW GRID
```

may produce a perspective view of the grid as shown in Fig 10.

In generating data through Formian, one often displays various parts of the generated data graphically for checking purposes. This is quite useful, since visual inspection is found to be an easy way of ensuring the correctness of the data.

NODE NUMBERING

The interconnection pattern of the grid has been formulated using normat coordinates. This is a convenient mode of operation when formulating the interconnection pattern. However, if the information about the list of elements is to be used for structural analysis, then it is normally required that the description is in terms of joint numbers. Therefore, it would be necessary to modify formices such as T, B and D and reproduce them in terms of a suitable joint numbering scheme. To achieve this, one may begin by writing a formex that

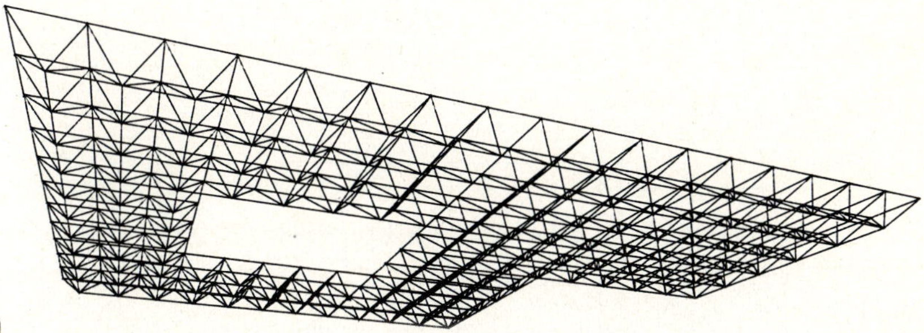

Fig 10

represents all the joints of the structure relative to the normat of Fig 2 with the joints listed in exactly the same order as one would want them to be numbered. Such a formex is referred to as a 'numerant'.

For example, if the joint numbering scheme is required to be as (partially) shown in Fig 11, then a numerant may be written as

N = lux(G # H)¦J

where

J = rin(1,17,2)¦J1 # rin(2,7,2)¦[35,1,2]

and where

J1 = rin(2,12,2)¦[1,1,2] # rin(2,11,2)¦[2,2,1]

where G and H are formices representing the joint positions in the empty regions of the top and bottom layers of the grid as discussed earlier. The joint positions represented by some parts of the above formulation are shown in Fig 12, where a plus sign is used to represent a top layer joint position and a dot is used to represent a bottom layer joint position.

With the numerant in hand, one may employ a 'dictum function' to transform a formex which is relative to a normat into a formex which is in terms of a joint numbering scheme. In the case under consideration, one may write

DT = dic(N)¦T,

DB = dic(N)¦B

and DD = dic(N)¦D

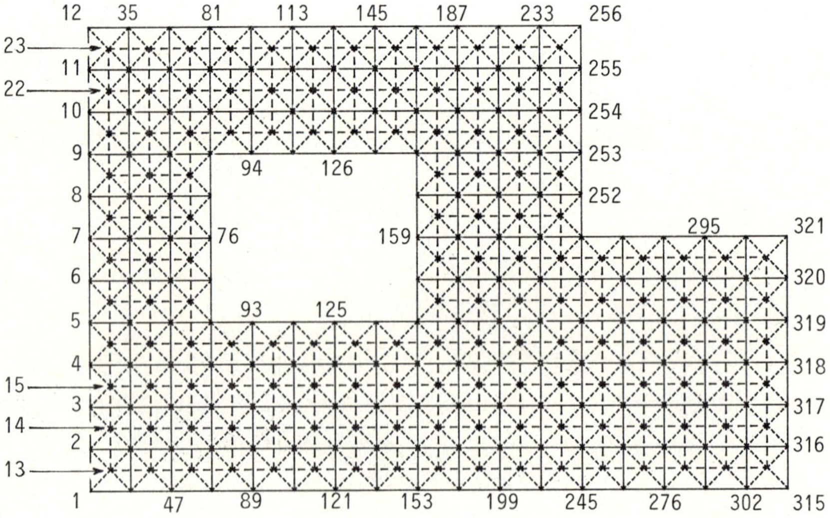

Fig 11

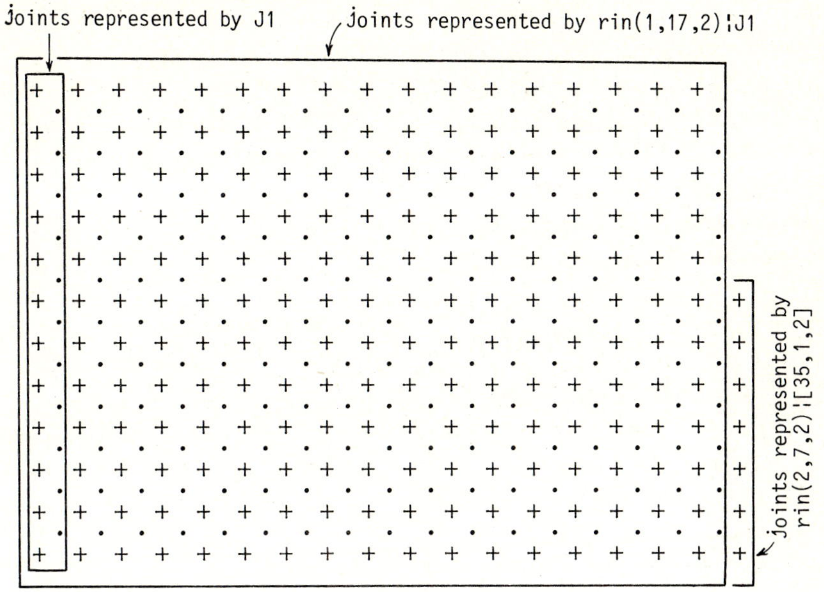

Fig 12

where DT, DB and DD are formices, respectively, representing the top layer elements, the bottom layer elements and the diagonal elements of the grid in terms of the joint numbering scheme of Fig 11. The construct

dic(N)

is a dictum function where N is the numerant obtained above and 'dic' is an abbreviation for the word 'dictum'.

LOAD AND SUPPORT DATA

The first loading case consists of vertical concentrated loads applied at all the top layer joints of the grid. The positions of these loads may be represented by a formex listing all the top layer joints. Such a formex, relative to the normat of Fig 2, may be written as

L1 = lux(G)¦L

where

L = rinid(18,12,2,2)¦[1,1,2].

Here, G is the formex representing all the top layer joint positions that are situated in the empty regions.

The second loading case consists of vertical concentrated loads applied at the bottom layer joints along the line indicated by arrows in Fig 1. The positions of these loads, relative to the normat of

Fig 2, may be given by the formex

L2 = rin(1,17,2)¦[2,6,1].

The structure under consideration involves two types of supports as explained at the beginning. The first support type consists of a single vertical constraint and the positions of supports of this type are indicated by little hollow circles in Fig 1. A formex representing these positions, relative to the normat of Fig 2, may be written as

S1 = {[2,2,1], [2,12,1], [2,22,1], [6,8,1],
[6,18,1], [12,2,1], [12,22,1], [24,2,1]
[24,12,1], [24,22,1], [34,2,1], [34,22,1]}.

Here the formex is written without the aid of any functions. In fact, it is not necessary to use functions for formex representations in all occasions and when a problem is simple enough then one may write the required formex explicitly.

The second support type provides complete translational constraint in all directions and the positions of supports of this type, which are indicated by solid circles in Fig 1, may be represented by the formex

S2 = {[18,8,1], [18,18,1]}.

One may now obtain formices representing the load and support positions in terms of the joint numbering scheme of Fig 11 by writing

DL1 = dic(N)¦L1,
DL2 = dic(N)¦L2,
DS1 = dic(N)¦S1
and DS2 = dic(N)¦S2

where N is the numerant derived earlier.

JOINT COORDINATES

In order to carry out the analysis of the grid, it would be necessary to provide information regarding the actual coordinates of the joints relative to a suitable global coordinate system such as the Cartesian coordinate system x-y-z shown in Fig 1. To provide this information, one may use a set of rules, referred to as 'coordinate specification', for transforming the normat coordinates into actual coordinates.

In the case under consideration, the coordinate specification may be written without difficulty. Thus, if U1, U2 and U3 are the normat coordinates of a typical joint and x, y and z are the corresponding actual coordinates, then x = 1.275 U1,

y = 1.275 U2

and z = 2.050 U3

where x, y and z are given in meters.

There are a number of standard coordinate specifications that may be selected and made 'current' in Formian and the above type of coordinate specification is included in the standard ones. When using Formian, at any given moment there is a coordinate specifiation which is current and any transformation of normat coordinates into actual coordinates will be performed using this current coordinate specification.

DATA SUBMISSION

Once the process of generation of various parts of data is complete, one may employ a SUBMIT statement to transmit the data for analysis with the information arranged in a predetermined manner using the concept of a 'plenix', as explained in Ref 3. In the particular case under consideration, the statement may take the form

SUBMIT {N,{DT,DB,DD},MA,{DS1,DS2},SA,{DL1,DL2},LA}

where the construct that follows the keyword 'SUBMIT' is an example of a plenix providing a structured form of presenting the data and where each one of the parts that are separated by commas is referred to as a panel. Each panel in the above plenix has a specific role in providing information for the analysis as explained below:

(1) The first panel, that is N, is a formex listing the normat coordinates of all the joints of the structure arranged in exactly the same order as the joints are required to be numbered. Formian will transform this formex into a list of joint coordinates using the current coordinate specification. The total number of joints in the structure need not be given explicitly since this can be deduced from the number of entries in N.

(2) The second panel, that is {DT,DB,DD}, is a sequence of formices each of which represents the members of a particular type. The formices in the panel are to be in terms of the joint numbering scheme implied by N. The number of different types of members and the number of members of each type may be obtained from the information contained in the panel and therefore it is not required to specify these numbers explicitly. Also, the band-width for the stiffness matrix will be determined by obtaining the maximum difference between the joint numbers of the individual members.

(3) The third panel, that is MA (standing for Member Array), is an array having as many rows as there are formices in the second panel, with each row containing information about the axial rigidity relating to a type of member. In the case under consideration, this panel may be of the form

$$\begin{bmatrix} EA1 \\ EA2 \\ EA3 \end{bmatrix}$$

where A1 is the cross-sectional area for the top layer members, A2 is the cross-sectional area for the bottom layer members, A3 is the cross-sectional area for the diagonal members and E is the modulus of elasticity of the material. In order to determine A1, A2 and A3, it may be necessary to repeat the analysis a number of times and for each such analysis A1, A2, A3 and E should have some values assigned to them.

(4) The fourth panel, that is {DS1,DS2}, is a sequence of formices each of which represents the constrained joints of a particular type. These formices are to be given relative to the joint numbering scheme implied by N. The number of different types of supports and the number of supports of each particular type are not required to be given explicity since these numbers may be determined from the information contained in the panel.

(5) The fifth panel, that is SA (standing for Support Array), is an array with as many rows as there are formices in the fourth panel, with each row providing information about a type of constrained joint. In the case under consideration, SA may be of the form

$$\begin{bmatrix} 0 & 0 & 1 \\ 1 & 1 & 1 \end{bmatrix}$$

where the first row indicates freedom of movement in the x and y directions and a constraint in the z direction and the second row indicates that all the three degrees of freedom are constrained.

(6) the sixth panel, that is {DL1,DL2}, is a sequence of formices each of which represents the load positions for a particular loading case relative to the joint numbering scheme implied by N. The number of different types of loaded joints and the number of loaded joints of each type may be determined from the information contained in the panel and therefore it is not necessary to specify these numbers explicitly.

(7) The seventh panel, that is LA (standing for Load Array), is an array with as many rows as there are formices in the sixth panel, with

each row containing information about a type of loaded joint. In the case under consideration, LA may be of the form

$$\begin{bmatrix} 0 & 0 & -8.50 \\ 0 & 0 & -14.8 \end{bmatrix}$$

where the first row specifies a load of 8.5 units in the negative z direction and the second row specifies a load of 14.8 units in the negative z direction.

The constitution of the above plenix suits most simple structural analysis problems. However, this constitution may be readily altered to cater for more complex situations.

An important feature of Formian is that it allows data to be modelled to suit any type of existing program. Thus, one starts by preparing the data along the lines indicated in the paper and the information conveyed through a SUBMIT statement may then be refashioned by a supplementary routine to fit a given program or package.

Finally, it should be realised that only a limited number of elements relating to the concepts of formex algebra, Formian and plenices are discussed in the paper. These elements allow the general attitude in algebraic data generation to be demonstrated. However, the full use of the concepts will provide far more capabilities for the operation.

REFERENCES

(1) Nooshin H, Formex Configuration Processing in Structural Engineering, Elsevier Applied Science Publishers, London, 1984.

(2) Disney P and El-Labbar O, An Introduction to Formian, Proceedings of the Third International Conference on Space Structures, ed H Nooshin, Elsevier Applied Science Publishers, London, 1984, PP 158-65.

(3) Haristchian M, Formex and Plenix Structural Analysis, Proceedings of the Third International Conference on Space Structures, ed H Nooshin, Elsevier Applied Science Publishers, London, 1984, PP 105-9.

TRAGLAST UND FORMÄNDERUNG VON RAUMFACHWERKPLATTEN IN ANSCHAULICHER NÄHERUNG

Max A.M. Herzog
Beratender Ingenieur
Rohrerstrasse 3
5000 Aarau (Schweiz)

Stützenfreie Grundrisse bis zu 10000 m² werden häufig mit plattenförmigen Raumfachwerken überdacht. Ihre genaue Berechnung ist selbst mit dem Computer aufwendig. Aus diesem Grund wird ein anschauliches Näherungsverfahren vorgestellt, das zumindest für Angebots- und Kontrollrechnungen genügt. Die Nachrechnung von fremden Modellversuchen und einem Prototypversuch, sowie von einem bekannten Schadensfall beweist die ausreichende Wirklichkeitsnähe des einfachen Handverfahrens.

Roofs covering areas up to 10000 m² without interior columns are frequently designed as plane space trusses. Their exact analysis is very expensive even with help of a computer. For this reason a transparent approximate method is presented that is sufficient at least for tender and control calculations. The posterior analyses of model tests and a prototype test by other researchers as well as of a known failure prove the simple hand method to be adequately realistic.

1 Einleitung

Bei stützenfrei zu überdachenden Grundrissen bis etwa 10000 m² werden an Stelle von Seilnetztragwerken und Hängeschalen /1/ häufig plattenförmige Raumfachwerke ausgeführt. Die genaue Berechnung dieser Tragwerke mit einigen tausend Stäben /2/ ist sehr aufwendig. Für die Angebotsbearbeitung und als Kontrollberechnung genügen auch bescheidene Handverfahren, wenn ihre Ergebnisse mit den ausgeführten Messungen an Modellen und Prototypen oder mit nachgerechneten Schadensfällen übereinstimmen.

2 Näherungsberechnung

Die grundlegende Näherung besteht darin, dass an Stelle des wirklichen Raumfachwerks eine Ersatzplatte berechnet wird. Die Gurtkräfte ergeben sich mit Hilfe der Netzhöhe aus den Biegemomenten und die Strebenkräfte folgen aus den Querkräften. Für Angebots- und Kontrollberechnungen genügt die Kenntnis dieser Schnittkräfte am Ort der grössten Beanspruchungen (Gurtkräfte im Feld und Strebenkräfte bei den Stützen). Die Ermittlung der Formänderungen (Durchbiegungen) und der Traglast erfolgt nach der Theorie drillweicher Platten /3/, weil Raumfachwerke i.a. keine nennenswerte Drillsteifigkeit besitzen.

3 Modellversuche

Von Schmidt, Morgan u.a. sind verschiedene Modellversuche an Halboktaeder-Tetraeder-Strukturen ausgeführt worden, die eine Kontrolle des vorgeschlagenen Näherungsverfahrens ermöglichen.

3.1 Umfangsgelagertes Raumfachwerk mit versagenden Druckgurten

Aus der Knicklast des Druckgurts (Bilder 1a und 1b) mit Rohrquerschnitt aus Aluminium ($\lambda = 305/3,99 = 77$ und $E = 41,1$ kN/mm²)

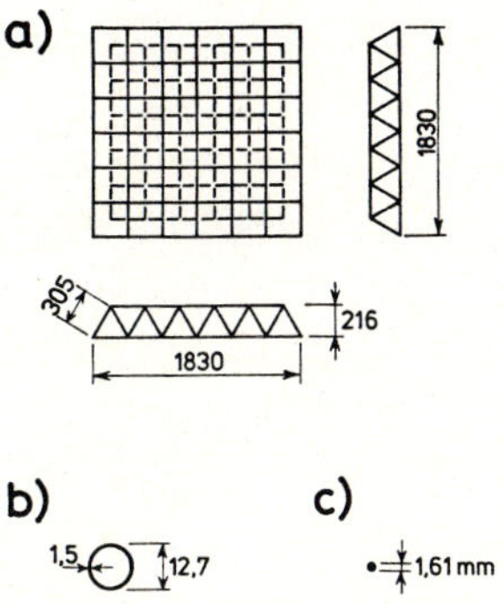

Bild 1: Umfangsgelagertes Raumfachwerk (nach /4, 5, 6/)

$$N_k = \sigma_k A = 0,0561.55,4 = 3,11 \text{ kN} \qquad (1)$$

und dem Feldbiegemoment der drillweichen quadratischen Platte

$$M = P/16 \qquad (2)$$

ergibt sich die Gurtkraft für den höchstbeanspruchten Druckgurt einer Tragrichtung zu

$$N = aM/h \qquad (3)$$

Bei bekannter Knicklast des Druckgurts erhalten wir die gesuchte Traglast des Raumfachwerks zu

$$P_u = 16\ N_k h/a = 16.3,11.0,216/0,305 = 35,2 \text{ kN} \qquad (4)$$

Gemessen wurde in drei Versuchen i.M. $P_u = (33,0 + 34,3 + 36,5)/3 = 34,6$ kN (= 98 % der rechnerischen Traglast).

Aus dem elastischen Biegemoment eines gedachten "Trägers" mit je einem Ober- und Untergurt im Achsabstand a = 305 mm

$$M' = aP_u/16 = 305.35,2/16 = 671 \text{ kNmm} \qquad (5)$$

und dem Trägheitsmoment dieses "Trägers"

$$I = \Sigma Ay^2 = 2.55,4(216/2)^2 = 1294000 \text{ mm}^4 \qquad (6)$$

ergibt sich die Durchbiegung ohne Gelenkspiel der Knotenpunkte zu

$$f_o = \frac{5}{48}\frac{ML^2}{EI} = \frac{5.671.1830^2}{48.45,5.1,294.10^6} = 4,0 \text{ mm} \qquad (7)$$

Das angenommene Gelenkspiel von 2 x 0,25 mm je Knoten entspricht einer Stabdehnung von $\varepsilon = 0{,}5/305 = 1{,}64.10^{-3}$ und einer Spannung von $\sigma_G = E\varepsilon = 45{,}5.1{,}64.10^{-3} = 74{,}6$ N/mm². Die verbesserte Durchbiegung in Feldmitte beträgt dann

$$f = f_o(\sigma_k + \sigma_G)/\sigma_k = 4{,}0.\frac{56{,}1 + 74{,}6}{56{,}1} = 9{,}3 \text{ mm} \qquad (8)$$

Gemessen wurden 10,6 mm.

3.2 Umfangsgelagertes Raumfachwerk mit versagenden Zuggurten

Dieses Raumfachwerk unterschied sich vom ersten nur dadurch, dass die Zuggurte (Bilder 1a und 1c) aus weichem Eisendraht Ø 1,61 mm bestanden, dessen Fliesslast $N_f = 0{,}57$ bis 0,72 kN und dessen Bruchlast $N_u = 0{,}76$ kN betrug. Die Traglast ergibt sich dann gemäss Gl.(4) zu $P_u = 16.0{,}76.0{,}216/0{,}305 = 8{,}6$ kN. Gemessen wurde die Traglast im ersten Versuch zu 21,8 kN und im zweiten zu 17,8 kN. Das Mittel der Messwerte liegt also um 130 % über dem Rechenwert. Das Versagen des Raumfachwerks erfolgte durch Knicken von vier Obergurten. Die rechnerische Traglast betrüge für diesen Fall (vier von insgesamt sechs Obergurten knickten) in Analogie zum Abschnitt 3.1 $P_u = 35{,}2.4/6 = 23{,}5$ kN (gemessen i.M. 19,8 kN = 84 % des Rechenwerts). Wie ist dieses beobachtete Verhalten zu erklären? Der auf den Umfangsstützen liegende Randzuggurt bestand aus demselben Aluminiumrohr wie die Druckgurte mit der Fliesslast

$$N_f = \sigma_f A = 0{,}294.55{,}4 = 16{,}3 \text{ kN} \qquad (9)$$

Mit wachsender Durchbiegung (unter der Traglast i.M. L/f = 1830/50 = 37) wurde im Raumfachwerk eine Druckmembranwirkung /7/ aktiviert, deren Horizontalschub vom überbemessenen Randzuggurt übernommen wurde. Infolge der veränderten Dehnungsverhältnisse im Raumfachwerk konnten die beiden Randobergurte jeder Tragrichtung keine Druckkräfte übertragen. Bei den Versuchen wurden in ihnen sogar kleine Zugkräfte gemessen (vgl. /5/ S. 138, Fig. 10).

Mit den unterschiedlichen Schwerlinienabständen des Druckgurtes (Aluminiumrohr) $y_o = 31$ mm und des Zuggurtes (Eisendraht) $y_u = 185$ mm erhalten wir die Biegesteifigkeit des gedachten "Trägers" nach Gl.(6) zu $EI = 45{,}5.55{,}4.31^2 + 205.2{,}02.185^2 = 16{,}4.10^6$ kNmm². Aus dem elastischen Biegemoment eines "Trägers" gemäss Gl.(5) M' = 305.8,6/16 = 164 kNmm ergibt sich die Durchbiegung ohne Gelenkspiel der Knotenpunkte nach Gl.(7) zu $f_o = \dfrac{5.164.1830^2}{48.16{,}4.10^6} = 3{,}5$ mm und mit Berücksichtigung des Gelenkspiels zu f = 8,9 mm. Da die Zuggurte ins Fliessen geraten waren und sich eine Druckmembranwirkung eingestellt hatte, waren die gemessenen Durchbiegungen (Versuch 1: f = 53 mm und Versuch 2: f = 47 mm) rund sechsmal so gross wie die für elastisches Verhalten berechneten.

4 Grossversuch

Zur Überprüfung der Modellversuche wurde von Schmidt, Morgan u.a. auch ein Grossversuch /6, 8/ ausgeführt. Das Raumfachwerk (Bild 2) bestand aus einer Halboktaeder-Halbkuboktaeder-Struktur, deren wirtschaftliche Überlegenheit später in /2/ nachgewiesen worden ist. Aus den im Versuch gemessenen Fliesslasten der Zuggurte $N_f=(69 + 66 + 66 + 79 + 64)/5 = 68{,}8$ kN und den Knicklasten der Druckgurte $N_k = -(107 + 67 + 82)/3 = -85{,}3$ kN erhalten wir die rechnerische Traglast des Raumfachwerks mit versagendem Zuggurt nach Gl.(4) zu $P_u = 16.\frac{68{,}6}{2}.\frac{0{,}970}{1{,}372} = 390$ kN (gemessen 375 kN nach /8/ bzw. 365 kN nach /6/, also i.M. 95 % des Rechenwerts). Der Faktor 2 im Nenner erfasst die Richtungsabweichung des Zuggurts vom Druckgurt. Die Summe der gemessenen Druckgurtkräfte $N = -(10 + 49 + 107 + 60 + 68 + 0) = -294$ kN für die Laststufe P = 374 kN stimmt mit dem rechnerischen Wert $N = - 6.68{,}8/\sqrt{2} = -291$ kN für die Laststufe P = 375 kN sehr gut überein.

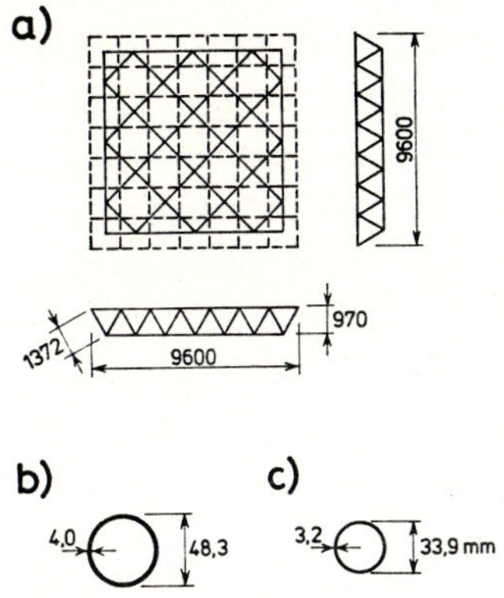

Bild 2: Umfangsgelagertes Raumfachwerk (nach /6, 7/) b) Druckgurt, c) Streben und Zuggurt

Aus dem grössten Biegemoment eines "Trägers" $M' = (68{,}8/\sqrt{2}).0{,}970 = 47{,}2$ kNm und seinem gedachten Trägheitsmoment gemäss Gl.(6) $I = 5{,}5.35^2 + 3{,}1.62^2 = 18600\ cm^4$ folgt die Durchbiegung ohne Gelenkspiel nach Gl.(7) zu $f_o = \frac{5.4720.960^2}{48.21000.18600} = 1{,}16$ cm. Durch das angenommene Gelenkspiel von 2 x 0,25 mm vergrössert sich die Mittendurchbiegung des gedachten "Trägers" auf f = 1,72 cm. Weil die Zuggurte teilweise ins Fliessen gerieten, ist der Rechenwert nur halb so gross wie die gemessene Durchbiegung von 3,6 cm.

5 Einsturz des Hartford Coliseums

Der in der Fachliteratur /9, 10, 11/ gut beschriebene Einsturz des auf nur vier Stützen gelagerten Raumfachwerks über dem Civic Center Coliseum von Hartford in Connecticut, USA, am 18. Januar 1978 (Bild 3) hat Anlass zu verschiedenen Untersuchungen /12, 13/ und Forschungsarbeiten /14, 15/ gegeben. Durch die besondere Ausbildung der Hilfsstrebenanschlüsse zwischen den Obergurten und den Hauptstreben der Halboktaeder-

Bild 3: Luftbild des eingestürzten Coliseumdaches (nach /15/ S. 217)

Tetraeder-Struktur mit randparallelen Gurten (Bilder 4, 5, 6) waren die Haupttragelemente in den Zwischenpunkten nicht starr sondern nur elastisch nachgiebig gestützt. Ihre Knicklänge lag also zwischen 9150/2 = 4575 und 9150 mm. Da die Bemessung der Raumfachwerkstäbe mit s_k = 4575 mm erfolgt war, lag in diesem Umstand eine Ursache des Einsturzes.

Der ursprünglichen Bemessung lagen die folgenden Vertikallasten /9/ zugrunde

- Schnee	30 psf = 1,46 kN/m²	(19 psf = 0,93 kN/m²)
- Dachfläche	20 " = 0,98 "	(20 " = 0,98 ")
- Eigenlast	18 " = 0,88 "	(26 " = 1,27 ")
- Summe	68 psf = 3,32 kN/m²	(65 psf = 3,18 kN/m²).

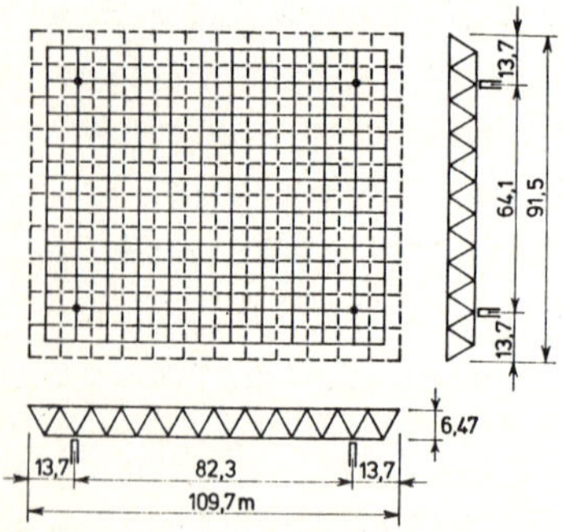

Bild 4: Punktgestütztes Raumfachwerk des Hartford Coliseums

Die Klammerwerte entsprechen der tatsächlichen Belastung beim Einsturz. Aus dem grössten Biegemoment in Längsrichtung

$$M_{xo} = 3{,}18.9{,}15\left(\frac{82{,}3^2}{8} - \frac{13{,}7^2}{2}\right) = 21900 \text{ kNm}$$

$$M_{xr} = 1{,}15.21900 = 25200 \text{ kNm (Rand)}$$

$$M_{xm} = 0{,}91.21900 = 19900 \text{ kNm (Mitte)}$$

erhalten wir mit Hilfe der Netzhöhe h = 6,47 m die grössten Gurtkräfte zu $N_{max} = \pm \frac{25200}{6{,}47} = \pm 3900$ kN. Bei einer Querschnittsfläche des grössten Druckstabes von A = 4.85,2 = 341 cm² beträgt die höchste Druckspannung σ = 3900/341 = - 11,5 kN/cm². Die wirkliche Stabschlankheit liegt zwischen den Extremwerten 9150/9,22 = 99 > λ > 4575/9,22 = 50. Die entsprechenden Knickspannungen ergeben sich für den verwendeten Stahl ASTM 572 (~ St 52) mit der Fliessgrenze σ_f = 35 kN/cm² gemäss Eurocode (Knickspannungslinie c) zu σ_k = 14,0 bzw. 26,2 kN/cm². Wir erkennen, dass einerseits die höchste auftretende Druckspannung stets unter den rechnerischen Extremwerten lag und anderseits die rechnerische Sicherheit im ungünstigsten Fall nur 14,0/11,5 = 1,22 betrug.

Es ist daher zu vermuten, dass der Einsturz durch die bereits bei der Montage beobachteten (Engineering News Record vom 6. April 1978, S. 9) Vorverformungen der Druckstäbe eingeleitet wurde. Unter der Voraussetzung einer geometrischen Imperfektion von 9150/300 = 30,5 mm betrüge das Biegemoment in Stabmitte M = 3900.0,0305 = 119 kNm.

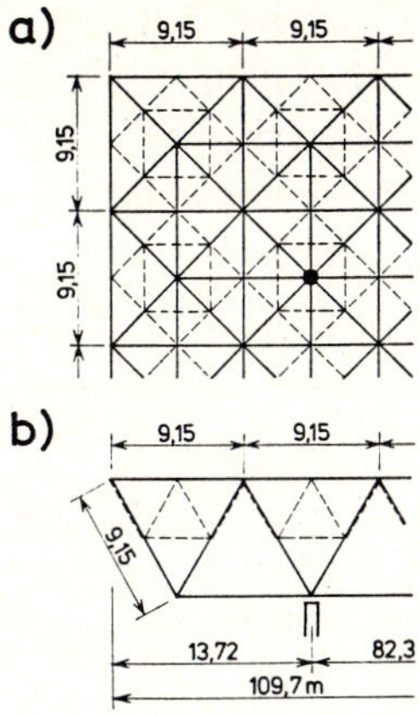

Bild 5: Struktur des Raumfachwerks mit Hilfsstreben (gestrichelt)

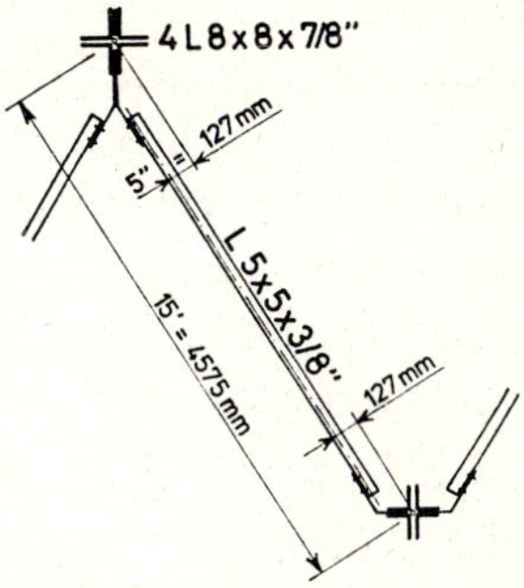

Bild 6: Hilfsstrebenanschlüsse mit abgekanteten Knotenblechen t = 3/8" = 9,5 mm

Die neue Faustformel /16/

$$\frac{N}{N_{pl}} + \frac{M}{1,1\ M_{pl}} \leq 1 - 1,2\ \frac{N}{N_{ki}} \qquad (10)$$

liefert für den höchstbeanspruchten Druckgurt des Raumfachwerks $\frac{3900}{11920} + \frac{119}{1,1.810} > 1 - 1,2.\frac{3900}{7230}$ entsprechend 0,461 > 0,352. Mit anderen Worten, die Beanspruchung liegt im Falle einer geometrischen Imperfektion von $s_k/300$ bereits über der Traglast eines normalen inneren Druckgurtes in Randnähe.

Mit Hilfe des Trägheitsmoments für einen gedachten "Träger" aus je einem Ober- und Untergurt nach Gl.(6) $I = 341.274^2 + 251.373^2 = 60,6.10^6$ cm^4 erhalten wir die Durchbiegung in Längsrichtung gemäss Gl.(7) zu

$$f_x = \frac{5.2,52.10^6.8230^2}{48.21000.60,6.10^6} = 14,0 \text{ cm}$$

und in Querrichtung auf analogem Weg zu f_y = 8,5 cm. Die gesamte Mittendurchbiegung betrüge daher ohne Gelenkspiel in den Knoten des Raumfachwerks (alle Stäbe waren HV-verschraubt) f_m = 14,0 + 8,5 = 22,5 cm. Im Januar 1973 hatte eine Kontrollmessung unter Eigenlast f_m = 21,3 cm und im April 1975 (ohne Schnee) f_m = 31,7 cm an Stelle der rechnerisch erwarteten von 21,6 cm ergeben.

6 Folgerungen

Erstens ist es für überschlägige Angebots- und Kontrollberechnungen möglich, plattenförmige Raumfachwerke von beliebiger Struktur mit einem anschaulichen Näherungsverfahren zutreffend zu analysieren. Zweitens sollte als Knicklänge der Druckstäbe stets ihre volle Systemlänge angesetzt werden. Drittens sollte für die Bemessung jener Wert als praktisch ausnutzbare Traglast (Grenzlast) angesehen werden, bei dem entweder das Reissen des ersten Zuggliedes oder das Knicken des ersten Druckgliedes auftritt, weil sich die Druckmembranwirkung erst bei so grossen Durchbiegungen einstellt, bei denen die normale Nutzung des Raumfachwerks kaum mehr möglich ist. Viertens können die Durchbiegungen im Gebrauchszustand bei bekanntem Gelenkspiel zuverlässig vorausgesagt werden.

Literatur

/1/ Herzog, M.: Elementare Betrachtung der Tragwirkung orthogonal vorgespannter Seilnetzdächer und Betonhängeschalen. IASS Symposium Dortmund 1984, Bd. 1, S. 91 - 112

/2/ Scheer, J. und Koep, H.: Zur Optimierung von Raumfachwerken. Bauingenieur 57 (1982) S. 27 - 33

/3/ Herzog, M.: Verteilungszahlen für die Berechnung orthotroper Platten unter gleichförmig verteilter Last nach der Streifenmethode. Schweizerische Bauzeitung 72 (1954) S. 457 - 458

/4/ Schmidt, L.C., Morgan, P.R. und Clarkson, J.A.: Space trusses with brittle-type strut buckling. Proceedings ASCE 102 (1976) ST 7, S. 1479 - 1492

/5/ Schmidt, L.C., Morgan, P.R. und Clarkson, J.A.: A space truss with yielding tensile chords. Civil Engineering Transactions 20 (1978) S. 134 - 141

/6/ Schmidt, L.C., Morgan, P.R. und Hanaor, A.: Ultimate load testing of space trusses. Proceedings ASCE 108 (1982) ST 6, S. 1324 - 1335

/7/ Herzog, M.: Die Membranwirkung in Stahlbetonplatten nach Versuchen. Beton- und Stahlbetonbau 71 (1976) S. 270 - 275

/8/ Schmidt, L.C., Morgan, P.R., O'Meagher, A.J. und Cogan, K.: Ultimate load behaviour of a full-scale space truss. Proceedings ICE, Part 2, 69 (1980) S. 97 - 109

/9/ Engineering News Record 200 (1978) Nr. 4, S. 8 - 9, Nr. 6, S. 9 - 10, Nr. 11, S. 13, Nr. 14, S. 9 - 11, Nr. 23, S. 5, Nr. 25, S. 36 - 37 und Nr. 26, S. 13

/10/ Nowak, B.: Schadensfälle in den USA als Folge heftiger Schneestürme. Stahlbau 47 (1978) S. 377 - 378

/11/ Koep, H.: Diskussionen über den Einsturz eines dreilagigen Raumfachwerks in Hartford, Conn., USA. Bauingenieur 55 (1980) S. 29 - 30

/12/ Lev Zetlin Associates, Inc.: Report on the engineering investigation concerning causes of the collapse of the Hartford Coliseum space truss roof on January 18, 1978. New York, June 12, 1978

/13/ Committee to investigate the Coliseum roof failure: Report. City of Hartford, Conn., July 13, 1978

/14/ Smith, E.A. und Epstein, H.J.: Hartford Coliseum roof collapse - Structural collapse sequence and lessons learned. Civil Engineering ASCE 52 (1980) Nr. 4, S. 59 - 62

/15/ Loomis, R.S., Loomis, R.H., Loomis, R.W. und Loomis R.W.: Torsional buckling study of Hartford Coliseum. Proceedings ASCE 106 (1980) ST 1, S. 211 - 231

/16/ Herzog, M.: Eine neue Faustformel für schlanke Stahlstützen unter ausmittigem Druck. Stahlbau 53 (1984) S. 281 - 282

PRACTICAL DYNAMIC ANALYSIS OF TRANSLATIONAL SHELLS

A. Samartín
Professor, University of Santander
E.T.S. de Ingenieros de Caminos, Canales y Puertos
39.004-Santander SPAIN
J. Martínez
Dr. Civil Engineer. Ministry of Public Works
28.003-Madrid SPAIN

1.- INTRODUCTION

This paper presents a practical approach for Levy type solutions of natural frequencies of translational shells. An extension to the study of forced vibrations is outlined.

2.- GENERAL EQUATIONS

The curved plate shallow linear elastic thin shell theory is assumed. The hypothesis and main assumptions of the theory can be seen in (1) and (2).

In the following only results arrived at in (3) are presented here. The positive directions of body forces, stress-resultants, stress-couples as well as the positive directions of the middle surface displacements are shown in figures 1, 2, 3 and 4 respectively.

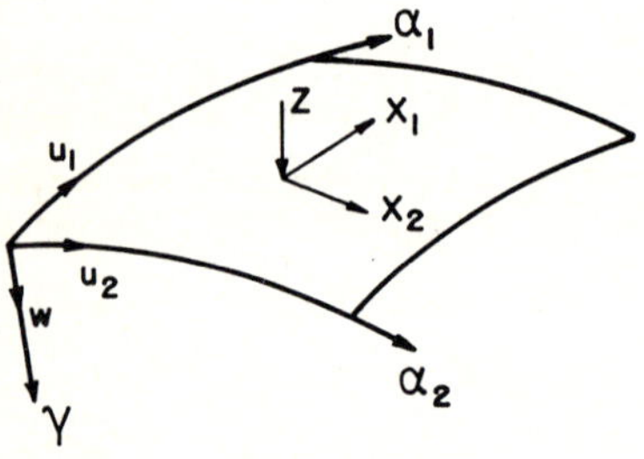

FIGURE 1.- BODY FORCES

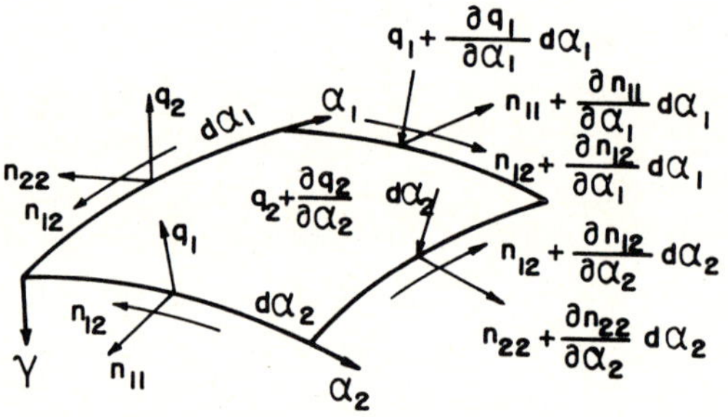

FIGURE 2.- STRESS - RESULTANTS

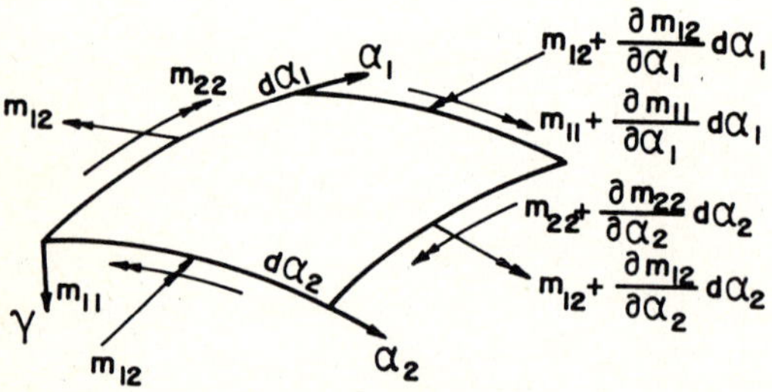

FIGURE 3.- STRESS - COUPLES

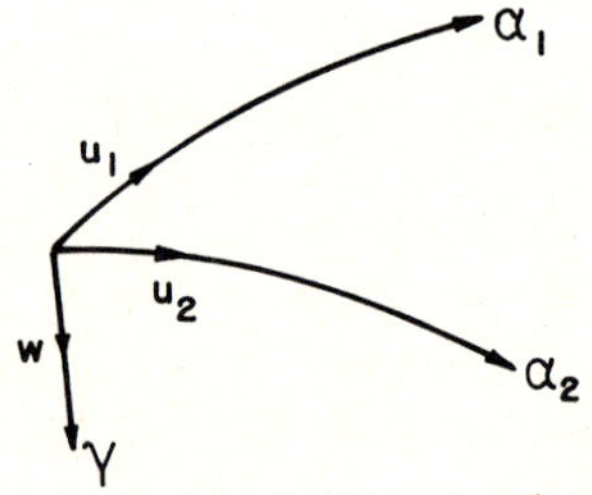

FIGURE 4.- ORTHOGONAL CURVILINEAR COORDINATES AND DISPLACEMENTS

Using the above assumptions the equations of equilibrium of a shell element become:

$$N_{11,1} + N_{21,2} + X_1 = 0$$

$$
\begin{aligned}
&N_{12,1} + N_{22,2} + X_2 = 0 \\
&N_{11,1} + N_{21,2} - Q_1 = 0 \\
&N_{12,1} + M_{22,2} - Q_2 = 0 \\
&Q_{1,1} + Q_{2,2} + N_{11}K_1 + N_{22}K_2 + Z = 0
\end{aligned}
\qquad (2.1)
$$

In order to derive the governing equations, the following relations are set up (2):

-Strain/displacements

$$
\begin{aligned}
e_{ij} &= \tfrac{1}{2}(u_{i,j} + u_{j,i} - 2K_{ij}w) \\
k_{ij} &= -w_{,ij}
\end{aligned}
\qquad (2.2)
$$

(i,j=1,2) and the Einstein convention is used.

$K_{ii}=K_i$ and $K_{ij}=0$, if $i\neq j$, because the lines $(\alpha_1\alpha_2)$ are curvature lines.

-Stress-resultants/strain

$$
\begin{aligned}
N_{ij} &= K\left[(1-\nu)e_{ij} + \nu\delta_{ij}\, e_{rr}\right] \\
M_{ij} &= D\left[(1-\nu)k_{ij} + \nu\delta_{ij}\, k_{rr}\right] \quad \text{(Einstein convention)}
\end{aligned}
\qquad (2.3)
$$

where $K = \dfrac{Eh}{1-\nu^2}$, $D = \dfrac{Eh^3}{12(1-\nu^2)}$ and δ_{ij} is the Kronecker delta.

-Stress function. Definition

$$
N_{ij} = (-1)^{i+j} \cdot \emptyset_{,ij} - \delta_{ij}\int X_i \, d\alpha_i \quad \text{(Not summed)} \qquad (2.4)
$$

With the introduction of $\emptyset$, the equations of equilibrium (2.1) and (2.2) are identically satisfied.

-Stress-resultants/displacements

$$
\begin{aligned}
M_{11} &= -D(w_{,11} + \nu w_{,22}) \\
M_{22} &= -D(w_{,22} + \nu w_{,11}) \\
M_{12} &= -D(1-\nu)w_{,12} = M_{21} \\
Q_1 &= -D\nabla^2 w_{,1} \\
Q_2 &= -D\nabla^2 w_{,2} \\
R_1 &= Q_1 + M_{12,2} = -D\left[w_{,111} + (2-\nu)w_{,122}\right] \\
R_2 &= Q_2 + M_{21,1} = -D\left[w_{,222} + (2-\nu)w_{,112}\right]
\end{aligned}
\qquad (2.5)
$$

-Governing Equations

Introducing these relations into the equations of equilibrium and the compatibility equations, the governing equations are derived. The compatibility equations can be set up using the static-kinematic analogy of Goldenweizer (see 1).

$$
\begin{aligned}
&\frac{1}{Eh}\nabla^4\emptyset + \nabla_K^2 w = \frac{1}{Eh}\left[\int X_{2,11}d\alpha_2 + \int X_{1,,2}d\alpha_1 + \nu(X_{2,2} + X_{1,1})\right] \\
&D\nabla^4 w - \nabla_K^2\emptyset = -K_1\int X_1 d\alpha_1 - K_2\int X_2 d\alpha_2 + Z
\end{aligned}
\qquad (2.6)
$$

where ∇^2 is the Laplacian operator $\frac{\partial^2}{\partial\alpha_1^2} + \frac{\partial^2}{\partial\alpha_2^2}$ and

$$\nabla_K^2 \equiv K_2 \frac{\partial^2}{\partial\alpha_1^2} + K_1 \frac{\partial^2}{\partial\alpha_2^2}$$

3.- DESCRIPTION OF THE LEVY PROCESS. STATIC CASE

As an introduction to the dynamic problem a matrix formulation of the Levy solution for the static case is briefly presented. Details can be seen in references (3) and (4).

The Levy solution corresponds to the following boundary conditions, along the edges $\alpha_1=0$ and $\alpha_1=l_1$ (rectangular planform):

$N_{11}=0$, $u_2=0$, $w=0$ and $M_{11}=0$ (normal gable)

The solution is of the form

$$w(\alpha_1\alpha_2) = \sum_{m=1}^{\infty} w^{(m)}$$

$$\phi(\alpha_1\alpha_2) = \sum_{m=1}^{\infty} \phi^{(m)}$$

where

$$w^{(m)} = \left[w_0^{(m)} + w_c^{(m)}\right] \sin\lambda_m\alpha_1$$

$$\phi^{(m)} = \left[\phi_0^{(m)} + \phi_c^{(m)}\right] \sin\lambda_m\alpha_1$$

$$w_0^{(m)}(\phi_0^{(m)})\sin\lambda_m\alpha_1 \quad \text{and} \quad w_c^{(m)}(\phi_i^{(m)})\sin\lambda_m\alpha_1$$

are called the particular solutions and the complementary solutions respectively.

$$\lambda_m = \frac{m\pi}{l_1}$$

The particular solutions are obtained from the Fourier expansion coefficients of the external load (X_1, X_2 and Z) and therefore they depend on the particular distribution of the loading.

The complementary solution is the general solution of the homogeneous equations (2.6), i.e.

$$\frac{1}{Eh}\nabla^4\phi + \nabla_k^2 w = 0$$
$$D\nabla^4 w - \nabla_k^2 \phi = 0 \qquad (3.1)$$

If $K_1 \neq K_2$, the Ambartsumyan function W defined as

$$w = \nabla^4 W$$
$$\phi = -Eh\nabla_k^2 W \qquad (3.2)$$

is introduced in (3.1). The first equation of (3.1) is identically satisfied and the second becomes:

$$\nabla^8 W + \frac{12(1-\nu^2)}{k^2}\nabla_k^4 W = 0 \qquad (3.3)$$

For the case $K_1=K_2=K$ (spherical shallow curved plate) Mishonov (5) introduced the auxiliary function W defined as:

$$\nabla^2 w = \nabla^2 W$$
$$\nabla^2\phi = -Eh\, K\, W \qquad (3.4)$$

in order to avoid the suppresion of solutions. The second equation (3.1) becomes in this case.

$$\nabla^4 W + \frac{12(1-\nu^2)}{k^2} K^2 W = 0 \tag{3.5}$$

The solution procedure for this special case (3.5) is very similar to the one presented for $K_1 \neq K_2$ (3.3) and therefore only this last procedure will be presented here.

Expanding W in Fourier series, the m-th term is $W=F(\alpha_2)\sin\lambda\alpha_1$ and substituting into (3.3) yields:

$$\left(\frac{d^2}{d\alpha_2^2} - \lambda^2\right)^4 F(\alpha_2) + \frac{12(1-\nu^2)}{k^2}\left(K_1 \frac{d^2}{d\alpha_2^2} - K_2\lambda^2\right)^2 F(\alpha_2) = 0$$

The solution of the above ordinary differential equation with constant coefficients is:

$$F(\alpha_2) = \sum_{i=1}^{2} \exp(-r_2\alpha_2)(A_{i1}\cos s_i\alpha_2 + A_{i2}\sin s_i\alpha_2) + \sum_{i=1}^{2} \exp(-r_i\beta_2)(B_{i1}\cos s_i\beta_2 + B_{i2}\sin s_i\alpha_2) \tag{3.6}$$

where A_{ij}, B_{ij} (i,j=1,2) are eight arbitrary constants

$\beta_2 = l_2 - \alpha_2$

$\pm r_i \pm s_i\sqrt{-1}$ (i=1,2) are the eight complex roots of the characteristic equations, i.e.

$$r_1 = \sqrt{\frac{a_1 + \sqrt{a_1^2\ b_1^2}}{2}} \qquad s_1 = \sqrt{\frac{-a_1 + \sqrt{a_1^2 + b_1^2}}{2}}$$

$$r_2 = \sqrt{\frac{a_2^2 + \sqrt{a_2^2 + b_2^2}}{2}} \qquad s_2 = \sqrt{\frac{-a_2 + \sqrt{a_2^2 + b_2^2}}{2}}$$

where if $K_1 < K_2$

$$a_1 = \lambda^2 + \mu\sqrt{(-K_1^2 + \Delta)/2}$$
$$b_1 = \mu\left[(K_1^2 + \Delta)/2 - K_1\right]$$
$$a_2 = \lambda^2 - \mu\sqrt{(-K_1^2 + \Delta)/2}$$
$$b_2 = \mu\left[(K_1^2 + \Delta)/2 + K_1\right]$$

if $K_1 > K_2$

$$a_1 = \lambda^2 + \mu\sqrt{(-K_1^2 + \Delta)/2}$$
$$b_1 = \mu\left[(K_1^2 + \Delta)/2 + K_1\right]$$
$$a_2 = \lambda^2 - \mu\sqrt{(-K_1^2 + \Delta)/2}$$
$$b_2 = \mu\left[(K_1^2 + \Delta)/2 - K_1\right]$$

$$\mu = \frac{\sqrt{3(1-\nu^2)}}{k}$$
$$\Delta = \sqrt{K_1^4 + 4(K_1-K_2)^2\lambda^4/\mu^2}$$

The results of interest of a shell analysis can be collected in a vector $\underline{R}$ defined as follows:

$$\underline{R}=(u_1, u_2, w_1, w_{,1}, w_{,2}, N_{11}, N_{22}, N_{12}, M_{11}, M_{22}, M_{12}, Q_1, Q_2, R_1, R_2)^T$$

where each element is a function of α_2 and varies along the direction α_1 as $\sin\lambda\alpha_1$, except $u_1, w_{,1}, N_{12}, M_{12}, Q_1$ and R_1 which vary as $\cos\lambda\alpha_1$.

For the complementary solution this vector $\underline{R}$ is denoted by $\underline{R}_c$ and it can be expressed in matrix form as follows:

$$\underline{R}_c = \underline{G}\left[\underline{C}_1\ \underline{P}(\alpha_2)\ \underline{A} + \underline{C}_2\ \underline{P}(\beta_2)\ \underline{B}\right] \tag{3.7}$$

where $\underline{G}$ is a 15x8 matrix shown in table 1.

TABLE I (matrix G)

	1	2	3	4	5	6	7	8
1	$-(K+\nu K_2)\lambda^3$	.	$-K_2\lambda$ $+(2+\nu)K_1\lambda$	.	.	.	.	.
2	.	$K_1\lambda^2$ $-(2+\ )K_2\lambda^2$	.	$K_2+\nu K_1$	.	.	.	.
3	λ^4	.	$-2\lambda^2$	.	1	.	.	.
4	λ^5	.	$-2\lambda^3$	.	λ	.	.	.
5	.	λ^4	.	$-2\lambda^2$	.	1	.	.
6	.	.	$K^2\lambda^2 Eh$	.	$-K_1 Eh$	.	.	.
7	$-K_2\lambda^4 Eh$	.	$K_1\lambda^2 Eh$	.	.	.	.	.
8	.	$-K_2\lambda^3 Eh$	.	$K_1\lambda Eh$	.	.	.	.
9	$\lambda^6 D$	.	$-(2+\nu)\lambda^4 D$	.	$(1+2\nu)\ {}^2D$	.	$-\nu D$	.
10	$\nu\lambda^6 D$	.	$-(1+2\nu)\lambda^4 D$	.	$(2+\nu)\lambda^2 D$	.	$-D$	.
11	.	$-(1-\nu)\lambda^5 D$	.	$2(1-\nu)\lambda^3 D$	.	$-(1-\nu)\lambda D$	.	.
12	$\lambda^7 D$	.	$-3\lambda^5 D$	.	$3\lambda^3 D$	.	$-\lambda D$	.
13	.	$\lambda^6 D$	.	$-3\lambda^4 D$	.	$3\lambda^2 D$	.	$-D$
14	$\lambda^7 D$	.	$-(4-\nu)\lambda^5 D$	.	$(5-2\nu)\lambda^3 D$	.	$-(2-\nu)\lambda D$	.
15	.	$(2-\nu)\lambda^6 D$	.	$-(5-2\nu)\lambda^4 D$	.	$(4-\nu)\lambda^2 D$	.	$-D$

$\underline{C}_1 = \left[\underline{C}_{11},\ \underline{C}_{12}\right]$ dimension 8x4

$\underline{C}_2 = \left[\underline{C}_{21},\ \underline{C}_{22}\right]$ dimension 8x4

$\underline{C}_{1i} = \left\{(-)^k \rho_i^k \cos k\varphi_i,\ (-)^k \rho_i^k \sin k\varphi_i\right\}$ dimensión 8x2

$\underline{C}_{2i} = \left\{\rho_i^k \cos k\varphi_i,\ \rho_i^k \sin k\varphi_i\right\}$ dimensión 8x2

$k = 0,1,2,\dots,7$

$\rho_i = (r_1^2+s_i^2)^{1/2}$; $\varphi_i = tg^{-1}\left[\dfrac{s_i}{r_i}\right]$

$$\underline{p}(\alpha_2) = \begin{bmatrix} p_{11}(\alpha_2) & p_{12}(\alpha_2) & 0 & 0 \\ -p_{12}(\alpha_2) & p_{11}(\alpha_2) & 0 & 0 \\ 0 & 0 & p_{21}(\alpha_2) & p_{22}(\alpha_2) \\ 0 & 0 & -p_{22}(\alpha_2) & p_{21}(\alpha_2) \end{bmatrix} \quad \text{dim 4x4}$$

$p_{i1}(\alpha_2) = \exp(-r_i\alpha_2)\cos s_i\alpha_2$

$$p_{i2}(\alpha_2) = \exp(-r_i\alpha_2)\sin s\alpha_2$$

$$\underline{A} = \left[A_{11}, A_{12}, A_{21}, A_{22}\right]^T \qquad \text{dimension 4x1}$$

$$\underline{B} = \left[B_{11}, B_{12}, B_{21}, B_{22}\right]^T \qquad \text{dimension 4x1}$$

The arbitrary constants A_{ij}, B_{ij} are obtained by imposing the boundary conditions along the edges $\alpha_2=0$ and $\alpha_2=l_2$, to the total solution

$$W = \sum_{m=1}^{\infty}(W_0+W_c)\sin\lambda_m\alpha_1$$

where W_0 is the particular solution depending on the type of loading applied on the shell structure.

The boundary conditions can be quite arbitrary but they are restricted here to be a linear combination of knematic and static boundary conditions. Namely, the kinematic boundary conditions are specified values for u_1, u_2, w and $w_{,2}$ and the static boundary conditions are specified values for N_{12}, $\bar{N}_{22}$, R_1 and M_{22}. By selection of the corresponding rows of the matrix $\underline{G}$ the submatrices $\underline{G}_d$ and $\underline{G}_p$ can be constructed and the boundary conditions can be collected in the vectors $\underline{d}$ and $\underline{p}$, displacements and forces vectors, defined as follows:

$$\underline{d} = \underline{d}_0 + \underline{G}_d\left[\underline{C}_1\,\underline{P}(\alpha_2)\,\underline{A} + \underline{C}_2\,\underline{P}(\beta_2)\underline{B}\right]$$
$$\underline{p} = \underline{p}_0 + \underline{G}_p\left[\underline{C}_1\,\underline{P}(\alpha_2)\,\underline{A} + \underline{C}_2\,\underline{P}(\beta_2)\underline{B}\right] \tag{3.8}$$

where $\underline{d}_0$ and $\underline{p}_0$ are the boundary conditions corresponding to the particular solutions By imposing the values $\alpha_2=0$ ($\beta_2=l_2$) and $\alpha_2=l_2$ ($\beta_2=0$) a linear system of 8 simultaneous equations is derived and the constants $\underline{A}$ and $\underline{B}$ can be found.

4.- FREE VIBRATIONS. NATURAL FREQUENCIES

For convenience with the above formulation only small normal vibrations will be studied. It could be possible to extend the Levy process to general vibrations in all directions, but in this case, the formulation in three differential equations in u_1, u_2, and w would be more convenient.

The boundary conditions are taken to be time-independent and homogeneous. The natural frequencies are obtained by assuming in the governing equations (2.6) the following loads:

$$X_1 = X_2 = 0 \qquad \text{and} \qquad Z = -\rho h\frac{\partial^2 w}{\partial t^2}$$

where ρh is mass per unit area.

Introducing the Ambartsumyan function $\bar{W}$ the free vibrations are governing by the equation:

$$\nabla^8\bar{W} + \frac{Eh}{D}\nabla_k^4\bar{W} + \rho\frac{h}{D}\nabla^4\frac{\partial^2\bar{W}}{\partial t^2} = 0 \tag{4.1}$$

To find the natural frequencies w let

$$\bar{W} = W(\alpha_1,\alpha_2)e^{i\omega t} \qquad (i = \sqrt{-1}) \tag{4.2}$$

and the equation (4.1) becomes:

$$\nabla^8 W + \frac{12(1-\nu^2)}{k^2}\nabla_k^4 W - p^2\nabla^4 W = 0 \tag{4.3}$$

where $p^2 = \frac{\rho h}{D}\omega^2$

The solution is of the form $W(\alpha_1\alpha_2) = \sum_{m=1}^{\infty} W^{(m)}(\alpha_1\alpha_2)$ where

$$W^{(m)}(\alpha_1\alpha_2) = F^{(m)}(\alpha_2) \; . \; \sin\lambda_m\alpha_1, \quad \lambda_m = \frac{m\pi}{l_1}$$

For clarity, the subscript m is dropped in all the following formulas.

The equation (4.3) is now

$$\left[\frac{d^2}{d\alpha_2^2}-\lambda^2\right]^4 F(\alpha_2) + \frac{12(1-\nu^2)}{h^2}\left[K_1\frac{d^2}{d\alpha_2^2} - K_2\lambda^2\right]^2 F(\alpha_2) -$$

$$- p^2\left(\frac{d^2}{d\alpha_2^2} - \lambda^2\right)^2 F(\alpha_2) = 0$$

An ordinary differential equation with constant coefficients and whose characteristic equation is:

$$\left[y^2 - \lambda^2\right]^4 + \frac{12(1-\nu^2)}{h^2}\left[K_1 y^2 - K_2\lambda^2\right]^2 - p^2\left[y^2-\lambda^2\right]^2 = 0$$

And with a change of variable

$$y^2 - \lambda^2 = z$$

$$z^4 + \frac{12(1-\nu^2)}{h^2}\left[K_1 z - (K_1-K_2)\lambda^2\right]^2 - p^2 z^2 = 0$$

or

$$z^4 + \left[\frac{12(1-\nu^2)}{h^2} K_1^2 - p^2\right]z^2 + \frac{24(1-\nu^2)}{h^2} K_1(K_1-K_2)\lambda^2 z +$$

$$+ \frac{12(1-\nu^2)}{h^2} \; . \; (K_1-K_2)^2\lambda^4 = 0$$

The roots of this equation can be obtained algebrically for each value of p. In the general case, there are n distint real roots y_i of multiplicity n_i and m pair of complex roots $r_i \pm s_i\sqrt{-1}$ of multiplicity m_i. Then

$$\sum_{i=1}^{n} n_i + \sum_{i=1}^{m} m_i = 4$$

The general solution $F(\alpha_2)$ is:

$$F(\alpha_2) = \sum_{i=1}^{n}\sum_{j=0}^{n_i-1}\left[\exp(-y_i\alpha_2)A_{ij}\alpha_2^j + \exp(-y_i\beta_2)B_{ij}\beta_2^j\right] +$$

$$\sum_{i=1}^{m}\sum_{j=0}^{m_i-1}\left[\exp(-r_i\alpha_2)(A'_{ij}\alpha_2^j \cos s_i\alpha_2 + B'_{ij}\alpha_2^j \sin s_i\alpha_2) + \right. \tag{4.4}$$

$$\left. + \exp(-r_i\beta_2)(A''_{ij}\beta_2^j \cos s_i\beta_2 + B''_{ij}\beta_2^j \sin s_i\beta_2)\right]$$

Using a similar procedure as the one described in the previous section, the following results can be derived:

$$\underline{R} = \underline{G}\left[\underline{C}_1\,\underline{P}(\alpha_2)\,\underline{A} + \underline{C}_2\,\underline{P}(\beta_2)\,\underline{B}\right] \tag{4.5}$$

where the matrices $\underline{C}_1$, $\underline{C}_2$ and $\underline{P}(\alpha_2)$ are now dependent on the roots nature and therefore, in general different to the ones defined in (3.7).

The kinematic and static boundary conditions can be expressed as:

$$\begin{aligned} \underline{d} &= \underline{G}_d\left[\underline{C}_1\,\underline{P}(\alpha_2)\,\underline{A} + \underline{C}_2\,\underline{P}(\beta_2)\,\underline{B}\right] \\ \underline{p} &= \underline{G}_p\left[\underline{C}_1\,\underline{P}(\alpha_2)\,\underline{A} + \underline{C}_2\,\underline{P}(\beta_2)\,\underline{B}\right] \end{aligned} \tag{4.6}$$

In addition to the previous stated assumption of time independence and homogenity, the boundary conditions will be further restricted to the following form:

$$\underline{K}_d \begin{bmatrix} \underline{d}_1 \\ \underline{d}_2 \end{bmatrix} + \underline{K}_p \begin{bmatrix} \underline{p}_1 \\ \underline{p}_2 \end{bmatrix} = \underline{0} \tag{4.7}$$

where $\underline{K}_d$ and $\underline{K}_p$ are two 8x8 diagonal matrices such that $K_d+K_p=I$ (unit matrix) $\underline{d}_1$ ($\underline{p}_1$) and $\underline{d}_2$ ($\underline{p}_2$) are the corresponding matrices obtained from (4.6) for the values $\alpha_2=0$ and $\alpha_2=l_2$ respectively.

Therefore equation (4.7) becomes:

$$\left\{ \underline{K}_p \begin{bmatrix} \underline{G}_p \underline{C}_1 \underline{P}_0 & \underline{G}_p \underline{C}_2 \underline{P}_1 \\ \underline{G}_p \underline{C}_1 \underline{P}_1 & \underline{G}_p \underline{C}_2 \underline{P}_0 \end{bmatrix} + \underline{K}_d \begin{bmatrix} \underline{G}_d \underline{C}_1 \underline{P}_0 & \underline{G}_d \underline{C}_2 \underline{P}_1 \\ \underline{G}_d \underline{C}_1 \underline{P}_1 & \underline{G}_d \underline{C}_2 \underline{P}_0 \end{bmatrix} \right\} \begin{bmatrix} \underline{A} \\ \underline{B} \end{bmatrix} = \underline{0} \tag{4.8}$$

i.e. a system of eight linear simultaneous equations, whose non-trivial solutions are given when

$$\det \left\{ \underline{K}_p \begin{bmatrix} \underline{G}_p \underline{C}_1 \underline{P}_0 & \underline{G}_p \underline{C}_1 \underline{P}_1 \\ \underline{C}_p \underline{C}_1 \underline{P}_1 & \underline{G}_p \underline{C}_2 \underline{P}_0 \end{bmatrix} + \underline{K}_d \begin{bmatrix} \underline{C}_d \underline{C}_1 \underline{P}_0 & \underline{G}_d \underline{C}_2 \underline{P}_1 \\ \underline{G}_d \underline{C}_1 \underline{P}_1 & \underline{G}_d \underline{C}_2 \underline{P}_0 \end{bmatrix} \right\} = 0 \tag{4.9}$$

This transcendental equation (4.9) in p^2 gives all the natural frequencies corresponding to m or number of half-waves along the α_1 direction.

An upper and a lower bound can be found by indirect method for every root of the equation (4.2-3) and by trial and error these bounds can be made to approach one another to any desired degree of accuracy. Further details are given in reference (6) where a computer program in FORTRAN IV language is presented.

For every root of (4.9) there exists a set of non-trivial values of A_i and B_i (i=1,..4) satisfying the homogenous simultaneous equations (4.8) and representing the corresponding modes of vibration of the shell. These can be designated w_{mn}, where m is the number of the harmonic along the α_1 directions and n is the root number, ordered according to increasing values.

The approach just outlined represents a formidable task even with the help of a digital computer. Numerical difficulties and large demand of computer time are normally encountered in the analysis. In the reference (6) this problem is treated in detail.

5.- ALTERNATIVE METHOD

The analysis of the free vibrations of translational shells can be carried out by means of the following numerical alternative approach. The shell is divided into a set of N adjacent shell elements by the following coordinate lines $\alpha_2=0$, l_2/N, $l_2/N \cdot 2$, ..., $l_2/N(N-1)$, $l_2/N \cdot N$. For each of these shell elements and each Fourier term m, it is possible to obtain the stiffness matrix, That means, a relationship between the vectors

$$\underline{p} = \begin{bmatrix} \underline{p}_1 \\ \underline{p}_2 \end{bmatrix} \quad \text{and} \quad \underline{d} = \begin{bmatrix} \underline{d}_1 \\ \underline{d}_2 \end{bmatrix}$$

of the following kind:

$$\begin{bmatrix} \underline{p}_1 \\ \underline{p}_2 \end{bmatrix} = \underline{k} \begin{bmatrix} \underline{d}_1 \\ \underline{d}_2 \end{bmatrix} \tag{5.1}$$

The stiffness matrix $\underline{k}$ can computed in the static situation, from the expresions

(3.8), assuming $\underline{d}_0=\underline{0}$ and $\underline{p}_0=\underline{0}$, i.e. no applied loading exists. In this case the following equations are derived:

$$\begin{bmatrix} \underline{d}_1 \\ \underline{d}_2 \end{bmatrix} = \begin{bmatrix} \underline{G}_d\, \underline{C}_1 & \underline{G}_d\, \underline{C}_2\, \underline{P}(l_2) \\ \underline{G}_d\, \underline{C}_1\, \underline{P}(l_2) & \underline{G}_d\, \underline{C}_2 \end{bmatrix} \begin{bmatrix} \underline{A} \\ \underline{B} \end{bmatrix} = \bar{G}_d \begin{bmatrix} \underline{A} \\ \underline{B} \end{bmatrix} \tag{5.2}$$

$$\begin{bmatrix} \underline{p}_1 \\ \underline{p}_2 \end{bmatrix} = \begin{bmatrix} \underline{G}_p\, \underline{C}_1 & \underline{G}_p \underline{C}_2 \underline{P}(l_2) \\ \underline{G}_p\, \underline{C}_1\, \underline{P}(l_2) & \underline{G}_p \underline{C}_2 \end{bmatrix} \begin{bmatrix} A \\ B \end{bmatrix} = \bar{G}_p \begin{bmatrix} \underline{A} \\ \underline{B} \end{bmatrix}$$

By elimination of the constants $\underline{A}$, $\underline{B}$ between the two matrix equations (5.2) the stiffness matrix is obtained:

$$\underline{k} = \underline{G}_p\, \underline{G}_d^{-1} \tag{5.3}$$

Similarly the mass matrix $\underline{m}$ of the shell element can be defined as:

$$\begin{bmatrix} \underline{p}_1 \\ \underline{p}_2 \end{bmatrix} = -\underline{m} \begin{bmatrix} \ddot{\underline{d}}_1 \\ \ddot{\underline{d}}_2 \end{bmatrix} \tag{5.4}$$

where $\underline{m}$ is computed using the standard matrix procedure by the formula:

$$\underline{m} = \frac{l_1}{2} \int_{l_2\, i/N}^{l_2(i+1)/N} \underline{N}^T\, \underline{\rho}\, \underline{N}\, d\alpha_2 \tag{5.5}$$

where the shape vector $\underline{N}$ is given by the expression

$$\underline{d} = \left[\underline{N}_1 \quad \underline{N}_2\right] \begin{bmatrix} d_1 \\ d_2 \end{bmatrix} = \underset{\sim}{N} \begin{bmatrix} \underline{d}_1 \\ \underline{d}_2 \end{bmatrix} \tag{5.6}$$

and therefore, the 4x8 matrix $\underline{N}$ is

$$\underline{N} = \left[\underline{G}_d\, \underline{C}_1\, \underline{P}(\alpha_2), \quad \underline{G}_d\, \underline{C}_2\, \underline{P}(\beta_2)\right] \underline{G}_d^{-1} \tag{5.7}$$

The mass matrix $\underline{\rho}$ is:

$$\underline{\rho} = \text{diag}\left[\rho_{u_1}, \rho_{u_2}, \rho_w, \rho_\theta h^2\right] h \tag{5.8}$$

If only normal vibrations are considered, then

$$\rho_{u_1} = \rho_{u_2} = \rho_\theta = 0 \; ; \qquad \rho_w = \rho \tag{5.9}$$

The integral (5.5) can be numerically evaluated or simply by using a lumped mass technique.

By assembling of the matrices (5.3) and (5.5) of each shell element the discretized dynamic equilibrium equations for the total shell can be derived.

$$\underline{M}\, \ddot{\underline{D}} + \underline{K}\, \underline{D} = \underline{0} \tag{5.10}$$

where $\underline{D} = \begin{bmatrix} \underline{d}_1 \\ \underline{d}_2 \\ \vdots \\ \underline{d}_N \\ d_{N+1} \end{bmatrix}$; $\underline{M} = \sum \underline{m}$ and $\underline{K} = \sum \underline{k}$

The equation (5.10) corresponds to undamped free vibrations of the shell. It can be solved using the standard procedures. Damping effects can be introduced by assuming an orthogonal mass matrix (8).

The extension to the study of forced vibrations can be carried out by introducing the applied loading in the dynamic equilibrium equations. That can be acomplished by the computation for each shell element of the nodal equivalent force vector $\underline{p}^0$, as follows:

$$p^0 = -\int_0^{l_2} \underline{N}^T \underline{g}\, d\alpha_2 \qquad (5.11)$$

where the applied loading vector $\underline{g}$ is described in the following way

$$\underline{g} = \left[X_1,\ X_2,\ Z,\ 0\right]^T$$

The equilibrium equations become now

$$\underline{M}\ \ddot{\underline{D}} + \underline{K}\ \underline{D} = \underline{P}^0 \qquad (5.12)$$

where $\underline{P}^0 = \sum \underline{p}^0$ and again the standard techniques of analysis of dynamic discrete systems can be applied (9).

6.- EXAMPLES

(a) The following cylindrical shell roof has been studied using the approach described in section 4:
Mean diameter: 6 ft, thickness: 0.5 in and effective length between the end supports 8 ft.

It is assumed $E=30x10^6\ lb/m^2$; $\nu=0.3$ and $\rho g=0.283\ lb/m^3$ for steel. This example was solved by the Rayleigh Ritz method in (7), for m=number of axial halfvawes =1:

Results from (7)	91	104	117	139	cycles/sec
Authors results	86	98	114	130	cycles/sec

(b) Simply supported concrete box bridge.

The bridge cross section is represented in figure 5a. The bridge span is 35.00 m. The concrete properties are:

$E = 3 \times 10^6\ tm^{-2}$, Poisson ratio $\nu = 0.20$, $\rho g = 2.5\ tm^{-3}$ and η (relative damping) = 0,02.

The folded plate structure is shown in figure 5b.

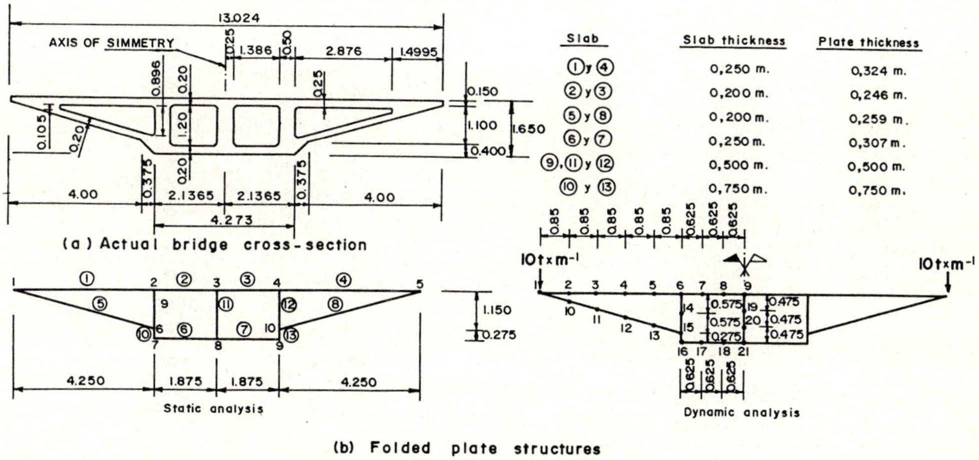

FIGURE 5.- BRIDGE_ CROSS-SECTION

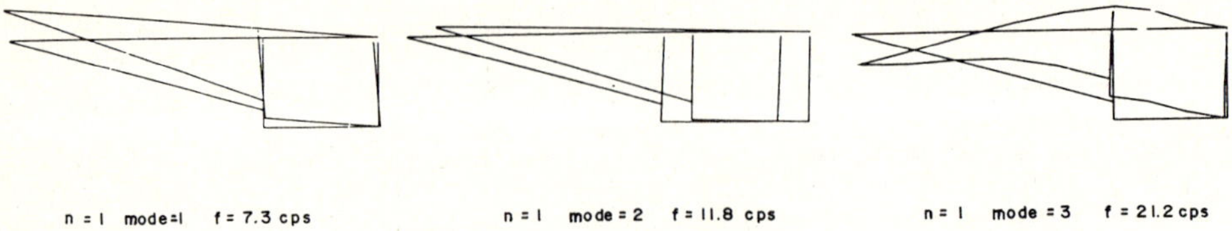

(a) Unsymmetric vibration modes

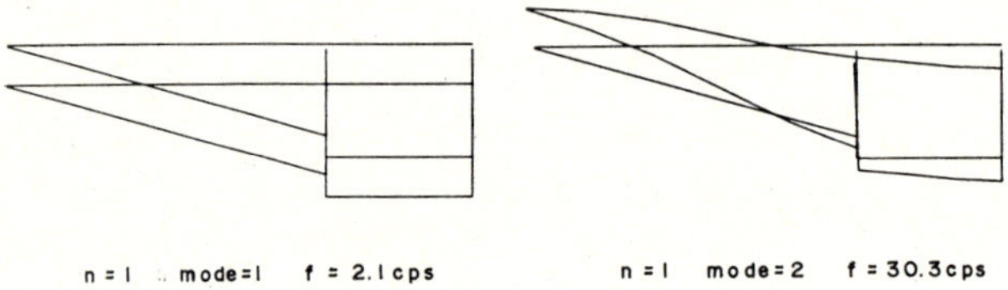

(b) Symmetric vibration modes

FIGURE 6.- MODES OF VIBRATION

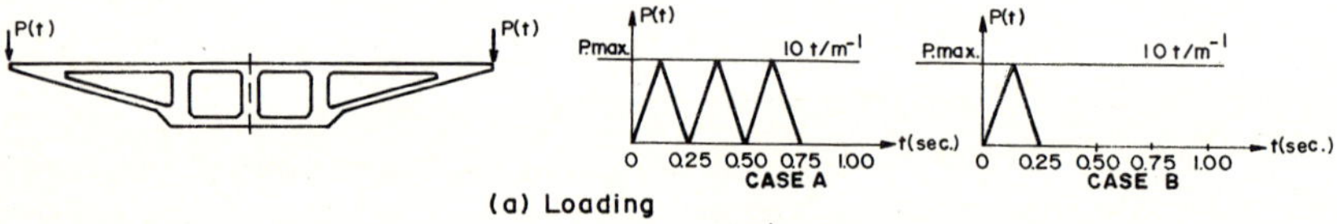

(a) Loading

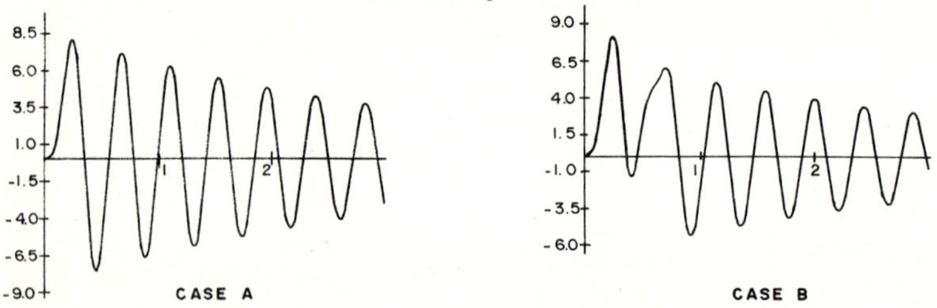

Edge 1.- Vertical deflection

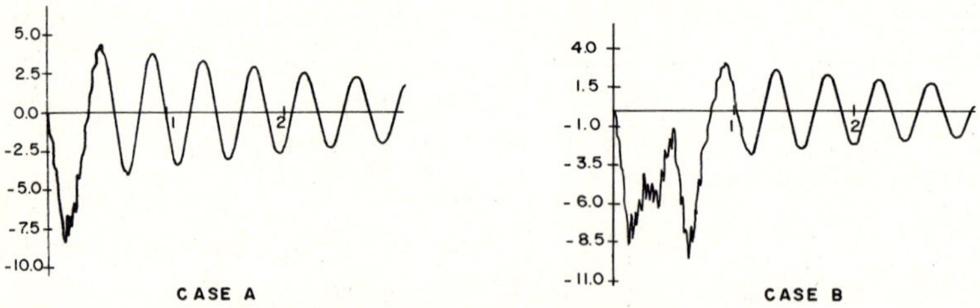

Edge 9.- Transversal bending moment

FIGURE 7.- DYNAMIC UNIFORM LOADING

In figure 6 the first natural frequencies and corresponding modes at central span cross-section for the symmetrical and unsymmetrical vibrations are shown. These values can be compared with the ones obtained by the elementary beam theory as follows:

Flexural mode: $f_F = \frac{\quad}{2l^2}\left(\frac{EI}{A}\right)^{1/2} = 2.35 \text{ sec}^{-1}$

Torsional mode: $f_T = \frac{1}{2} \frac{GJ}{I_p l^2} = 7.62 \text{ sec}^{-1}$

The corresponding results given by the shell theory are: f_F=2.10 and f_T=7.30 sec^{-1}.

If the dynamic load is deterministic known a time history analysis can be carried out. In figure 7 the results for a deflection at node 1 and the bending moment at node 9, are given.

If a vertical earthquake loading is known as a pseudovelocity response spectra (figure 8) the results obtained by application of the root-mean square (RMS) and sum of absolute values (ABS) rules are for two nodes:

	RMS	ABS	
Node 1: Vertical deflection:	0.164	0.165	m
Node 9: Transversal moment:	0.95	1.25	mt/m

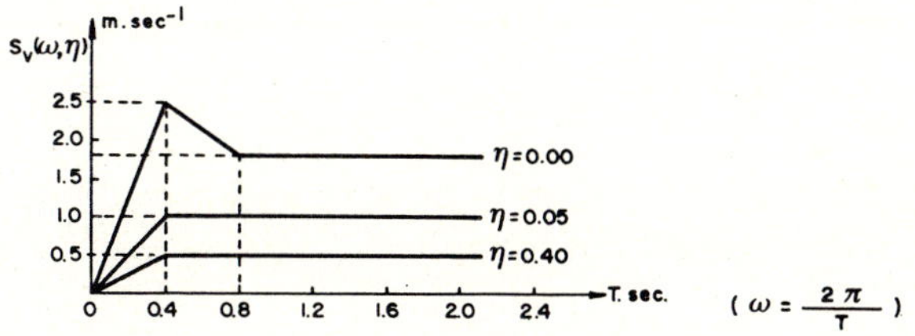

FIGURE 8.- PSEUDO-VELOCITY RESPONSE SPECTRA

7.- CONCLUSIONS

The Levy solution of analysis of translational shells can be extended to dynamic situations. However important difficulties (large computer runs and numerical inestabilities) may appear in this approach. An alternative method based also in the Levy solution technique is presented . The method can efficiently handle rather general and complicated systems of shells structures.

8.- REFERENCES

(1) GOLDENWEIZER, A.L. "Theory of Elastic Thin Shells". Pergamon Press, Oxford, 1961 (English translation of 1953 Russian Edition).

(2) VLASOV, V.Z. "General Theory of Shells and its Application in Engineering". N.A.S.A. TTF -99, 1964.

(3) SAMARTIN, A. and MUNRO, J. "Dynamic Analysis of Translational Shells". Imperial College. CSTR, 67/2. October 1967.

(4) BOUMA, A.L. "Some Applications of the Bending Theory Regarding Doubly-curved Shells". International Union of Theoretical and Applied Mechanics. Proceedings, Symposium on the Theory of Thin Elastic Shells, Delft, 1959.

(5) MISHONOV, M. "On the Theory of Shallow Shells". 691, 1958.

(6) OLIVEIRA, J.D.G. "Dynamic Analysis of Translational Shells". Ph. D. Thesis University of London. June, 1972.

(7) WARBURTON, G.B. "The Dynamical Behaviour of Structures". Pergamon Press, 1964.

(8) CLOUGH and PENZIEN "Dynamic of Structures". McGraw Hill. 1975.

(9) SAMARTIN, A. and MARTINEZ, J. "Dynamic Analysis of Translational Shells". IASS Congress. Alma Ata. Sept. 1977.

DYNAMIC EXPERIMENTS AND EARTHQUAKE OBSERVATION OF RETICULATED SINGLE-LAYER DOMES

T. Tanami and Y. Hangai
Institute of Industrial Science
University of Tokyo
7-22-1 Roppongi, Tokyo 106, JAPAN

1. INTRODUCTION

The importance of considering the dynamic treatments of spatial structures at the time of design has been pointed out from the various points of view[1], based on, for example, the numerical results of large seismic responses for a spherical shell due to vertical earthquake motions[2]. Hence it is desirable to accumulate more materials for a better understanding of the dynamic behaviours of spatial structures.

Based on the above considerations, the purpose of the present paper is to investigate the dynamic behaviours of reticulated single-layer shallow domes by means of response observations of a reticulated single-layer model under the natural earthquakes as well as the free vibration tests and the shaking table tests for another model with the same configuration and dimensions.

2. DOME MODELS

Fig.1 gives the layout of the model, which is a reticulated single-layer shallow dome with triangular unit. The model, with the base diameter(a) of 3m, is composed of 156 members connecting at 61 joints. The rise (h) and the rise-to-span ratio (h/a) of the model are 40.2cm and 0.134, respectively. The member lengths are distributed into 7 kinds between 34.2cm and 46.6cm, and the mean member length is 37.8cm. All the members are of constant cross section and are pin-connected at their joints (see Fig.2), which are assumed not to transmit any bending moments or torques in the numerical analysis.

As properties of the member, which is made of brass, the cross-sectional area, the moment of inertia of cross-section, the specific gravity and Young's modulus are $1.13cm^2$, $0.102cm^4$, 8.47 and $0.73 \times 10^6 kgf/cm^2$, respectively.

3. FREE VIBRATION TESTS

Fig.3 shows the first twelve natural frequencies and the corresponding mode shapes, which are obtained by the free vibration analysis, in which the model is idealized as a space truss. The first two mode shapes are asymmetrical and the first axisymmetrical mode shape appears in the third. Fig.4 shows a histogram of distribution of natural frequencies. In this figure, it is seen that a lot of natural frequencies exist in the vicinity of 100Hz.

Let us proceed to the free vibration test. A model is set on the square base (3.5m x 3.5m), which is composed of I-shaped steel members, with the height of 30cm, being on the shaking table. Fig.5 shows the installed positions of 11 accelerometers. One (ch.1) is on the shaking table for recording the input acceleration, and three accelerometers (ch.2 to ch.4) are set on the base. The remainders, which are installed at the joints, are used for recording time histories of response acceleration in the normal direction.

Free vibration was caused by cutting a string which supported the weight of 8kgf at the crown (ch.5) or the eccentric position (ch.8) as shown in Fig.5 by circles. These two types, which may be used for generating the axisymmetrical and the asymmetrical mode shapes, are represented by "Central Load" and "Eccentric Load", respectively, in the following.

Figs.6 and 7 denote the time histories for "Central Load" and "Eccentric Load", respectively. As shown in Fig.6, the accelerations of the central joint (ch.5) and the surrounding joints move the contrary direction, that is, have the inverse phases. This means that the mode shape is axisymmetric in the case of "Central Load", and it's understood to be combination of the axisymmetric mode shapes shown in Fig.3. In the case of "Eccentric Load", ch.7 and ch.9 have the same phases to ch.8, but in the contrary, ch.6 and ch.10 have the inverse phases to ch.8. From this relation, it can be seen that this mode shape is asymmetric.

Figs.8 and 9 shows the Fourier spectra of response acceleration and spectrum ratios based on the Fourier spectra of the loading positions, at which the weight is supported before vibration, for both the cases of "Central Load" and "Eccentric Load", respectively. As compared with the Fourier spectra of the loading positions, in the case of "Eccentric Load", the prominent components exist in the neighbourhood of 40Hz, which is lower than the result of 60Hz in "Central Load". By the comparison with the numerical results given in Fig.3, these components (40Hz and 60Hz) are considered to correspond to the frequencies of local vibration, that is generated by the forced deformation for the free vibration. The spectrum ratio in the case of "Central Load" shows that the axisymmetrical component exists in the domain of over 90Hz, which corresponds to the natural frequency of the third mode shape in Fig.3. In the case of "Eccentric Load", one of the prominent components exists between 70Hz and 80, which are considered to represent the natural frequencies of asymmetric modes.

4. SHAKING TABLE TEST

Shaking table tests are carried out in order to investigate the earthquake responses of the same model by using the up-and-down excitations of various earthquake records. In this chapter, let us describe the results used the natural earthquake wave recorded at the Chiba Experiment station of the Institute of Industrial Science, University of Tokyo on October 4, 1985 (see[3]). when exciting, the time scale of input data is condensed in one-fifth of real time.

Fig.10 shows the time histories of the acceleration, in which ch.1 represents the input acceleration on the shaking table. Since the base is also a structure, the input acceleration is slightly amplified (see ch.2 and ch.4). Hence, we considered the acceleration on the base as the input excitation for the model. The maximum accelerations on the base and the model are 705gal and 2330gal respectively. Hence the amplification ratio is 3.3 for this input.

Fig.11 and 12 give the spectra and the spectrum ratios, obtained by making the ratio of each spectrum to ch.2 on the base, respectively. The spectra have the two well-separated peaks in the vicinity of 40Hz and 80Hz, which corresponds well to the results given in Fig.9. But, the peaks of the spectrum ratios may be considered to show the posititions of natural frequencies of the model. If we observe Fig.12, these spectra have roughly the characteristics of both spectrum ratios given in Figs.8 and 9.

5. RESPONSE OBSERVATION UNDER REAL EARTHQUAKES

A photograph in Fig.13 shows the model for the response observation under the natural earthquakes, which is set on the second floor of a reinforced concrete tower. The tower is located at the Chiba Experimental Station, and has been used for the observation of dynamic soil-structure interaction during the earthquakes [3].

The observation of response accelerations of the model was begun on March 15, 1986 by using 9 accelerometers. Data can automatically be recorded by digital recorders at a sampling interval of 1/200 second, when an earthquake occurs.

6. CONCLUDING REMARKS

In order to investigate the dynamic behaviours of reticulated single-layer domes, both free vibration tests and shaking table tests were tried by using the small model. Tests show that the model has the

asymmetrical and the symmetrical mode shapes in the vicinity of 75Hz and over 90Hz, respectively. And it is shown that these values correspond to the numerical results obtained by the free vibration analysis. Also, it is obtained from the shaking table tests that the input acceleration is amplified by about 3.3, in which the time scale of input data is condensed in one-fifth of real time.
In order to accumulate more data under the real earthquakes, the earthquake observation of a reticulated single-layer model was planned. In this paper, the outline of this plan was introduced.

REFERENCES

[1] IASS Working Group on Spatial Steel Structures, "Analysis, Design and Realization of Space Frames, a State-of-the-art Report," Bulletin of the IASS, h. 84/85, Vol.XXV-1/2, 1984.

[2] Kato, S., Yoshikawa, K. and Nagahama, T., "A Note on the Seismic Response of a Spherical Shell due to U-D Motions," Research Report No.5 of Structural Division, School of Regional Planning, Toyohashi University of Technology, 1981 (in Japanese).

[3] Yamagami, T. and Hangai, Y., "Observation of Dynamic Soil-Structure Interaction of Reinforced Concrete Tower," Bulletin of Earthquake Resistant Structure Research Center, Institute of Industrial Science, University of Tokyo, No.19, 1986.

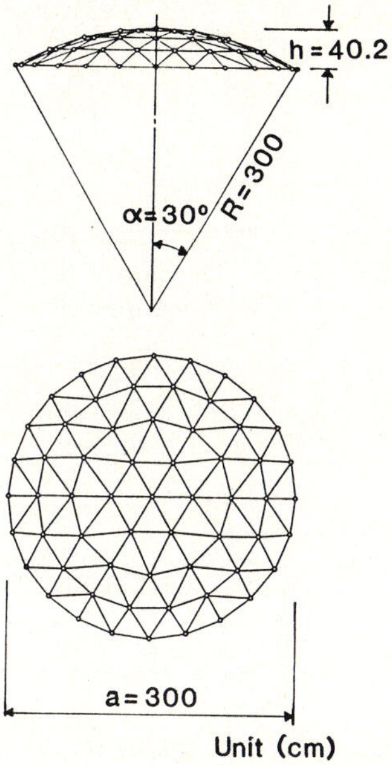

Fig.1:
Reticulated Single-Layer Model

(a) TEST MODEL

(b) JOINT

Fig.2: Photograph of Model and Joint

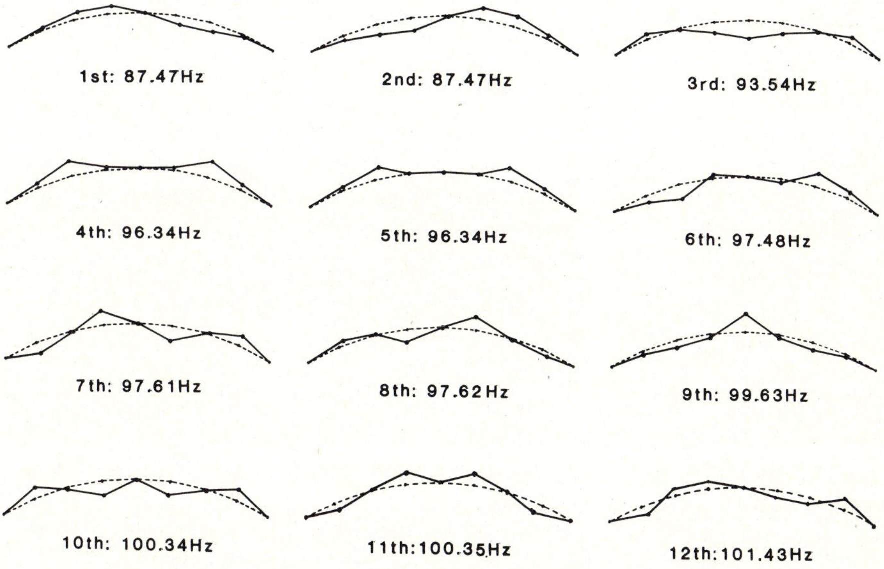

Fig.3: Natural Frequencies and Mode Shapes

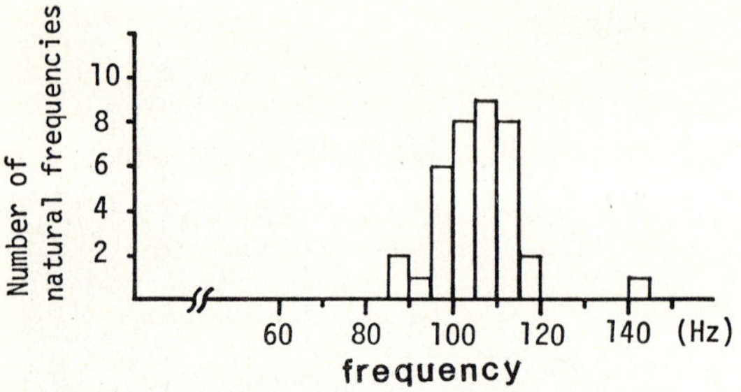

Fig.4: Histogram of Distribution of Natural Frequencies

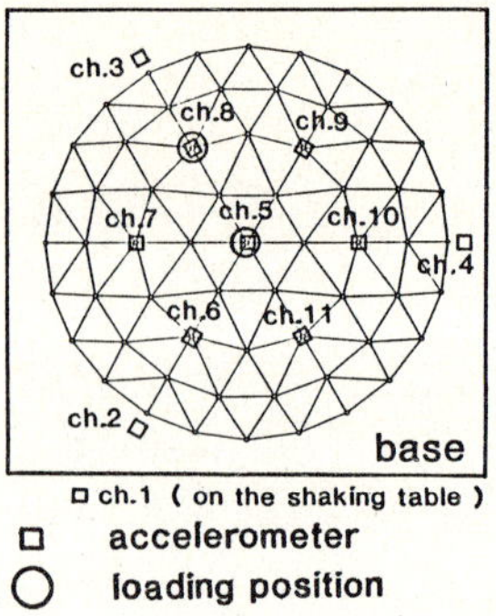

Fig.5: Installed Positions of Accelerometers

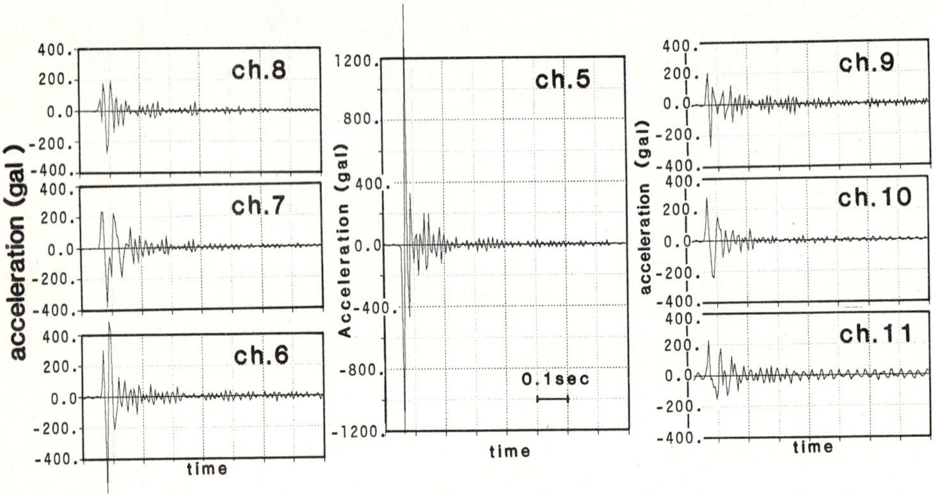

Fig.6: Time Histories of Acceleration in the case of "Central Load"

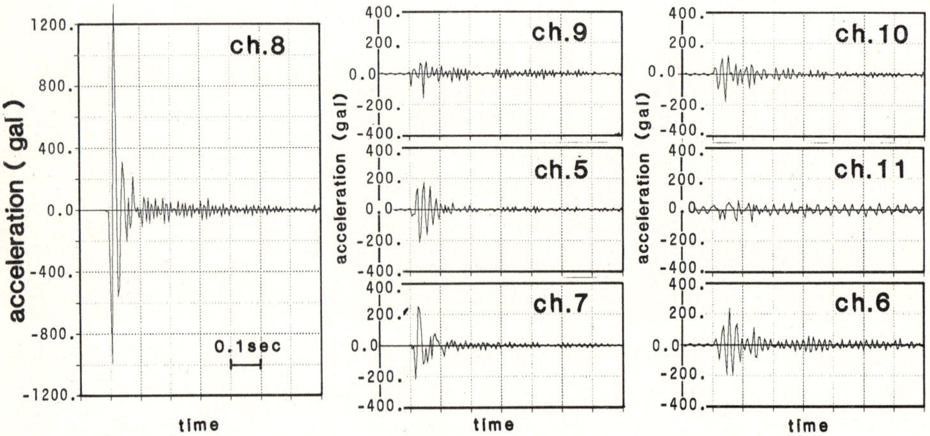

Fig.7: Time Histories of Acceleration in the case of "Eccentric Load"

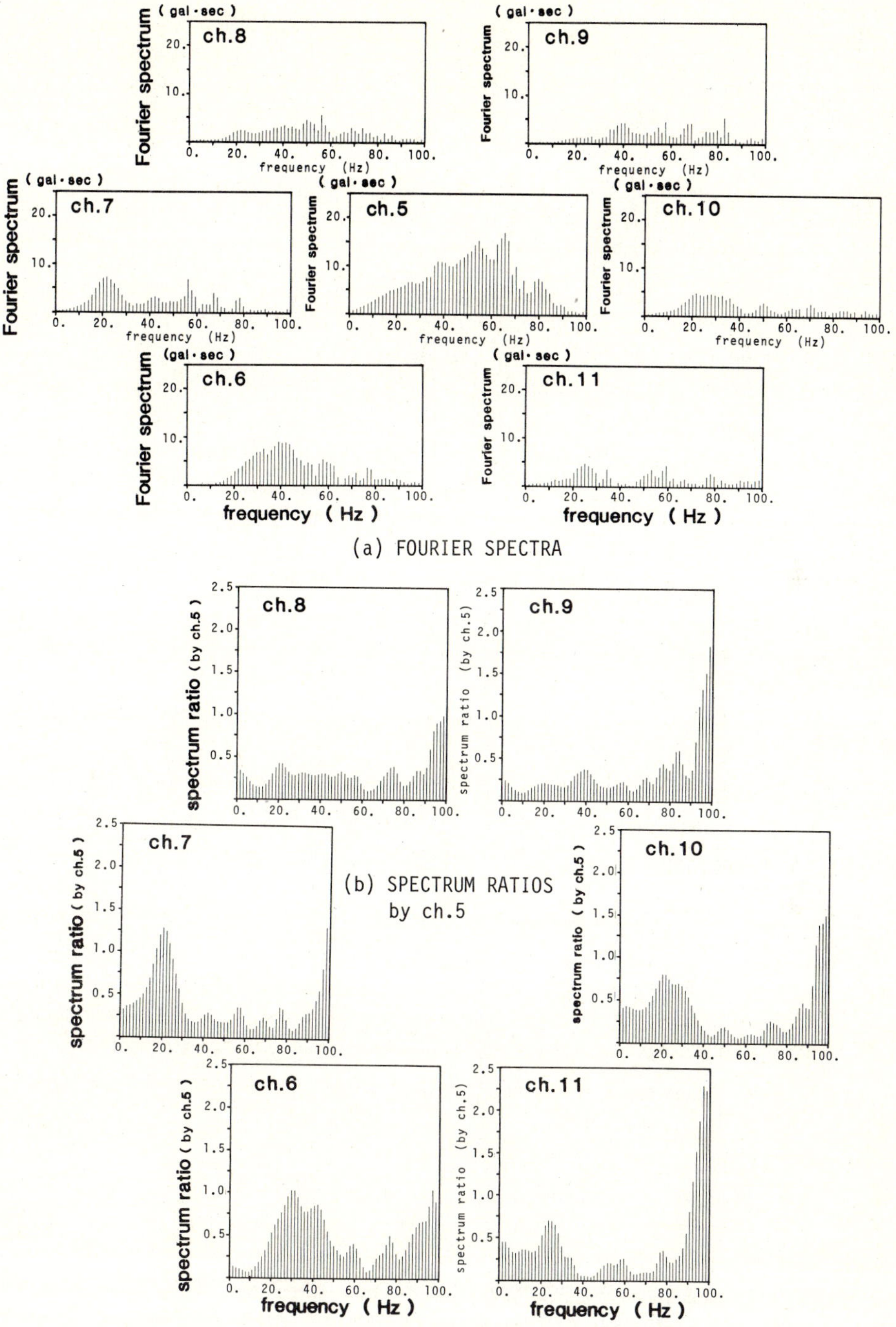

Fig.8: Fourier Spectra and Spectrum Ratios in the case of "Central Load"

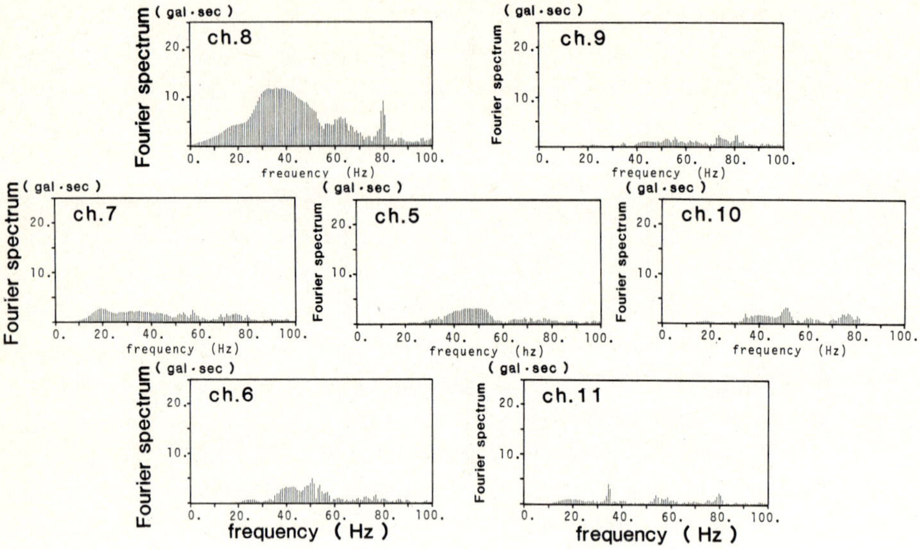

(a) FOURIER SPECTRA

(b) SPECTRUM RATIOS by ch.8

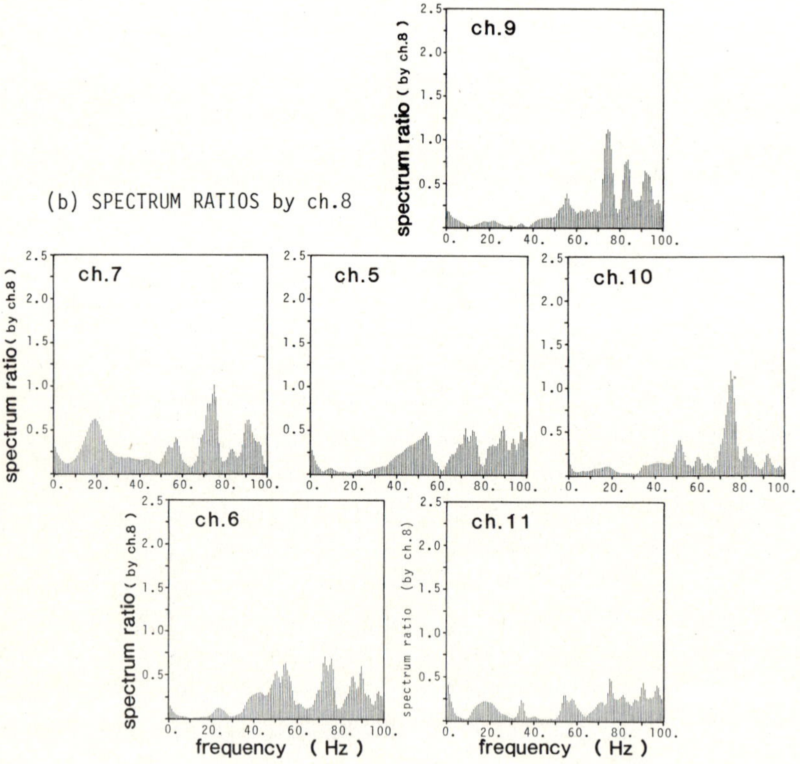

Fig.9: Fourier Spectra and Spectrum Ratios
in the case of "Eccentric Load"

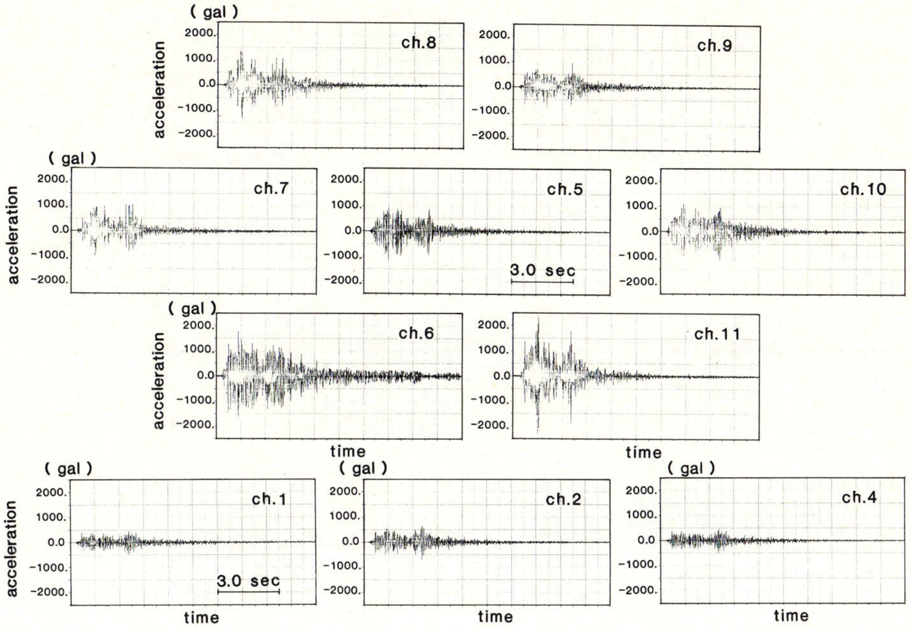

Fig.10: Time Histories of Acceleration in the Shaking Table Test

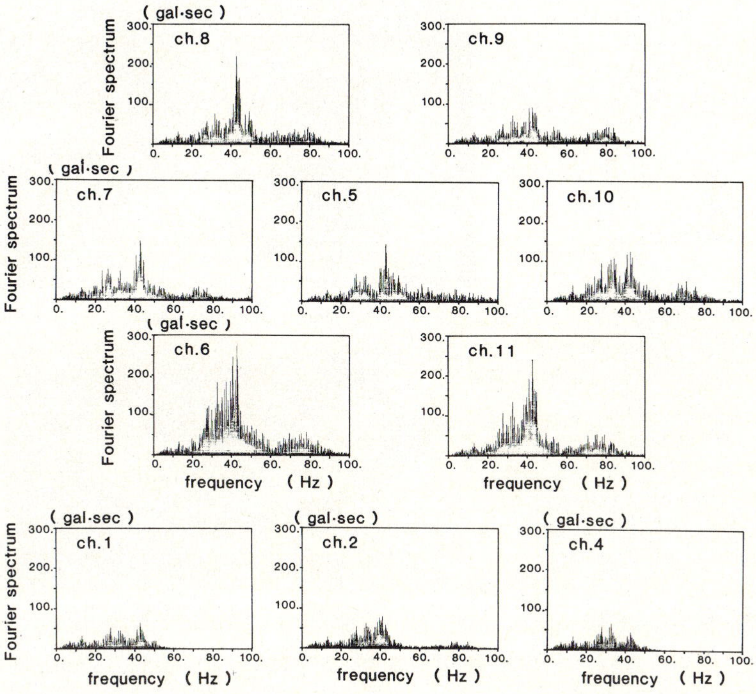

Fig.11: Fourier Spectra in the Shaking Table Test

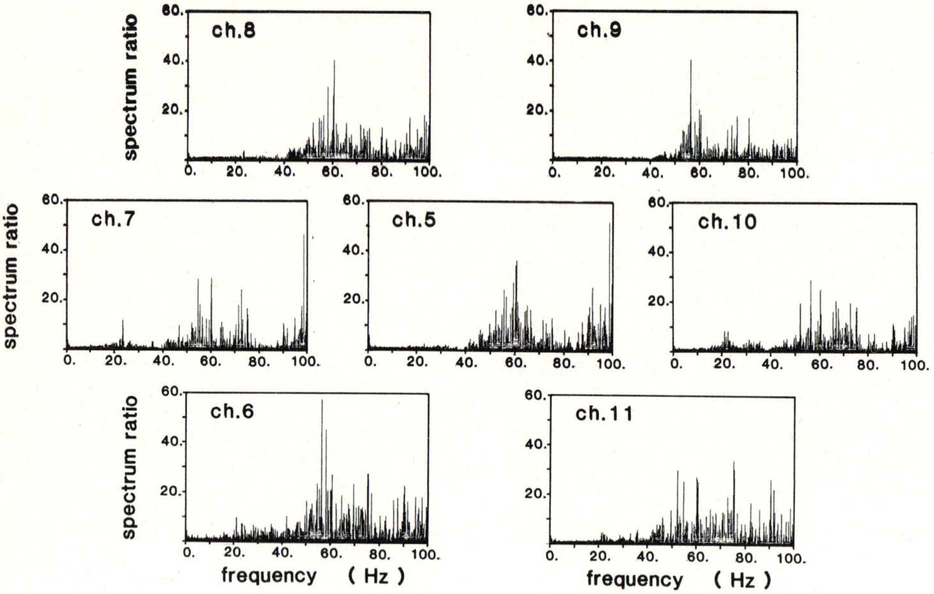

Fig.12: Spectrum Ratios in the Shaking Table Test

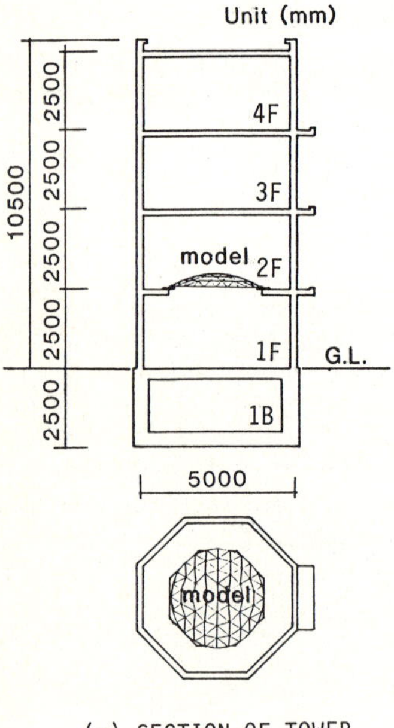

(a) SECTION OF TOWER

(b) PHOTO OF TOWER

(c) PHOTO OF OBSERVATION MODEL

Fig.13: Model installed on the Second Floor of R. C. Tower

ATMOSPHERIC TURBULENCE AND ITS CONSEQUENCES ON THE DYNAMIC BEHAVIOUR OF SHELL STRUCTURES

J. Monbaliu
K.U.Leuven, Hydraulics Laboratory
de Croylaan 2
3030 HEVERLEE - BELGIUM

1. INTRODUCTION

It is well known that wind loads on structures should not be underestimated. The collapse of many structures, e.g. tall masts, chimneys, oil storage tanks [1] and cooling towers [2] only give an indication that the process of wind loading is not fully understood. To predict the behaviour of a structure, the structural parameters as mass, stiffness and damping are not sufficient. Also knowledge about the nature of the loading conditions is necessary. In the case of wind loading, a first step towards understanding is looking at the vertical variation of the mean and fluctuating component of the wind speed. Then, the turbulent spectrum of the wind provides a description of the frequency composition of the load input.
In a second step, knowing the wind climate and therefore the design velocity at the site of the construction, the wind loading condition can be estimated. A quasi-steady approach is most often used to get pressure fluctuations out of ambient velocity fluctuations. Although this way of working has considerable success, there are limitations to its applicability. The wind-structure interaction complicates the understanding of the load input. For many constructions, wind tunnel simulation is still the only feasible way of determining the wind loads. But also here it is very important that the modelling is done with appropriate scaling parameters.

2. DESCRIPTION OF THE WIND FIELD

The wind velocity is composed of a mean velocity component $\bar{u}$ and fluctuating components u', v', w'. The averaging time for the mean velocity is from a meteorological point of view about 1 hr. For engineering purposes shorter averaging times are often used (e.g. 10 minutes). Strong winds are normally of the greatest interest for structural applications. In this case one can assume that the thermal effects in generating the turbulence are small compared to the mechanical (friction) effects [3].
For the vertical variation of the wind speed in uniform terrain Harris & Deaves recommend [5] :

- for the mean component : $\bar{u}(z)/u_* = k^{-1}\,[\ln(z/z_o) + a_1\,(z/h)]$ (1)

$$h = u_*/\beta f \qquad (2)$$

where z = height above the ground

$\overline{u}(z)$ = mean wind speed at height z

k = von Karman constant (= 0.4)

u_* = friction velocity

z_o = roughness length

a_1 = 5.75 : constant obtained from experiments

h = height of the boundary layer

β = 6 : constant

f = coriolis parameter ($2\,\omega \sin\lambda$ where λ = latitude and ω rotational speed of the earth)

- for the turbulent component : $\sigma_u(z)/u_* = 2.63\,\eta[0.538 + 0.090\ln(z/z_o)]^{\eta^{16}}$ (3)

where $\sigma_u(z) = \sqrt{T^{-1}\int_o^T u'^2(z)\,dt}$ and $T \simeq 1\text{hr}$

$\eta = (1 - z/h)$

The height of the boundary layer h in eq. (2) corresponds with the depth of the mixed layer defined in [3] for strong winds. Within this layer atmospheric properties are well mixed. Values for the roughness length z_o can be found in the ESDU publication [6], also taken over in [3]. Combination of eq. (1) and eq. (2), for assumed value of z_o and known velocity $\overline{u}(z)$, will give a velocity profile.
Eq. (3), although complex in appearance, implies that the turbulence intensity decreases almost linearly with height unless you are close to the ground.
For strong winds, below 150 m of height, eq. (1) and eq. (3) can in practice be simplified to [4] :

$$u(z)/u_* = k^{-1}\ln(z/z_o) \tag{4}$$

and

$$\sigma_u(z) = 2.5\,u_* \tag{5}$$

Not only the vertical variation of the wind speed, but also the frequency composition of the turbulent eddies in the wind, is of interest. A link between the frequency (f) and the size of an eddy (L_{ux}) is made through the mean velocity :

$$\overline{u}/f = L_{ux} \tag{6}$$

The most frequently used mathematical formulation for the spectrum of the longitudinal velocity component, are the expressions by Davenport [7] and by ESDU [6] :

Davenport formulation $f\,S_{uu}(f)/\sigma_u^2 = 2\,x^2/3(1 + x^2)^{4/3}$ (7)

ESDU formulation $f\,S_{uu}(f)/\sigma_u^2 = 4\,n_u/(1 + 70.8\,n_u^2)^{5/6}$ (8)

where $x = L\,f/\overline{u}(z)$ and $n_u = L_{ux}\,f/\overline{u}(z)$

Note that $\sigma_u^2(z) = \int_0^\infty S_{uu}(f)\,df$

The length scales L and L_{ux} can be determined from the frequency or wavelength of the peak of the spectrum. For measured spectra this peak is not always very well defined because of the greater statistical uncertainty of the low frequency components. It is therefore far easier to fit the inertial subrange, which has a -2/3 slope and extend this line to where $f\,S_{uu}(f)/\sigma_u^2 \equiv 1$ (on a log-log scale) [8]

$$L = 0.544\ (\bar{u}(z)/f)_1 \text{ and } L_{ux} = 0.039\ (\bar{u}(z)/f)_1 \tag{9}$$

An illustration of this procedure is given in Figure 1 (from [8])

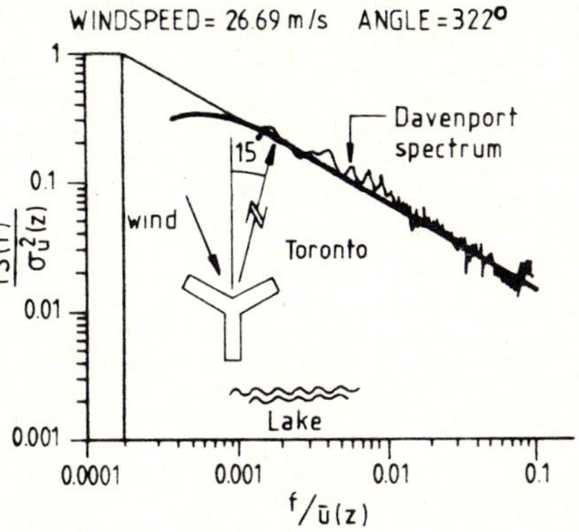

Fig. 1 : Example of velocity spectrum

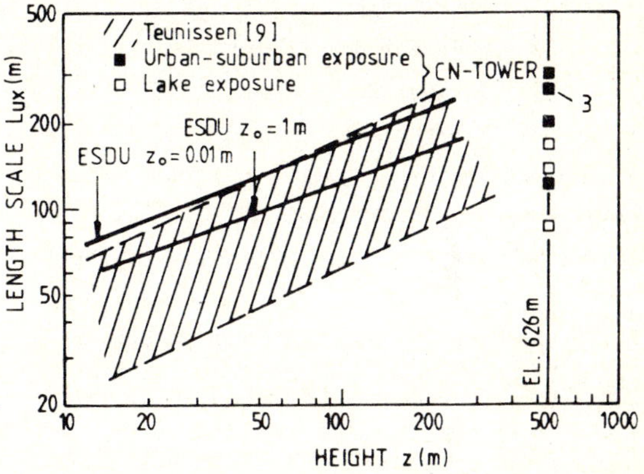

Fig. 2 : Compilation of longitudinal length scales of turbulence

This gives L = 2700 m and L_{ux} = 200 m. Physically L_{ux} can be seen as the eddy size where most of the energy is concentrated. Therefore it will be an important parameter in relation to the size of the structure under consideration.

A compilation of the variation of the longitudinal length scale of turbulence with height was done by Teunissen [9], here given in Figure 2. Added to Teunissen's work are measured length scales at the top of the CN-Tower in Toronto, Canada collected by the author [8]. Also the suggested values by ESDU [6] are given.

Other heights than the top level height at the CN-Tower were instrumented, but for those heights the interference of the tower was clearly significant [10]. Because the measured length scales at those heights are not representative for the ambient longitudinal velocity spectrum, they are not included here.

There is one more thing which needs attention and that is the spatial correlation of the turbulence. It is obvious that the correlation is equal to unity if point 1 coincides with point 2. And it is also not difficult to see that when point 1 and 2 are very far apart, the correlation tends to zero. In general, the correlation function can be seen as composed out of individual frequency components. Then a narrow band correlation can be written as [7] :

$$R_{12}(f) = S_{12}(f)/\sqrt{S_{11}(f)\,S_{22}(f)} \tag{10}$$

The squared coherence (Coh) is here defined as the sum of squares of the in-phase and out of phase components of the above correlation function [7]. The square root of the above definition is then used to indicate the coherence. The model normally used for the squared coherence is an exponential function. Values for the squared coherence are [3] :

- vertical separation : $Coh\ (f, \Delta z) = \exp\ (-a_z\ f\ \Delta z/\bar{u})$
- lateral separation : $Coh\ (f, \Delta y) = \exp\ (-a_y\ f\ \Delta y/\bar{u})$ (11)

where $a_z = 12 + 11\ \Delta z/\bar{z}$

$a_y = 12 + 11\ \Delta y/\bar{z}$

$\bar{z}$ = mean height

Δz = vertical separation

Δy = lateral separation

Up to now only the longitudinal velocity component has been described. Although the lateral and vertical component of the velocity are important as well for the dynamic excitation of structures (e.g. chimneys, long span bridges), it is outside the scope of this paper to comment upon them. The reader is herefore referred to [3].

3. FORMULATION OF THE STRUCTURAL BEHAVIOUR

Constructions, such as tall buildings, chimneys or cooling towers, experience aerodynamic forces generated both by the upstream flow impinging on the structure, as well as by the wake behind the structure. In addition, aeroelastic forces may be generated by the motion of the structure through the air [11]. A linear elastic system can be simulated mathematically :

$$M\ddot{X} + C\dot{X} + KX = F_T\ (t) + F_W\ (t) + A\ (X, \dot{X}, \ddot{X}, \ldots) \tag{12}$$

where M = mass parameter

C = damping parameter

K = stiffness paremeter

$F_T\ (t)$ = turbulent forces in the oncoming flow (function of time)

$F_W\ (t)$ = forces in the wake (function of time)

$A\ (X, \dot{X}, \ddot{X}, \ldots)$ = forces due to aeroelastic effects

$X, \dot{X}, \ddot{X}$ = vectors of displacement, velocity and acceleration

Eq. (12) is quite general and may, in principal, define a MDOF system. Eq. (12) then represents a set of coupled differential equations and the response must be obtained by integrating these equations simultaneously. However, after modal decomposition, uncoupled equations are considered. And to make it even easier a SDOF system is taken. It is more convenient to work in the frequency domain and eq. (12) can be written as [11] :

$$\left[\left(1 - \frac{\omega^2}{\omega_o^2} - a^2\right) + 2\ i\ \left(\frac{\omega}{\omega_o}\right)\ (\zeta_s + \zeta_a)\right]\ X\ (\omega) = \frac{F_T\ (\omega) + F_W\ (\omega)}{K} \tag{13}$$

where $\omega_o = \sqrt{K/M}$

$\omega = 2\pi f$

$a^2 = A(\omega)/K$: aerodynamic stiffness ratio; is normally small compared to unity and therefore neglected

$\zeta_a = -A^{*}(\omega)/2M\omega_o$: aerodynamic damping

$\zeta_s = C/2M\omega_o$: structural damping

To obtain eq. (13), the aeroelastic forces A $(X, \dot{X}, \ddot{X}, \ldots)$ are assumed to be a linear combination of X and $\dot{X}$. This is justified when the amplitudes of motion are small. $A(\omega)$ and $A^{*}(\omega)$ then define the in-phase and out of phase component of the motion dependent forces. Let us now investigate the different terms in eq. (13) in relation to the wind field described above.

3.1. Responce due to turbulence in the approaching flow (F_T)

To keep the discussion short, only the along wind direction will be considered. To simplify the matter even more, consider an axisymmetric construction (e.g. oil storage tank, ...) and describe it by z and θ (Figure 3)

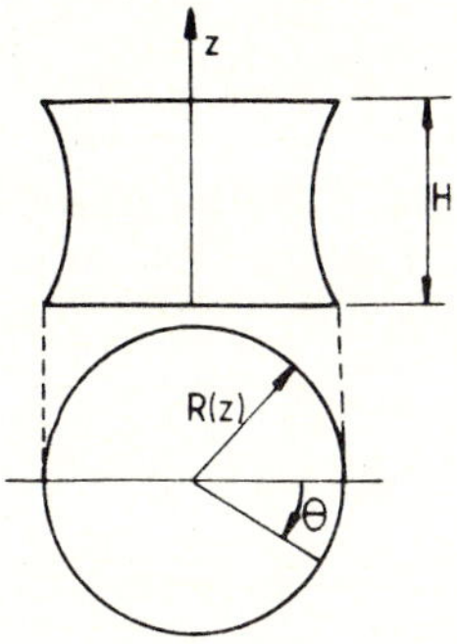

Fig. 3 : Used notations for axisymmetric construction

The spectrum of the generalized force in the along wind direction due to the atmospheric turbulence can be written as

$$S_{GF_T}(t) = \int_0^H \int_0^H \int_0^{2\pi} \int_0^{2\pi} S_p(z, z'; \theta, \theta'; f)\, \mu_i(z, \theta)\, \mu_i(z', \theta')\, R^2 \cos\theta \cos\theta' dz dz' d\theta d\theta' \quad (14)$$

where $S_p(z, z', \theta, \theta', f)$: cospectrum of the pressure

$\mu_i(z, \theta)$: assumed mode shape for mode i

It is the cospectrum of the pressure which is quite complicated to deal with. We have to derive if from velocity measurements upstream of the structure and then relate the velocity fluctuations to pressure fluctuations. The most common approach, the quasi-steady approach, is to write the fluctuating pressure p' as :

$$p' = \rho \bar{u} u' C_p(z, \theta)$$

where ρ : air density

$C_p(z, \theta)$: mean pressure coefficient (function of z and θ)

Then the cospectrum of the pressure can be rewritten as

$$S_p(z, z', \theta, \theta'; f) = \rho^2 \bar{u}(z)\bar{u}(z')\, C_p(z, \theta)\, C_p(z', \theta')\sqrt{S_{uu}(z) \cdot S_{uu}(z')}\sqrt{Coh(\Delta z, \Delta\theta, f)}$$

One can then in eq. (15) substitute eq. (4) for $\bar{u}(z)$, eq. (8) for $S_{uu}(z)$ and an equivalent formula to eq. (11) for the Coh.[12] . This can then be put into eq. (14). The calculation of the integral is obviously complicated but, with modern computing techniques, reasonably straightforward. An example of an analysis of a cooling tower random response, using the quasi-steady aerodynamic theory and Davenport's spectrum, is given in [12]. A finite element formulation was adopted and both the direct matrix inversion and the modal superposition method were used.

Unfortunately, there are a few complications. Can we relate the ambient velocity fluctuations to pressure fluctuations ? How do we determine the pressure coefficients C_p ? Are the C_p values independent of wind speed ? How reliable are wind speed predictions ? What about vortex shedding ? Some of these questions will be commented upon in the rest of this paper.

It is probably the easiest to start with the pressure coefficients. They relate the mean wind speed to the mean pressure. Ideally the C_p values should be obtained from full-scale experiments [13] . However full-scale measurements are expensive, time consuming and a lot of technical difficulties are involved. Therefore most C_p values are obtained from wind tunnel experiments. It might be worth noting that in case of an open construction, such as a cooling tower, the inside pressure coefficient has to be subtracted from the outside pressure coefficient to obtain the C_p value used in the equations. Inside pressure coefficients are essentially constant and negative, but can vary in magnitude depending on the circumstances (i.e. venting, flow or no flow inside ...). Also the dependence of the pressure coefficients of structures with circular cross-section on Reynolds number (i.e. velocity) is well known [14].

For the relationship between velocity fluctuations and pressure fluctuations, it seems reasonable to assume that on the windward face of the structure the pressure fluctuations follow the velocity fluctuations. This is so in the case of large scale of turbulence ($L_{ux} >> 2R$). The longitudinal fluctuations are seen as changes in the mean wind speed, and lateral fluctuations are seen as changes in the wind direction. Hunt [15] comes to the same observation that at low frequencies and large scale of turbulence, the pressure spectrum should follow the velocity spectrum. At high frequencies or L_{ux} the same order or smaller than 2R, the fluctuating pressure should fall more rapidly than the upstream turbulence. An indication of the fall in the fluctuating stagnation pressure on a circular cylinder when $L_{ux}/2R$ decreases is given in [16] and sketched here in Figure 4.

So, at higher frequencies the dynamic load on the structure will be exagerated. For angles away from the stagnation zone (say $\theta = 40°$), the quasi-steady theory fails to

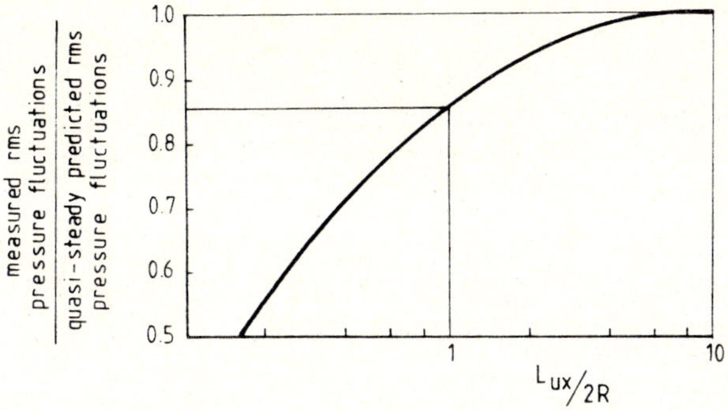

Fig. 4 : Comparison between measured and quasi-steady predicted pressure fluctuations

predict pressure fluctuations from the knowledge of ambient turbulence (for an illustration on a circular cylinder, see [16]). It is only when the length scale of turbulence L_{ux} is greater than about 10 times the diameter and where the fluctuations due to incident turbulence dominate those due to vortex shedding, that quasi-steady theory yields a satisfactory result. This is illustrated in [16] by the comparison between measured fluctuating pressure distribution and the quasi-steady theory on a reinforced concrete chimney. The aspect ratio was about 10, $L_{ux}/2R$ about 7.5 and relatively weak vortex shedding was suggested by the pressure spectra. To conclude, one can say that when the diameter of the structure is smaller than 20 m, the limitations of the quasi-steady theory are minimal.

3.2. Response due to turbulence in the wake (F_W)

In the wake of the structure, the effect of the incident turbulence appears to be quite weak. Pressure fluctuations are in a first approximation independent of the incident turbulence [16]. Whithout explicitely stating, the self induced turbulence due to vortex shedding and separation of the flow has been introduced. Although this self induced turbulence is affected by the incident turbulence, the incident turbulence cannot be seen as the cause of it. The turbulence in the wake is not always as well organized as a regular vortex shedding pattern may suggest, but a force in the along wind direction at a frequency about twice the vortex shedding frequency is sometimes noticable. However, across wind vibration and ovaling of e.g. chimneys, are well known.

3.3. Aerodynamic damping (ζ_a)

One term in eq. (13) was introduced because of the motion of the body. It was called the aerodynamic damping term because the forces connected to it are proportional to the velocity. Along-wind aerodynamic damping has been shown to agree quite closely with theoretical estimates based on quasi-steady aerodynamics [7]. For a given mode it can be expressed as (for very small amplitudes of vibration) :

$$\zeta_a = \frac{\pi f_i \int_0^{2\pi} \int_0^H [\rho\, C_p(z,\theta)\, \bar{u}(z) + \frac{\rho\, \bar{u}^2(z)}{2} \frac{\partial C_p(z,\theta)}{\partial u(z)}]\, \mu(z,\theta) \cos\theta\, R\, dz\, d\theta}{(2\pi f_i)^2 M_i} \qquad (16)$$

Normally the term $\frac{\rho\, \bar{u}^2(z)}{2} \frac{\partial C_p(z,\theta)}{\partial \bar{u}(z)}$ can be ignored except when the C_p values change rapidly with velocity (think on the magnitude of the drag coefficient of a circular cylinder in subcritical, transcritical ans supercritical Reynolds number range), and when the geometry of the cross-section is dependent on the wind speed (deformable cross-section). In the drag direction, the aerodynamic damping force is opposing the motion and is extracting energy from the vibration of the structure (positive aerodynamic damping). Although this aerodynamic damping can be significant, it is quite often "neglected" because the structural damping is not so well known either.
The response in the lift direction will not be discussed here but it might be of interest that aerodynamic instabilities such as galloping are due to negative damping forces. Due to the motion of the body a force in the direction of the motion is introduced. Also in eq. (16), it is assumed that the flow is smooth. The effect of turbulence is most noticable in the across-wind direction, where the spectrum of the lift forces is broadened and flattened due to an ever changing velocity [11].

4. WIND TUNNEL TESTING

It is by now clear that for light, flexible structures and/or structures of unusual shape, the aerodynamics are not always easy to describe mathematically. Therefore designers will often use a wind tunnel to model the turbulent boundary layer and measure the behaviour of a scaled down version of their structure. In this way it is not only possible to determine the effect of atmospheric turbulence on the dynamic behaviour of the structure, but also the influence of nearby constructions or obstacles can be investigated (see e.g. J. Armitt about the Ferrybridge C power station [2] and B.J. Vickery and A. Daly about the buffeting effect of two closely spaced cylinders [17]). Also, most building codes indicate to test unusual structures in an appropriate simulated flow, unfortunately without specifying what is appropriate.
Surry [18] gives an interesting outline on the consequences of modelling the flow. The reader is hereby referred to this interesting paper, but a few points will be mentioned here. Surry says that the pressures on a structure are function of the far field, the near field and the self field. In the far field simulation, it is relatively easy to model a desired mean speed profile and longitudinal turbulence intensity distributions. Usually only the longitudinal component is modelled, while the other components are assumed to be reasonably correct. However, the main difficulty is modelling the length scale of turbulence, i.e. the energy frequency distribution.

This length scale of turbulence is an important factor when the correlation of loads over large areas is important, e.g. for flexible structures as membrane structure. Mean and local unsteady loads on the other hand, seem to be less sensitive to the modelling of the boundary layer parameters. The effect of the near field (other surrounding structures) or of the self field, can in these cases be dominant.
For the resonant response in the along wind direction, mismatching of scales might distort the dynamic response considerably. Although the overall (mean + dynamic) response might still be sufficiently accurate, the predicted accelerations might be far off. There is little known about the effect of distortion of the scale for the across-wind response, but the indications are that it is at least as sensitive as the along wind response. Therefore, scales of turbulence should be maintained better than a factor 2 for aeroelastic tests [18]. Even wind tunnel testing is not the answer to all problems. A good understanding of the forcing mechanisms prior to testing is necessary. It seems therefore reasonable to incorporate modelling into a total reliability approach. Not only the uncertainty about the wind field, but also the modelling uncertainty can be treated with statistical means [19].

5. CONCLUSION

The atmospheric boundary layer is described by means of the vertical variation of the mean velocity and the turbulence intensity. The frequency composition of the energy in the wind field can be expressed mathematically by the Davenport or ESDU spectrum formulation. The longitudinal length scale of turbulence is an important parameter. The coherence function was introduced so that the spectrum of the generalized force in the along wind direction could be obtained. It was necessary to find the relationship between the fluctuating velocities and the fluctuating pressures. A quasi-steady approach was used and its limitations were discussed. Also, the aerodynamic damping was shown to be significant.
The uncertainty involved in estimating the different parameters, the complexity of the flow-structure interaction and the aerodynamic influence of nearby structures, often force us to rely on wind tunnel simulation techniques to predict the overall loads. But also in the wind tunnel care has to be taken to model the flow accurately, certainly for relatively ligth and flexible structures.

ACKNOWLEDGEMENTS

Discussions with and comments by Prof. J. Berlamont and Prof. G. De Roeck are gratefully acknowledged.

REFERENCES

1. Holroyd, R.J., "On the Behaviour of Open-Topped Oil Storage Tanks in high Winds, part I, Aerodynamic Aspects", Jnl. of Wind Eng. and Ind. Aerodynamics, 12, pp. 329-352, 1983.
2. Armitt, J., "Wind Loading on Cooling Towers", Jnl. of Structural Division, ASCE, Vol. 106, No. ST3, March, 1980.
3. Panofsky, H.A. and Dutton, J.A., "Atmospheric Turbulence Models and Methods for Engineering Applications", John Wiley & Sons, 1984.
4. Panofsky, H.A. "Wind Structure in Strong Winds below 150 m", Wind Engineering, Vol. 1, No. 2, 1977.
5. Harris, R.I. and Deaves, D.M., "The Structure of Strong Winds", paper 4, Wind Engineering in the Eighties, Proc. of the CIRIA conference, Nov. 1980.
6. ESDU (Engineering Science Data Unit), "Characteristics of Atmospheric Turbulence near the Ground", part II, Single Point Data for Strong Winds (Neutral Atmosphere), ESDU Item, No. 74301, ESDU, London, U.K., 1974.
7. Davenport, A.G., "Wind Structure and Wind Climate", Safety of Structures under Dynamic Loading, Trondheim, June 1977.
8. Monbaliu, J., "Full-Scale Observations at the CN-Tower, Toronto, Canada", M.E.Sc. Thesis, Faculty of Engineering Science, University of Western Ontario, London, Ontario, 1985.
9. Teunissen, H.W., "Characteristics of the Mean Wind and Turbulence in the Planetary Boundary Layer", UTIAS Review No. 32, University of Toronto, 1970.
10. Monbaliu, J., Dance, P.G., Isyumov, N. and Davenport, A.G., "Properties of the atmospheric Boundary layer Measured at the CN-Tower, Toronto, Canada", Fifth U.S. National Conference on Wind Engineering, Lubbock, Texas, U.S.A., Nov. 1985.
11. Davenport, A.G., "The Influence of Turbulence on the Aeroelastic Responses of tall Structures to Wind", IAHR/IUTAM, Karlsruhe, 1979.
12. Yang, T.Y. and Kapania, R.K.,"Finite Element Random Response Analysis of Cooling Towers", Jnl. of Engineering Mechanics, Vol. 110, No. 4, April, 1980.
13. Sollenberger, N.J., Scanlan, R.H. and Billington, D.P., "Wind Loading and Response of Cooling Towers", Jnl. of Structural Division, ASCE, Vol. 106, No. St. 3, March, 1980.
14. Isyumov, N., Abu-Sitta and Davenport, A.G., "Approaches to the Design of Hyperbolic Cooling Towers against the dynamic Action of Wind and Earthquake", IASS Symposium on Hyperbolic Cooling Towers, Brussels, May, 1971.
15. Hunt, J.C.R., "Turbulent Velocities near and Fluctuating Surface Pressures on Structures in turbulent Wind", Proc. of fourth Int. Conf. on Wind Effects on Buildings and Structures, pp. 309-320, Heathrow, 1975.
16. Basu, R., "Across-Wind Response of slender Structures of circular Cross-Section to atmospheric Turbulence", Ph.D.Thesis, Faculty of Engineering Science, University of Western Ontario, London, Ontario, 1982.
17. Vickery, B.J. and Daly, A., "Wind Tunnel Modelling as a Means of Predicting the Response of Chimneys to Vortex Shedding", Int. Conf. on Design against Wind-Induced Failure, Bristol, England, Jan., 1984.
18. Surry, D., "Consequences of Distortion in the Flow including Mismatching of Scales and Intensities of Turbulence". Int. Workshop on Wind Tunnel Modelling Criteria and Techniques, N.B.S., Gaithersburg, Maryland, April, 1982.
19. Davenport, A.G., "Contributions to the Reliability of Structures Under Wind Loading", Boundary Layer Wind Tunnel Laboratory, U.W.O., London, Ontario, 1982.

NUMERICAL ANALYSIS OF CONCRETE PEDESTAL OF ANTENNA BUILDING AT LESSIVE, BELGIUM

D. Van Gemert and C. Vanoverschelde
K.U.Leuven - Reyntjens Laboratory
de Croylaan 2

3030 Heverlee - BELGIUM

P. Wouters
B.E.C. n.v.
Louisalaan 228 bus 3

1050 Brussel - BELGIUM

J. Decock
S.A.I.T. Electronics n.v.
Steenweg op Ruisbroek 66

1190 Brussel - BELGIUM

ABSTRACT

The search of an appropriate structure lay-out for the supporting structure of the telecommunications antenna at Lessive is discussed. The combination of radio-electrical, mechanical and aesthetic requirements has led to a framed three-dimensional structure.

A dynamic analysis of the building was executed. The results are compared with experimental results, obtained from a pulsation test on the real structure.

1. INTRODUCTION

Lessive III is the name of a terrestrial telecommunication station at Lessive, Belgium, Fig. 1.

Fig. 1 : Terrestrial station Lessive III

The parabolic antenna of the station must continuously be pointed at the telecommunications satellite to enable a reliable radio-electric contact. For that purpose the movements of the antenna are commanded by two motors, which determine the displacements of the antenna around two orthogonal axes : the elevator axis and the azimuth axis, Fig. 2.

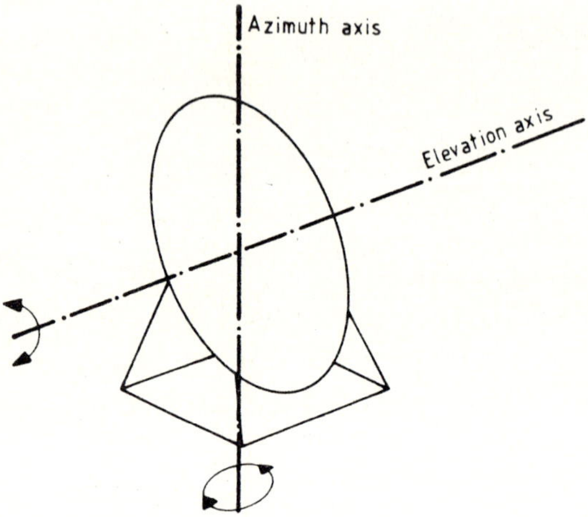

Fig. 2 : Definition of elevator and azimuth axis

The specifications for the deformational behaviour of the building were severe. The most important requirements are summerized hereafter. The static and dynamic loads, executed on the foundation building by the antenna, are listed in Table 1. The antenna has a diameter of 18 m.

Load condition	R_1	R_2	R_3	R_4	V	H	H max
Static	25	25	15	15	4	-	-
130 km/H Wind	31	36	1	12	4	15	-
	10	9	31	30	4	-	31
200 km/H Wind	23	43	0	11	6	8	-
Horizontal Seismic (0.3 g)	50	19	0	9	6	25	-

Unit : Ton

Notes : (1) R_1-R_4 : Vertical loads transmitted through antenna structure wheels (refer to Fig. 3)

(2) V, H : Vertical and horizontal loads on the antenna pedestal center (refer to Fig. 3)

(3) 130 km/H Wind : Antenna survival wind velocity in any position

(4) 200 km/H Wind : Antenna survival wind velocity in stow position

Table 1 : Static and dynamic loads from antenna

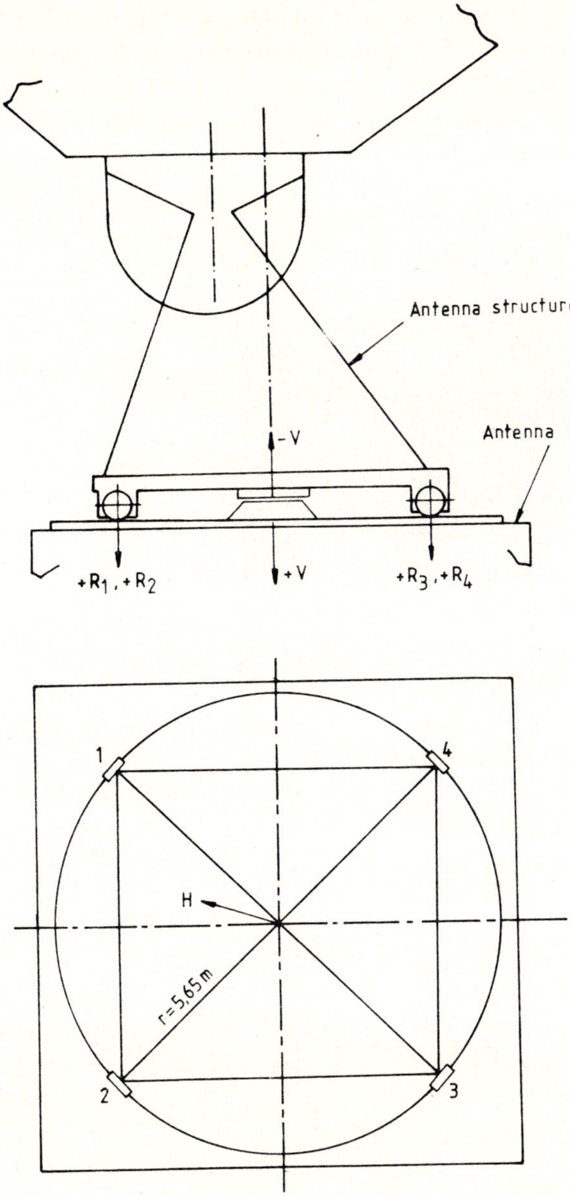

Fig. 3 : Definition of loads on antenna pedestal

The position of the loads is clarified in Fig. 3. Under these loads the vertical deflection at the center of the roof had to be smaller than 2 mm. The lowest natural frequency of the antenna building with soil had to be greater than 2 Hz.
The Lessive telecommunication site has also touristic importance, and therefore the

design office B.E.C. got considerable freedom in the structural conception of the foundation building, in order to make a building with lightness and cosiness, attractive to visitors and tourists.

The designer has chosen for a three dimensional framed structure, composed of six frames, interconnected with rings, Fig. 4.

Fig. 4 : Three dimensional framework concept

In cooperation with the Reyntjens Laboratory, the designer has studied different structural concepts, and through the numerical evaluation of these, the final concept, meeting all the requirements, has grown.

2. FIRST CONCEPT : NO INTERNAL COLUMNS

The dimensions of the framework are given in Fig. 5.

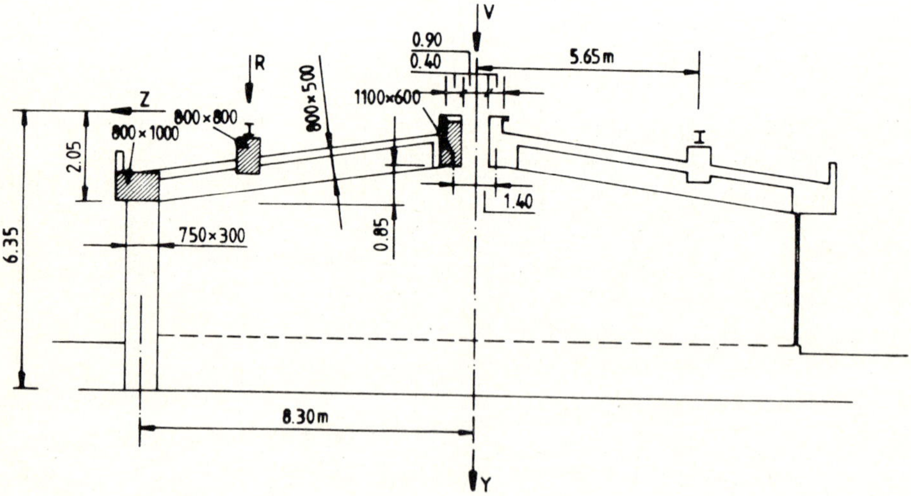

Fig. 5 : Framework without internal columns

A plan of the structure, indicating the most unfavourable positions of the antenna, is given in Fig. 6.

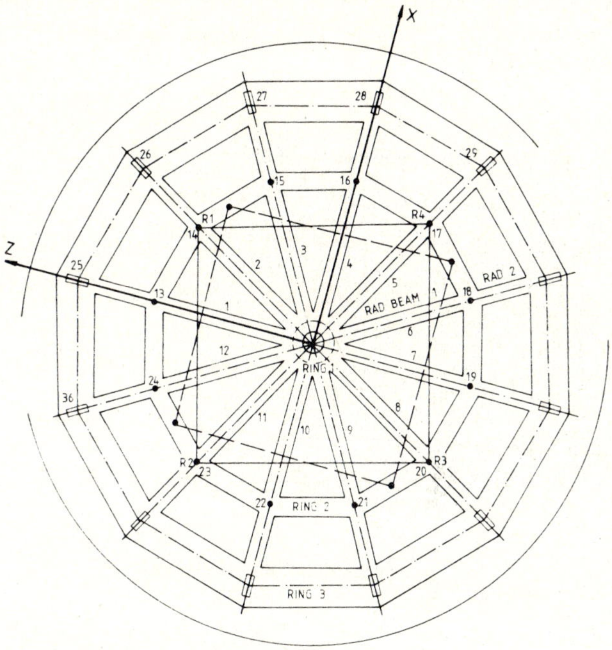

Fig. 6 : Plan of structure

The structure is composed of

- three circular beams : RING 1, 2, 3
- 12 radial beams : RAD 1 and RAD 2
- 12 columns

The analysis was executed with the finite element computer program SAP IV. To check the possibility of a prefabricated construction procedure, where all structural elements could be prefabricated, and assembled on the site, the calculations were made with the connections between the columns and the outer circular beam RING 3 assumed to be either fixed or articulated. The deflections in the respective cases were calculated to be 6,8 mm and 9,5 mm. These deflections were too big to be acceptable for the good functioning of the antenna.

3. SECOND CONCEPT : FRAME WITH INTERNAL COLUMNS

To overcome the excessive deflections without being obliged to make the structure too heavy and clumsy, and as a consequence unattractive to visitors, it was decided to add a row of columns to support the wheel carrying circular beam RING 2. A typical frame element is shown in Fig. 7. The idealised frame is shown in Fig. 8. The eccentricity of the beam RING 2 with respect to the axis of the radial beams 1 and 2 is introduced in the analysis by means of short connection beams. The deflection at the center of the roof was now limited to 1,97 mm. This maximum deflection would occur at a horizontal seismic load on the antenna equal to 0,3 g, combined with the dead weight of the total structure.

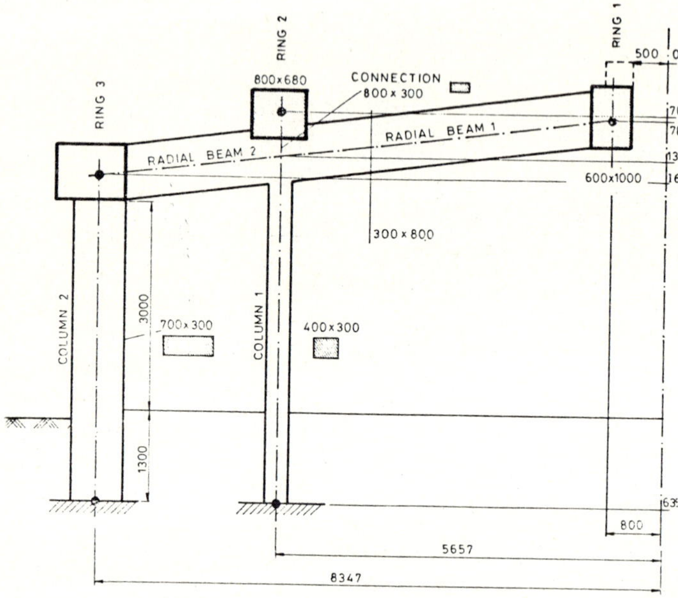

Fig. 7 : Typical frame element

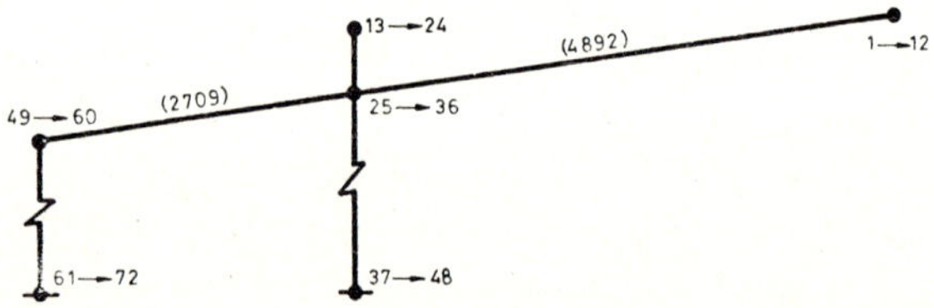

Fig. 8 : Idealised frame

The natural frequency of the foundation structure was calculated as well. Different boundary conditions were applied to get a complete set of significant natural frequencies. The results are summarized in Table 2.

The results indicate that the structure is too flexible in torsion. With a framed structure it was impossible to reach the required lowest natural frequency of 2 Hz.

Assumptions	circular frequency ω (rad/s)	f Hz	T s
1. Idealised structure 3-dim. frame			
1.1. Rotational vibration around vertical Y-axis	0,62	0,10	10,2
1.2. Translational vibration	0,82	0,13	7,62
2. Nodes 49 to 60 fixed in X-Z-plane			
-vibration in Y-direction	3,99	0,64	1,57
3. Torsional stiffness of columns x 10^{10}			
-vibration in Y-direction	3,45	0,55	1,82

Table 2 : Summary of natural frequencies

4. FINAL CONCEPT : FRAMED STRUCTURE WITH STIFFENING WALLS

To meet the required lowest natural frequency of 2 Hz four reinforced concrete diaphragms were introduced into the structure. The diaphragms are placed as indicated in Fig. 9. The thickness of the reinforced concrete walls is 140 mm. Their height was reduced, to leave an opening of 0,75 m for the windows beneath the circular beam RING 3.

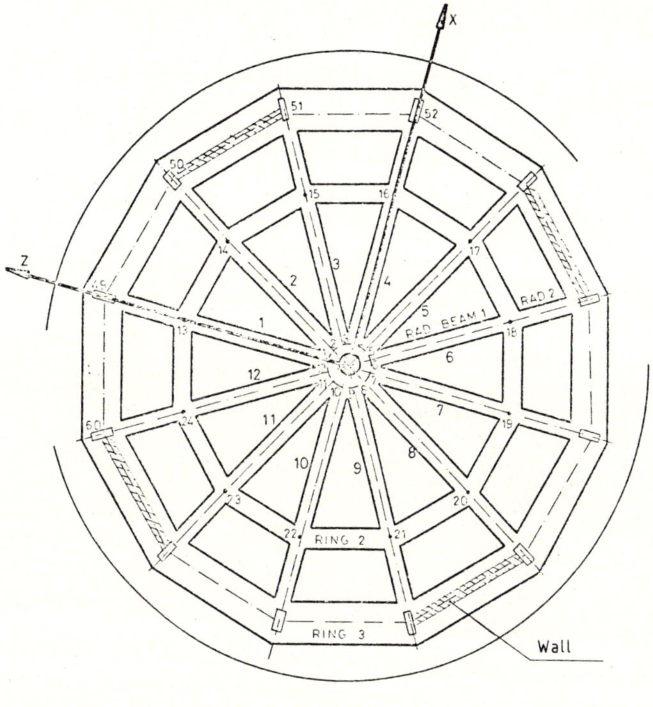

Fig. 9 : Position of stiffening diaphragms

The response of the structure was calculated under the action of

- a unit horizontal load
- a unit vertical load
- a unit twisting moment around the vertical axis (twisting)
- a unit bending moment around a horizontal axis (rocking).

The spring constants were calculated by the relation force/displacement. The calculation of the natural frequencies were executed, taking into account the soil-structure interaction. For the calculation of foundation stiffnesses the formulas, given in 'Fundamentals of Earthquake Engineering', by N.M. Newmark and E. Rosenblueth were used. The structure was now idealised as shown in Fig. 10.

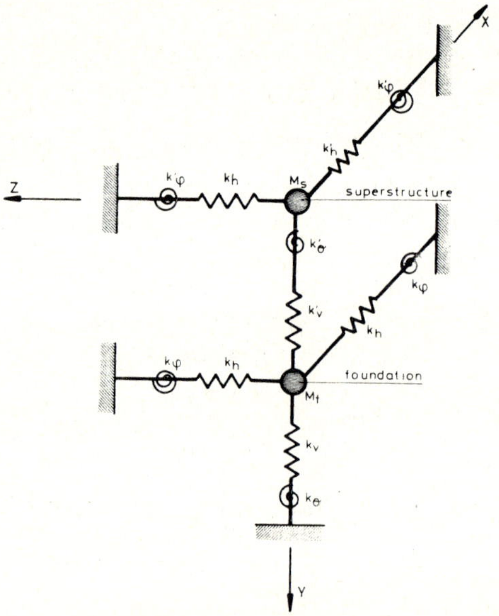

Fig. 10 : Idealised structural dynamic system

For the masses and mass moments of inertia we calculated :

- foundation M_f = 200.000 kg
  I_φ = 5,384 E12 kg.mm^2
  I_θ = 1,077 E13 kg.mm^2
- superstructure M_s = 340.000 kg
  I_φ = 8,236 E12 kg.mm^2
  I_θ = 1,5945 E13 kg.mm^2

The idealisation provides a structure with 12 degrees of freedom.
The four lowest natural frequencies were calculated, with the results

1. Horizontal displacement	frequency	7,118	period (sec)	0,141
2. Horizontal displacement		7,118		0,141
3. Rocking		17,03		0,059
4. Rocking		17,03		0,059

All these are movements of the superstructure, the foundations do not take part in the movements.

From the analysis we conclude that the introduction of four stiffening walls gave the antenna foundation building the desired rigidity. The lowest natural frequency is 7,1 cycles/sec.

5. EXPERIMENTAL MEASURING OF THE NATURAL FREQUENCY OF THE LESSIVE III - EARTH STATION

The earth station must be continuously pointed at the telecommunications satellite with a preciseness of about one tenth of a degree. For this purpose, it disposes of a so-called enslaving system of the master engines for the shifting of the antenna as per two orthogonal rotation axes, the elevation and the azimuth axes, Fig. 2. The pursuing system is meant to get exterior parasite effects under control, as for instance the drifting out of position of the satellite, the attenuation of the signal received by the hydrometeors (snow, rain ...) the wind, the seismic effects, the elasticity of the antenna support.

The system corrects the effects of low frequency disturbances (wind, seism), but it may not respond any further when there is a risk of entering into resonance with the natural frequencies of the mechanical system, tc avoid damage of the structure.

With the object of measuring the systems response in frequency, a push-and-pull movement around an axis is imported to the antenna and the real movement of the whole structure is measured by the decoders installed on the antenna. The decoders indicate the relative shifting between the moving part and the reference support, which is :

- azimuth : - moving part = antenna
  - reference support = building
- elevation : - moving part = elevation structure
  - reference support = building + azimuth structure.

The natural frequency of the fixed support is, per definition, the frequency at which, at equal loading, the deformation is the greatest. At this frequency, the relative shifting between the moving parts and the support will therefore be less important because the enslaving system doesn't react anymore. This shifting can be recorded. Consequently the resonance frequencies of the supports are the anti-resonance frequencies of the pursuing system. For ease of measuring, a speed instruction is injected and the relative shifting speed is recorded. The command signal is injected to the motors as a sine-shaped tension with a constant amplitude and a frequency fluctuating continuously between 0,1 and 4,2 Hz, Fig. 11.

The speed response is an electric signal proportional to the relative speed recorded by tachometric dynamos. At the resonance frequency the relative speed is minimum.

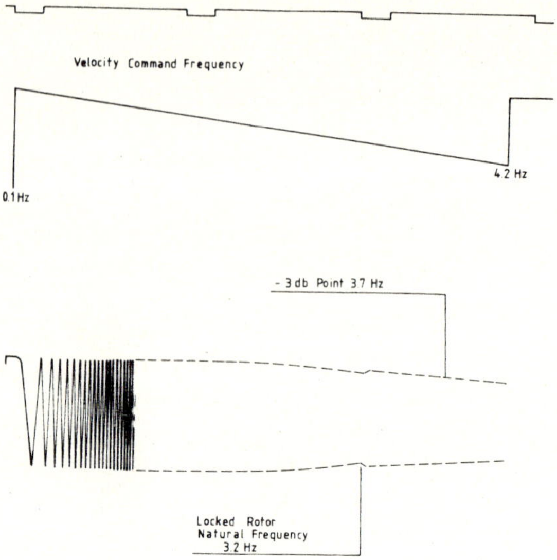

Fig. 11 : Velocity command frequency and relative speed response

This minimum is observed at the recorded response, Fig. 11.
The results for the station were : - azimuth pulsation : 3,2 Hz
- elevation pulsation : 3,7 Hz.

The accuracy of these measurement is poor, but sufficient to proove that the structure meets the required 2 Hz resonance frequency. The correspondance with the calculations, however, is even poorer. The results indicated a lowest natural frequency of 7,1 Hz, corresponding to an horizontal displacement. The experiments on the actual building indicate a lowest frequency in torsion. This can be explained by the fact that in the analysis all the connections are ideal, perfect fixations. In reality the connections are not as rigid, resulting in lower natural frequencies. A further element can be the difference in the masses, because the calculations were made without the antenna, whilst the experiments are executed on the building with antenna installed.

6. CONCLUSION

A numerical analysis enables the designer to adapt or adjust the structure lay-out to meet the specifications. However, the few experimental results, obtained in a real size vibration test, show that especially with respect to dynamic analysis the calculated results must be interpreted and handled with precaution.

ON THE NONLINEAR BEHAVIOR OF R/C FRAME STRUCTURES UNDER SEISMIC LOADS

W.B. Krätzig, K. Meskouris
Lehrstuhl KIB III, Ruhr-Universität Bochum
4630 Bochum 1, W. Germany

1 Introduction

A widely accepted design philosophy for buildings in seismically active regions stipulates that weak earthquakes with high frequency of ocurrence should cause little or no damage to the structure, whereas strong earthquakes might well cause a substantial degree of damage as long as collapse with ensuing loss of life is avoided. To this effect, strength and ductility of aseismic structures should be sufficiently high to allow the structure to survive a number of loading cycles without critical deterioration. In most design codes, special provisions are included to ensure ductile performance of the structure and to minimize the danger of sudden collapse. Since, however, linear analyses are unable to predict the nonlinear behavior of the structure under strong seismic excitation, a rational design concept should compare the available ductility reserves of the structure with ductility demands evaluated for appropriate ground motions by nonlinear time-history calculations. The results of such calculations allow a critical appraisal of the structure's weak points (e.g. accumulation of inelastic deformation in a certain story, hinging in columns, high shear regions in beams) and suggest preferable design alternatives.

R/C frame and wall/frame buildings should ideally be idealized by spatial models for dynamic analysis, in order to automatically include biaxial bending and torsional effects. However, since aseismic engineering principles dictate that buildings subject to earthquakes should be more or less symmetric, it is common practice in this case to use two-dimensional models in the principal directions and to separately consider torsion loading, biaxial bending and multi-dimensional earthquake excitation. In this paper, two-dimensional frame models are employed for the analysis, but the basic principles apply equally well to 3D structural models.

2 Ductility balance method

The ductility balance method is an analysis method for assessing available earthquake safety in a structural system. Since it focuses attention on a structure's weak points, e.g. where ductility demands are perilously near or even above the available supply, it can also serve as a valuable tool in the design process, if appraisal and/or strengthening of planned, existing or damaged buildings is desired. The principle is simply a comparison of local nonlinear damage with the available ductility reserves of the system. Ductility is mostly expressed through maximum ductility factors, defined as ratios of maximum allowed (on the supply side) or maximum reached (on the demand side) deformations to their respective elastic limits.

On the supply side, section ductility factors η_Q, defined as ratios of the maximum allowable curvature κ_u of a R/C section to its elastic limit κ_f are especially suited for the quantification of ductility reserves. Moment-curvature diagrams are best evaluated by computer as functions of the mechanical and geometrical section properties. As an example, Fig. 1 shows M-κ - curves for a 20 by 30 cm rectangular section (B 35 concrete, BSt 500 S steel) with variable tensile rein-

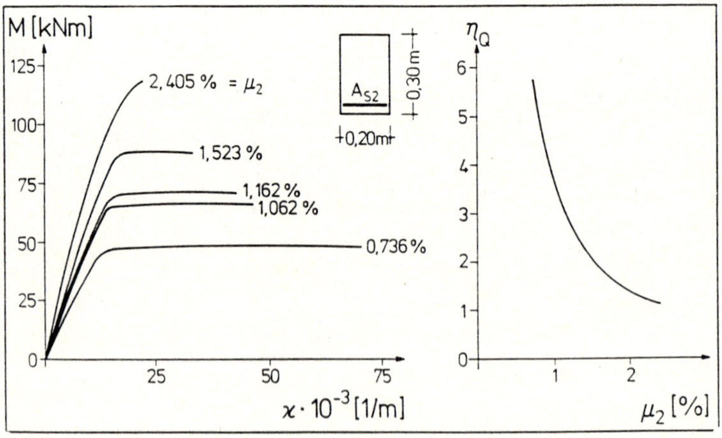

Fig. 1. - Influence of tensile reinforcement on curvature ductility

forcement. It can be seen that with increasing reinforcement the section's moment carrying capacity increases while sectional ductility decreases. Fig. 2 depicts the influence of compressive axial loads (expressed in terms of the balance point load P_b) on the ductility of the same section as in Fig. 1, only with symmetric double reinforcement. The marked decrease of ductility for high compressive stresses is the reason for the limits imposed in many building codes (e.g. in DIN 4149) on the allowable axial loads in columns.

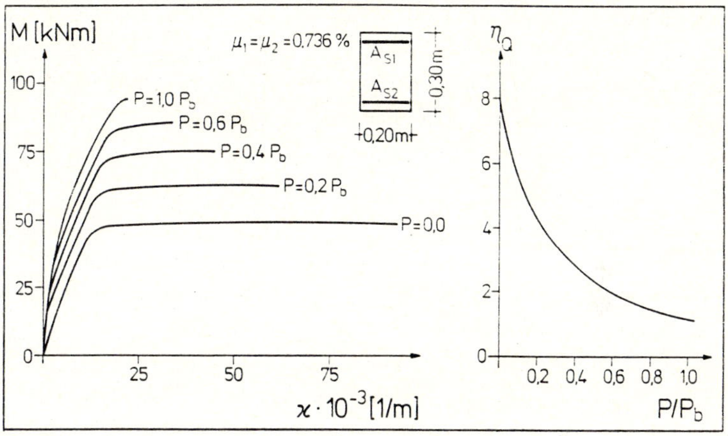

Fig. 2. - Influence of axial load on curvature ductility

On the demand side, we must evaluate the evolving local damage in the structure subject to several severe ground motions. The main difficulty lies in accurately describing the complex, stiffness-and-strength-degrading, hysteretic behavior of concrete structures under high cyclic loads. Nonlinear structural models can be simple or quite complex. Examples for the former are Saiidi and Sozen's Q-model /3/ as a single degree of freedom idealisation and generalized shear beam type models /4/. Although these are quite successful in predicting overall inelastic behavior, they cannot accurately describe local (sectional) damage. At the other extreme, fiber models yield abundant information on structural behavior at prohibitive computational cost for large or even medium-sized buildings. As a "middle of the road"-alternative, point hinge member models in which R/C beams and columns are idealized by elastic members with nonlinear rotational springs at their end sections (single component models) /5/ /6/ are frequently employed. The nonlinear hysteretic behavior of these springs follows predefined rules, mostly of Takeda-type /7/. Having determined maximum plastic rotations, section ductility demands can be evaluated by assuming plastic zone lengths (roughly one-half of beam depth). Currently, we also employ single-component point hinge models as described above /10/, while a more refined idealization based on section properties is being developed. In it, finite length of plastic zones is considered as well as bond slip and cracking, also no pre-defined set of hysteretic rules is necessary at element level.

Evaluated ductility demands, also expressed as section ductilities η_Q, can now be compared with the available supply. Danger of collapse is excluded if demand values range well below their supply counterparts everywhere in the structure. Of course, a more detailed picture of structural damage can be readily acquired by evaluating additional damage indicators, such as cumulative ductilities, dissipated energies, stiffness reduction factors etc. /8/.

Because of the stochastic character of earthquakes, this investigation should be carried out with three to five different, measured or artifically generated ground motions, preferably of different durations. Special properties of the site (type of subsoil, probable epicentral distance and earthquake intensity) should be considered in choosing these ground motions. If, for example, we are dealing with a near-field site, ground motions should resemble the Pacoima dam accelerogram of the 1971 San Fernando earthquake with its characteristic long-wave acceleration pulses. If no measured accelerograms are readily available, artificially generated ground motions can be used. They can be synthesized as spectrum-compatible accelerograms or through the filtering of white noise /9/, in which case filter parameters can be adjusted to represent local site conditions.

3 Example

Application of the above method will now be demonstrated by an example. Fig. 3 shows a R/C frame of a six-story building that was designed according to DIN 4149 and, separately, according to the CEB code. Materials were B 25 concrete and BSt 500 steel, the frames were 6m apart.

Earthquake loads were given by design response spectra which, in both cases, featured a constant spectral acceleration of 1.0 m/s^2 for periods up to 0.45 s. The adjoining branches for $T > 0.45$ s differed, however, in that the exponent was equal to -0.8 for DIN 4149 spectrum and equal to -2/3 for the CEB spectrum. For the frame fundamental period of 0.9 s this meant a 10% difference in spectral accelerations.

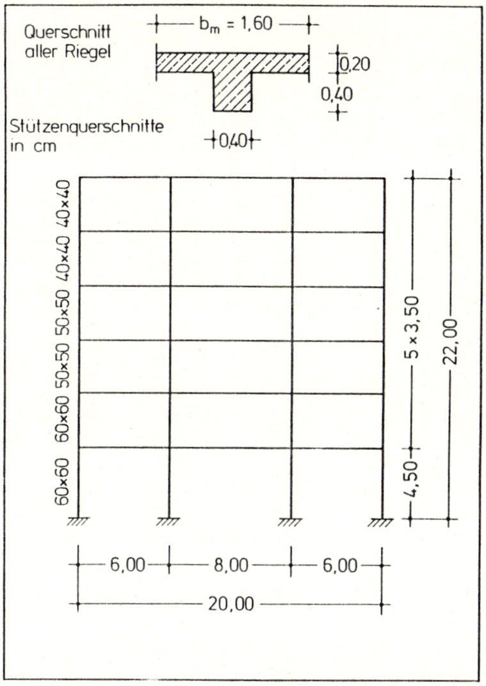

Fig. 3. - R/C frame

Due to the low seismic loading, static load combinations governed the design almost everywhere. In order to comply with DIN 4149 provisions for seismically loaded R/C structures (section 9.2) additional longitudinal reinforcement was necessary in almost all columns, reduction of reinforcement in view of surplus sectional area being explicitly excluded. Also, additional stirrups were provided in the interior columns of the lower four stories and extra longitudinal bars reinforced the underpart of beam/column joints. In the CEB design, in order to comply with ductility level II specifications, exterior columns in the upper four stories and interior columns in the upper two stories required additional longitudinal reinforcement, as well as some beam/ column connections.

Ductility supply expressed as curvature ductility η_Q, has been evaluated for both design variants and is depicted in Fig. 4. Ultimate curvature κ_u was here defined by spalling of the concrete cover and buckling of the reinforcing bars.

As a first step for the evaluation of the nonlinear behavior of both design variants, static limit load analyses were conducted for horizontal loading in form of an inverted triangle. In our experience, such analyses are quite useful for preliminary assessment of hinging

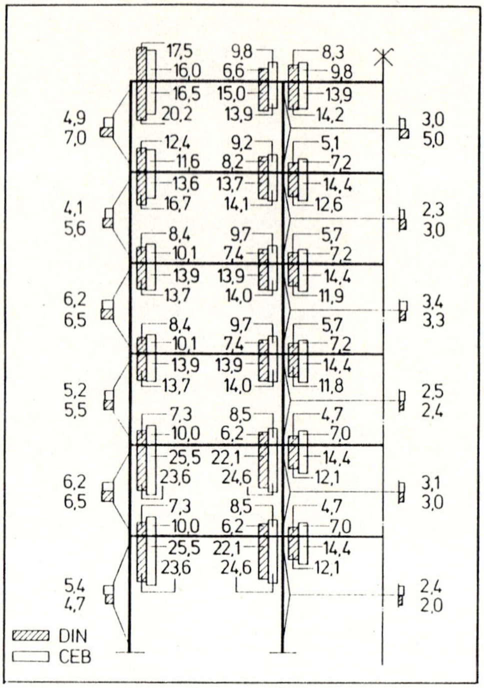

Fig. 4. - Ductility supply, η_0

patterns in the structure. Fig. 5 shows resulting load-deflection curves as well as the limit state hinge configurations of both variants. System ductility is lower for the DIN design, which also exhibits more unwanted hinge formation in columns, especially in the 5th story.

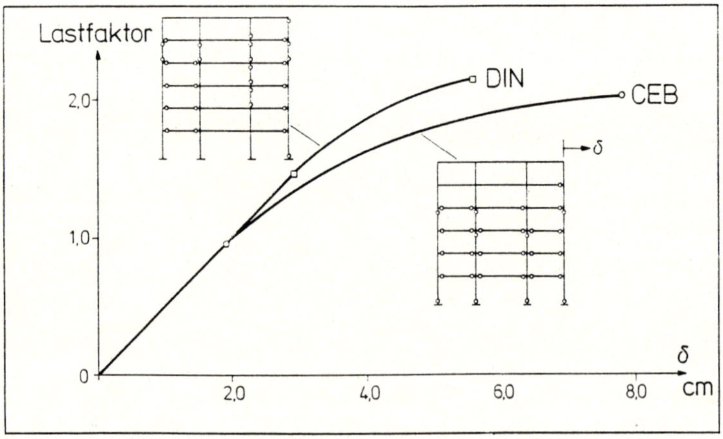

Fig. 5. - Static load-deflection diagrams

For the nonlinear dynamic analysis three spectrum compatible motions with durations between 6 s and 8 s as well as a recorded accelerogram were used. The latter was a component of the April 15th, 1979 Montenegro earthquake, recorded in Petrovac, with 0.45 g peak ground acceleration and almost 30 s duration. In order to simulate typical earthquakes, the synthetic motions were scaled by 3 and the recorded motion by 0.25. Corresponding ARIAS intensities of the scaled accelerograms, defined as time integrals of squared acceleration over the entire duration ranged from 1.4 m^2/s^3 to 1.7 m^2/s^3. Fig. 6 shows response spectra of all four scaled ground motions together with the utilized smooth design spectra according to DIN 4149 and to CEB Model Code.

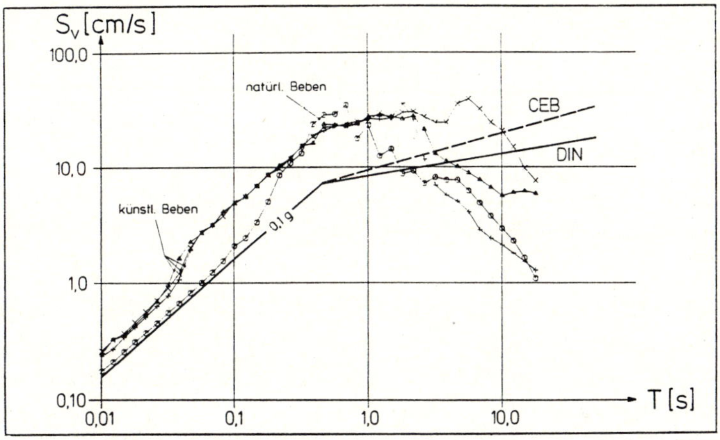

Fig. 6. - Design spectra and spectra of amplified ground motions

Some results of the investigation can be seen in Figs. 7 and 8. In Fig. 7, 5th story drift time histories of both design variants subject to the 6 s synthetic ground motion are shown. A certain degree of plastic damage accumulation is evident for the DIN design. Fig. 8 depicts maximum reached curvature ductilities for all four motions and both design variants. Comparison with ductility supply, Fig. 4, shows no danger of collapse of either variant for ground motions of the assumed intensity, that is with peak ground accelerations 3-4 times higher than those implied in the design spectrum. The evaluated ductility demands were, by the way, essentially the same for the different accelerograms used. Fig. 8 also points out the structural members whose additional strengthening would drastically enhance seismic safety, e.g. 5th story columns in the DIN design variant.

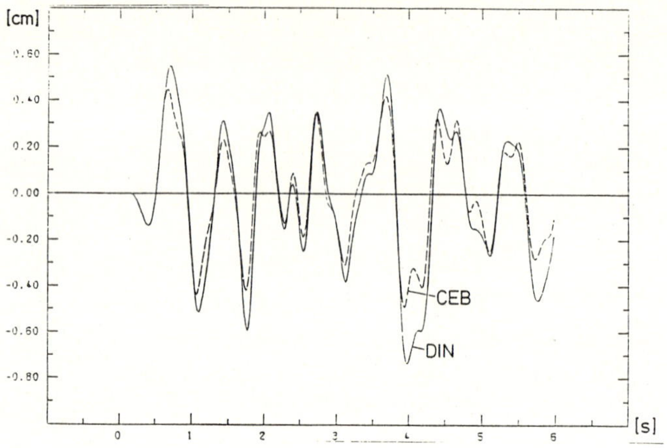

Fig. 7. - Relative displacement of the 5th floor

4 Conclusions

The explicit ductility balance approach is a powerful tool for analysing seismic safety of building structures. It requires more or less sophisticated models for the structure as well as for the loading, but it furnishes realistic insights in the structure's nonlinear behavior and thus focuses attention on its weak points. Ongoing research deals with the influence of the mathematical model parameters on the computed nonlinear response and also with the suitable choice of sample ground motions for the specific site. With computing power becoming increasingly cheaper with the advent of micro-computers, it is hoped that such powerful analysis tools will soon find their way in everyday design practice.

The authors wish to thank their co-workers, Dipl.-Ing. A. Elenas, Dipl.-Ing. L. Heiny and Dipl.-Ing. I. Meyer for their assistance in preparing this paper. The continuing financial support of the DFG (Sonderforschungsbereich 151) is gratefully acknowledged.

5 References

/1/ CEB Model Code for Seismic Design of Concrete Stuctures. Bulletin d'Information No. 165, April 1985

/2/ Entwurf Eurocode Nr. 8, Gemeinsame einheitliche Regeln für Bauwerke in Erdbebengebieten. Bericht EUR 8850, 1984

/3/ Saiidi, M., Sozen, M.A. : "Simple Nonlinear Seismic Analysis of R/C Structures". Journ. Struct. Div., ASCE, Vol. 107, ST5, May 1981

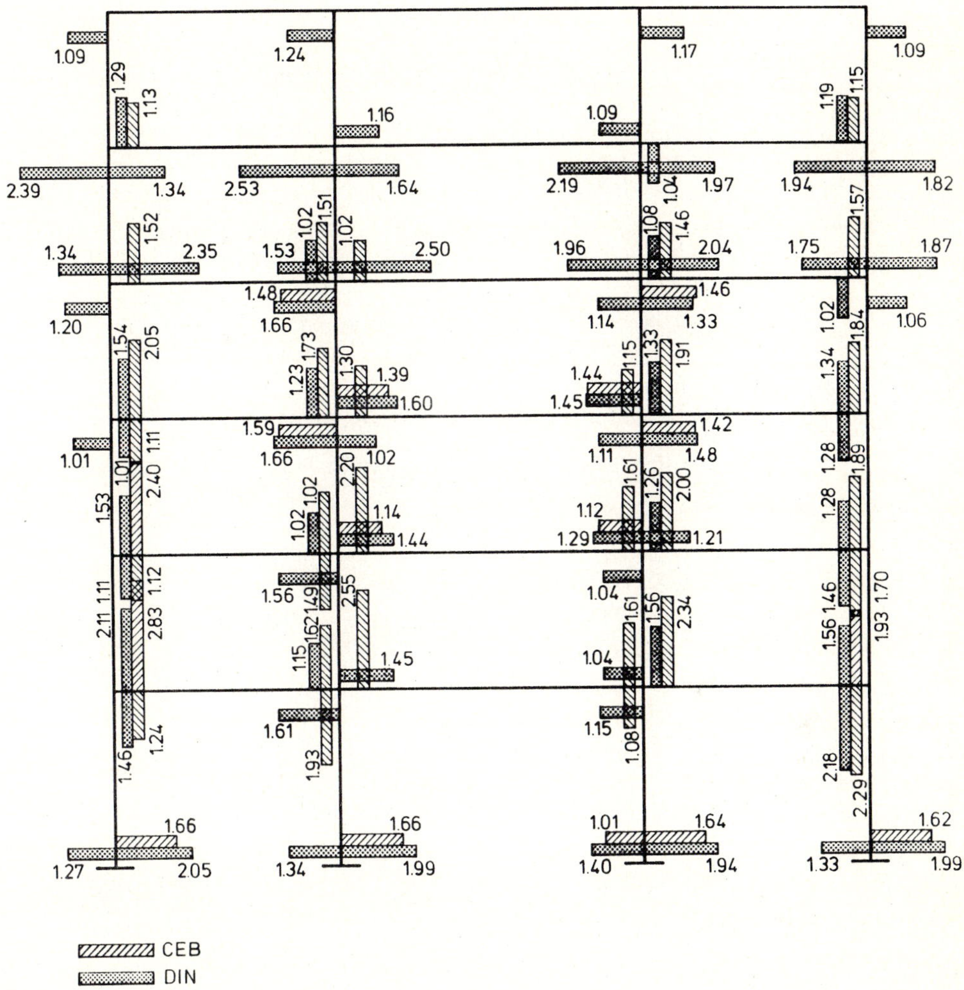

Fig. 8. - Ductility demands, η_Q

/4/ Park, Y.J., Ang, A.H.-S., Wen, Y.K. : "Seismic Damage Analysis of Reinforced Concrete Buildings". Journ. Struct. Eng., ASCE, Vol. 111, No. 4, April 1985

/5/ Otani, S. : "SAKE - A Computer Program for Inelastic Response of R/C Frames to Earthquakes". Struct. Research Series No. 413, University of Illinois, Urbana, November 1974

/6/ Powell, G.H. : "Supplement to Computer Program DRAIN-2D", University of California, Berkeley, August 1975

/7/ Takeda, T., Sozen, M.A., Nielsen, N.N. : "Reinforced Concrete Response to Simulated Earthquake". Journ. Struct. Div., ASCE, Vol. 96, ST12, December 1970

/8/ Banon, H., Biggs, J.M., Irvine, H.M. : Seismic Damage in Reinforced Concrete Frames. Journ. Struct. Div., ASCE, ST), September 1981, S. 1713 - 1728

/9/ Meskouris, K. : Nichtlineare Rahmenmodelle unter simulierten Erdbebenlasten. VDI-Bericht Nr. 419, 1981, S. 119 - 127

/10/ Hohlsiepe, U. : Beitrag zur Untersuchung des nichtlinearen Tragverhaltens ebener Stahlbetonrahmen unter seismischen Einwirkungen. Mitteilung Nr. 85-3, Institut für Konstruktiven Ingenieurbau / SFB 151, Ruhr-Universität Bochum, Mai 1985

Dynamic Analysis of Liquid-Filled Tanks including Plasticity and Fluid Interaction - Earthquake Effects -

J. Eibl and L. Stempniewski
Universität Karlsruhe, Institut für Massivbau und Baustofftechnologie
Postfach 6380, D-7500 Karlsruhe 1, Germany

1. Introduction - Problem

Tanks enclosing highly explosive or dangerous fluids are needed for purposes of energy supply as well for the chemical industry. In Germany containers with a capacity of more than 100 000 m^3 are being planned.

With regard to public safety such tanks also have to be designed against catastrophic external loads of dynamic character such as earthquakes, external explosions, impact and air plane crashes. While with small tanks very rough investigations may be tolerated from safety aspects as well as from economical consideration these structures have to be studied very thoroughly.

The following contribution which reports on the fluid-structure interaction in a tank under earthquake excitation may help to solve this problems. In this article emphasis lies however more on modelling of the fluid-structure interaction in an elastoplatic steel structure than on the type of dynamic excitation. The latter can easily be exchanged in favour of other actions as e.g. explosion loads or impact.

2. Available Knowledge

Finite elements have previously been developed for the numerical analysis of fluid-structure interaction. Along the lines of derivation the used methods can be categorized into three approaches:

- "added mass" approximation
- "Eularian" formulation
- "Lagrangian" formulation

In the "added mass" approach some fraction of the fluid mass is added to the structural model. So that spatial discretisation of the fluid is avoided. In the mid 1950's, *Housner*[1,2] formulated a mechanical model for estimating liquid response in seismically excited rigid, rectangular and cylindrical tanks. He divided the hydrodynamic pressure of the contained liquid into two components; the impulsive pressure caused by the accelerated portion of the liquid within the tank and the convective pressure resulting from that part of the fluid, which sloshes in the tank. This "added mass" approach is suitable for predimensioning, because it can be done fast and easily. But for a more accurate analysis in particular for nonlinear problems the whole fluid-structure model must be analyzed numerically, normally by means of F.E. methods.

In the "Eulerian" approach the behaviour of the fluid is formulated in terms of a pressure potential at the finite element nodes. The spatial (Eularian) coordinates for the fluid must be coupled with the material (Lagragian) coordiantes of the container. ***Belytschko et al.*** [3] and ***Donea et al.*** [4] realized the interface conditon by means of the "arbritary Lagrangian-Eularian" (ALE) code. With this code a steady transition from "Eularian" to "Lagragian" coordinates is possible. But it requires a special-purpose computer program. Also, in this code coupled equations lead to unsymmetrical equations with large bandwith.

In the "Lagrangian" approach, the behaviour of the fluid is described in terms of the displacements at finite element nodes. Hence, this element typ automatically satisfies the requirements of compatibility at the interface nodes between fluid and structure. The major advantage of this approach is that these elements can be readily be incorperated in a general-purpose computer program. Therefore the "Lagragian" approach was used in our investigation.

3. Aim of Investigation

As already mentioned in our study the interaction between dynamic load, tank wall and inner fluid is emphasized. This nonlinear dynamic response of liquid filled tank poses a very challenging task to numerical analysts.

In a first step only the material nonlinearities and free surface sloshing are treated. Geometrical nonlinearity, soil-structure interaction including lift-up are under work but are not yet subject of this article.

4. Method of Investigation

4.1 Isoparametric Fluid Element

The stiffness coefficients within the F.E. formulation are derived from the principle of virtual displacements. The formulation accounts only for infinitesimal displacements and strains including fluid compressibility. "Hourglass" modes resulting from the reduced integration of the element terms are eliminated by introducing the following rotations [7]

$$\omega_{ij} = \frac{1}{2}(u_{i,j} - u_{j,i})$$

An important behaviour of the fluid is its ability to displace without change in volume. In tanks this movement appears in form of sloshing waves where parts of the fluid experience large vertical displacements of low frequency at the surface of the fluid with nearly no volume change.

The resulting expression for the total internal virtual work A_i including "hourglass" control and free surface sloshing are:

$$A_i = \int_V \delta \underline{\epsilon}^T \, \underline{p} \, dV + \int_V \delta \underline{\omega}^T \, \underline{\underline{C}} \, \underline{\omega} \, dV + \int_S \delta \underline{u}_S^T \, \rho \, g \, \underline{u}_S \, dS$$

Where $\underline{\epsilon}$ are the strains, $\underline{p}$ is the pressure, $\underline{u}_S$ are the free surface displacments, ρ is the fluid density and g is the acceleration due to gravity.

The Matrix $\underline{\underline{C}}$ consists of:

$$\underline{\underline{C}} = \begin{vmatrix} E_H & & \\ & E_H & \\ & & E_H \end{vmatrix}$$

and E_H being the "hourglass"-modul.

4.2 Isoparametric Shell Element

The tank wall is approximated by isoparametric shell elements with independent degrees of freedom for rotation and translation . The constitute relations are given in the frame of plasticity using the kinematical yield surface of Mises[5] with isotropic strain hardening. In principle the yield surface is a cylinder in the main stress spaces the diameter of which is encreasing with the applied plastic work.

$${}^tF = \frac{1}{2} {}^t s_{ij} {}^t s_{ij} - \frac{1}{3} ({}^t\sigma_y)^2$$

The ${}^t s_{ij} = {}^t\sigma_{ij} - \frac{1}{3} {}^t\sigma_{mm} \delta_{ij}$ are the deviatoric stresses with ${}^t\sigma_{mm} = \sum {}^t\sigma_{mm}$, t characterizing the instant of time.

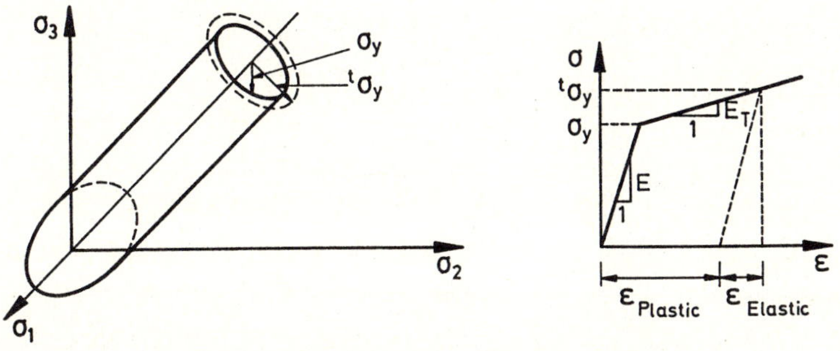

Figure 1: Von Mises yield condition with isotropic hardening [5]

4.3 Isoparametric Contact Element

Furthermore a contact element is applied to impose inequality constraints between nodes along the interface. At the boundary of the fluid, only normal stresses are to be transferred to the tank wall, while displacements non perpendicular to the wall surface are possible without stress transfer. To incorporate this physical fact *Liu* [8] coupled fluid and tank with special spring elements. We chose a special 16-node isoparametric solid element with local fictitious stress-strain relations $\underline{\underline{C}}_L$. The total internal virtual work A_i for this element typ is calculated by

$$A_i = \int_V \delta \underline{\epsilon}^T \; \underline{\underline{T}}^T \; \underline{\underline{C}}_L \; \underline{\underline{T}} \; \underline{\epsilon} dV$$

with

$$\underline{\underline{C}}_L = \begin{vmatrix} 0 & & & & & \\ & E & & & & \\ & & 0 & & & \\ & & & 0 & & \\ & & & & 0 & \\ & & & & & 0 \end{vmatrix}$$

$\underline{\underline{T}}$ representing a matrix that transform the stress-strain law from the local coordinate system to the global cartesian coordinate system.

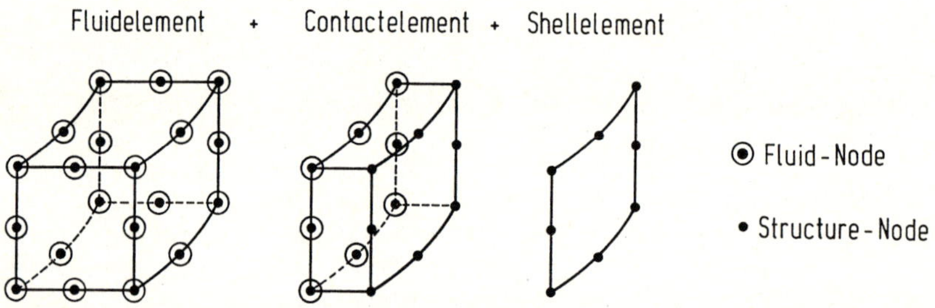

Figure 2: Isoparametric fluid-, contact- and shell element

5. Results

5.1 Solution of the Matrix Equation of Motion

The resulting matrix equation of fluid-structure interaction excluding damping is:

$$\underline{\underline{M}}\,^{t+\Delta t}\ddot{\underline{u}}^{(i)} + {}^{t}\underline{\underline{K}}_{NL}\,\Delta\underline{u}^{(i)} = {}^{t+\Delta t}\underline{R} - {}^{t+\Delta t}\underline{F}^{(i-1)}$$

where $\underline{\underline{M}}$ is the mass matrix , $\underline{\underline{K}}$ is the nonlinear stiffness matrix and $\underline{u}$ is the vector of nodal displacements. $\underline{R}$ is an external load vector, which may include body forces and external pressure loads. Vector $\underline{F}$ represents the nodal forces corresponding to the element stresses.

Vector $\underline{R}$ may consists of

$$\underline{R}(t) = \underline{\underline{M}}\,\underline{a}(t) + \underline{F}_P(t) + \underline{F}_S(t)$$

where vector $\underline{a}$ contains the nodal accelerations during an earthquake and $\underline{F}_P$ is the vector of external pressure. Vector $\underline{F}_S$ is to represent the free surface waves :

$$\underline{F}_S(t) = \int_S \rho\, g\, \underline{u}_S(t) dS$$

Here $\underline{u}_S$ are the free surface displacements in the direction of gravity, where index S applies to the free surface.

For the solution of this matrix equation of motion Newmark's integration method is used. Resulting from the material nonlinearity of the tank wall, equilibrium iterations have to be performed at each time step.

5.2 Steel tank subjected to seismic motion

The effects of "material-nonlinear tank wall" and "free surface sloshing" are very strong if the tank is subjected to seismic motion. Unsymmetric pressure loading can lead to the well known "elephant footing" of the tank.

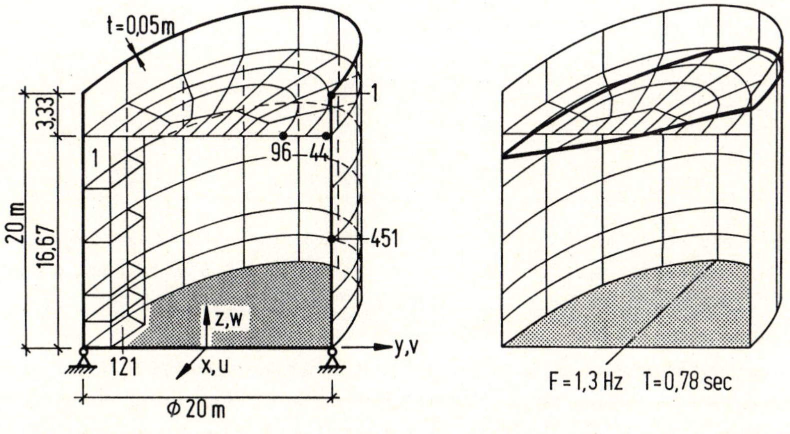

Figure 3: Dimensions, configuration and first sloshing mode of the tank

Dimensions and configuration of an investigated tank are given in fig. 3, while fig. 4 shows the applied earthquake acceleration. As an arbitrary chosen example the nine seconds lasting modified N-S component of the Vrancha earthquake - Bucharest - was taken.

Material properties of the tank and fluid were chosen as follows:

Tank: $E = 2.1 * 10^8 KN/m^2$, $\nu = 0.3$, $\sigma_y = 2.4 * 10^5 KN/m^2$, $E_T = 0$

Fluid: $E = 1 * 10^6 KN/m^2$, $E_H = 1 * 10^{10}$, $\rho = 1KNsec^2/m^4$

The Newmark parameters are $\delta = \frac{1}{2}$, $\alpha = \frac{1}{4}$, ΔT=0.02sec.

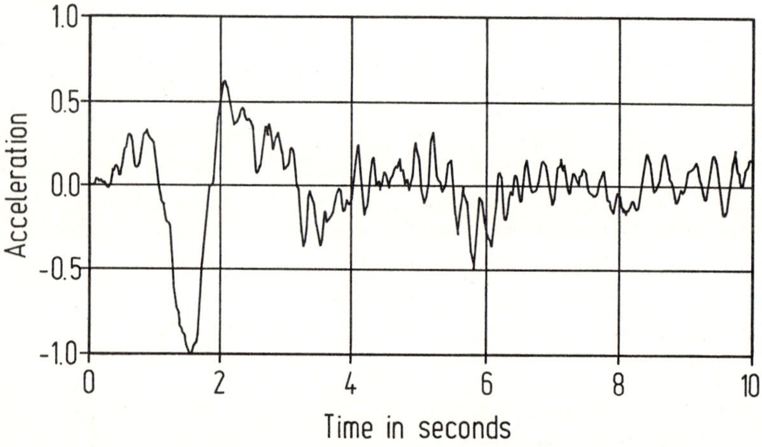

Figure 4: N-S horizontal acceleration of the Vrancha earthquake

Plots of the sloshing motion at the free surface are shown in fig. 5. It gives the vertical displacements at the free surface nodes 44 and 96 with a period of about 0,8 sec. The maximum wave height is about 75 cm. The displacements of the tank wall at the node 1 and node 451 in tank mid-height are shown in fig. 6. Plots of the total pressure at fluid elements 1 and 121 are given in fig. 7. The much higher pressure at tank top is due to superposition of the sloshing pressure and the impulsive pressure. The free surface motions are shown in fig. 8 at several time steps. Since the free surface motion is much larger than the tank deformation, plots of the tank are depicted separatly in fig. 9 because of clear representation.

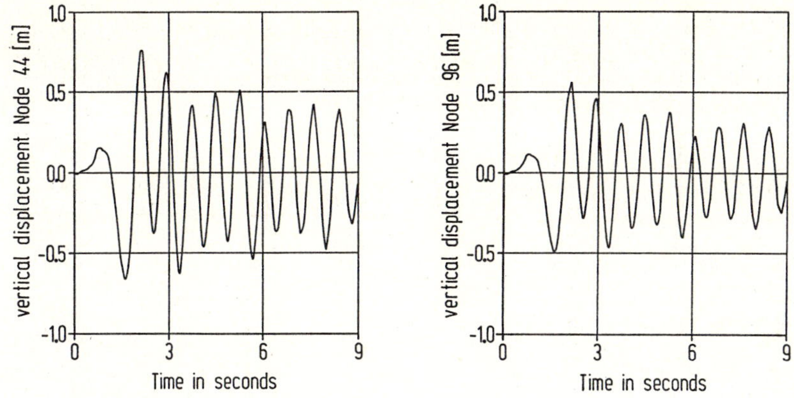

Figure 5: Sloshing of the free surface

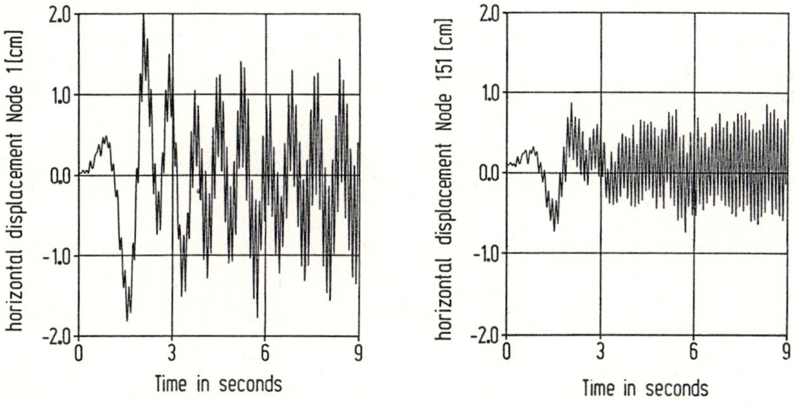

Figure 6: Tank wall displacements at a) tank top b) tank mid-height

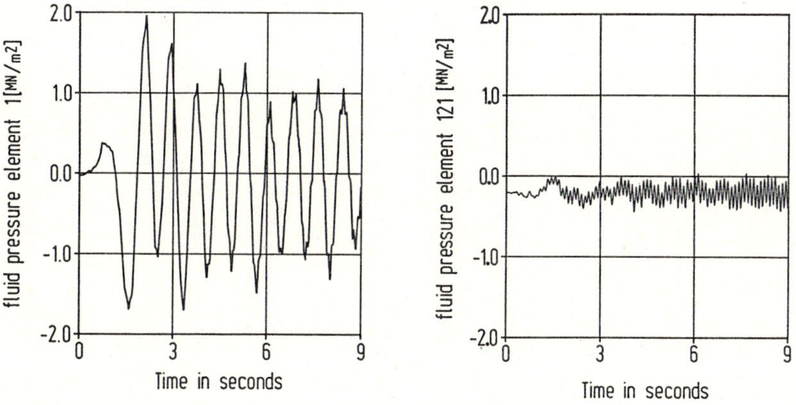

Figure 7: Total fluid pressure at a) tank top b) tank bottom

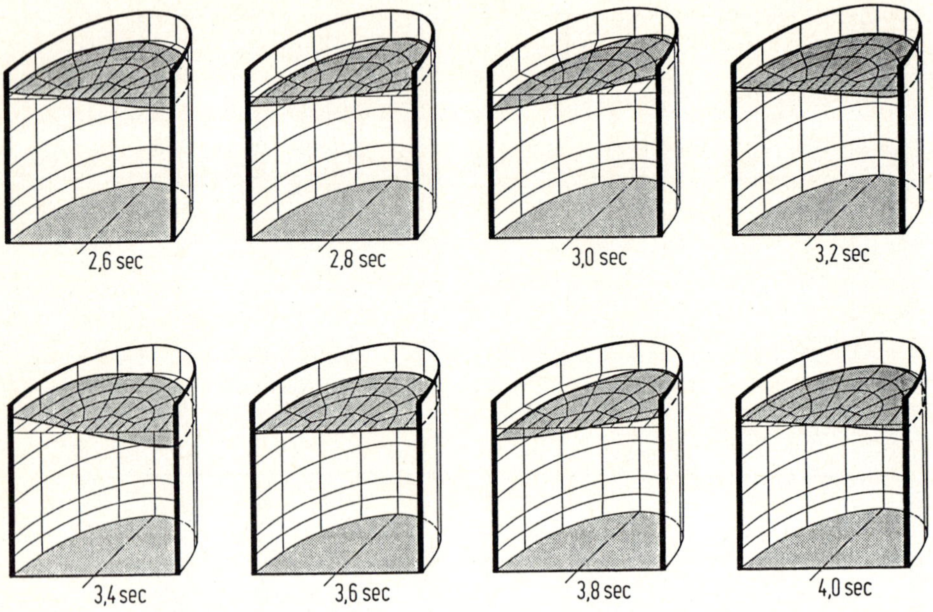

Figure 8: Free surface motion

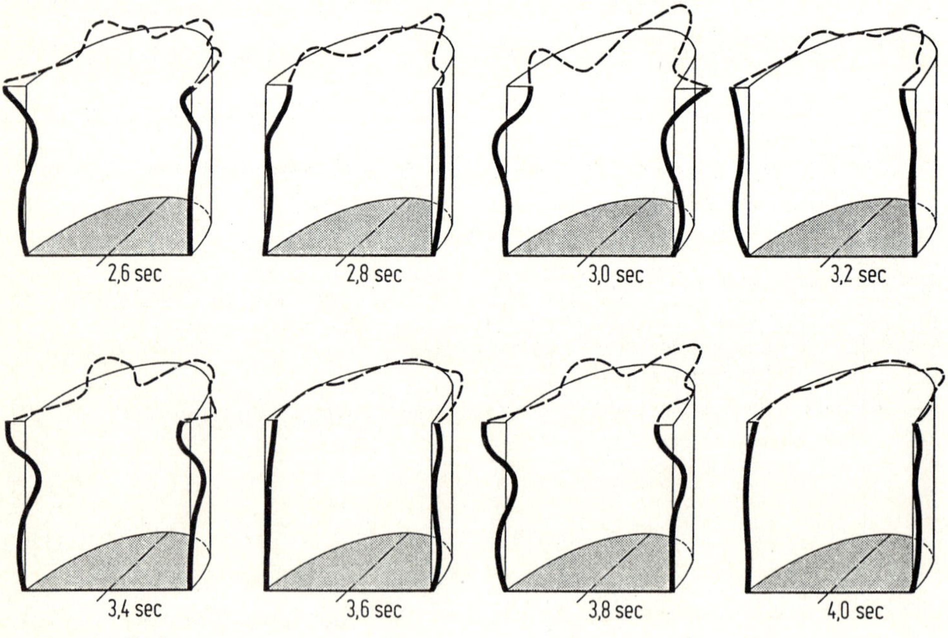

Figure 9: Tank response

The behaviour of the vessel may be sketched as follows:
The tank first plastifies at the top, with resulting membran hoop forces. Then it plastifies at the bottom.
The overturning moment caused by the sloshing motion tries to lift off the tank on one tankside and initiates axial compression forces on the other side - compare the same time steps in fig. 8 and fig. 9 - which may initiate buckling or the so called "elephant footing". Only because the tank is anchored the "elephant foot" isn't so distinct as if it might be in case of an unanchored bottom.

6. Summary

In this paper, a numerical tank model was developed treating thoroughly the fluid-structure interaction and the nonlinear elasto-plastic material behaviour of the tankwall. The example of a seismic excitation is to demonstrate its application.

7. References

[1] *Housner, G.W.*, "Dynamic pressure on accelerated Fluid containers", Bulletin Seis. Soc. America, Vol.47, No. 1, 1957, pp.15-35
[2] *Housner, G.W.*, "The dynamic behaviour of water tanks", Bulletin Seis. Soc. America, Vol.53, No. 1, 1963, pp.381-387
[3] *Belytschko, T., Flanagan, D.P., Kennedy, J.M.*, "Finite element methods with user-controlled meshes for fluid- structure interaction", Computer Methods in applied mechanics and engineering 33(1982) 669-688
[4] *Donea, J., Giuliani, S., Halleux, J.P.*, "An arbitrary lagrangian-eulerian finite element method for transient dynamic fluid-structure interaction", Computer Methods in applied mechanics and engineering 33(1982) 689-723
[5] *Bathe, K.-J.*, "Finite element procedures in engineering analysis", Englewood Cliffs, New Jersey 1982
[6] *Zienkiewicz, O.C.*, "Methode der finiten Elemente", Wien: Hanser 1984
[7] *Wilson, E.L., Khalvati, M.*, "Finite elements for the dynamic analysis of fluid-solid systems", International Journal for numerical Methods in engineering, Vol.19,1657-1668(1983)
[8] *Liu, W.K.*, "Development of finite element procedures for fluid-structure interaction", EERL-80-6,California Institute of Technology, Pasadena, Calif., Aug., 1980
[9] *Eibl, J., Stempniewski, L.*, "Flüssigkeitsbehälter unter äußerem Explosionsdruck" Forschungsbericht Institut für Massivbau und Baustofftechnologie, Universität Karlsruhe, 1986 (in preparation)

STATIC AND DYNAMIC ANALYSIS OF TENSEGRITY SYSTEMS

R.MOTRO,S.NAJARI,P.JOUANNA

Laboratoire de Génie Civil
Université des Sciences et Techniques du Languedoc
34060 Montpellier-France

Although Tensegrity Systems have an undeniable attraction for builders, they are hardly used at all because of the lack of accurate quantitative knowledge of their structural behaviour. The present article concerns the experimental and theoretical study of Tensegrity Systems and deals with static and dynamic analysis. Theoretical study is suitable for any kind of Tensegrity System; experimental work has been developped on the smallest tensegrity module, so-called "Simplex".

1 DEFINITION AND MODELS FOR THE STUDY OF TENSEGRITY SYSTEMS

1.1 Definition and scope of the study

Running together the words "tensional" and "integrity", R.B. FULLER defined a new type of system as follows:

"A Tensegrity System is established when a set of discontinuous compressive components interact with a set of continuous tensile components to define a stable volume in space."

The scope of the study is limited by two others complementary conditions: the longitudinal axis of the constituent elements are rectilinear; each node of the reticulated system is incident to one member in compression and one only.

1.2 Characteristic property of Tensegrity Systems

One of the basic features of these systems is that they must form a stable volume in space in the absence of any external action and according to the first order theory, equilibrium of the nodes is given by the following matrix equation:

$$[K].\{D\} = \{W\} \tag{1}$$

$[K]$ matrix of rigidity of the system, $\{D\}$ displacement vector of the nodes, $\{W\}$, external vector action which is identically zero giving:

$$[K].\{D\} = 0 \tag{2}$$

For self-stressing to occur, the determinant of the matrix of rigidity must be zero in order to obtain a solution other than zero vector for $\{D\}$. In this case, the components of the system are either compressed or under tension with the overall stresses in equilibrium. Tensegrity Systems therefore belong to the Critical Configuration Class of the Reticulated Space Systems. Most calculation associated with them include the effect of geometrical non linearity and that of the terms of self-stress. The form of such systems is not defined "a priori": the spatial position of nodes must satisfy the zero condition of the determinant of $[K]$. In order to fit the definition proposed by R.B.Fuller, two conditions have to be fulfilled:

* topology: relational structure must meet certain criteria concerning spatial stability (necessary condition on the number of cables incident to a node)

* form: the spatial coordinates of the nodes are determined by the fact that Tensegrity Systems are critical configurations. Form-finding presupposes knowledge of the relational structure.

1.3 Models for the study of Tensegrity Systems

Full design study of selfstressed reticulated space structures (Tensegrity Systems) can be carried out in three phases:

* determination of the relational structure of the system (list of nodes and their connections). Use of graph theory for a system with n nodes ("n" being an even number) leads to the construction of relational structure which fit the definition.

* calculation of the geometry of the system corresponding to the possibility of self-stressing; this geometry is not known and can be determinated by the application of dynamic relaxation. (Ref. 1)

* analysis of mechanical behaviour: this analysis is carried out for the static and dynamic aspects. For the first, the mechanical behaviour of the tensegrity systems assuming static loading was studied. A mathematical model was used which takes second order effects and possible prestressing into account. In the case of a basic module, the validity of the results was shown by tests carried out on a prototype. In the second case, identification of the dynamic characteristics of the systems studied was achieved using what is known as the "inappropriate" method; it was applied with an harmonic excitation of the systems. A second prototype, with the same geometry as the first made it possible to complete the theoretical analysis and to check certain results.

2 PRESENTATION OF THE MATHEMATICAL MODELS

2.1 Static analysis

The mathematical model chosen for the calculation is based on the displacement method, the theory of which has been widely expounded. This method leads to the solving of linear equations of the following form:

$$\{W_r\} = [K_{Sr}] \cdot \{D_r\} \tag{3}$$

in which $\{W_r\}$ is the vector of actions at the nodes- the components of unknown reactions in supporting directions.

$\{D_r\}$ is the displacement vector sought.

$[K_{Sr}]$ is the matrix of rigidity of the structure.

In practice, deformations are considerable in reticulated systems comprising cables and it is necessary to take into account the non linearity effects of geometrical origin. Thus while with a "first order" calculation, only the equilibrium of the structure in its initial geometry is given, in a second order method, it is necessary to ensure the equilibrium of the structure in its deformed state under the corresponding load. Taking into account the second order terms leads to putting the matrix of rigidity into the following form:

$$[K_{Sr}] = [K_{LD}] + [K_S] \tag{4}$$

$[K_{LD}]$ is called the "large displacement matrix". The value of its terms depends on the rigidity of the members and present node coordinates.

$[K_S]$ the so-called "initial stress matrix" depends simultaneously on actual stress values in members and actual node coordinates.

As the value of $[K_S]$ varies with the load applied, it is necessary to linearise the system of equations using a Newton Raphson type incremental method in order to solve (3). The total load {W} is applied progressively. The state of equilibrium in the deformed position with the total load is sought for each increase {ΔW}. It is considered that the system is in equilibrium if all the components of the vector of the loads applied correspond to the vector of the balanced loads to whithin x% (the percentage of error is fixed at the beginning of the calculation).(See Ref.1)

2.2 Dynamic analysis

2.21 Introduction

Dynamic identification of the structures consists of determining the physical parameters involved in the mathematical modelling of their vibratory behaviour.

2.22 Direct problem

a/ Expression of the transfer functiun

It is assumed the modelling parameters p_i are known, the purpose of the direct problem is to determine the relation between the excitation noted {P*} and the response in the form of displacement {y*} , which in fact means determining the form of the transfer functiun F(j$\tilde{\omega}$), $\tilde{\omega}$ being the excitation pulsation. Thus:

$$\{y^*\} = [F(j\tilde{\omega})] \, . \, \{P^*\} \tag{5}$$

The equations for the movement of discrete systems with "n" degrees of freedom in time space are as follows:

$$[M].\{\ddot{y}(t)\} + [B].\{\dot{y}(t)\} + [K].\{y(t)\} = \{P(t)\} \tag{6}$$

$[M]$, $[B]$, and $[K]$ "mass","damping", and "stifness" matrices (constant of order n)
$\{y(t)\}$, $\{\dot{y}(t)\}$, $\{\ddot{y}(t)\}$ "n" dimension column vectors: displacements, velocity and acceleration of the nodes in the structure.

$[M]$ and $[K]$ are defined, positive matrices for stable structures. The eigen values of the undamped free system are the eigen pulsations ω_k. With mode "k", all points in the structure vibrate in phase with pulsation ω_k (Ref. 2). It is preferable under these conditions to use a modal representation on the basis of the eigen mode shape vectors $\{\Phi_k\}$ of the associated undamped system:

$$[M].\{\ddot{y}(t)\} + [K].\{y(t)\} = \{0\} \tag{7}$$

Supposing:

$$\{y\} = [\Phi].\{q\} \qquad (8)$$

There follows:

$$[\mu].\{\ddot{q}(t)\} + [\beta].\{\dot{q}(t)\} + [\delta].\{q(t)\} = [\Phi]^T.\{P(t)\} \qquad (9)$$

$[\mu]$ generalised real matrix of mass (diagonal)

$[\delta]$ generalised real matrix of rigidity (diagonal)

$[\beta]$ full generalised damping matrix

$[\Phi]$ eigen shape mode matrix

In the case of harmonic excitation:

$$\{P(t)\} = \{P^*\}.\exp(j\tilde{\omega}t) \qquad (10)$$

$\ddot{q}$, $\dot{q}$ and q are harmonics of pulsation $\tilde{\omega}$ (Ref.3) and working in frequency space gives:

$$\{y^*\} = [\Phi].(-\tilde{\omega}^2.[\mu] + j\tilde{\omega}.[\beta] + [\delta])^{-1}.[\Phi]^T.\{P^*\} \qquad (11)$$

The transfer functiun matrix is:

$$[F(j\tilde{\omega})] = [\Phi].(-\tilde{\omega}^2.[\mu] + j\tilde{\omega}.[\beta] + [\delta])^{-1}.[\Phi]^T \qquad (12)$$

b/ <u>Case in which $|\beta|$ is diagonal (Basile's hypothesis)</u>

In this case $[F(j\tilde{\omega})]$ is expressed in the form of a sum of the contributions of each eigen mode. The transfer functiun between an excitation point M_v and the response M_u will take the form:

$$F(M_u, M_v, j\tilde{\omega}) = \sum_k \left(\frac{A_k(M_u, M_v)}{p - p_k} + \frac{\bar{A}_k(M_u, M_v)}{p - \bar{p}_k} \right) \qquad (13)$$

A_k and $\bar{A}_k$ are the residues and are purely imaginary; p_k and $\bar{p}_k$ are the poles which are the roots of the equation:

$$p^2.\mu_k + p.\beta_k + \delta_k = 0 \qquad (14)$$

They can be expressed in functiun of the reduced damping α_k defined by:

$$\beta_k = 2.\alpha_k.\omega_k.\mu_k \qquad (15)$$

c/<u>Case in which $[\beta]$ is not diagonal</u>

When there is a viscous coupling between the various vibration modes, ($[\beta]$not being diagonal) and when the systems have low dissipation, the transfer functiun $[F(j\tilde{\omega})]$ can be expressed as a development in series of complex eigen functiuns.The following can therefore be written:

$$F(M_u, M_v, j\tilde{\omega}) = \sum_{k=1}^{n} \left(\frac{A_k(M_u, M_v)}{j\tilde{\omega} - p_k} + \frac{\bar{A}_k(M_u, M_v)}{j\tilde{\omega} - \bar{p}_k} \right) \qquad (16)$$

The residues A_k and $\bar{A}_k$ are no longer purely imaginary. Their imaginary part is determined only by the characteristics of mode "k" of the associate conservative system,whereas their real part involves the non diagonal coefficients of $|\beta|$.

d/ Expression of $F(M_u, M_v, j\tilde{\omega})$ near a eigen pulsation ω_r.

If it is supposed in the case of dissipative systems that the real parts $Re(A_k(M_u,M_v))$ are small infinitesimals of the 1st order, and that the pulsation of excitation is close to one of the eigen pulsations ω_r, it is possible to write, postulating:

$$A_k(M_u,M_v) = A'_k - j.A''_k \qquad (17)$$

$$F(M_u,M_v,j\tilde{\omega}) = H'_r.\exp(j.\frac{A'_r}{A''_r}) + \sum_{\substack{k \\ k\neq r}} H''_k \qquad (18)$$

for $\tilde{\omega} \cong \omega_r$

H'_r corresponds to the case in which the mode "r" is the only one involved (diagram is the same as a system with a single degree of freedom: circle arc). The term $j.A'_r/A''_r$ corresponds to the rotation caused by the terms of the coupling.

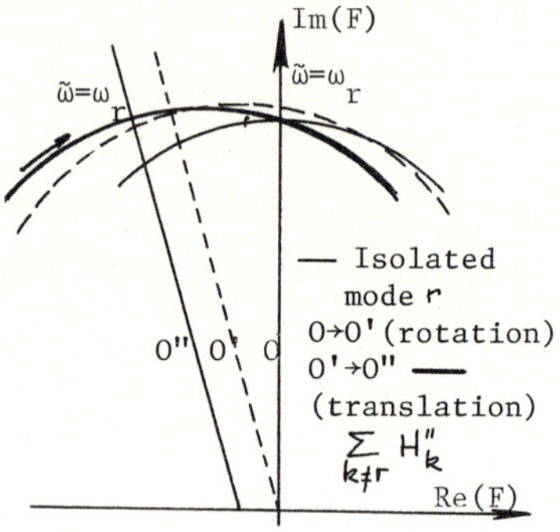

Fig.2 Diagram of the transfer functiun $F(M_u,M_v,j\tilde{\omega})$ near ω_r.

Translation $\sum_k H''_k$ (for $k\neq r$) is determined by the characteristics of inertia and rigidity of modes "k" for $k\neq r$.

2.23 Inverse problem

Although it is easy to determine the transfer functiun when the dynamic characteristics of the structure are known (direct problem), it is much more difficult to determine these dynamic characteristics when a limited number of tranfer functiuns is known (inverse problem). This inverse problem consists of determining the dynamic characteristics of a structure by comparing experimental and theoretical transfer functiuns . The rational fraction which is most compatible with the measured values is determined by mathematical smoothing method. According to the possibility of writing:

$$F(j\tilde{\omega}) = P(j\tilde{\omega}) / Q(j\tilde{\omega}) \qquad (19)$$

the breakdown into single elements is explained in functiun of the modal vibration characteristics of the structure. The smoothing method needs precautions to be taken in the choice of methods for matrix inversion and search for eigen values and eigen vectors. Attention must also be paid to the choice of degree for numerator and denominator of the rational fraction. The algorithm selected for the search of eigen values and eigen vectors is based on a method of simultaneous iterations .(Ref.4)

The breakdown of $F(M_u,M_v,j\tilde{\omega})$ reveals a set of mathematical parameters (poles p_k and residues $A_k(M_u,M_v)$ the determination of which leads to the physical parameters

The following are thus obtained:

- eigen pulsations $\omega_k = \text{Im}(p_k)$
- the ratio: generalised damping over generalised mass:

$$\beta_{kk}/\mu_k = -2.\ \text{Re}(p_k) \qquad (20)$$

- μ_k results of knowledge of the imaginary part of the residue A_k and so it is for eigen shapes $\phi_k(M_u)$.
- the terms β_{kr} result from a test in which the excitation is known and applied at a single point M_v; the responses are measured at all points.

3 TESTS

3.1 Geometry of the prototype

In both the cases examined -static and dynamic- the prototype comprises 9 cables and three struts (Fig.3a and 3b).Cables are of equal length. The ratio between the length of the struts and the length of the cables is 1.468, corresponding to practically zero-selfstressing.

Fig.3a. Static analysis prototype

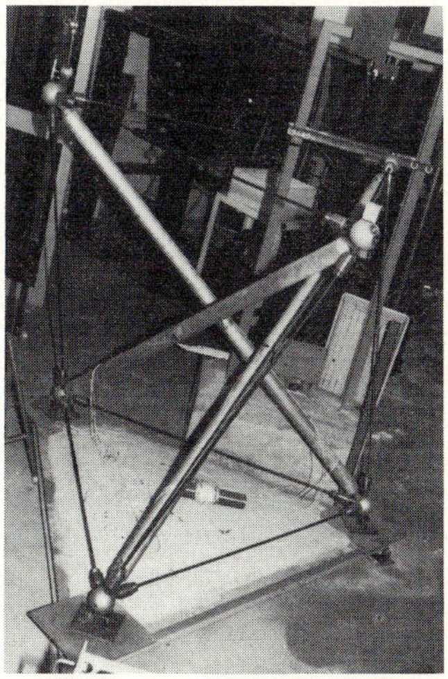

Fig.3b Dynamic analysis prototype (Mero nodes and struts)

3.2 Results of the static study

3.21 Test description

*Prototype: cables had an apparent diameter of 5.7mm;struts have an outside diameter of 42.4 mm and were 2.6 mm thick and 2 m long. The connection nodes had to provide good articulation for the members. They were easy to assemble and dismantle and also held the devices to tighten the cables and the parts to enable loading (upper nodes. See Fig.3a).

* <u>Supports</u> : the support devices were such that 6 degrees of freedom were fixed A bearing thrust block permitted movement in the plane of one node, a needle slide controlled the one directional movement of the second; the third node was fixed.

* <u>Loading</u> : Transmission of actions to the upper nodes was realised by a system of weights, levers and cables. This type of loading enables considerable displacement of the upper nodes while maintaining the actions constant. The loading schedule planned five steps each of 100 daN for each of the three upper nodes; this is a "one-dimensional compression test".

* <u>Measurements</u> : x,y and z displacements of all the nodes were determined by topometric measurements using three theodolites forming a triangular datum base. The accuracy obtained was very satisfactory, but this system required measurements which took a great deal of time.

3.22 <u>Results</u>

The results shown are for a model whose self-stressing is close to zero; they were compared to the theoretical results obtained with the mathematical model (§2.1)

Fig.4 Nodes displacements (OZ direction)

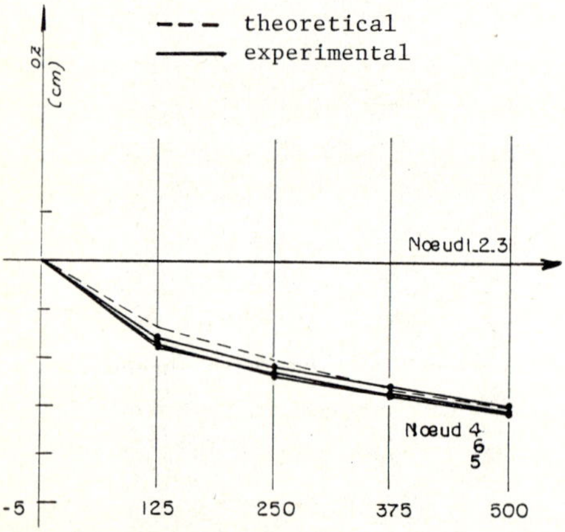

Fig.5 Forces in struts

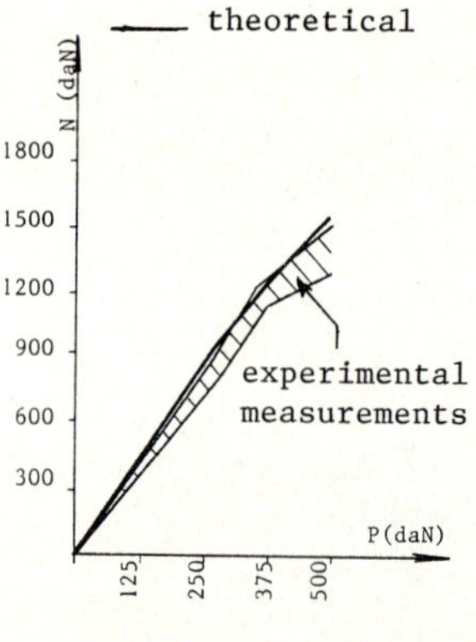

Fig.6 <u>Forces in cables</u>

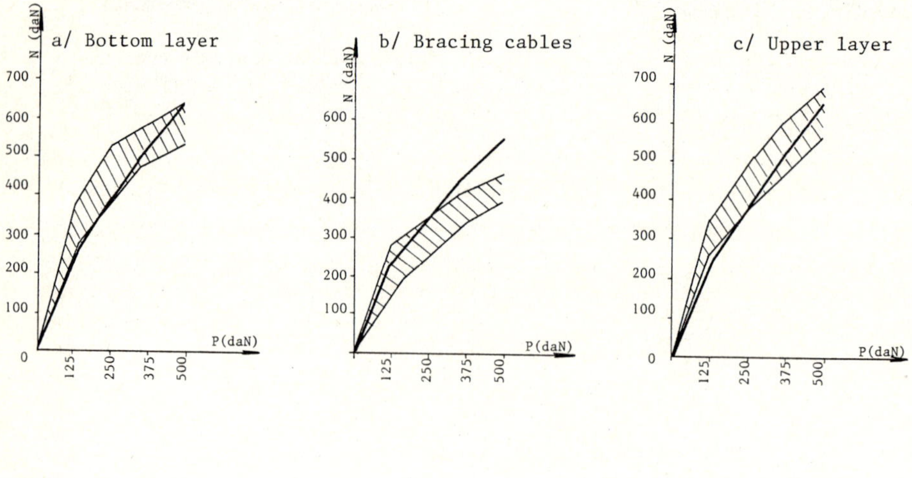

* Displacements : Vertical displacements were in conformity with the theoretical description (Fig.4). They are coherent for the three upper nodes, showing that the parallelism of the upper and lower triangles was maintained in the deformation.

The final deflection is almost identical to the theoretical value. The increasing rigidity of the system can be observed when the loads are applied; this was caused mainly by the tension which is created in the cables. The deviation observed at the start of loading is a result of not taking into account in the theoretical calculation the dead load of the system , whose effect diminished later, and the effect of the taking up of the mechanical play. Rotation of the two upper and lower triangles was in the order of 7° both in the experiment and in theory. Notable reduction of all displacements could be expected if self-stressing were not zero (this was confirmed by the theoretical results obtained with this hypothesis with five stages of self-stressing).

* Stresses : the stresses supported by the struts were some three times greater than those in the cables which nevertheless accumulated nearly 85% of the internal deformation energy: the struts play the role of'following'supports for the system of cables. Deviations between theory and experiment was observed, particularly for the bracing members. These resulted from a lack of accuracy in the execution of the connection modes of these cables with the nodes; this lack of accuracy affected the value of the angle of incidence of these members with the horizontal plane. It should also be noted that parasitic friction occured at the end of the test in the slide enabling unidirectional displacement of one of the support nodes.

Overall comparison of the four forces gaphs shows transfer of forces compared to theory from bracing cables and struts to the layer cables. This explains why the theoretical and experimental results obtained for the displacements agreed: theoretical and experimental rigidities agree in testing under one-directional compression. The overall rigidity thus obtained can then be used to construct a model of the behaviour of self-stressed reticulated space systems made up by the addition of elements similar to the prototype.

* Concluding remarks : in addition to the validation of the theoretical model, the test carried out demonstrated the "feasibility" of structural systems of an unusual type. The knowledge acquired during the test led us to use adapted industrial type assembly nodes (Mero system) for the new test.

3.3 Results of the test to determine the dynamic characteristics

The prototype possesses cables with an apparent diameter of 9 mm; the cables are 7-stranded and each strand is 0.6 mm in diameter. Tubes with an outside diameter of 60 mm and 1.8 mm thick were supplied by the manufacturer "Mero" with nodes whose holes threaded according to the angles of incidence of the Simplex, take the struts and anchorage tensioning apparatus for the cables.

Determination of the dynamic characteristics of the prototype is carried out by harmonic excitation of a node and measurement of the response of the other nodes. Details of the experimental procedure is provided in Ref. 5 . The electronic excitation device used was able to develop a force of up to 40 N (in a range of 1 to 500Hz) Displacements were measured using a proximity detector with a response of 0.15V/mm.

The transfer functiuns were measured using harmonic excitation by frequency scannings. Two measurements were made: a first coarse reading in which a frequency scan was applied to the whole range to be studied, and a second finer reading in the frequency ranges in which the eigen frequencies were located. The location of the natural frequencies is obtained using a criterion of maximum amplitude. Fig. 7 illustrates the results obtained: the plotting of the module and of the argument of the transfer functiun between excitation and one of the displacements shows the existence of six eigen modes in the 0 to 100 Hz: 6, 10, 14.5, 23, 49, and 69 Hz. The results obtained guarantee the independence of the eigen frequencies in relation to the points of measurement of the response.

It is therefore possible to make a more accurate seep around the resonance frequency (Fig.8); by entering in the complex plan the ../..

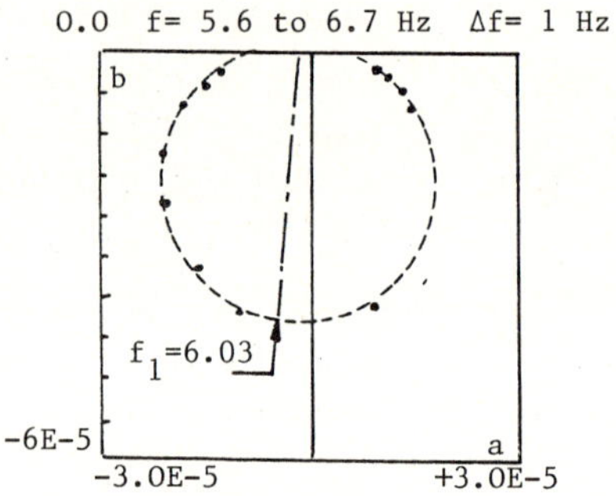

• measured points
a= Re(y/p) m/N
b= Im(y/p) m/N

Fig.8 Transfer functiun near eigen frequency f_k= 6Hz

Fig.7 Module of the transfer functiun (y/p)

y displacement in z direction of an upper node

p excitation force at node 4 in direction Oz

• measured points

+ calculated points

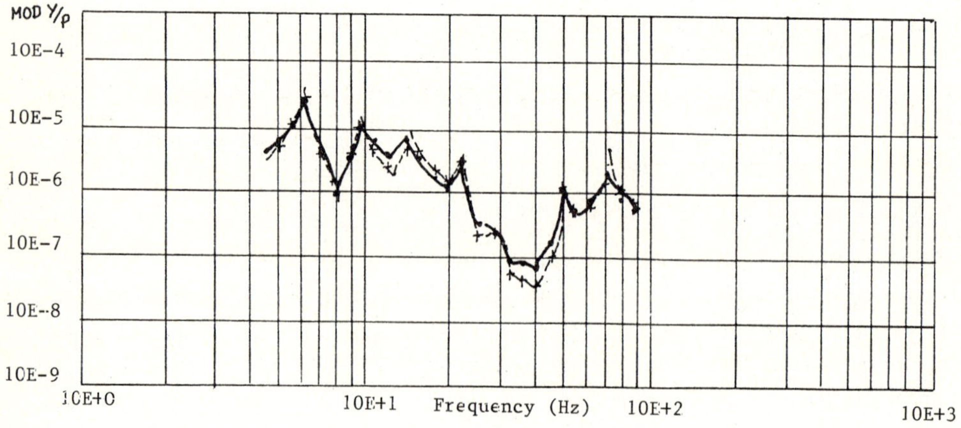

the polar diagram of the transfer functiun measured, taking in account the frequency parameter, the characteristic circular arc of the single degree of freedom system is obtained. The contribution of the other eigen modes and of the other non diagonal terms of the damping matrix is revealed by the displacement (translation plus rotation) of the circular arc. Comparison with the theoretical values entered on the same diagram is satisfactory (Fig. 8)

CONCLUSION

The two studies presented made it possible to gain better knowledge of the static behaviour and dynamic characteristics of tensegrity systems. The mathematical models can be used with any degree of self-stressing; the tests were carried out with practically zero self-stressing and should be continued with increasing degrees of self-stress. Coherence between the theoretical and experimental results can be considered to be good.

This theoretical and experimental work brings information which was hiterto been lacking and whose integration in the overall study described at the beginning of this text should make it possible to construct very light structures combining compression and tension members to make systems which do not require the usual anchor blocks that go with tension structures.

REFERENCES

1 MOTRO R. Formes et Forces dans les Systèmes Constructifs. Cas des Systèmes Réticulés Spatiaux Autocontraints. Thèse d'Etat. Montpellier 1983

2 VALID R. La mécanique des milieux continus et le calcul des Structures. Editions Eyrolles. 1977. p132.

3 DE LARMINAT P., THOMAS Y. Automatique des Systèmes linéaires. Tome 1: Signaux et Systèmes Flammarion 1975. §2.5,p.239

4 LALANNE M., TROMPETTE P., CHAMBAUD M.T. Recherche des valeurs propres et vecteurs propres par itération simultanée. Rapport D.R.M.E. Fév. 1972.

5 NAJARI S. Identification par analyse harmonique de Structures en Génie Civil. Application de la méthode non appropriée et comparaison à la méthode appropriée. Thèse de 3° Cycle. Toulouse Décembre 1985.

STABILITY OF THIN-WALLED BEAMS - A GENERAL THEORY

Yeong-Bin Yang
Department of Civil Engineering
National Taiwan University, Taipei, Taiwan
China

INTRODUCTION

The classic theory of thin-walled beams was developed by Vlasov during the 1930s (1961). This theory is more general than the Euler-Bernoulli-Navier theory of solid beams, in that the cross sections are allowed to warp nonuniformly along the axis of a member. In addition to the St. Venant torque caused by uniform warping, two actions are generated by variation of warping from cross section to cross section of a bar. One is the torque generated by resistance to warping deformation and the other is the bimoment that results from self-equilibrating direct stresses. The presence of such actions influences not only the linear but also the nonlinear behavior of beams.

A thin-walled beam may buckle by flexure, by torsion, or by a combination of flexure and torsion, which depends upon the the type of loadings and the support conditions. In an eigenvalue analysis, the buckling shape and the associated critical load may be determined solely from the governing differential equations and the prescribed boundary conditions. In the present discussions, although reference will be made also to a potential energy expression or to an element stiffness equation, one recognizes that for conservative systems, both of them are closely related to the differential equations and they contain just as much information as that of the associated differential equations.

In principle, the differential equations are subjected to some theoretical restrictions that characterize the capability or incapability of each theory. For instance, the buckling equations derived by Vlasov (1961) consider only the effect of axial compression and bending moments. Similar limitations are present in the theories derived by Goodier (1941), Bleich (1952), and Timoshenko and Gere (1961). The effect of applied torques on stability of bars was considered approximately by Barsoum and Gallagher (1970) in their potential energy expression. The same effect was considered by Chen and Atsuta using an equilibrium approach (1977).

It is interesting to note that in Yoo's (1980) potential energy expression, the effect of an applied bimoment on the stability of thin-walled bars was included, while the effect of an applied torque was neglected. Recent efforts on the research of thin-walled beams include the work of Gjelsvik (1981). Although his final expressions of the buckling equations turn out to be similar to those of Vlasov's, the effect of warping across the thickness of the shell is duly taken into account. Using an updated Lagrangian approach, Yang and McGuire (1986) established an element stiffness matrix for thin-walled beams of bisymmetrical sections. This formulation allows the inclusion of all types of loading.

This paper intends to derive the differential equations of equilibrium for the stability of thin-walled beams of an arbitrary open cross section. As the equations are derived for a deformed bar, various nonlinear effects due to the geometric change are considered. Of interest is the investigation of the instability of a bar subjected to an applied torque, whether semitangential or quasitangential, with various boundary conditions.

NONLINEAR THEORY OF THIN-WALLED BEAMS

In a geometrically nonlinear analysis, the equations of equilibrium are written for a body in the deformed configuration. Based on the principle of virtual displacements, the equation of equilibrium of the deformed body can be expressed as:

$$\int_{\bar{V}} \tau_{ij}\, \delta e_{ij}\, d\bar{V} = \int_{\bar{S}} t_k\, \delta u_k\, d\bar{S} \tag{1}$$

if the body force is neglected, where τ_{ij} = the Cauchy stress tensor, e_{ij} = the linear or Almansi strain tensor, t_k = the vector of surface tractions, δu_k = a variation of the displacement components, and $\bar{V}$ = the volume and $\bar{A}$ = the surface area of the deformed body. In the above equation, it is assumed that all quantities occur in the deformed state and are measured in the same state.

Using a Lagrangian description, the equilibrium expression of (1) can be transformed to an equation referring to the initial undeformed configuration (Bathe 1982):

$$\int_V S_{ij}\, \delta\varepsilon_{ij}\, dV = \int_S t_k\, \delta u_k\, dS \tag{2}$$

where S_{ij} = second Piola-Kirchhoff stress tensor, ε_{ij} = Green-Lagrange strain tensor, t_k = surface tractions, V = volume, and S = surface area of the undeformed body.

<u>Three-Dimensional Beam</u>. In the present formulation, a right-handed coordinate system x, y, and z will be used such that x coincides with the centroidal axis, and y and z coincide with the principal axes of the cross section in the initial configuration (Fig. 1). Based on the engineering theory of a three-dimensional beam, there are only three independent stress components and three independent strain components, i.e.,

$$S_x = S_{11}, \quad S_{xy} = S_{12}, \quad S_{xz} = S_{13}, \tag{3}$$

and

$$\varepsilon_x = \varepsilon_{11}, \quad \gamma_{xy} = 2\varepsilon_{12}, \quad \gamma_{xz} = 2\varepsilon_{13}; \tag{4}$$

the other components of the stress and strain being assumed to vanish. As a result, the equilibrium expression of a deformed three-dimensional beam can be obtained from Eq. 2 as:

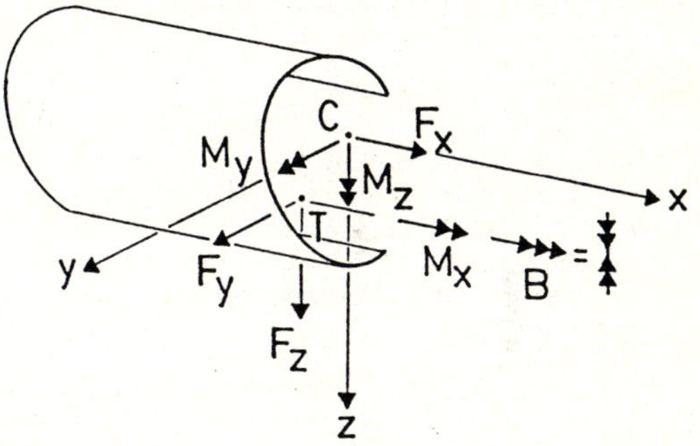

Fig. 1 Coordinates of a Thin-Walled Bar

$$\int_V [S_x\, \delta\varepsilon_x + S_{xy}\, \delta\gamma_{xy} + S_{xz}\, \delta\gamma_{xz}]\, dV = \int_S t_k\, \delta u_k\, dS \tag{5}$$

In a large-displacement analysis, the strain-displacement relations are:

$$\begin{aligned} \varepsilon_{xx} &= u_{x,x} + \tfrac{1}{2}(u_{x,x}^2 + u_{y,x}^2 + u_{z,x}^2) \\ \gamma_{xy} &= u_{x,y} + u_{y,x} + u_{x,y}\, u_{x,x} + u_{y,y}\, u_{y,x} + u_{z,y}\, u_{z,x} \\ \gamma_{xz} &= u_{x,z} + u_{z,x} + u_{x,z}\, u_{x,x} + u_{y,z}\, u_{y,x} + u_{z,z}\, u_{z,x} \end{aligned} \tag{6}$$

where, u_x, u_y, and u_z = displacements along the x, y, and z axes, respectively, of an arbitrary point on the section contour of the deformed bar. Substitution of the above relations into Eq. 2 yields:

$$\int_V S_x \,\delta u_{x,x}\,dV + \int_V [S_{xy}\,\delta(u_{x,y} + u_{y,x}) + S_{xz}\,\delta(u_{x,z} + u_{z,x})]\,dV + \frac{1}{2}\int_V S_x \,\delta(u_{y,x}^2 + u_{z,x}^2)\,dV$$

$$+ \int_V [S_{xy}\,\delta(u_{x,y}u_{x,x} + u_{z,y}u_{z,x}) + S_{xz}\,\delta(u_{x,z}u_{x,x} + u_{y,z}u_{y,x})]\,dV = \int_S t_k \,\delta u_k \,dS \tag{7}$$

where it is noted that $u_{y,y} = u_{z,z} = 0$ according to the engineering beam assumptions. In arriving at Eq. 7, the term containing the square of the derivative of u_x in the axial strain has been neglected as usual (Washizu 1975).

<u>Sectional Displacements and Stress Resultants.</u> Vlasov's (1961) assumptions for thin-walled beams include: 1) the beam is shallow, 2) the section is thin-walled, 3) there is no in-plane distortion of the cross section, and 4) the shearing deformation of the profile is negligible. According to the hypothesis of in-plane rigid cross sections, the transverse displacements u_y and u_z of an arbitrary point, N, (Fig. 2), may be written as

$$u_y = u_{yT} - (z-a_z)\theta_x, \qquad u_z = u_{zT} + (y-a_y)\theta_x \tag{8}$$

if the angle of twist θ_x is reasonably small, where (a_y, a_z) = location and (u_{yT}, u_{zT}) = transverse displacements of the shear center, T, respectively. The longitudinal displacement u_x of point N is found by utilizing the hypothesis concerning the absence of shearing strain in the profile (Vlasov 1961):

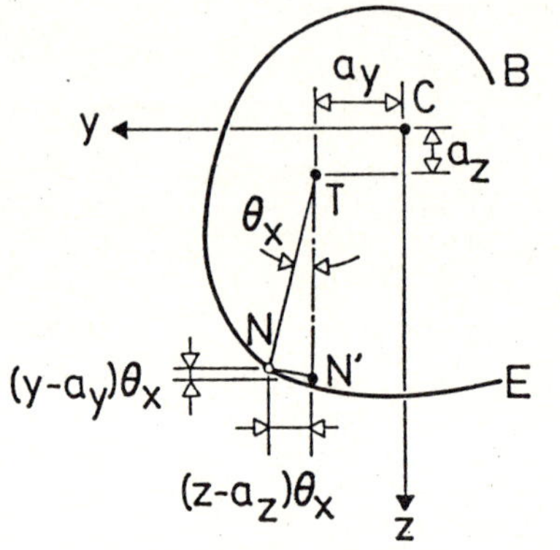

Fig. 2 Displacement of a Point on the Section Contour

$$u_x = u_{xC} - zu'_{zT} - yu'_{yT} - \omega\theta'_x \tag{9}$$

where u_{xC} = axial displacement of the centroid, C, and ω = normalized sectorial area. The fourth term on the right-hand side of Eq. 8 determines the part of displacement arising as a result of sectorial warping, which does not obey the law of plane sections.

The stress resultants on a cross section of a member are referred to the initial state of that cross section. For a thin-walled beam, they are:

$$F_x = \int_A S_x \,dA, \quad F_y = \int_A S_{xy}\,dA, \quad F_z = \int_A S_{xz}\,dA,$$

$$M_x = \int_A [S_{xz}\,(y-a_y) - S_{xy}\,(z-a_z)]\,dA \tag{10}$$

$$M_y = \int_A S_x \,z dA, \quad M_z = -\int_A S_x \,y dA, \quad B = -\int_A S_x \,\omega dA$$

where A = cross sectional area. The bimoment B, an additional bar force occurring in thin-walled bars, has the dimension of moment multiplied by length.

<u>Virtual Work Due to Linear Axial Strain.</u> Using Eq. 9 for the longitudinal displacement and Eqs. 10 for stress resultants, one obtains the first integral in Eq. 7 as follows:

$$\int_V S_x \,\delta u_{x,x}\,dV = \int_0^L (F_x \,\delta u'_{xC} - M_y \,\delta u''_{zT} + M_z \,\delta u''_{yT} + B \,\delta\theta''_x)\,dx \tag{11}$$

where L = length of the beam. In a linearized theory, the stress resultants can be expressed in terms of the displacements by the same relations as in the linear theory:

$$F_x = EAu'_{xC}, \quad M_y = -EI_y u''_{zT}, \quad M_z = EI_z u''_{yT}, \quad B = EC_w \theta''_x \tag{12}$$

where I_y, I_z = moments of inertia about the y and z axes, and and C_w = warping constant. Consequently, the first integral in Eq. 7 reads:

$$\int_V S_x \delta u_{x,x} dV = \frac{1}{2} \int_0^L [EA\delta(u'^2_{xC}) + EI_y \delta(u''^2_{zT}) + EI_z \delta(u''^2_{yT}) + EC_w \delta(\theta''^2_x)]dx \tag{13}$$

Clearly, this equation expresses the variation of the strain energy of the thin-walled beam that is associated with the axial, flexural, and warping deformations.

Virtual Work Due to Linear Shear Strains. Similarly, substitution of Eqs. 8 for u_y and u_z into the second integral of Eq. 7 results in the following expression:

$$\int_V [S_{xy} \delta(u_{x,y} + u_{y,x}) + S_{xz} \delta(u_{x,z} + u_{z,x})]dV$$

$$= \int_0^L \int_A \{S_{xy} [-(z - a_z) - \frac{\partial\omega}{\partial y}] + S_{xz} [(y - a_y) - \frac{\partial\omega}{\partial z}]\} \delta\theta'_x \, dAdx \tag{14}$$

Because the shear strains due to flexural bending and nonuniform torsion are neglected, the shear strains in the above equation are caused by the St. Venant torque alone. Such strains may be given as (Timoshenko and Goodier 1970):

$$S_{xy} = -G\theta'_x [(z - a_z) + \frac{\partial\omega}{\partial y}], \quad S_{xz} = G\theta'_x [(y - a_y) - \frac{\partial\omega}{\partial z}] \tag{15}$$

Substitution of the above shear stresses into Eq. 14 yields:

$$\int_V [S_{xy} \delta(u_{x,y} + u_{y,x}) + S_{xz} \delta(u_{x,z} + u_{z,x})]dV = \frac{1}{2} \int_0^L GJ\delta(\theta'^2_x)dx \tag{16}$$

where J = torsional constant, which can be recognized as (Love 1944):

$$J = \int_A \{[(z - a_z) + \frac{\partial\omega}{\partial y}]^2 + [(y - a_y) - \frac{\partial\omega}{\partial z}]^2\}dA \tag{17}$$

The total strain energy, in the variational form, is the sum of those given in Eqs. 13 and 16, which turns out to be identical with the one used in Bleich (1952), Barsoum and Gallagher (1970), and Yoo (1980).

Virtual Work Due to Nonlinear Axial Strain. Substituting Eqs. 8 for the transverse displacements into the third integral on the left-hand side of Eq. 7, and making use of the expressions for stress resultants (Eqs. 10), one obtains:

$$\frac{1}{2} \int_V S_x \delta(u^2_{y,x} + u^2_{z,x})dV$$

$$= \frac{1}{2} \int_0^L F_x \delta[u'^2_{yT} + u'^2_{zT} + 2a_z u'_{yT} \theta'_x - 2a_y u'_{zT} \theta'_x]dx + \frac{1}{2} \int_0^L \bar{K}\delta(\theta'^2_x)dx$$

$$- \int_0^L M_y \delta(u'_{yT} \theta'_x)dx - \int_0^L M_z \delta(u'_{zT} \theta'_x)dx \tag{18}$$

where the Wagner coefficient $\bar{K}$ is:

$$\bar{K} = \int_A S_x [(z-a_z)^2 + (y-a_y)^2]dA \tag{19}$$

For a thin-walled beam, four actions may contribute to the axial stress (Vlasov 1961):

$$S_x = \frac{F_x}{A} + \frac{M_y z}{I_y} - \frac{M_z y}{I_z} - \frac{B\omega}{C_w} \tag{20}$$

Accordingly,

$$\bar{K} = F_x r^2 + M_y \beta_z - M_z \beta_y - B\beta_w \tag{21}$$

where the following notation is introduced:

$$r^2 = \frac{I_y + I_z}{A} + a_y^2 + a_z^2, \qquad \beta_z = \int_A z(y^2+z^2)dA/I_y - 2a_z$$

$$\beta_y = \int_A y(y^2+z^2)dA/I_z - 2a_y, \qquad \beta_w = \int_A \omega(y^2+z^2)dA/C_w \tag{22}$$

Equation 18 represents the potential of the applied loads in the variational form that is associated with the axial stress. It should be mentioned that expressions similar to Eq. 18 were used in Yoo's (1980) analysis.

Virtual Work Due to Nonlinear Shear Strains. By performing the following steps to the fourth and last integral on the left-hand side of Eq. 7: 1) substituting the expressions for the displacements (Eqs. 8 and 9), 2) neglecting the higher order terms, and 3) utilizing the stress resultant expressions (Eqs. 10), one obtains:

$$\int_V [S_{xy}\, \delta(u_{x,y}\, u_{x,x} + u_{z,y}\, u_{z,x}) + S_{xz}\, \delta(u_{x,z}\, u_{x,x} + u_{y,z}\, u_{y,x})]dV$$

$$= - \int_0^L F_y\, \delta(u'_{yT}\, u'_{xC})dx - \int_0^L F_z\, \delta(u'_{zT}\, u'_{xC})dx + \int_0^L F_y\, \delta(u'_{zT}\, \theta_x)dx$$

$$- \int_0^L F_z\, \delta(u'_{yT}\, \theta_x)dx + \int_0^L (\alpha M_x + F_z\, a_y)\delta(u'_{zT}\, u''_{yT})dx$$

$$- \int_0^L [(1-\alpha)M_x - F_y\, a_z]\delta(u'_{yT}\, u''_{zT})dx \tag{23}$$

where

$$\alpha = \int_A S_{xz}\, y dA \,/\, M_x = 1 - \int_A (-S_{xy}\, z)dA \,/\, M_x \tag{24}$$

The parameter, α, as a function of x, depends not only on the warping boundary conditions, but also on the shape of the cross section of a bar.

Equation 23 gives the potential of the applied loads in the variational form. With regard to this equation, several points can be made here.

1) The terms containing M_x, $F_y a_z$, and $F_z a_y$ represent the potential of torsion in a variational form. Their presence makes possible the study of the torsional-flexural buckling of a bar under the action of an applied torque. Specifically, the terms with $F_y a_z$ and $F_z a_y$ are generated as a result of transformation of a torque from the shear center axis to the centroidal axis, which may be appreciated from the differential equations to be presented later.

2) The terms containing $(u'_{yT}\, \theta_x)$ and $(u'_{zT}\, \theta_x)$ represent another component of the torsional-flexural interaction. They may be examined together with the terms containing $(u'_{yT}\, \theta'_x)$ and $(u'_{zT}\, \theta'_x)$ in Eq. 18.

3) The terms with $(u'_{yT}\, u'_{xC})$ and $(u'_{zT}\, u'_{xC})$ result from the interaction between axial and flexural deformations.

No expression similar or equivalent to Eq. 23 has been presented in Vlasov, Timoshenko, Goodier, Bleich, or Yoo's formulations. Thus, these formulations are by themselves unable to account for the effects stated above. In particular, they can not be used to detect the instability of a bar under the action of a torque. Note that such an effect has been considered by Barsoum and Gallagher (1970) in an approximate sense in that the terms with $F_y a_z$ and $F_z a_y$ were overlooked. Using an equilibrium approach, this effect was later considered by Chen and Atsuta (1977). The same effect was also considered by Yang and McGuire (1986) for a bisymmetrical section in their finite element analysis.

Virtual Work Equation of Equilibrium for a Deformed Bar. Combining the expressions for the variation of strain energy in Eqs. 13 and 16, and those for the potential in Eqs. 18 and 23, one obtains from Eq. 7 the equation of equilibrium for a deformed thin-walled bar:

$$\frac{1}{2}\int_0^L [EA\delta(u_{xC}'^2) + EI_y\,\delta(u_{zT}''^2) + EI_z\,\delta(u_{yT}''^2) + EC_w\,\delta(\theta_x''^2) + GJ\delta(\theta_x'^2)]dx$$

$$+\frac{1}{2}\int_0^L F_x\,\delta[u_{yT}'^2 + u_{zT}'^2 + 2a_z\,u_{yT}'\,\theta_x' - 2a_y\,u_{zT}'\,\theta_x']dx + \frac{1}{2}\int_0^L \bar{K}\delta(\theta_x'^2)dx$$

$$-\int_0^L M_y\,\delta(u_{yT}'\,\theta_x')dx - \int_0^L M_z\,\delta(u_{zT}'\,\theta_x')dx - \int_0^L F_y\,\delta(u_{yT}'\,u_{xC}')dx$$

$$-\int_0^L F_z\,\delta(u_{zT}'\,u_{xC}')dx + \int_0^L F_y\,\delta(u_{zT}'\,\theta_x)dx - \int_0^L F_z\,\delta(u_{yT}'\,\theta_x)dx$$

$$+\int_0^L (\alpha M_x + F_z\,a_y)\delta(u_{zT}'\,u_{yT}'')dx - \int_0^L [(1-\alpha)M_x - F_y\,a_z]\delta(u_{yT}'\,u_{zT}'')dx$$

$$= \int_0^L (q_x\,\delta u_{xC} + q_y\,\delta u_{yT} + q_z\,\delta u_{zT} + m_x\,\delta\theta_x)dx$$

$$+ [F_x^*\,\delta u_{xC} + F_y^*\,\delta u_{yT} + F_z^*\,\delta u_{zT} + M_x^*\,\delta\theta_x - M_y^*\,\delta u_{zT}' + M_z^*\,\delta u_{yT}' + B^*\,\delta\theta_x']_0^L \tag{25}$$

in which m_x = distributed torque, and q_x, q_y, and q_z = distributed forces. The axial force (q_x) is acting at the centroid, and the torque (m_x) and the transverse forces (q_y, q_z) are acting about and at the shear center, respectively. The forces with a superscript of "*" on the right-hand side of Eq. 25 denote the concentrated nodal loads.

Nonlinear Differential Equations for Thin-Walled Beams. Integrating Eq. 25 by parts and admitting the arbitrary nature of virtual displacements: δu_{xC}, δu_{yT}, δu_{zT}, $\delta\theta_x$, one obtains a set of nonlinear differential equations for the thin-walled beam:

$$EAu_{xC}'' - (F_y\,u_{yT}')' - (F_z\,u_{zT}')' = -q_x \tag{26a}$$

$$EI_z\,u_{yT}'''' - (F_x\,u_{yT}')' - (F_x\,a_z\,\theta_x')' + (F_z\,\theta_x)' + (M_y\,\theta_x')' + (F_y\,u_{xC}')'$$
$$+ (M_x\,u_{zT}'')' - (F_y\,a_z\,u_{zT}'')' + (F_z\,a_y\,u_{zT}')'' = q_y \tag{26b}$$

$$EI_y\,u_{zT}'''' - (F_x\,u_{zT}')' + (F_x\,a_y\,\theta_x')' - (F_y\,\theta_x)' + (M_z\,\theta_x')' + (F_z\,u_{xC}')'$$
$$- (M_x\,u_{yT}'')' + (F_y\,a_z\,u_{yT}')'' - (F_z\,a_y\,u_{yT}'')' = q_z \tag{26c}$$

$$EC_w\,\theta_x'''' - GJ\theta_x'' - (\bar{K}\theta_x')' + M_y\,u_{yT}'' - (F_x\,a_z\,u_{yT}')' + M_z\,u_{zT}''$$
$$+ (F_x\,a_y\,u_{zT}')' = m_x \tag{26d}$$

Several points can be drawn from the above equations:

1) Equations 26 state the equilibrium of a thin-walled bar in the deformed configuration, in which the geometric nonlinearities due to various types of loads, including the axial force, shear forces, bending moments, and torque, has been considered.

2) Two types of interaction between bending and torsion need be distinguished. One is the influence of the bending moments (M_y and M_z) on the torsional deformation, and, therefore, on the torsional equation of equilibrium. The other is the influence of the torque (M_x) on the flexural equilibrium equations. Both effects are essential to the study of instability of a thin-walled bar subjected either to a bending moment, a torque, or their combinations.

3) The terms: $F_y u_{yT}''$ and $F_z u_{zT}''$ in Eq. 26a, $F_y u_{xC}''$ in Eq. 26b, and $F_z u_{xC}''$ in Eq. 26c, are usually neglected in practice.

4) Since the torque and bending moments are acting about the axes of the shear

center and of the centroid, respectively, they must be referred to the same axis when writing a flexural or torsional equilibrium equation. The terms with the a_y and a_z in the above equations are generated by the transformation of torques or bending moments from one axis to the other.

5) The bimoment contribution to stability is contained in the Wagner coefficient $\bar{K}$ given in Eq. 21.

As a by-product of the variational procedure, general natural boundary conditions may be obtained for each degree of freedom at both ends of the beam:

$$F_x^* = EAu'_{xC} - F_y u'_{yT} - F_z u'_{zT}$$

$$F_y^* = - EI_z u'''_{yT} + F_x u'_{yT} - (M_y - F_x a_z)\theta'_x - F_z \theta_x - F_y u'_{xC}$$
$$- (M_x - F_y a_z + F_z a_y)u''_{zT}$$

$$F_z^* = - EI_y u'''_{zT} + F_x u'_{zT} - (M_z + F_x a_y)\theta'_x + F_y \theta_x - F_z u'_{xC}$$
$$+ (M_x - F_y a_z + F_z a_y)u''_{yT}$$

$$M_x^* = - EC_w \theta'''_x + GJ\theta'_x + \bar{K}\theta'_x - (M_y - F_x a_z)u'_{yT} - (M_z + F_x a_y)u'_{zT} \quad (27)$$

$$M_y^* = - EI_y u''_{zT} + [(1 - \alpha)M_x - F_y a_z]u'_{yT}$$

$$M_z^* = EI_z u''_{yT} + (\alpha M_x + F_z a_y)u'_{zT}$$

$$B^* = EC_w \theta''_x$$

The linearized differential equations (Eqs. 26) along with the natural boundary conditions (Eqs. 27) can be used as a valid basis for a large displacement analysis of thin-walled beams.

TORSIONAL-FLEXURAL INSTABILITY OF BARS WITH AN APPLIED TORQUE

Thin-walled beams that are subjected to the action of an axial compression and/or bending moments may buckle in the form of flexure, torsion, or a combination of flexure and torsion. Such problems of instability of thin-walled beams have been thoroughly investigated by many researchers (Goodier 1941, Timoshenko and Gere 1961, Vlasov 1961). In contrast, very little work has been done on the instability of a bar under the action of a torque. In this section, an attempt is made to study the instability of a bar under the action of a torque. For illustration, a bisymmetrical section will be considered. For the present case, the differential equations may be reduced from Eq. 26 with the subscripts "T" dropped as

$$EI_z u''''_y + Mu'''_z = 0 \quad \text{and} \quad EI_y u''''_z - Mu'''_y = 0 \quad (28)$$

Two types of conservative torques will be considered here, the semitangential and the quasitangential torque (Ziegler 1977, Yang and McGuire 1986). If a torque (M) is semitangential, two moment increments, $Mu'_{yL}/2$ and $Mu'_{zL}/2$, acting about the y and z axes, respectively, may be generated by the torque at the free end undergoing buckling deformations (Fig. 3). It follows that Eq. 28 may be integrated to yield

$$EI_z u''_y + Mu'_z = \frac{1}{2} Mu'_{zL} \quad \text{and} \quad EI_y u''_z - Mu'_y = - \frac{1}{2} Mu'_{yL} \quad (29)$$

or

$$u_y'' + m_2 u_z' = \frac{1}{2} m_2 u_{zL}' \quad \text{and} \quad u_z'' - m_1 u_y' = -\frac{1}{2} m_1 u_{yL}' \tag{30}$$

with

$$m_1 = \frac{M}{EI_y} \quad \text{and} \quad m_2 = \frac{M}{EI_z} \tag{31}$$

The boundary conditions for the column are:

$u_y = 0$, $u_y' = 0$, $u_z = 0$, and $u_z' = 0$, at $x = 0$

and

$u_y' = u_{yL}'$, and $u_z' = u_{zL}'$, at $x = L$.

By differentiating the first of Eqs. 30 once and making use of the second equation, one obtains a single differential equation as:

$$u_y''' + m^2 u_y' = \frac{1}{2} m^2 u_{yL}' \quad \text{where} \quad m^2 = m_1 m_2 \tag{32}$$

of which the general solution is

$$u_y = c_1 \cos mx + c_2 \sin mx + c_3 + \frac{1}{2} u_{yL}' x \tag{33}$$

Similarly, the solution for u_z is

$$u_z = d_1 \cos mx + d_2 \sin mx + d_3 + \frac{1}{2} u_{zL}' x \tag{34}$$

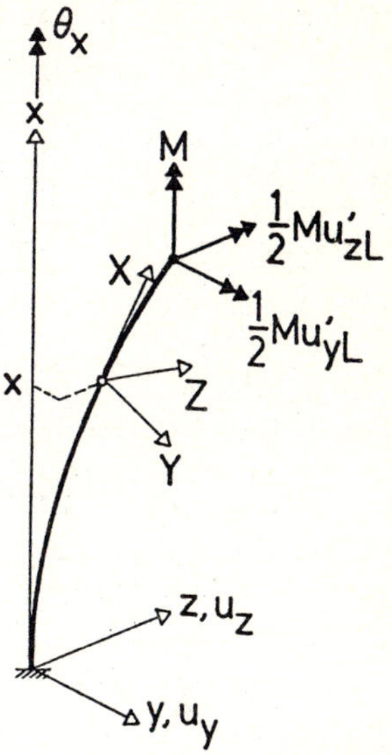

Fig. 3 Buckling of a Column Subject to a Torque

In the reduction of the two-variable system (Eqs. 30) to the single-variable system (Eq. 32), two extraneous coefficients have been introduced. This problem can be resolved by substituting the solutions for u_y and u_z into the original equations and making sure that they are identically satisfied. For instance, by substituting the solutions for u_y and u_z into the first of Eqs. 30, one obtains

$$d_1 = -c_2 \frac{m}{m_2} \quad \text{and} \quad d_2 = c_1 \frac{m}{m_2} \tag{35}$$

Thus u_z may be rewritten as:

$$u_z = -c_2 \frac{m}{m_2} \cos mx + c_1 \frac{m}{m_2} \sin mx + d_3 + \frac{1}{2} u_{zL}' x \tag{36}$$

The solutions for u_y and u_z (Eqs. 33 and 36) must satisfy the boundary conditions. By substituting the expressions for u_y and u_z into the six boundary conditions, one obtains a set of linear equations as:

$$\begin{bmatrix} 1 & 0 & 1 & 0 & 0 & 0 \\ 0 & m & 0 & 0 & 1/2 & 0 \\ -m \sin mL & m \cos mL & 0 & 0 & -1/2 & 0 \\ 0 & -m/m_2 & 0 & 1 & 0 & 0 \\ m_1 & 0 & 0 & 0 & 0 & 1/2 \\ m_1 \cos mL & m_1 \sin mL & 0 & 0 & 0 & -1/2 \end{bmatrix} \begin{Bmatrix} c_1 \\ c_2 \\ c_3 \\ d_3 \\ u_{yL}' \\ u_{zL}' \end{Bmatrix} = \begin{Bmatrix} 0 \\ 0 \\ 0 \\ 0 \\ 0 \\ 0 \end{Bmatrix} \tag{37}$$

Further, by letting the determinant of the linear equations equal to zero, one obtains the characteristic equation:

$$\frac{1}{2} m m_1 (1 + \cos mL) = 0 \tag{38}$$

which is satisfied whenever

$$\cos mL = -1 \tag{39}$$

The critical load is given by $mL = \pm \pi$, namely,

$$M_{cr} = \pm \frac{\pi E}{L}(I_y I_z)^{1/2} \tag{40}$$

The two signs correspond to the fact that the sense of the torque is arbitrary. The result can be written in a formula similar to Greenhill's,

$$M_{cr} = \pm \frac{k\pi E}{L}(I_y I_z)^{1/2} \tag{41}$$

For the present case, the numerical factor k has the value of 1.

Next, assume the torque is quasitangential. In this case, only a moment increment Mu'_{yL}, acting about the y axis, will be generated during buckling at the free end. Thus, the differential equations become

$$u''_y + m_2 u'_z = 0 \quad \text{and} \quad u''_z - m_1 u'_y = - m_1 u'_{yL} \tag{42}$$

The same boundary conditions may be used except that $u'_z = u'_{zL}$ at $x = L$ is no longer needed. Following the same procedure, one obtains the characteristic equation

$$mm_1 \cos mL = 0 \tag{43}$$

from which

$$mL = \pm \pi/2 \tag{44}$$

This solution corresponds to $k = 1/2$ in the formula of Eq. 41.

The problem can be solved in a similar manner for a bisymmetrical shaft with four other types of end conditions (Fig. 4). Complete results are listed below. It is interesting to note that for the special case of a circular shaft ($I_y = I_z = I$), the present solutions coincide exactly with Ziegler's (1977).

Torques	Critical Load Factor k				
	Type 1	Type 2	Type 3	Type 4	Type 5
Semitangential	2.861	2.000	1.000	2.168	1.564
Quasitangential	2.861	2.000	0.500	1.576	1.000

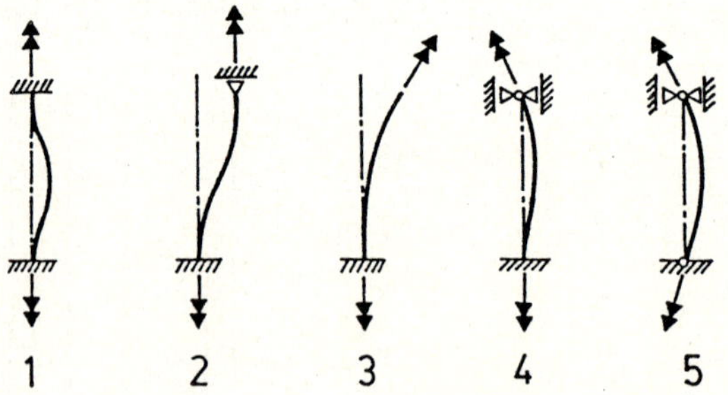

Fig. 4 Columns with Various Boundary Conditions

SUMMARY AND CONCLUSIONS

In the large displacement analysis of a thin-walled bar, the equations of equilibrium are written for the bar in the deformed configuration. This results in the inclusion of instability effects associated with various types of loading in the governing differential equations. Using such equations, one is able to study various torsional-flexural buckling problems of a thin-walled bar. In this paper, the torsional-flexural buckling of a bar under the action of a torque is examined.

ACKNOWLEDGEMENT

This research is sponsored by the National Science Council of the Republic of China under Grant No. NSC-74-0414-P-002-21.

APPENDIX. - REFERENCES

Barsoum, R. S., and Gallagher, R. H., (1970). "Finite Element Analysis of Torsional-Flexural Stability Problems," International Journal for Numerical Methods in Engineering, Vol. 2, pp. 335-352.

Bathe, K.-J., (1982). Finite Element Procedures in Engineering Analysis, Prentice-Hall, Inc., Englewood Cliffs, New Jersey.

Bleich, F., (1952). The Buckling Strength of Metal Structures, McGraw-Hill Book Co., Inc., New York, N.Y.

Chen, W. F., and Atsuta, T., (1977). Theory of Beam-Columns, Vol. 2, Space Behavior and Design, McGraw-Hill, N.Y.

Gjelsvik, A., (1981). Theory of Thin Walled Bars, John Wiley and Sons, New York, N.Y.

Goodier, J. N., (1941). "The Buckling of Compressed Bars by Torsion and Flexure," Cornell University Engineering Experiment Station Bulletin, No. 27, Ithaca, N.Y.

Love, A. E. H., (1944). A Treatise on the Mathematical Theory of Elasticity, 4th ed., Dover Publication, N.Y.

Timoshenko, S. P., and Gere, J. M., (1961). Theory of Elastic Stability, 2nd ed., McGraw-Hill Book Co., Inc., N.Y.

Timoshenko, S. P., and Goodier, J., (1970). Theory of Elasticity, 3rd ed., McGraw-Hill Book Co., Inc., N.Y.

Vlasov, V. Z., (1961). Thin-Walled Elastic Beams, 2nd ed., Moscow, 1959; English translation, Israel Program for Scientific Translation, Jerusalem, Israel.

Washizu, K., (1975). Variational Methods in Elasticity and Plasticity, 2nd ed., Pergamon Press, Oxford, England.

Yang, Y. B., and McGuire, W., (1986). "A Stiffness Matrix for Geometric Nonlinear Analysis," Journal of Structural Engineering, ASCE, Vol. 112, No. 4.

Yoo, C. H., (1980). "Bimoment Contribution to Stability of Thin-Walled Assemblages," Computers and Structures, Vol. 11, pp. 465-471.

Ziegler, H., (1977). Principle of Structural Stability, 2nd ed., Birkhauser Verlag, Basel und Stuttgart, West Germany.

GEOMETRICALLY AND PHISICALLY NON–LINEAR ANALYSIS OF PLANE FRAMES SENSITIVE TO IMPERFECTIONS

M.A. Giżejowski
Department of Metal Structures, Warsaw Technical University, ul. Armii Ludowej 16, 00–637 Warsaw, Poland

1. Introduction

It has long been recognized that in practise the response of frame type structures can be significantly affected by both geometric and inelastic material properties [1]. Recently it also has been proved [2] that initial imperfections such as eccentricities of axial loads, crookednesses of frame elements, initial nodal displacements, stray of yield stress within the cross sections as well as residual stresses have to be included in the analysis of frames sensitive to the stability effects as it is widely utilized in the analysis of individual compression members. The above imperfections are always present in the real frames and may highly reduce their maximum load carrying capacity estimated by means of the linear buckling theory or limit analysis theory. Accordingly the necessity of providing the numerical methods and the effective implements for the analysis of both nonlinear and imperfect framed structures is self–evident. It can be done by use of computer orientated methods of nonlinear structural mechanics.

This contribution deals with the nonlinear in–plane analysis of metal frames which is able to consider any imperfection to give more correct value of the real structure failure load. It includes the effects of both gradual penetration of yielding in the cross sections and the spread of inelastic zones along the length of elements. The fundamental problem equations are expressed in terms of displacements increments and numerically solved via FEM (finite element method). The NRIT (Newton–Raphson interpolation technique) is used step by step to follow the load–deflection characteristic of structures due to the prescribed loading program. This technique was combined with the GT(gradient test) to search for the limit point and to follow the full response of structure both in the prelimit and the postlimit ranges. The concept of MPCS (multi–pointed cross section) is utilized to substitute the continuous functions of strain and stress by sets of their discrete values. It enabled to take into consideration the unloading effect of ealier plastificated fibres and arbitrary residual stress pattern.

2. Strain–displacement relationship for the beam–column element

Let us consider the motion of an elemental length Δz of the beam–column element in a fixed Cartesian co–ordinate system, as shown in Fig. 1. The z axis coincides with the mid–height line of the cross section and is the reference line to describe the displacements of arbitrary point M as the functions of their values for the reference line point 0. Geometrical imperfections are recognized as small distrubances for the perfect state of structure. Undeformed position of imperfect structure beam –column elements is treated as unstressed configuration. Hence it is convinient to relate all kinematic variables of imperfect element to the reference configuration, being initial one for elements of

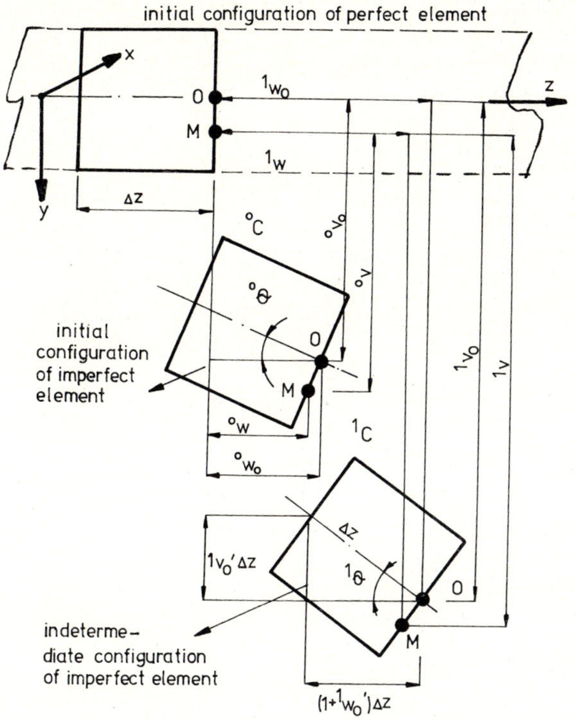

Fig. 1

perfect structure. Let us denote $^o\bar{q}_k$ as displacement magnitude defined in the initial configuration ^oC of imperfect structure and $^1\bar{q}_k$ – its value corresponding to the arbitrary intermediate configuration ^1C on the loading path. Thus we can write down

$$^1\bar{q}_k = {}^o\bar{q}_k + \bar{q}_k \tag{1}$$

where $\bar{q}_k$ – effective value of arbitrary point M displacement measured from the unstressed configuration.

Arbitrary point M axial and transverse displacements $\bar{q}_k$ are denoted w and v as well as their values for reference line point O – w_o and v_o, respectively.

To establish the equilibrium equations the following assumptions are made

i) (y,z) is the plane of symmetry of the elements and the surface tractions. The surface tractions may be reduced to the concentrated generalized forces located at the ends of frame elements and distributed ones acting along their length.

ii) Euler–Bernouilli kinematic hypothesis takes place which states that initially plane sections remain plane and undistorted during the deformation.

iii) Sections originally normal to the z axis will remain normal to the deflected reference axis of the elements after deformation; shear deformation and influence of shear stresses on plastification are neglected.

Making use of assumptions ii) iii) permits the displacements ^1v and ^1w of an arbitrary point M on an element cross section to be expressed in terms of the displacements of the beam–column reference axis. Thus, referring to Fig. 1, these displacements can be defined as

$$\begin{aligned} ^1v &= {}^1v_o - y\,[1 - \cos({}^1\theta)]\,, \\ ^1w &= {}^1w_o - y\sin({}^1\theta). \end{aligned} \tag{2}$$

By differentiating Eqs. 2 we obtain

$$\begin{aligned} ^1v' &= {}^1v'_o - y({}^1\theta')\sin({}^1\theta), \\ ^1w' &= {}^1w'_o - y({}^1\theta')\cos({}^1\theta). \end{aligned} \tag{3}$$

where the symbol ()' is used herein to denote z–derivatives.

To evaluate the term $({}^1\theta')$, we differentiate the expression for the rotation angle of the elemental length Δz

$$^1\theta\,' = (\text{arc tg}\,\frac{^1v'_o}{1 + {^1w'_o}})' = [\,(1 + {^1w'_o})(^1v''_o) - (^1v'_o)(^1w''_o)]\; c_o^2,\; c_o = \frac{1}{e} \tag{4}$$

in which $e^2 = 1 + 2e_o$ and e_o is the Green strain measure of the reference axis, calculated from the initial configuration for perfect element. From the kinematics of individual element we can also put down

$$\sin(^1\theta) = {^1v'_o}\, c_o, \quad \cos(^1\theta) = (1 + {^1w'_o})\, c_o. \tag{5}$$

Since TLD (total Lagrangian discription) is used, Eqs. 3 connected with Eqs. 4 and 5 incorporate fully geometrically nonlinear behaviour.

Substituting Eqs. 4 and 5 into Eq. 3 yields the expressions for components of strain–displacement equation. They permit the strain to be determined throughout the beam–colum element in terms of displacement of the reference axis. Retaining all nonlinear terms allows the strain–displacement equation to be written as

$$^1\varepsilon_z = {^1w'} + 0{,}5[(^1w')^2 + (^1v')^2] = \sum_{i=0}^{2} (^1\varepsilon_{zi}) = \sum_{i=0}^{2} [(-1)^i c_i y^i e_i\,]. \tag{6}$$

The values of c_i and e_i are given in Tabl. 1

Tabl. 1. Parameters of the strain–displacement equation

Sub-script i	Parameters c_i	Parameters e_i
1	1	$^1w'_o + 0{,}5[(^1w'_o)^2 + (^1v'_o)^2]$
2	$\frac{1}{(1+2e_o)^{3/2}}$	$(1 + {^1w'_o})^3(^1v''_o) - (1+{^1w'_o})^2(^1w''_o)(^1v'_o) + (1+{^1w'_o})(^1v'_o)^2(^1v''_o) - (^1w''_o)(^1v'_o)^3$
3	$\frac{1}{2(+2e_o)^3}$	$(1+{^1w'_o})^4(^1v''_o) - 2(1+{^1w'_o})^3(^1w''_o)\;(^1v'_o)(^1v''_o) + (1+{^1w'_o})^2\,(^1v'_o)^2\,[(^1w''_o)^2 + (^1v''_o)^2\,]\; -2(1+{^1w'_o})(^1w''_o)(^1y'_o)^3(^1v''_o) + (^1w''_o)^2(^1v'_o)^4$

3. Total equilibrium equation for the beam–column element

Considering the equilibrium of the element at configuration ^1C, the principle of virtual displacements gives [5]

$$\int_{^oV} {^1_oS_{ij}}\, \delta\, {^1_o\varepsilon_{ij}}\; {^od\vartheta} = {^1R} \tag{7}$$

where ^1R is the external virtual work expression due to the surface tractions with components $^1_ot_k$ and body forces with components $^1_of_k$

$$^1R = \int_{^oA} {^1_ot_k}\, \delta\, \bar{q}_k\; {^oda} + \int_{^oV} {^o\rho}\, {^1_of_k}\, \delta\, \bar{q}_k\; {^od\vartheta}. \tag{8}$$

In Eqs. 7 and 8, $\delta\,\bar{q}_k$ is a (virutal) variation in the current displacement components $\bar{q}_k$; $\delta\,{^1_o\varepsilon_{ij}}$ is a (virtual) variation in the Cartesian components of the Green–Langrange strain tensor in the configuration ^1C refered to the initial configuration ^oC of imperfect element and $^1_oS_{ij}$ are the Cartesian components of the 2nd Piola–Kirchhoff stress tensor in the configuration ^1C and measured in the configuration ^oC.

In the problem under consideration we have only one non–zero component of the strain and the stress tensors (see assumption iii)). Thus, the total strain at any point M on the element, due to axial and transverse displacements measured from the axis of the imperfect element in the initial configuration, takes the form

$$ {}^1_0\varepsilon_z = {}^1\varepsilon_z - \varepsilon_z^\star + \varepsilon_R \tag{9}$$

where ε_R is the residual strain, time–independent as well as self – equilibrated i.e.

$$\int_A \varepsilon_R\, y^i\, da = 0, \qquad i = 0,1,2 \tag{10}$$

and $\varepsilon_z^\star$ is given by Eq. 6 in which ${}^1\varepsilon_{zi}$, c_i, e_i have to be replaced by $\varepsilon_{zi}^\star, c_i^\star$ and $e_i^\star$, respectively. Further–more all displacements (and their derivatives) left hand superscripts 1 have to be changed for 0 when we describe $\varepsilon_z^\star$ by Eq. 6.

Noting that displacements products distinguished by left hand superscript 0 are time–independent, the total strain given by Eq. 9 is only function of four time–dependent quantities i.e. first and second derivatives of axial and transverse displacements. Thus the variation in strain $\delta({}^1_0\varepsilon_z)$ can be obtained from Eq. 9 remembering that w_o, w_o'', v_o' and v_o'' may vary independently. It results in

$$\delta({}^1_0\varepsilon_z) = \delta({}^1\varepsilon_z) = A_1\delta w_o' + A_2\delta w_o'' + A_3\delta v_o' + A_4\delta v_o'', \tag{11}$$

where

$$A_j = \sum_{i=0}^{2} a_{ji}\, y^i, \qquad j = 1,2,3,4 \tag{12}$$

and

$$a_{1i} = \frac{\partial\, {}^1\varepsilon_{zi}}{\partial w_o'}, \quad a_{2i} = \frac{\partial\, {}^1\varepsilon_{zi}}{\partial w_o''}, \quad a_{3i} = \frac{\partial\, {}^1\varepsilon_{zi}}{\partial v_o'}, \quad a_{4i} = \frac{\partial\, {}^1\varepsilon_{zi}}{\partial v_o''}. \tag{13}$$

Assuming further that configuration ${}^0C$ is geometrically fairly close to the initial configuration of perfect element, the left hand side of Eq. 7 can be transformed to

$$INT = \int_V {}^1_0\sigma_z\, \delta({}^1_0\varepsilon_z)d\vartheta = \int (A_{1n}\delta w_o' + A_{2n}\delta w_o'' + A_{3n}\delta v_o' + A_{4n}\delta v_o'')dz \tag{14}$$

where:

$$A_{jn} = \sum_{i=0}^{2} a_{ji}\, n_i \tag{15}$$

and n_i are stress resultants defined as

$$n_i = \int_A ({}^1_0\sigma_z)y^i\, da. \tag{16}$$

Eq. 14 is utilized to describe the work done by generalized nodal forces [4]. Integrating Eq. 14 by parts enables to set up the differential equilibrium equations as the Euler–Langrange equations of the functional and the associated boundary conditions, namely

$$INT = INT_o + {}^1J, \tag{17}$$

$$INT_o = \left|(A_{1n} - A'_{2n})\,\delta w_o + A_{2n}\delta w_o' + (A_{3n} - A'_{4n})\delta v_o + A_{4n}\delta v_o'\right|_o^l, \tag{18}$$

$${}^1J = \int_l (A'_{1n}\delta w_o - A''_{2n}\delta w_o' + A'_{3n}\delta v_o - A''_{4n}\delta v_o')dz. \tag{19}$$

The stress resultants contained in the integration brackets of Eq. 18 are the specified values of resultant stress at the ends of the element. Equivalent generalized nodal forces are defined in Tabl. 2.

Tabl.2. Natural boundary conditions and generalized element distributed loads

Variation of independent variable	Associated generalized nodal force (Natural boundary condition)		Associated generalized element distributed load	
	Linear component	Nonlinear component	Linear component	Nonlinear component
$\delta\, w_o$	${}^1Q_z$	Q_{nz}	${}^1p_z$	0
$\delta\, w'_o$	0	Q_{nzy}	0	m_{zyq}
$\delta\, v_o$	${}^1Q_y$	Q_{ny}	${}^1p_y$	0
$\delta\, v'_o$	${}^1Q_{yz}$	Q_{nyz}	$-{}^1m_{yz} = -{}^1p_z y_p$	m_{yzq}

The components ${}^1Q_i({}^1Q_z, {}^1Q_y$ and ${}^1Q_{yz})$ are statically equivalent to the n_o, n'_1 and n_1, respectively calculated at nodes of elements. The components Q_{ni} of nonlinear part of natural boundary conditions are additionally functions of n'_2. The quantities n'_1 and n'_2 are not considered as generalized stresses. They are celled nodal reactions because they are only needed for equilibrium. Their values may be easily calculated from the equilibrium conditions for the elements when generalized nodal forces and the element surface tractions are known.

To establish the total external virtual work, the components ${}^1_o t$ of surface tractions are reduced to the distributed p_z an p_y loads, acting in the plane (y, z) at the distance of y_p from the reference axis. Assuming ${}^1_o\rho \equiv 0$, Eq. (13) may then be replaced by

$$ {}^1R = \int_l ({}^1p_z\, \delta\, {}^1w\, |_{y\,=\,y_p} + {}^1p_z\; \delta\; {}^1v\, |_{\,y\,=\,y_p})dz. \tag{20}$$

Applying the introduced kinematic hypothesis (see assumption ii) and iii)) we can put down

$$ \begin{aligned} \delta\, {}^1w &= \delta\, w_o + y\, c_1({}^1v'_o)(1 + {}^1w'_o)\, \delta w'_o - y\, [c_o - c_1({}^1v'_o)^2\,]\delta\, v'_o\,, \\ \delta\, {}^1v &= \delta\, v_o - y\, c_1({}^1v'_o)(1 + {}^1w'_o)\, \delta\, v'_o + y\, [\,c_o - c_1(1 + {}^1w'_o)^2]\delta\, w'_o\,. \end{aligned} \tag{21}$$

The components of ${}^1R$ are given in Tabl. 2, in which

$$ m_{zyq} = c_o - c_1(1 + {}^1w'_o)^2({}^1m_{zy}) + c_1(1 + {}^1w'_o)({}^1v'_o)({}^1m_{yz}), \tag{22}$$

$$ m_{yzq} = -c_1(1 + {}^1w'_o)({}^1v'_o)({}^1m_{zy}) + [\,1 - c_o + c_1({}^1v'_o)^2\,]\,({}^1m_{yz}) \tag{23}$$

and ${}^1m_{zy} = {}^1p_y y_p$ as well as ${}^1m_{yz} = {}^1p_z y_p$.

Thus the total external virutal work is given by

$$ EXT = EXT_o + {}^1R \tag{24}$$

where

$$ EXT_o = INT_o. $$

Equating the r.h.s. of Eqs. 14 and 24 the following variational total equilibrium equation is obtained

$$ \begin{aligned} &\int_l (A_{1n}\, \delta\, w'_o + A_{2n}\, \delta\, w''_o + A_{3n}\, \delta\, v'_o + A_{4n} \delta\, v''_o)dz - \int_l [\,{}^1p_z\, \delta\, w_o + m_{zyq} \delta\, w'_o + \\ &+ {}^1p_y\, \delta\, v_o + (m_{yzq} - {}^1m_{yz})\, \delta\, v'_o\,]\;\; dz - |\, ({}^1Q_z + Q_{zn})\, \delta w_o + Q_{zyn}\; \delta w'_o + \\ &+ ({}^1Q_y + Q_{yn})\, \delta\, v_o + (Q_{yzn} - {}^1Q_{yz})\; \delta v'_o\, |\,{}^l_o = 0\,. \end{aligned} \tag{25}$$

In the presented analysis all nonlinear terms of Eq. 25 are included. A variety of beam—column problem formulations may be now available depending on the degree to which Eq. 25 is approximated i.e. which terms are excluded in Eq. 25.

4. First level of discretization and the element incremental equilibrium equation

To obtain a set of algebraic equations of the problem, convinient for computational calculations, the reference axes of the frame members are assumed to be divided into finite elements with the eight DOF (degrees of freedom) in the points of their origin and end. The axial and transverse reference axis displacements w_o and v_o are approximated as a linear combination of four nodal generalized displacements q_{oi} and four standard cubic shape functions ϕ_i. The same base functions are used to describe the initial geometrical imperfections. In the nondimensional isoparametric element co–ordinate these functions are given e.g. in [5] .
They are chosen so that the introduced independent variables w_o and v_o given by the approximation are compatible with those of the adjacent elements

$$\bar{q}_{ok} = \langle\phi\rangle \{q_o\} \quad , \qquad \langle\phi\rangle = \langle \phi_1 , \phi_2 , \phi_3 , \phi_4 \rangle . \tag{26}$$

If the displacements, w_o and v_o are functions of the discrete set of K nodal displacements co–ordinates. q_{oi}, Eq. (25) may be rewritten as

$$^1\Psi_i = {}^1\psi_i - {}^1Q_i = 0 \tag{27}$$

in which

$$^1\Psi_i = \int_l \left(A_{1n} \frac{\partial w_o'}{\partial q_{oi}} + A_{2n} \frac{\partial w_o''}{\partial q_{oi}} + A_{3n} \frac{\partial v_o'}{\partial q_{oi}} + A_{4n} \frac{\partial v_o''}{\partial q_{oi}}\right)dz -$$

$$\int_l \left[{}^1p_z \frac{\partial w_o}{\partial q_{oi}} + m_{zyq} \frac{\partial w_o'}{\partial q_{oi}} + {}^1p_y \frac{\partial v_o}{\partial q_{oi}} + (m_{yzq} - {}^1m_{yz}) \frac{\partial v_o'}{\partial q_{oi}}\right] dz - Q_{ni}, \tag{28}$$

$i = 1, 2, \ldots , K.$

Eqs. 27 are phisically and geometrically nonlinear. To establish the element incremental equilibrium equations the solution technique arised from a Taylor´s series expansion is applied [6] . For any specified values of q_{oi}, Q_i and p_i, say $({}^1q_{oi}, {}^1Q_i, {}^1p_i)$, the values of $^2\Psi$ at a configuration ^2C, sufficiently close to ^1C, may be determined as

$$^2\Psi_i \approx {}^1\Psi_i + \frac{\partial\, {}^1\Psi_i}{\partial q_{ok}} \Delta q_{ok} + \left[\frac{\partial\, {}^1\Psi_i}{\partial\, {}^1p_i} + \frac{\partial^2 ({}^1\Psi_i)}{\partial\, {}^1p_i \partial q_{ok}} \Delta q_{ok} \right] \Delta p_i + \frac{\partial\, {}^1\Psi_i}{\partial\, {}^1Q_i} \Delta Q_i ,$$

$i = 1,2, \ldots , K.$ (29)

In Eq. 29 only one nonlinear term in increments is retained. To find an equilibrium position in configuration ^2C it is necessary that $^2\Psi_i$ be zero [7] .
The approximate element incremental equilibrium equation may now be written in matrix from as

$$[k_T] \{\Delta q_o\} = \{\Delta Q\} = \{Q\} - \{Q_R\} \tag{30}$$

where

$$k_{Tik} = \frac{\partial\, {}^1\Psi_i}{\partial q_{ok}} = \int_l \left(\frac{\partial A_{1n}}{\partial q_{ok}} \frac{\partial w'_o}{\partial q_{oi}} + \frac{\partial A_{2n}}{\partial q_{ok}} \frac{\partial w''_o}{\partial q_{oi}} + \frac{\partial A_{3n}}{\partial q_{ok}} \frac{\partial v'_o}{\partial g_{oi}} + \frac{\partial A_{4n}}{\partial q_{ok}} \frac{\partial v''_o}{\partial q_{oi}} \right) dz - \int_l \left(\frac{\partial m_{zyq}}{\partial q_{ok}} \frac{\partial w'_o}{q_{oi}} + \frac{\partial m_{yzq}}{\partial q_{ok}} \frac{\partial v'_o}{\partial q_{oi}} \right) \Big|_{{}^1p_i + \Delta p_i} dz, \tag{31}$$

$$Q_i = {}^1Q_i + \frac{\partial\, {}^1\Psi_i}{\partial\, {}^1Q_i} \Delta Q_i, \quad Q_{Ri} = \frac{\partial\, {}^1\Psi_i}{\partial\, {}^1p_i} \Delta p_i - {}^1\psi_i \tag{32}$$

and the summation convention is used for repeated indices. The matrix $[k_T]$ is called element tangential (or incremental) stiffness matrix. The vector $\{\Delta Q\}$ is called element unbalanced (or incremental) nodal forces and it is measured as difference between vector of applied generalized nodal forces $\{Q\}$ and vector of resisting forces $\{Q_R\}$. Each influence coefficient k_{Tik} of tangential stiffness matrix may be separated into constitutive and geometrical parts by decomposition of $\partial A_{jn}/\partial q_{ok}$

$$\frac{\partial A_{jn}}{\partial q_{ok}} \sum_{i=0}^{2} \left(a_{ji} \frac{\partial n_i}{\partial q_{ok}} + n_i \frac{\partial a_{ji}}{\partial q_{ok}} \right) = A^E_{jkn} + A^G_{jkn}. \tag{33}$$

First component of Eq. 33 gives rise to the constitutive stiffness matrix $[k_E]$, which contains conventional linear stiffness matrix $[k_e]$, and the geometrical stiffness matrix $[k_G]$, which contains conventional linear stability matrix $[k_s]$. The nonlinear components of $[k_E]$ and $[k_G]$ matrices are called initial displacement stiffness matrix $[k_q]$ and initial stress stiffness matrix $[k_\sigma]$, respectively. Further–more second integral of Eq. 31 gives rise to the additional component of tangential stiffness matrix which may be called surface traction matrix, $[k_p]$. Thus

$$[k_T] = [k_E] + [k_G] - [k_p] \tag{34}$$

and

$$[k_E] = [k_e] + [k_q], \quad [k_G] = [k_s] + [k_\sigma].$$

The component A^E_{jkn} of Eq. 33 is expressed in terms of stiffness resultants s_m as follows

$$A^E_{jkn} = \sum_{i=0}^{2} a_{ji} \left(\sum_{m=i}^{i+2} [s_m a_{k(m-i)}] \right) \tag{35}$$

where

$$s_m = \int_A E^\star y^m \, da \tag{36}$$

and $E^\star$ is the tangential modulus (depends on the current strain state). The history of the strain (loading, unloading) is described by means of the special parameters of physical equation for the given model of elastic–plastic material.

The matrices $[k_E]$ and $[k_G]$ may be p erformed in the same maner as

$$[k_m] = \begin{bmatrix} [k^{11}_m] & [k^{12}_m] \\ [k^{21}_m] & [k^{22}_m] \end{bmatrix}, \quad m = E, G \tag{37}$$

where because of symmetry must be $[k^{21}_m] \equiv [k^{12}_m]$ and the submatrices take the form

$$k^{11}_m = \int_l \left[\left(\langle\phi'\rangle \frac{\partial A^m_{1n}}{\partial w'_o} + \langle\phi''\rangle \frac{\partial A^m_{2n}}{\partial w'_o} \right) \{\phi'\} + \left(\langle\phi'\rangle \frac{\partial A^m_{1n}}{\partial w''_o} + \langle\phi''\rangle \frac{\partial A^m_{2n}}{\partial w''_o} \right) \{\phi''\} \right] dz, \tag{38}$$

$$[k_m^{12}] = \int_l [(<\phi'> \frac{\partial A_{1n}^m}{\partial v_o'} + <\phi''> \frac{\partial A_{2n}^m}{\partial v_o'}) \{\phi'\} + (<\phi'> \frac{\partial A_{1n}^m}{\partial v_o''} +$$

$$<\phi''> \frac{\partial A_{2n}^m}{\partial v_o''}) \{\phi''\}] dz, \qquad (39)$$

$$[k_m^{22}] = \int_l [(<\phi'> \frac{\partial A_{3n}^m}{\partial v_o'} + <\phi''> \frac{\partial A_{4n}^m}{\partial v_o'}) \{\phi'\} +$$

$$(<\phi''> \frac{\partial A_{3n}^m}{\partial v_o''} + <\phi''> \frac{\partial A_{4n}^m}{\partial v_o''}) \{\phi''\}] dz. \qquad (40)$$

The matrix $[k_p]$ may be expressed in the similar way

$$[k_p] = \begin{bmatrix} [k_p^{11}] & [k_p^{12}] \\ \text{symm.} & [k_p^{22}] \end{bmatrix} \qquad (41)$$

where

$$[k_p^{11}] = \int_l <\phi'> \frac{\partial m_{zyg}}{\partial w_o'} \Big|_{^1p_i + \Delta p_i} \{\phi'\} \; dz, \qquad (42)$$

$$[k_p^{12}] = \int_l <\phi'> \frac{\partial m_{zyg}}{\partial v_o'} \Big|_{^1p_i + \Delta p_i} \{\phi'\} \; dz, \qquad (43)$$

$$[k_p^{22}] = \int_l <\phi'> \frac{\partial m_{yzg}}{\partial v_o'} \Big|_{^1p_i + \Delta p_i} \{\phi'\} \; dz. \qquad (44)$$

5. Second level of discretization and the structure incremental equilibrium equation

One the expressions for w_o and v_o displacements are defined in terms of the q_{oi}, all components of the tangential stiffness matrix and the vector of resisting forces may be evaluated for elements, one by one and numerical integration scheme with respect to z co–ordinate may be carried out to form matrices $[k_T]$ and $\{\Delta Q\}$.The GIM (Gaussian integration method) is used herein to create these matrices. At each sampling points along the elements the stress and stiffness resultants have to be calculated due to their definitions given by Eqs. 16 and 36. Hence the additional integration scheme over the cross section is required at each Gauss point. To include sufficiently the influence of initial residual stresses and the strain history (stress–strain diagram is irreversible) the one layer MPCS concept is used. In this concept all flange and web cross section components are discretized by dividing into finite elements $\Delta A_f = t_f \Delta x$, $\Delta A_w = t_w \Delta y$. At each finite element the total strain then stress and tangential modulus are computed as the average values at its centroid. It yields the contribution of one point to the Gauss point stress and stiffness resultants. Summing over all points gives the stress and stiffness resultants at the considered Gauss point. Summing again over all sampling points the multiplication products of $[k_T]$ (and $\{Q_R\}$) components and appropriate weight factors gives the tangential stiffness matrix (and vector of resisting forces). The $\{Q_R\}$ vector may be substracted from the applied loads one to obtain the vector of unbalanced forces.

In such a way the components of Eq. 30 can be evaluated numerically for all elements. To analyse the complete structure it is necessary to perform consideration of equilibrium and compatibility at the nodes. Since the matrix representation of incremental equilibrium equations for a beam–column element is related to the local element co–ordinates, it can be directly used only for the analysis of beam–column system [5] when each element is aligned with the system. In the analysis of frames, Eq. (30) is transformed to the global frame co–ordinates as follows

$$([T]^T [k_T] [T]) \{\Delta r_o\} = \{R\} - [T]^T \{Q_R\} \qquad (45)$$

where [T] is transformation matrix such as given in [6],

$\{R\}$ is the vector of applied generalized nodal forces in frame co–ordinates.

The element stiffness matrices can now be assembled to form the structure incremental equilibrium equations

$$[K_T] \{\Delta r\} = \{\Delta R\} . \qquad (46)$$

In this process the force boundary conditions and displacement ones are incorporated automatically into the right and left hand sides of Eq. 46, respectively.

The vector of total displacements of the structure nodes are obtained by simple addition of increments

$$\{r\} = \sum_{m=1}^{N} \{\Delta r\} . \qquad (47)$$

The computer program based on the formulation presented above was elaborated and implemented on PDP computer. Some numerical examples were tested. They show that this method allowed for all kinds of nonlinearities is effective from the practical point of view because of its relatively small computer CPU time consuming.

References

[1] W. Merchant, The failure load of rigid–jointed frameworks as influenced by stability. The Struct. Engr. 32 , 185–191 (1954).

[2] M. Giżejowski, Non–linear analysis of frames sensitive to the instability effects. Proc. of Conf. Ultimate Load–Carrying Capacity of Metal Structures, Bratislava, CSSR (1986).

[3] K.J. Bathe and S. Bolourchi, Large displacement analysis of three–dimensional beam structures. Int. J. Num. Meth. Engng. 14 , 961–986 (1979).

[4] J.H. Argyris, B. Boni, W.Hindenland and M. Kleiber, Finite element analysis of two–and three – dimensional elasto–plastic frames – the natural approach. Comput. Meth. Appl. Mech. Engng. 35 , 221–248 (1982).

[5] M.Epstein and D.W. Murray. Large deformation in–plane analysis of elastic beams. Comput. Structures 6 , 1–9 (1976).

[6] M.H.El;Zanaty and D.W. Murray, Nonlinear finite element analysis of steel frames. J.Struct. Engng. 109 , 353–368 (1983).

[7] K.Washizu, Variational Methods in Elasticity and Plasticity. Pergamon Press, Oxford (1968).

NUMERICAL ANALYSIS IN THE VICINITY OF CRITICAL POINTS BY THE GENERALIZED INVERSE

Y. Hangai
Institute of Industrial Science
University of Tokyo
7-22-1 Roppongi, Tokyo 106, JAPAN

1.INTRODUCTION

The load-incremental method is one of the basic analytical methods for the numerical analysis of the geometrically nonlinear problem. However, at the critical points such as the limit point and the bifurcation point, the determinant of the coefficient matrix becomes zero, and a numerically unstable situation develops in the vicinity of critical points. Therefore, it is impossible to pursue the equilibrium path beyond the critical point by the load-incremental method.

To avoid such a situation, the displacement-incremental method [1], the perturbation method [2], etc. have been used. But these methods have also disadvantages for the numerical analysis of nonlinear multi-degree-of-freedom systems. The difficulty lies in the selection of an incremental parameter in the vicinity of the critical points.

The arc-length method has been used to remove the drawback mentioned above, and has been shown to be applicable to the pursuit of highly nonlinear equilibrium curves [3]. However, a nonlinear constraint concerning the arc-length has to be added to the ordinary equilibrium equations, so that the theoretical as well as the numerical procedures become complicated.

Based on the above considerations,"the generalized incremental method" was presented by means of the generalized inverse [4], and a systematic procedure to pursue the post-critical path was developed. In the paper, the rectangular matrix, which was constucted by adding the load vector to the tangent stiffness matrix, was used in the incremental interval, so that the existence condition of solution was not used advantageously. Hence, in this paper, the meaning of this condition will be examined in detail, and then an analytical method in the vicinity of critical points will be presented.

2.GENERALIZED INVERSE AND SOLUTIONS OF SYSTEMS OF LINEAR EQUATIONS

Let $\underline{A}$ be an m×n rectangular matrix, and let us adopt the following definition of the generalized inverse of $\underline{A}$, which is expressed by $\underline{A}^-$.

If a matrix $\underline{A}^-$ exists that satisfies the four conditions below, we shall call $\underline{A}^-$ the generalized inverse (or the Moore-Penrose generalize inverse) of $\underline{A}$.

$$(\underline{A}\,\underline{A}^-)' = (\underline{A}\,\underline{A}^-),\quad (\underline{A}^-\underline{A})' = (\underline{A}^-\underline{A})$$
$$\underline{A}\,\underline{A}^-\underline{A} = \underline{A},\quad \underline{A}^-\underline{A}\,\underline{A}^- = \underline{A}^- \tag{1}$$

where $\underline{A}'$ denotes the transpose of $\underline{A}$.

If $\underline{A}$ is a nonsingular square matrix, it is clear that $\underline{A}^{-1}$ satisfies all conditions of Eqs.(1). And it is proved[5]that for each matrix $\underline{A}$ there exists a unique matrix $\underline{A}^-$ that satisfies Eqs.(1): that is, each matrix $\underline{A}$ has a unique generalized inverse matrix.

Let us consider a system of m equations with n unknowns

$$\underline{A}\,\underline{x} = \underline{g} \tag{2}$$

where $\underline{A}$, $\underline{x}$ and $\underline{g}$ are an m×n matrix, an n×1 vector and an m×1 vector, respectively. Eq.(2) is called "consistent" if and only if

$$\underline{A}\,\underline{A}^-\underline{g} = \underline{g} \tag{3}$$

in which case Eq.(2) can have a solution, which may be written in the form for each and every vector $\underline{\alpha}$

$$\underline{x} = \underline{A}^-\underline{g} + (\underline{I} - \underline{A}^-\underline{A})\,\underline{\alpha} \tag{4}$$

where I is the n×n identity matrix. In the following, Eq.(3) is called the existence condition of solution. Let us denote the column vector of the matrix $(\underline{I} - \underline{A}^-\underline{A})$ by $\underline{a}_1$, $\underline{a}_2, \dots, \underline{a}_n$. If the rank of the matrix $(\underline{I} - \underline{A}^-\underline{A})$ is p, then there exists exactly p vectors represented by $a_1, \dots, a_p$ of these n vectors that are linearly independent. For such a case, each of the remaining n-p vectors is expressed as a linear combination of these p vectors. Then, the solution given by Eq.(4) has the form

$$\underline{x} = \underline{A}^-\underline{g} + \alpha_1\underline{a}_1 + \alpha_2\underline{a}_2 + \quad + \alpha_p\underline{a}_p \tag{5}$$

where $\alpha_1, \dots, \alpha_p$ are any set of scalars. When the rank of A is r, then p = n-r.

3.INCREMENTAL EQUATIONS FOR NONLINEAR EQUILIBRIUM EQUATIONS

For a discrete conservative system described by n displacement parameter D_i, i=1, ··· ,n and a load parameter Λ, let us consider the nonlinear equilibrium equation of the form

$$f_r(D_1, D_2, \dots, D_n, \Lambda) = f_r(D_i, \Lambda) = 0 \tag{6}$$

where r takes 1 through n. Eq.(6) is usually obtained by the Galerkin method, the finite element method, the finite difference method, etc. to the continuous system.

Let us assume that a point (D_i^0, Λ^0) represents a known point on the equilibrium path which satisfies Eq.(6), and, in addition, that a point (D_i, Λ) with increments of displacement parameters d_i and with the load parameter λ, represents another point on the equilibrium path in a distance from point (D_i^0, Λ^0), as shown in Fig.1. Then, the equilibrium equations take the following forms:

$$f_r(D_i^o, \Lambda^o) = 0 \tag{7}$$

and

$$f_r(D_i, \Lambda) = f_r(D_i^o + d_i, \Lambda^o + \lambda) = 0 \tag{8}$$

Taylor's expansion of Eq.(8) about the known point (D_i^o, Λ^o), and then deviation of Eq.(7) from the result give us the localized incremental equilibrium equations:

$$\left[d_i \frac{\partial}{\partial D_i} + \lambda \frac{\partial}{\partial \Lambda}\right] f_r^o + \frac{1}{2}\left[d_i d_j \frac{\partial^2}{\partial D_i \partial D_j} + 2 d_i \lambda \frac{\partial^2}{\partial D_i \Lambda} + \lambda^2 \frac{\partial^2}{\partial \Lambda^2}\right] f_r^o + \cdots = 0 \tag{9}$$

where the summation convention is used with respect to the subscripts i and j. Eq.(9) is the nonlinear incremental equation in terms of d_i and λ for the known point (D_i^o, Λ^o).

In order to solve Eq.(9), let the increments d_i and λ be functions of an arbitrary single parameter t, which have their origin at the known point, i.e., $d_i(0)=0$ and $\lambda(0)=0$, from which it follows that for $t=0$, $(D_i,\Lambda)=(D_i^o,\Lambda^o)$. The expansion of $d_i(t)$ and $\lambda(t)$ into Maclaurin series for $|t|<1$ leads to

$$d_i(t) = \dot{d}_i t + \frac{1}{2}\ddot{d}_i t^2 + \cdots \tag{10}$$

$$\lambda(t) = \dot{\lambda} t + \frac{1}{2}\ddot{\lambda} t^2 + \cdots \tag{11}$$

where dot means the derivative with respect to t.

If $\dot{d}_i$, $\ddot{d}_i$, ... and $\dot{\lambda}$, $\ddot{\lambda}$, ... are obtained, the increments (d_i and λ) on the equilibrium path can be determined for an arbitrary step length of the parameter t.

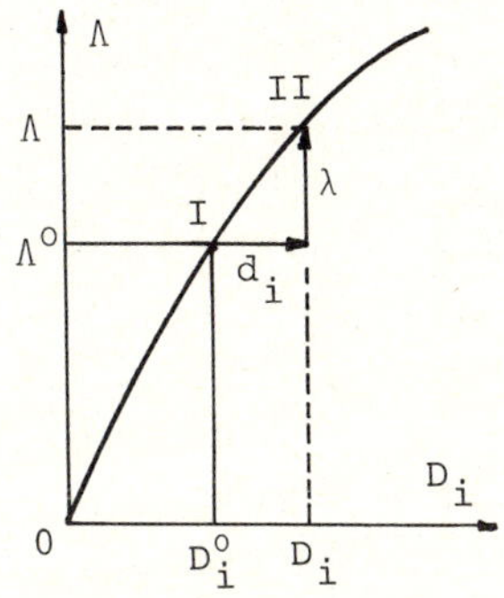

Fig.1: Equilibrium Path and Increments of Parameters

Introducing Eqs.(10) and (11) into Eq.(9), and then equating the coefficients of equal power of t, we obtain the following sequence of equations after the arrangement in the matrix form.

$$\underline{K}\,\dot{\underline{d}} + \underline{h}\,\dot{\lambda} = \underline{0} \tag{12}$$

$$\underline{K}\,\ddot{\underline{d}} + \underline{h}\,\ddot{\lambda} + \underline{r}(\,\dot{\underline{d}}\,,\,\dot{\lambda}\,) = \underline{0} \tag{13}$$

where the component of the vector r is represented as

$$r_r = f^o_{r,ij}\dot{d}_i\dot{d}_j + 2f^o_{r,i\lambda}\dot{d}_i\dot{\lambda} + f^o_{r,\lambda\lambda}\dot{\lambda}^2. \tag{14}$$

4.CLASSIFICATION OF CRITICAL POINTS

If the determinant of the coefficient matrix $\underline{K}$ is not zero, Eq.(12) can be solved to give the unique relation between the incremental displacements and the increment of load as

$$\dot{\underline{d}} = -\underline{K}^-\,\underline{h}\dot{\lambda}$$

in which $\underline{K}^-$ is coincident with $\underline{K}^{-1}$.

In the converse case, where the determinant of $\underline{K}$ is zero, a critical point exists at the point (D^o_i, Λ^o) on the equilibrium path. This case will be treated in the following. This section deals with the classification of critical points.

Since $|\underline{K}|$ =0 at the critical point, the rank of $\underline{K}$ must be (n-1) at the simple critical point. In this case, the existence condition of solution and the solution of Eq.(12) take the form respectively from Eqs.(3) and (4):

$$(\underline{I} - \underline{K}\,\underline{K}^-\,)\,\underline{h}\,\dot{\lambda} = 0 \tag{15}$$

and

$$\dot{\underline{d}} = \underline{K}^-\,\underline{h}\,\dot{\lambda} + (\underline{I} - \underline{K}^-\,\underline{K})\,\dot{\underline{\alpha}} \tag{16}$$

Since $\underline{K}$ is symmetric for a conservative system, $\underline{K}^-\underline{K}=\underline{K}\,\underline{K}^-$can be led by using Eq.(1), and then $(\underline{I}-\underline{K}\underline{K}^-)'=\underline{I}-\underline{K}^-\underline{K}$ holds. As stated in the second section, the matrix $(\underline{I}-\underline{K}^-\,\underline{K})$ has one linearly independent column vector, which is represented by $\underline{k}$, because the rank of this matrix is p=n-(n=1)=1. Then, Eqs.(15) and (16) become respectively

$$g\,\dot{\lambda} = 0 \tag{17}$$

where $g = \underline{k}'\underline{h}$ and

$$\dot{\underline{d}} = \underline{K}^+\,\underline{h}\dot{\lambda} + \dot{\alpha}_1\underline{k}. \tag{18}$$

At the critical point, Eq.(17) should hold for the existence of solution. Then we can classify critical points as limit points or bifurcation points, based on whether g is zero or not, as shown in Fig.2.

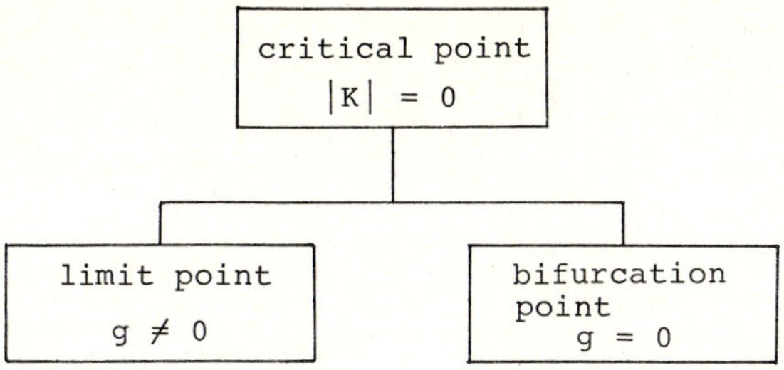

Fig.2: Clasification of Critical Points

In the case of the limit point, $\dot{\lambda}=0$ has to be satisfied according to both $g\dot{\lambda}=0$ and $g\neq 0$. However, for the bifurcation point, both cases hold, $\dot{\lambda}=0$ and $\dot{\lambda}\neq 0$. The symmetrical bifircation path corresponds to $\dot{\lambda}=0$, and the asymmetric bifurcation path appears for $\dot{\lambda}\neq 0$ [6].

5.INITIAL POST-CRITICAL MODE

For the limit point on the equilibrium path, $\dot{\lambda}=0$ holds. Hence, the solution of Eq.(18) takes the form

$$\underline{\dot{d}} = \dot{\alpha}_1 \underline{k} \tag{19}$$

In another form, Eq.(19) with $\dot{\lambda}=0$ gives us

$$\begin{Bmatrix} \underline{\dot{d}} \\ \dot{\lambda} \end{Bmatrix} = \begin{Bmatrix} \underline{k} \\ 0 \end{Bmatrix} \dot{\alpha}_1 \tag{20}$$

which shows that in the case of the limit point, the buckling mode consists of the linearly independent column vector of $(\underline{I}-\underline{K}^{-}\underline{K})$.

For the bifurcation point, the mode of load $(\underline{h})$ and $\underline{k}$ are orthogonal because of $g=0$. Also, $\underline{K}^{-}\underline{h}$ and $\underline{k}$ are orthogonal because $\underline{h}'(\underline{K}^{-})'(\underline{I}-\underline{K}^{-}\underline{K})=\underline{h}'(\underline{K}^{-})(\underline{I}-\underline{K}\underline{K}^{-})=0$ by the definition of the generalized inverse. Hence, we have two independent vectors such as

$$\begin{Bmatrix} \underline{\dot{d}} \\ \dot{\lambda} \end{Bmatrix} = \begin{Bmatrix} \underline{k} \\ 0 \end{Bmatrix} \dot{\alpha}_1 + \begin{Bmatrix} \underline{K}^{-}\underline{h} \\ 1 \end{Bmatrix} \dot{\alpha}_2 \tag{21}$$

which $\dot{\alpha}_1$ and $\dot{\alpha}_2$ are arbitrary parameters. Existence of two independent vectors for the solution corresponds to the existence of two distinct equilibrium paths emerging at the bifurcation point, though these paths may be represented by the combination of the above two vectors. To find out the way of combination is one of the objective of the post-bifurcation analysis, which will be presented in the next section.

6.POST-BIFURCATION ANALYSIS

In the case of limit points, the number of incremental parameter is one as understood in Eq.(20). Hence, post-critical paths can be calculated easily, for instance, by the displacement-incremental method[1]. However, for bifurcation points, the ratio of two parameters, $\dot{\alpha}_1$ and $\dot{\alpha}_2$ in Eq.(19) must be determined to pursue the post-bifurcation path. This can be done by using the second perturbation equation (13).

The existence condition of solution for Eq.(13) leads to

$$(\underline{I} - \underline{K}\,\underline{K}^{-})(\underline{h}\ddot{\lambda} + \underline{r}(\dot{\underline{d}}, \dot{\lambda})) = 0 \tag{22}$$

Since the linearly independent vector of the coefficient matrix of Eq.(22) is $\underline{k}$ and $g=\underline{k}'\underline{h}=0$ at the bifurcation point, then Eq.(22) can be written as

$$\underline{k}'\underline{r}\ (\dot{\underline{d}}, \dot{\lambda}) = 0 \tag{23}$$

Eq.(23) can finally be reduced, by introducing Eq.(21) into Eq.(14), to a quadratic equation of $\dot{\alpha}_1$ and $\dot{\alpha}_2$:

$$a(\dot{\alpha}_1)^2 + 2b\,\dot{\alpha}_1\dot{\alpha}_2 + c(\dot{\alpha}_2)^2 = 0 \tag{24}$$

from which the two ratios κ_1 and κ_2 , where $\kappa_i = \dot{\alpha}_2/\dot{\alpha}_1$, $i=1,2$, can be obtained. After all, the first approximation of the post-bifurcation path has the form

$$\begin{Bmatrix} \dot{\underline{d}} \\ \dot{\lambda} \end{Bmatrix}_i = \left(\begin{Bmatrix} \underline{k} \\ 0 \end{Bmatrix} + \begin{Bmatrix} \underline{K}^{-}\underline{h} \\ 1 \end{Bmatrix} \kappa_i \right) \dot{\alpha}_2 \tag{25}$$

where i=1 and 2. Thus, two incremental parameters can be reduced to one arbitrary incremental parameter, which enables us to pursue numerically the post-bifurcation paths.

7.CONCLUSION

An analytical procedure to pursue the equilibrium paths in the vicinity of critical points is presented in the paper by using the

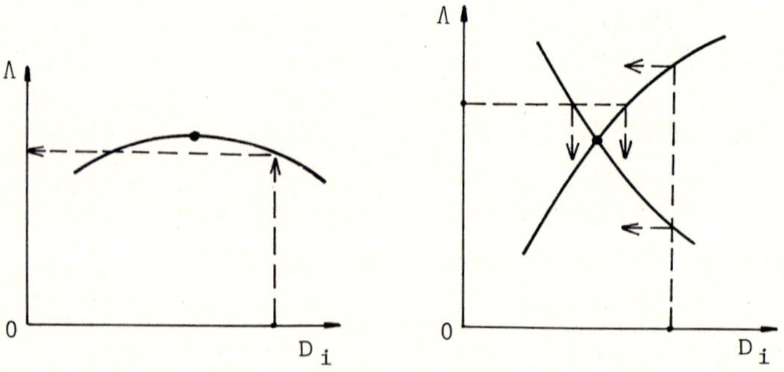

Fig.3: Limit Point and Bifurcation Point

generalized inverse. The present procedure has the following characteristics:

(1)The selection of the incremental parameters is not necessary,

(2)Critical points can be classified into limit points and bifurcation points by the existence condition of solution,

and

(3)Post-bifurcation paths can be automatically calculated by using the existence condition of solution for the second order perturbation equation.

When the geometrically nonlinear problems is analyzed, we face a troublesome situation of numeriacal instability caused by the zero-determinant of the coefficient matrix, especially in the vicinity of critical points. This paper presents one answer to this problem by means of the generalized inverse.

REFERENCES

[1] HANGAI, Y. and KAWAMATA, S., "Nonlinear Analysis of Space Frames and Snap-through Buckling of Reticulated Shell Structures," Proc. IASS Pacific Symposium on Tension Structures and Space Frames, 1972, pp.803-816.

[2] WALKER, A. C., "A Nonlinear Finite Element Analysis of Shallow Circular Arches," International Journal of Solids and Structures, Vol.5, 1969, pp.97-107.

[3] RIKS, E., "Some Computational Aspects of the Stability Analysis of Nonlinear Structures, Computer Methods in Applied Mechanics and Engineering, Vol.47, 1984, pp.219-259.

[4] HANGAI, Y., "Application of the Generalized Inverse to the Geometrically Nonlinear Problem," SM Archives, Vol.6, 1981, pp.129-165.

[5] GRAYBILL, F. A., "Introduction to Matrices with Application in Statistics," Wadsworth Publishing Company, 1969.

[6] THOMPSON, J. M. T., "Instabilities and Catastrophes in Science and Engineering," John Wiley & Sons, 1982.

Nonlinear computation of a barrel vault

Marcel VAN LAETHEM, Professor; Jacques de COEN,research engineer; Katholieke Universiteit Leuven (Louvain) Celestijnenlaan 200 A, B-3030 Leuven, Belgium

1. Introduction

A basic assumption for application of linear thin shell theory is that the ratio w/t should be tiny enough. Bringing a small contribution to the question "To what extent?" is our target.

2. Geometry

The beamless barrel vault spans 18m; [1]. Cross section: Figure 1. The origin of the xyz-system is the centre of curvature of a circular edge, the z-axis vertically upward directed and the x-axis in the plane of that edge. Half of the central angle θ , being zero in the summit and 34,2° at the edges, will be used in Figure 4. Two prestressing cables: $\theta=(34.2y/9)\cdot(2-y/9)$. Half a cable appears in each of the meshes in Figure 4: isometric projections of a quarter of the shell limited by the planes y=0.00, y=9.00m, $\theta=0.00°$ and $\theta=34.2°$.

3. Material properties

The shell material is linear elastic with $E=3000kN/cm^2$ and $\nu=0$, and weights $2.5t/m^3$. The cable material and the cross sectional area are not mentioned here. The reason will appear later on.

4. Boundary conditions

The supports are diaphragms acting as membranes along the curved edges: y=0.00m and y=18.00m. The straight edges are free.

5. Load and prestress - Linear and nonlinear

The load is $1.75kN/m^2$ (i.e. the self weight), the prestressing force 615kN and the friction neglected.

Only the geometrical nonlinearity has been taken under account, although the material nonlinearity would partly counterbalance its effect.

Prestress has been applied in ten steps of 61.5kN, without self weight, for theoretical comparison.

The implicit method has been used, adapting the stiffness matrix every time five iterations have been done. Iteration goes on till residual forces become less or equal to 0.001kN, residual displacements less or equal to 0.001cm.

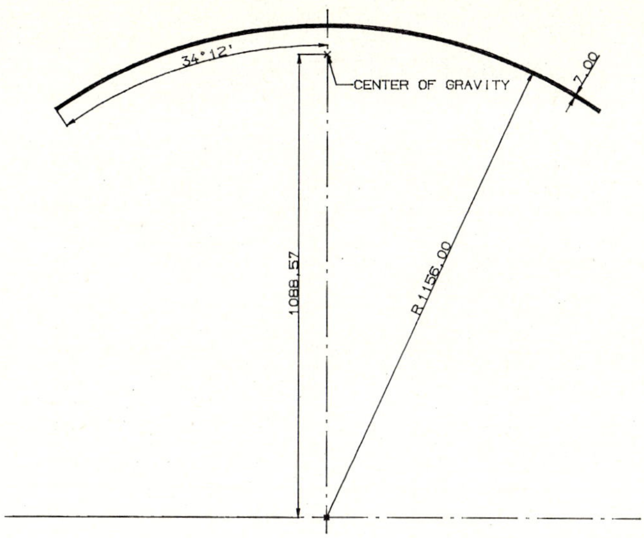

FIGURE 1 - ORIGINAL CROSS SECTION

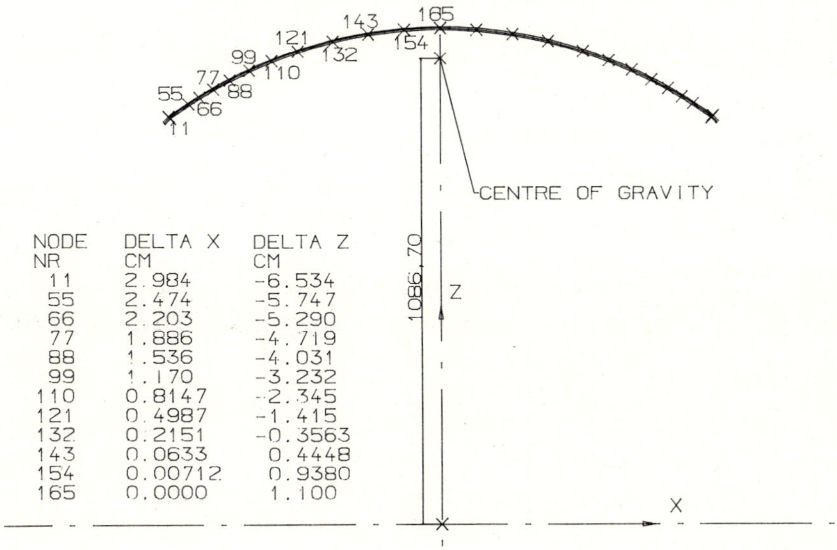

NODE NR	DELTA X CM	DELTA Z CM
11	2.984	-6.534
55	2.474	-5.747
66	2.203	-5.290
77	1.886	-4.719
88	1.536	-4.031
99	1.170	-3.232
110	0.8147	-2.345
121	0.4987	-1.415
132	0.2151	-0.3563
143	0.0633	0.4448
154	0.00712	0.9380
165	0.0000	1.100

FIGURE 2 - DEFORMED CENTRAL CROSS SECTION
SELF WEIGHT - LINEAR

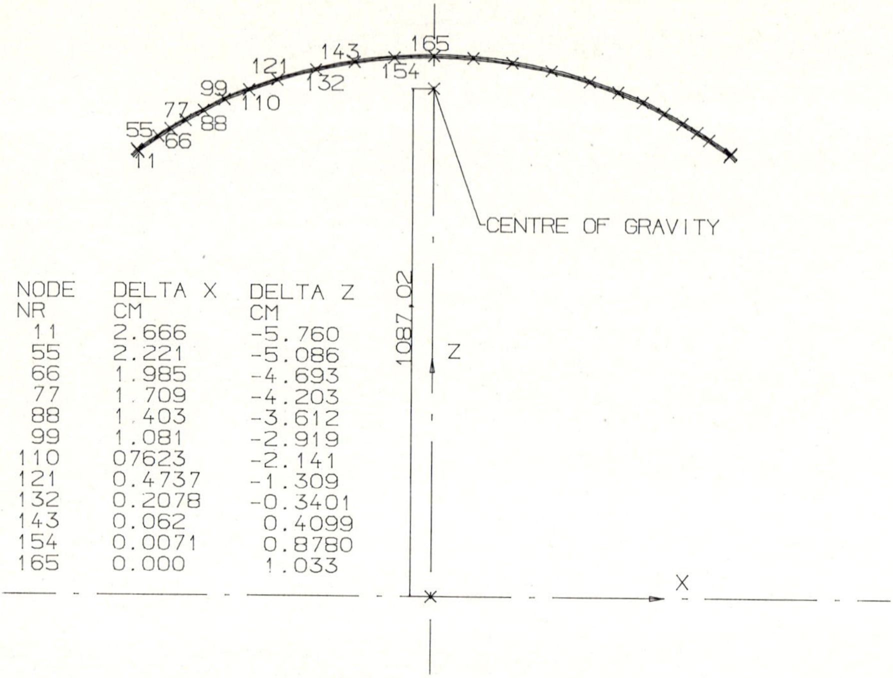

NODE NR	DELTA X CM	DELTA Z CM
11	2.666	-5.760
55	2.221	-5.086
66	1.985	-4.693
77	1.709	-4.203
88	1.403	-3.612
99	1.081	-2.919
110	07623	-2.141
121	0.4737	-1.309
132	0.2078	-0.3401
143	0.062	0.4099
154	0.0071	0.8780
165	0.000	1.033

FIGURE 3 - DEFORMED CENTRAL CROSS SECTION
SELF WEIGHT - NONLINEAR

6. Finite element meshes - Program code

The used program code is TITUS, marketed by FRAMATOME; [3].

Two meshes are considered; the first is shown in Figure 4.b), the second in 4.c).

Fig.4.b consists of 140 quadrilateral thin elements (code number in TITUS is 2004; elements which can be used only in linear analysis) each with four nodes, each node having six degrees of freedom. Those elements are resulting from a programmed mesh: the PROGRAM-routine meshes the area swept by a one dimensional initial mesh, as all the nodes of this are following a user-defined trajectory. Description of the elements: [5].

Modelling the prestressing cable has been done by using two-nodal-truss-elements. Fig.4.c consists of 224 thin triangular shell elements (code number in Titus 2003; elements wich can be used for nonlinear computations also) each with three nodes, each node having six degrees of freedom. This mesh is obtained by the routine TRIANGLE of TITUS.

7. Methods of introducing prestressing forces

In a previous paper [2] we used a fictive thermal load in a curved cable, excercising consequently a continuous lateral force on the shell itself. As the resulting structure displacements would invoke unwanted cable forces, we replaced the real cable characteristics by fictive ones. We reused this method in combination with the mesh 4.b, in which the cable trajectory is replaced by the polygon composed of the mentioned truss elements.

Using the other mesh (with polygon for cable form as well), the 615kN prestress force is applied in the nodes of this polygon, following its material sides.

8. Results

8.1 Deflection of the centres of gravity of the cross sections due to self weight only

As pointed out by Kvetoslav ZAHRADNIK [4], all the centers of gravity deflect like in any beam, but the summit raises while the edge points drop under self weight as appears from FIGURES 2 and 3, both compared with FIGURE 1.

The linear computation yields 1088.57 - 1086.70 = 1.87cm deflection. This result, obtained with the one mesh is confirmed by the result of the other, rectangular mesh: 1088.57 - 1086.90 = 1.67cm. The nonlinear computation yields 1088.57 - 1087.02 = 1.55cm. Thus the influence of the geometrical nonlinearity is 1.87 - 1.55 = 0.32cm.

The computation by hand yields 2.9cm with the formula $f=5pl^4/384EI$ in which $p=2415.25$ kg/m; $l=18$m, $EI=11.28\cdot10^6$ kN/m².

This influence of the shearing force appeared when we used modified boundary conditions with the other mesh.Letting the boundary nodes slide along the curved edge, except the summits, we obtained 2.14cm.

8.2 Deflection surfaces due to prestress only

The contours are shown in FIGURE 5.a, 6.a and 7.a. The linear computations compare rather well, but along to the nonlinear computations the edge displacement doubles: curves J in the three figures.

8.3 Transversal bending moments due to prestress only

Figure 4.a is comparising three curves, each of which is $M_\theta=f(\theta)$. The best value found for M_θ in $\theta=0$ in [2] with linear computations corresponds well with the two values shown in Figure 4.a for the linear computations, specially with the rectangular elements, result which is moreover nearest to the result of the nonlinear computations. As the material nonlinearity counteracts partly the effect of the geometrical nonlinearity, we estimate M_θ = 1kNm/m in the central summit point of the shell for prestress only.

The positive maximum disappears when nonlinearity is considered, while the negative maximum jumps from ..-0.8..-1.0 till -1.3. The latter was found in [4] but our best curve in [4] yielded -0.75 kN.

The contours of M_θ are shown in the FIGURES 5.c, 6.c, 7.c.

8.4 Normal stress resultants due to prestress only

The contours are shown in the FIGURES 5.b, 6.b and 7.b. Due to nonlinearity, shifts the extreme value from ...-8.5...-8.6... till -10 kN/cm.

8.5 Longitudinal bending moments

The contours are shown in the FIGURE 5.d, 6.d and 7.d. The nonlinearity makes shift the extreme value from ...-2.5...-2.6... till -3.996kN.

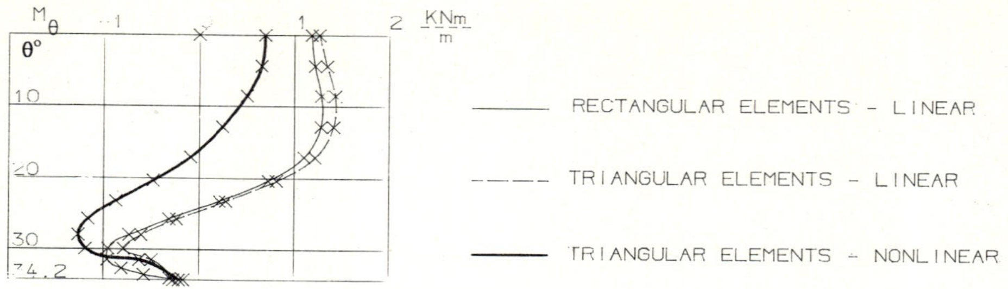

a) TRANSVERSAL BENDING MOMENTS M_ϑ ALONG THE CENTRAL CROSS SECTION DUE TO PRESTRESS ONLY - MESS WITH RECTANGULAR ELEMENTS AND MESH WITH TRIANGULAR ELEMENTS, THE LATTER ALSO FOR THE NONLINEAR PART

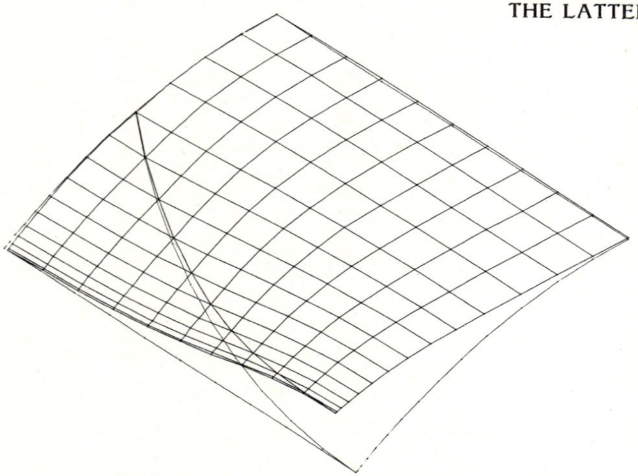

b) MESH WITH RECTANGULAR ELEMENTS FOR LINEAR COMPUTATIONS - MESH DEFORMED BY PRESTRESS ONLY

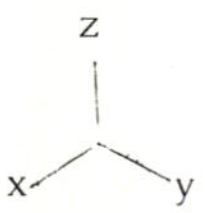

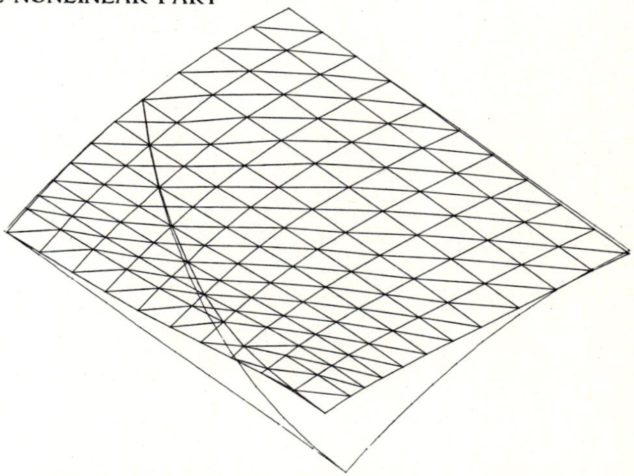

c) MESH WITH TRIANGULAR ELEMENTS FOR LINEAR AND NONLINEAR COMPUTATIONS - MESH DEFORMED BY PRESTRESS ONLY

FIGURE 4

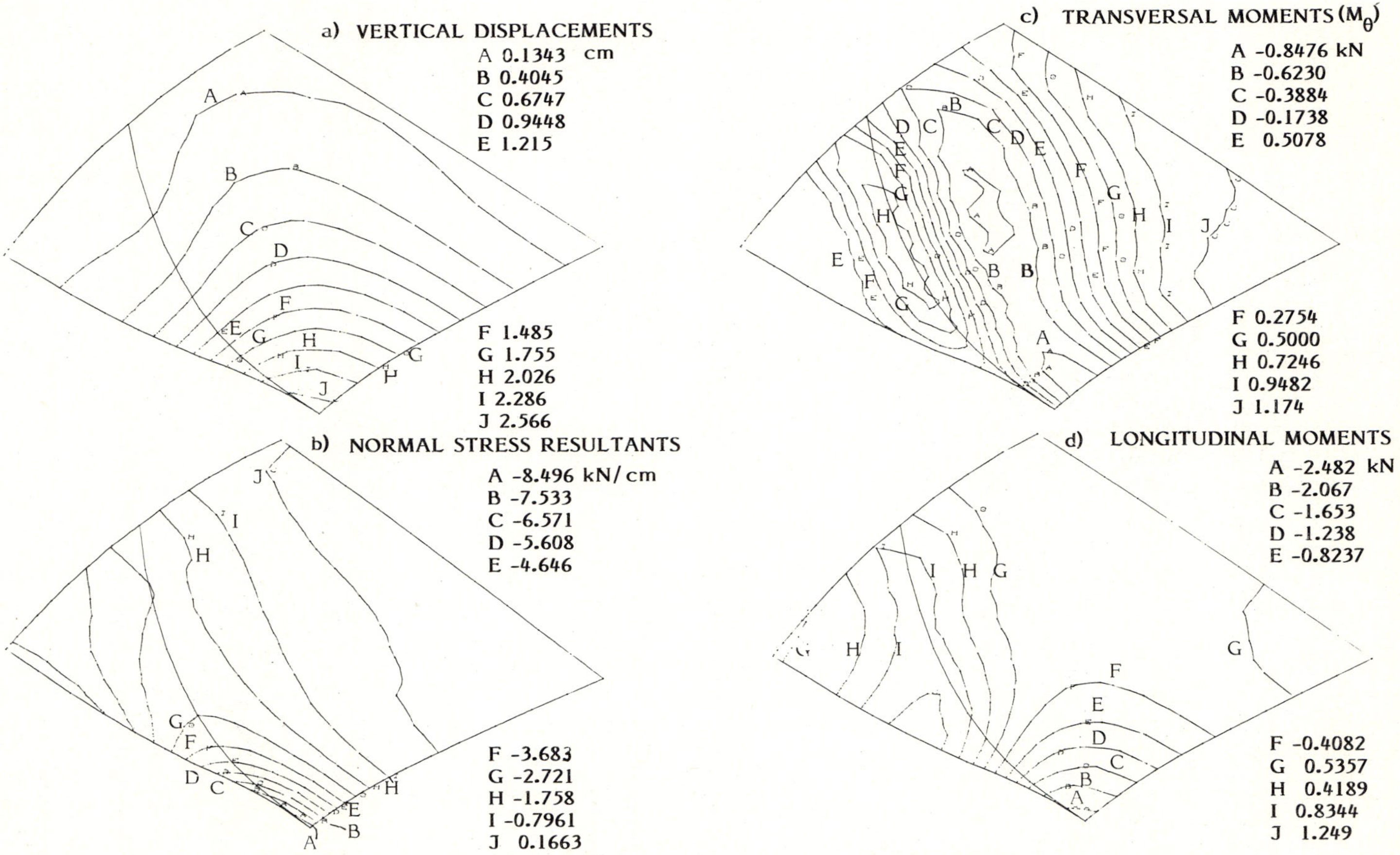

FIGURE 5 - LINEAR COMPUTATIONS WITH THE RECTANGULAR ELEMENTS

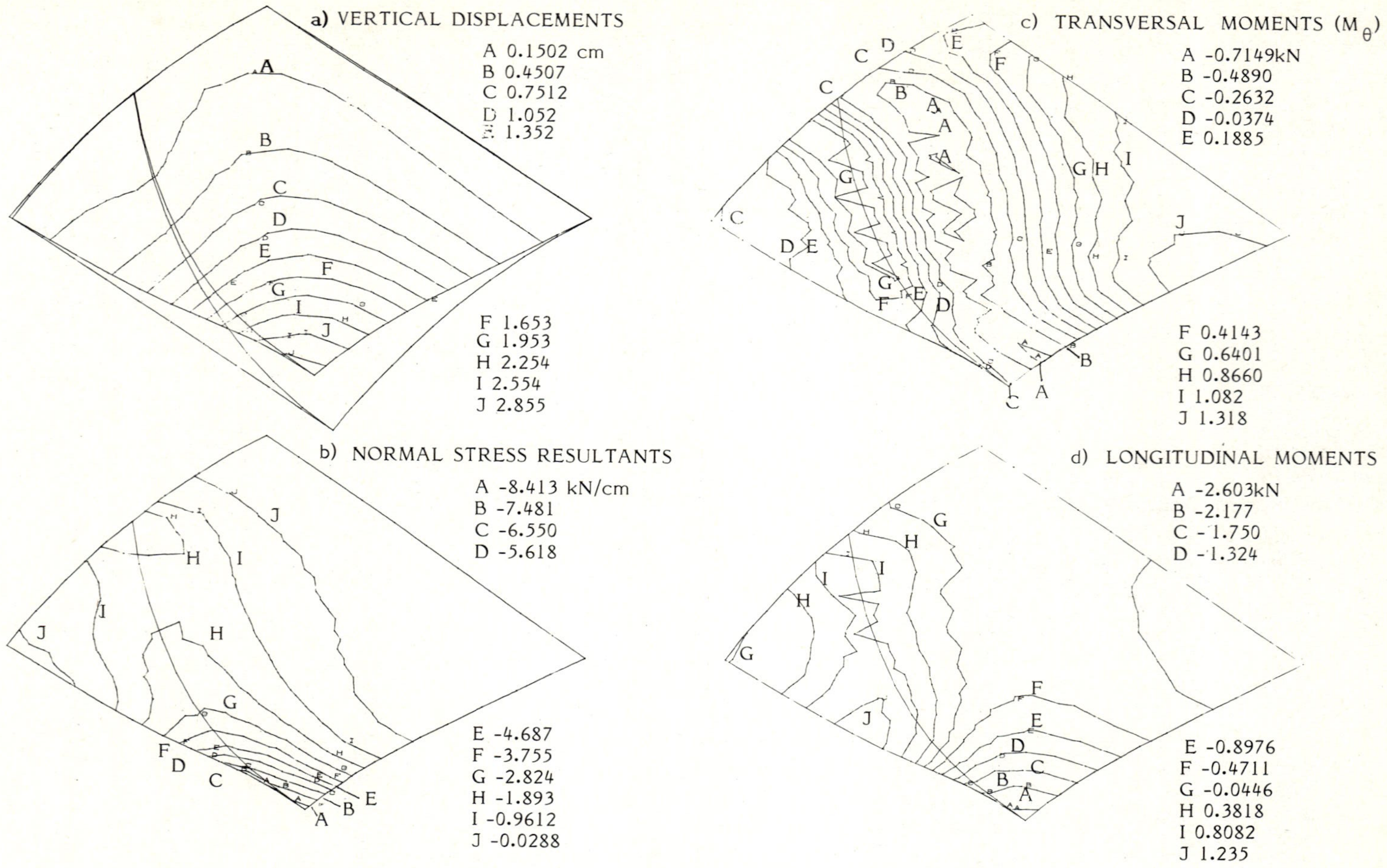

FIGURE 6 - LINEAR COMPUTATIONS WITH THE TRIANGULAR ELEMENTS

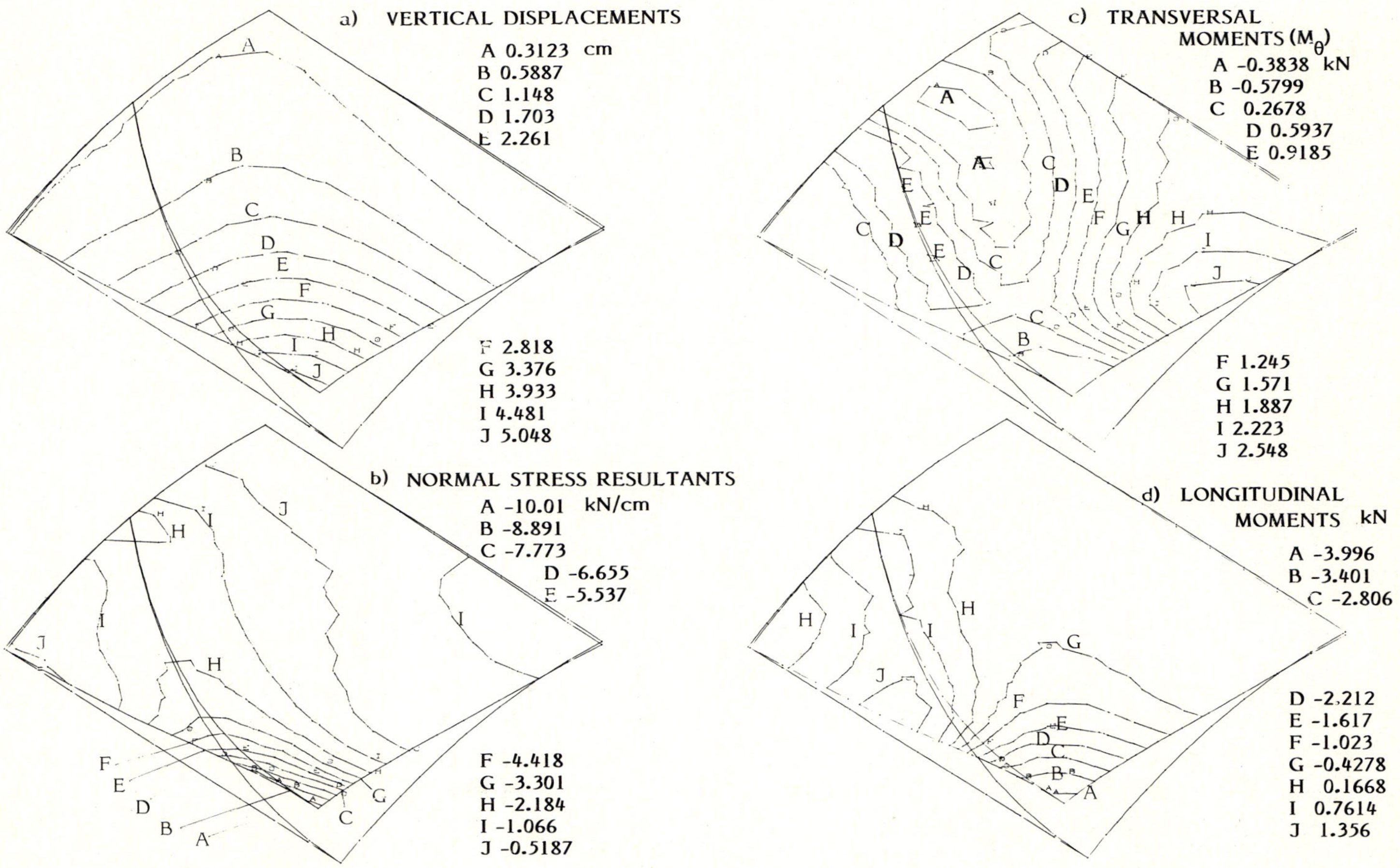

FIGURE 7 - NONLINEAR COMPUTATIONS WITH THE TRIANGULAR ELEMENTS

9. Conclusion

For prestressed, important shell structureswhich displacements are nearly equal to the shell thickness, nonlinear computations could be necessary for the secondary stresses. The simultaneous loading with self weight and prestress rests to be examined.

10. Acknowledgements

Dr.ir. Edgard BACKX (SCIA, B-3912 Herk-de-Stad, Belgium), drs.ir. Hans WYNENDAELE and ir. Peter VERHEYDEN (both of Dynamics Engineering (supporting TITUS), Ambachtenlaan 21, B-3030 Heverlee, Belgium) helped mainly in the use of TITUS.

ir. Carla GORDONS and ir. Jan CARREIN, assisted with the CAD-program CDM300 of KONGSBERG for composing the FIGURES 1, 2 and 3 and the computings in them.

Literature

[1] W.R. De Sitter, Theoretisch en experimenteel onderzoek naar het gedrag van een voorgespannen tonschaal, Heron 11, Nr.1 (1963), Technische Hogeschool Delft, Delft (The Netherlands).

[2] M. Van Laethem und J. de Coen, Löwen, Belgien; Vorspannung einer kurzen Kreiszylinderschale ohne Randträger, Osterreichische Ingenieurszeitschrift, Heft 9, Jg.23(1980).

[3] FRAMATOME - Service Calcul - BP13 - 71380 SAINT MARCEL-FRANCE.

[4] K. Zahradnik - Tests of beamless cylindrical shells - Symposium Simplified Computations, held by IASS in Brussels.

[5] O.C. Zienckiewicz - The Finite Element Method - McGraw-Hill, 1977; chapters 10 and 13.

A FINITE ELEMENT MODEL FOR THE NONLINEAR ANALYSIS OF REINFORCED CONCRETE SHELL STRUCTURES

by M. Cervera , A.J. Kent and E. Hinton.
Department of Civil Engineering, University College of Swansea, U.K.

SUMMARY

This paper presents computational models for simulating the behaviour of reinforced concrete structures which exhibit a combination of geometric and material nonlinearity. The nonlinear finite element model is described and methodes are presented for the solution of the nonlinear equilibrium equations using arc-length procedures. In order to examine the the performance of the finite element model and the solution procedures, examples are presented which involve material and geometric nonlinearity with snap-through and/or snap-back behaviour.

1. INTRODUCTION AND OBJECTIVES

The ultimate load behaviour of many reinforced concrete shell structures is dependent on the combination of geometric and materially nonlinear effects. Geometric nonlinearities in the strain displacement relationships become significant when the deflections are of the same order as the shell thickness. In such circumstances structural instability in the form of buckling and bifurcation, represented by limit points in the load-deflection response, may be in evidence.

The objective of this paper is therefore to describe the development of a finite element model which attempts to incorporate the above modes of behaviour thereby enabling the ultimate load responses of such shell structures to be predicted. The proposed numerical model has the following features :

a) a finite element shell formulation which permits large displacements and rotations,

b) a material model which includes nonlinear effects due to concrete cracking and crushing, and steel yielding,

c) a nonlinear equation solution procedure which is efficient and which can negotiate limit points in the structural response.

2. DESCRIPTION OF SHELL GEOMETRY AND BEHAVIOUR

The present work follows the shell formulation of Oliver and Onate [1] and allows for large displacements and rotations and also transverse shear. For further details and explicit forms of the matrices the reader is referred elsewhere, [1],[2].

The resulting element is an isoparametric degenerate shell element of the type originally developed by Ahmad et al. [3]. The geometry of the shell is thus described in terms of the position of its midsurface, as shown in Figure 1. The vector of nodal degrees of freedom consists of three orthogonal midsurface displacements and two independent rotation components.

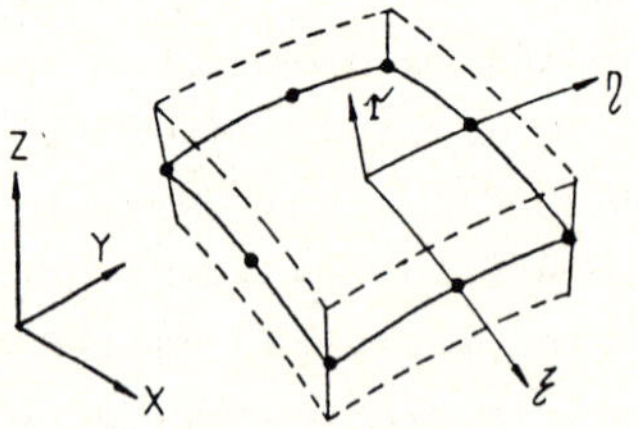

Figure 1 The degenerate shell element

A total Lagrangian formulation is used so that all quantities are referred to the original, undeformed configuration. These equations can be written in terms of the external applied loads $\underline{R}$ and the out of balance or residual forces, as

$$\Psi(\underline{a}) = \int_V \underline{B}^T \underline{\sigma}\, dv - \underline{R} = 0$$

The volume integrals appearing in the discretised equilibrium equations are performed numerically. They are evalated using product integration rules resulting from Gaussian quadratures for area integration and Gaussian or Lobatto rules for the through-thickness integration.

Deficiencies in some standard degenerate shell elements such as shear and membrane locking, the presence of spurious zero energy modes, and failure to pass the patch test are well known to the authors, necessitating the use of special integration techniques [4], and othe procedures. However, for the moderately thin situations encountered with reinforced shell concrete shell structures, the use of selective integration has proved to give reasonable results.

3. MODELLING OF THE MATERIAL BEHAVIOUR

Concrete is a highly complex, multi-phase material. However, in practice it is usually idealised as an equivalent continuum. The interaction between the reinforcement and the surrounding concrete further complicates the material idealisation used in the numerical model. Such idealisation should be able to represent compressive yielding and crushing and tensile cracking of the concrete, and also yielding of the reinforcing steel.

In the present model any point within the shell is assumed to be in a state of plane stress, the stress component normal to the midsurface being assumed to be zero. The incremental constitutive relations may be written as

$$\Delta\underline{\sigma} = \underline{D}\Delta\underline{\varepsilon}$$

where $\Delta\underline{\sigma}$ and $\Delta\underline{\varepsilon}$ are vectors containing increments of second Piola-Kirchhoff stresses and Green strains respectively. The constitutive matrix $\underline{D}$ is modified according to the material state. Typical failure envelopes and hardening curves for concrete are shown in Figure 2.

3.1 MODELLING OF CONCRETE IN COMPRESSION

In the present work, for concrete in biaxial compression a yield criterion is adopted which is defined in terms of the two stress invariants, as shown in Figure 2, [5]. The direction of plastic straining is defined by the normality condition of an associated flow rule. A parabolic hardening function is used in which cf_c' ($0<c<1$) defines the initial yield surface. Continuation of loading beyond the proportionality limit therefore results in a uniform expansion of the yield surface. Unloading inside the current yield surface is assumed to be elastic.

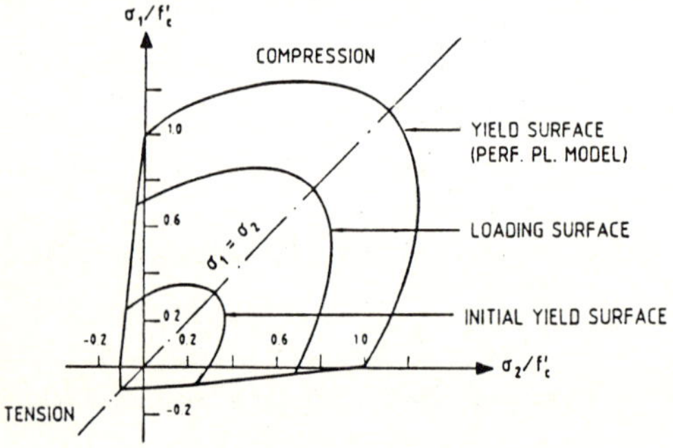

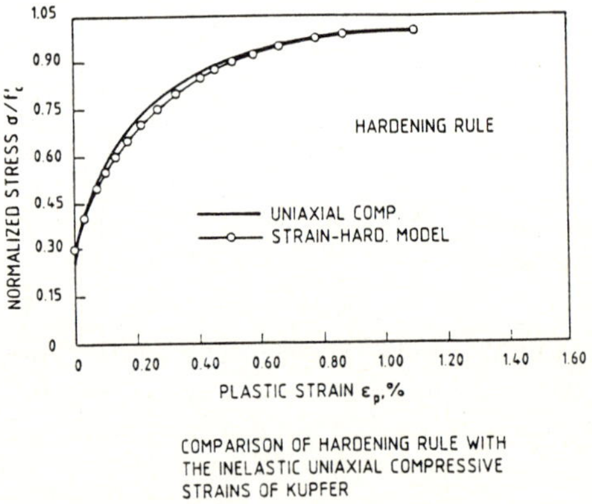

Figure 2 Failure envelope and hardening for concrete

Inelastic deformation in the biaxial compression zone continues until a failure surface defined in principal strain space is reached, at which stage the concrete is assumed to be crushed and all stresses are released.

3.2 MODELLING OF CONCRETE IN TENSION

Due to the comparatively low tensile strength cracking is a major source of material nonlinearity in concrete. However, the degree of scatter in experimental observations of the associated material parameters suggests that a fairly simple cracking criterion is appropriate. Here, the maximum tensile stress criterion is adopted. The 'smeared' approach is used for the numerical representation of the cracks. This averaging philosophy is consistent with the assumptions of continuum mechanics and the finite element method , and is the approach usually used for flexural systems in which the behaviour is not governed solely by a few dominant cracks.

In zones of biaxial tension or tension-compression, cracking is assumed to occur when a principal stress exceeds the tensile strength, f_t' . The cracks form in a vertical plane orthogonal to the offending principal stress.

Thereafter, the material associated with the cracked sampling point becomes orthotropic with the local material axes being defined in directions parallel and perpendicular to the crack plane. The subsequent loss of stiffness of the cracked concrete is modelled by a reduction of the Young's modulus and the shear modulus in the directions perpendicular and parallel to the crack respectively. These reductions are related to the normal tensile strain across the crack.

Subsequent opening and partial or full closing of the cracks is permitted and is monitored by the strain normal to the crack plane.

3.3 STRAIN-SOFTENING

Experimental and numerical evidence shows that the release of the stress normal to a crack follows a post-peak strain-softening curve. The process of redistribution of the stresses away from a damaged region is therefore a gradual one. The rate at which this stress release occurs is determined by the descending branch of the tensile stress-strain relationship, as shown in Figure 3.

In order to overcome the problem of mesh independency the present work adopts a strain-softening relation based on the fracture energy concept. With 'stiff' testing apparatus the fracture energy G_f may be obtained as a fundamental material property. This concept , and the definition of a characteristic length l_c associated with the sampling point provides an objective way of selecting the parameter α which defines the softening curve.

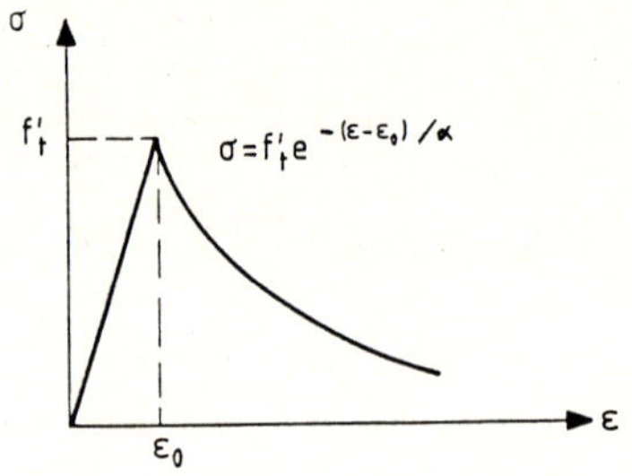

Figure 3 The strain-softening curve

3.4 MODELLING OF THE STEEL REINFORCEMENT

Each set of reinforcing bars is idealised by an equivalent smeared layer with uniaxial properties in the direction of the discrete bars. An elasto-plastic formulation is adopted which permits linear isotropic hardening and elastic unloading. Perfect bond is assumed to exist at the steel-concrete interface.

4. THE ARC-LENGTH SOLUTION PROCEDURE

The arc-length solution techniques of Riks [6] and Crisfield [7] allow limit points in the load-deflection response to be both pinpointed and successfully negotiated without major complications to the solution procedure.

In the present work the arc-length method based on a spherical path [7] is implemented in conjunction with a modified Newton-Raphson incremental-iterative procedure. With an automatic load incrementation scheme, this combination posseses the following advantages over traditional solution methods :

a) it has the ability to handle snap-through and snap-back situations,

b) it significantly improves the convergence characteristics throughout the entire loading range, and

c) it allows automatic and self-adaptive incrementation of loads.

The expression of the iterative displacement changes $d\underline{a}_i$ in terms of a change of the load parameter $d\lambda_i$ is fundamental to all arc-length procedures. The evaluation of this change in load, however, is dependent upon the nature of the convergence path followed by the iterative process. Here the convergence path is constrained to follow a spherical path in n-dimensional displacement space. This constraint is introduced by means of an extra equation in terms of the incremental displacements.

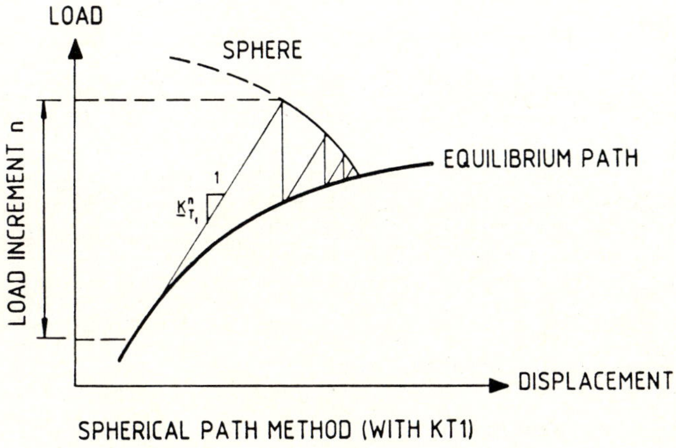

Figure 4 Spherical path method for a single degree of freedom

For a single degree of freedom system this reduces to a circular arc of radius l^n, as shown in Figure 4. The appropriate choice of $d\lambda_i$ from the resulting quadratic expression enables the iterative displacement change to be obtained and the solution procedure updated.

The new incremental length l^n is computed at the beginning of each increment, following the suggestion of Ramm [8], as a function of the previous number of required iterations, N_{n-1}, and a specified desired number of iterations, N_d, so that

$$l^n = l^{n-1} (N_d / N_{n-1})^{1/2}$$

This procedure automatically reduces the size of the increment in the regions where more nonlinear behaviour occurs. As suggested by Crisfield [7] the sign of successive load increments (originally assumed as positive) follows that of the previous one unless a change in the sign of the determinant of the tangential stiffness matrix is observed.

To avoid inappropriate increment sizes the magnitude of the load increments are referred to specified upper and lower limits. By relating these limits to the current stiffness parameter the 'grading' of the solution procedure in the areas of most rapid change can be significantly accelerated [9].

5. NUMERICAL APPLICATIONS

5.1 Example 1 - Thin cylindrical shell

This example is presented here to illustrate the ability of the proposed solution procedure to handle snap-through and snap-back behaviour.

The geometry and material properties of the shell are given in Figure 5. A symmetric quarter of the slab is idealised using four shell elements. The constrained boundary conditions are in the form of clamped longitudinal edges and free curved edges. The shell is subjected to a central point load.

The load-deflection curve for the central point is presented in Figure 5 and compared to the combined load and displacement incrementation solution presented by Sabir and Lock [10]. Horizontal and vertical tangents which occur at various stages of the load-deflection curve, present difficulties for traditional load and displacement incrementation methods respectively. They can be successfully overcome using the arc-length method.

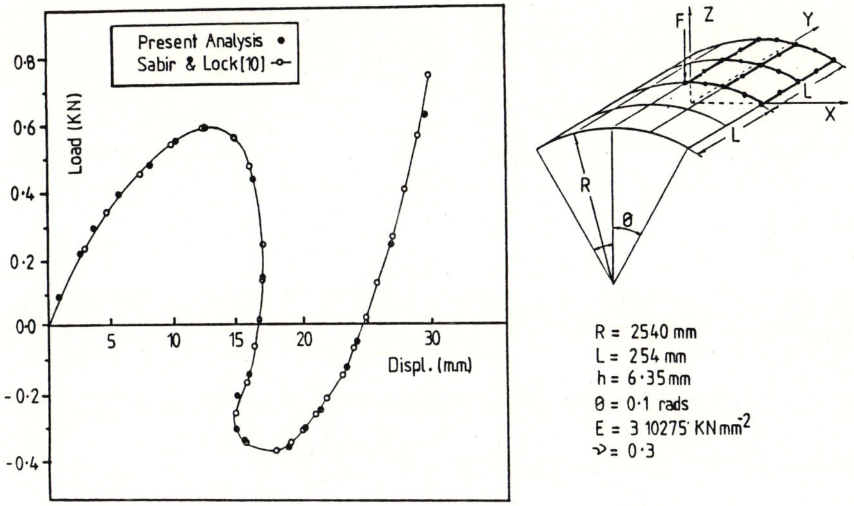

Figure 5 Example 1 - Thin cylindrical shell

5.2 Example 2 - Reinforced concrete hyperbolic paraboloid shell

In this example consideration is given to the reinforced concrete hyperbolic paraboloid shell previously studied by Mueller [11]. The saddle shell, shown in Figure 6, is initially loaded with a dead load (DL) of 150 lbs/ft^3 and then with a live load (LL) in multiples of 20 lbs/ft^2 . The mesh used to idealise a symmetric quarter of the shell is shown in Figure 6 and consists of 14 Heterosis elements : 10 to represent the shell and 4 to represent the edge beam.

Load-deflection curves for the highest point are presented in

Figure 6. The response is almost linear, up until failure.

The structure is analysed firstly using a local softening curve with an ultimate tensile strain of 0.00015. The ultimate load obtained compares well with the value given in reference 11 where the same model was used. The analysis is then repeated using the fracture energy concept with G_f = 0.25 N/mm. A much lower failure load is then obtained , thus showing the need for an objective method of defining the softening branch of the tensile model.

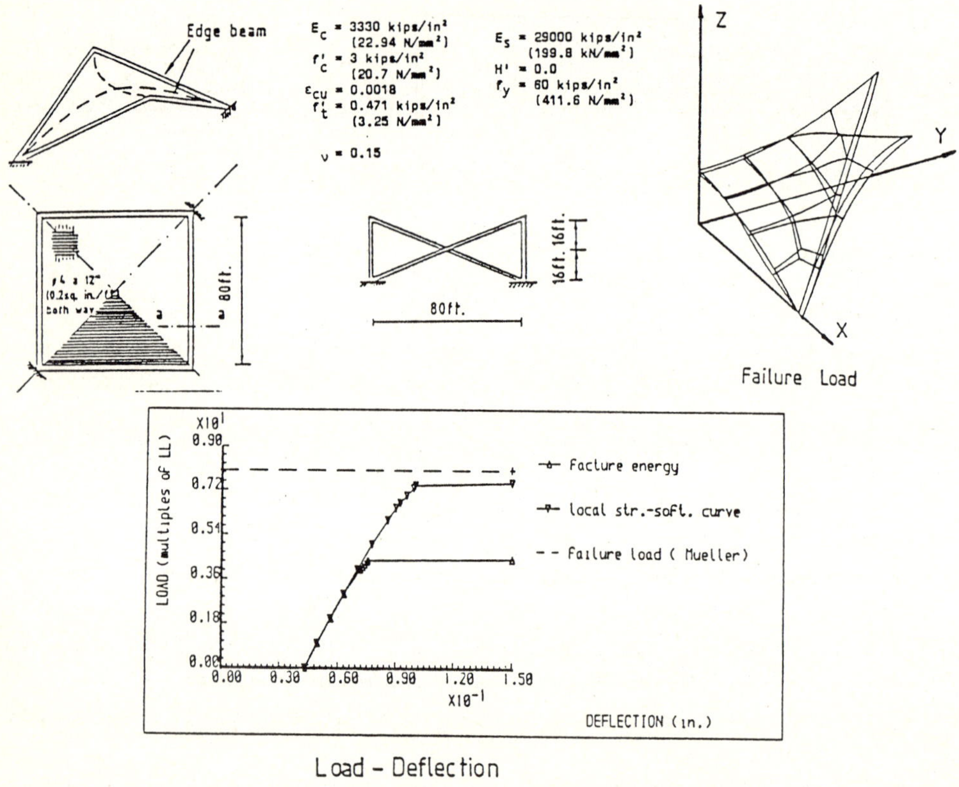

Figure 6 Example 2 - Hyperbolic paraboloid shell

5.3 Example 3 - Plate buckling problem

The plate chosen is No.11 tested by Swartz et al [12], which was reported to have buckled with one half wave in each direction. Details of the geometry and material properties of the plate are given in Figure 7. Again, due to its symmetrical nature only one quarter of the plate was considered. To initiate the buckling process a small transverse load of 1/1000th of the total axial load was applied at the centre. The load-deflection curves for this central point demonstrate the significance of the nonlinear geometric behaviour, which is shown in Figure 7.

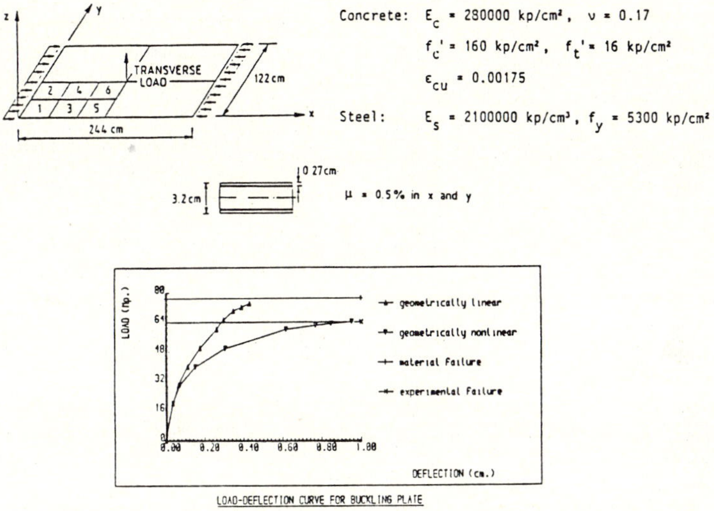

Figure 7 Example 3 - Plate buckling problem

5.4 Example 4 - Fully constrained reinforced concrete slab

This clamped rectangular reinforced concrete slab was tested experimentally by Park [12]. One quarter of the slab is idealised using six shell elements and is subjected to a uniformly distributed load.

The load-deflection response for the central point is presented in Figure 8 and is compared to the displacement incrementation analysis of Ganaba [14]. The initial compressive membrane action, which occurs because of the clamped boundary of the slab, precedes a clearly defined snap-through point, after which reinforcement in the slab acts as a tensile net. Cracks occur on both upper and lower surfaces near the boundary and at the centre respectively. The apparent 'phase' difference between the experimental and numerical results is attributed to the difficulty in obtaining a true clamped boundary condition in practice.

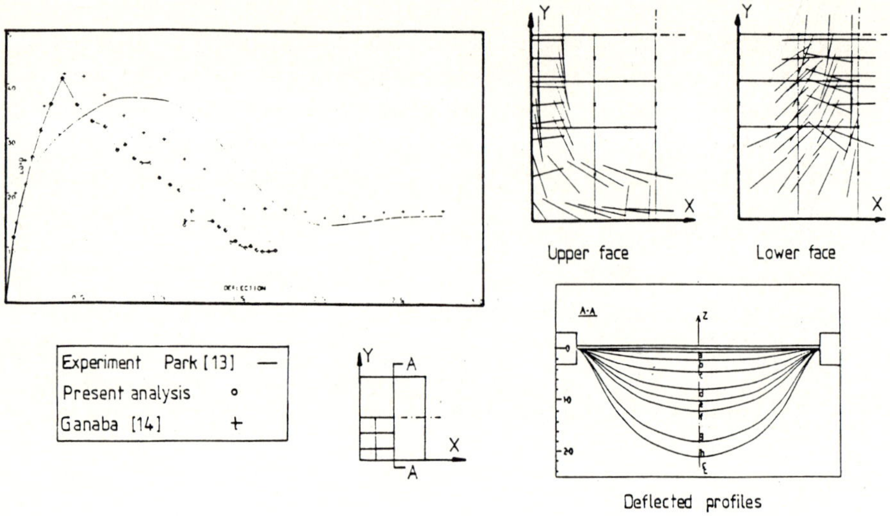

Figure 8 Example 4 - Constrained reinforced concrete slab

DISCUSSION AND CONCLUSIONS

It has been shown that the proposed numerical model is capable of analysing reinforced concrete shell structures which exhibit both geometric and material nonlinearity. Snap-through and snap-back analyses are also possible although the presence of alternative equilibrium paths for reinforced concrete structures can cause difficulties when using the arc-length method. Research is currently underway to investigate the behaviour of similar structures using three-dimensional elements and material representations.

REFERENCES

[1] Oliver J. and Onate E.
'A total Lagrangian formulation for the geometrically nonlinear analysis of structures using finite elements. Part two. Shell and plate

structures',
Int.J.Num.Meth.Engng., 20, 2253-2281, 1984.

[2] Cervera M.
'Finite element techniques for nonlinear static and transient dynamic analysis of reinforced concrete structures',
Ph.D.Thesis. University College of Swansea.
(to be submitted 1986)

[3] Ahmad S. Irons B.M. and Zienkiewicz O.C.
'Analysis of thick and thin shells by curved shell elements',
Int.J.Num.Meth.Engng., 2, 419-451, 1970.

[4] Huang H.C. and Hinton E.
'A nine noded lagrangian Mindlin plate element with enhanced shear interpolation',
Engineering Computations, 1, 369-379, 1984.

[5] Owen D.R.J. Figueiras J.A.
'Ultimate load analysis of reinforced concrete plates and shells including nonlinear effects',
Finite Element Software for Plates and Shells.
(ed. E. Hinton and D.R.J. Owen)
Pineridge Press, Swansea U.K. 1984.

[6] Riks E.
'An incremental approach to the solution of buckling problems.'
Int.J.Solids and Structures, 15, 524-551, 1972.

[7] Crisfield M.A.
'A fast solution procedure that handles snap-through',
Computers and Structures, 13, 55-62, 1982.

[8] Ramm E.
'Strategies for tracing nonlinear responses near limit points',
Nonlinear Finite Element Analysis in Structural Mechanics.
Springer, New York, 1984.

[9] Kent A.J.
'The use of arc-length solution techniques in the geometrically nonlinear finite element analysis of plate and shell structures',
Internal Report, Department of Civil Engineering, University College of Swansea, 1986.

[10] Sabir A.B. and Lock A.C.
'The application of finite elements to the large deflection geometrically nonlinear behaviour of shells.'
Variational Methods in Engineering, 7/54-7/65, Southampton University Press, 1972.

[11] Mueller G.
'Numerical problems in nonlinear analysis of reinforced concrete',
Report No.UC SESM 77-5 Department of Struct.Eng.Mech. 1977.

[12] Swartz S.E. Rosebrough V.H. and Berman M.Y.
'Buckling tests on rectangular concrete panels',
J.Amer.Conc.Inst., 71, 33-39, 1974.

[13] Park R.
'Ultimate strength of rectangular concrete slabs under short term loading with edges restrained against lateral movement',
Proc.Instn.Civ.Engng., 28, 125-150, 1964.

[14] Ganaba T.H.
'Nonlinear finite element analysis of plates and slabs',
Ph.D.Thesis, Department of Engng., University of Warwick, 1985.

THE BEHAVIOUR OF A STEEL CYLINDER UNDER THE INFLUENCE OF A LOCAL LOAD IN THE ELASTIC AND ELASTO-PLASTIC AREA

J. Spiekhout, N.V. Nederlandse Gasunie, Groningen
A.M. Gresnigt and G.M.A. Kusters, TNO-Institute for Building Materials and Structures, Rijswijk

SUMMARY

In this paper a calculation model for the relation between local load and indentation depth, including internal pressure, is given. The formulas have been derived by means of theory of shells and theory of plasticity. Via coefficients and a few interventions these theoretical formulas have been adapted to the results of tests and the results of finite element calculations. The calculation model thus obtained is capable of predicting the real local load - indentation depth diagram in a sufficiently accurate manner.

The finite element method calculations were carried out with the DIANA programme of IBBC-TNO. In these computer calculations elasto-plastic material behaviour, large displacement options and non-conservative gas pressure were included.

INTRODUCTION

The work described in this paper was carried out as part of a research programme on the subject of resistance of pipelines against mechanical damage. The cause of the mechanical damage was assumed to be a hydraulic excavator. For a case of indentation of a pipeline by the hydraulic excavator a model for the relation between local load and indentation depth was developed. Further, on the basis of fracture mechanics and burst tests, a calculation model was developed for the burst pressure of a pipe with a notched dent. In (10) a description of that calculation model is given.

In order to be as brief as possible, first the formulae are presented and then an explanation about the backgrounds of the formulae is given.

CALCULATION MODEL

Formulae

For the area of elastic deformation the following formula applies to the relation between dent depth H_F and force F:

$$H_F = 0{,}149\ F.R^3 \sqrt{f_{rr}}/(B.EI) \qquad (1)$$

where:

$B = 1{,}33\ R \sqrt{R/t} + b_1$
$f_{rr} = p_{cr}/(p_{cr} + p_i)$
$p_{cr} = 3\ EI/R^3$
$EI = E\ t^3/12$

in which:

R	-	pipe diameter (= D/2)	(mm)
t	-	wall thickness	(mm)
H_F	-	dent depth	(mm)
B	-	acting width of pipe wall	(mm)
b_1	-	width of denting object in axial direction	(mm) (mm)
F	-	denting force	(N)
p_i	-	internal pressure	(MPa)
p_{cr}	-	critical external pressure with regard to stability	(MPa)
f_{rr}	-	rerounding factor	
I	-	moment of inertia of pipe wall per unit length	(mm^3)
E	-	Young's modulus	$(= 2{,}1.10^5\ N/mm^2)$

The following formula has been derived for the relation between dent depth H_F and the force F in the area of plastic deformation:

$$F = F_p + F_{pi} \qquad (2)$$

and $F = F_m + F_{pi}$ for $F_p + F_{pi} > F_m + F_{pi}$ (3)

and $F_p = 1{,}15\ t^2.Re\ (1-0{,}75\ (\sigma_r/Re)^2).B_p.S/R$ (4)
$B_p = 0{,}80\ R \sqrt{R/t} + b_1$
$S = R/(R-0{,}35b_2)$
$\sigma_r = p_i.R/t$

$F_m = 2X.H_F.S/ \sqrt{H_F^2 + l^2}$ (5)
$X = 0{,}8\ l.Re.t$
$l = R \sqrt{R/t}/8$

$F_{pi} = 1{,}5\ H_F.p_i.B_{pi}$ (6)
$B_{pi} = 0{,}4R \sqrt{R/t}\ f_{rr} + b_1$

where:

B_p	-	acting width of pipe wall in plastic area due to bending in pipe wall	(mm)
B_{pi}	-	acting width of pipe wall in plastic area due to internal pressure	(mm)

b_2 - width of denting object perpendicular to the axial direction (mm)
F_p - denting force in plastic area due to bending in pipe wall (N)
F_{pi} - denting force in plastic area due to internal pressure (N)
F_m - denting force in plastic area due to extension forces in pipe wall (N)
σ_r - circumferential stress (N/mm^2)

Re - yield stress (N/mm^2)

S - correction factor for calculation of F_p

According to formula (1) there is a linear relation between force and dent depth in the region of elastic deformation. This region extends to load F_p. As may be derived from (4) F_p is not dependent on dent depth. According to (5) and (6) the influence of the extension forces (F_m) and the internal pressure (F_{pi}) on the force increase with the dent depth (H_F). It should also be noted that it takes a substantially greater force to dent a pipeline with internal pressure than one without internal pressure.

Background to formulae

The formulae for elastic deformation have been derived with shell theory for a pipe with two opposite forces. By applying $\sqrt{f_{rr}}$ in formula (1) instead of f_{rr} reasonable agreement with the results of denting tests was obtained.

The formulae for the plastic region are, in fact, semi-empirical. In the first instance the load bearing capacity is determined by the occurrence of plastic hinges in the circular cross-section (see fig. 1a and 1b). For the acting width of pipe wall an analoguous formula was chosen to the one derived for elastic deformation. By adaptation of the coefficients reasonable agreement with the results of denting tests was obtained.

As the dent depth increases, the influence of the "second-order" becomes greater. The contribution of the extension forces (F_m) was derived by means of equilibrium as illustrated in figure 1c. The contribution of the internal pressure (F_{pi}) was derived by means of virtual work, as may be seen from figure 1d. For the four hinge model in this figure, the coefficient in formula (6) is 2,0. Because in reality the deformations are somewhat different, a coefficient of 1,5 was chosen. For F_m and F_{pi} as for F_p, by adaptation of the coefficients in the expressions for X, S, 1, B_{pi} and B_p reasonable agreement with the results of tests was obtained.

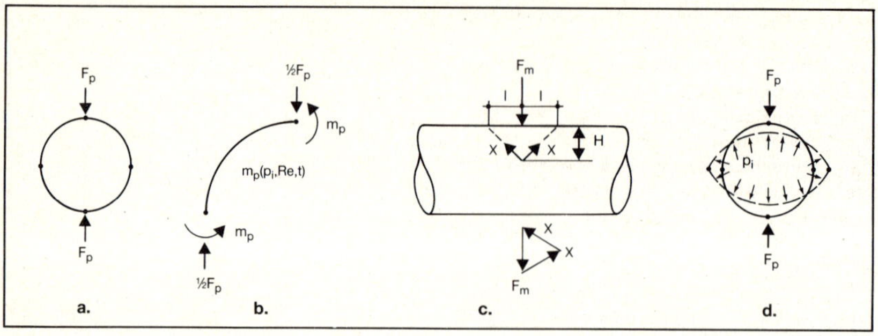

Fig. 1 Failure mechanism when denting.

Actually, when deformation occurs in the plastic region, the contributions to the force of the bending action in circumferential direction (F_p), of the extension forces (F_m), and of the internal pressure (F_{pi}) should be summated. However, as the membrane forces increase, the pipe wall absorbs less bending, so the share of bending tends to diminish with increasing dent depth. This effect is taken account of in a more or less qualitative manner by applying formulae (2) and (3).

The force/dent depth curves calculated on this principle appears to agree fairly well with the load/dent depth diagrams obtained from dent tests.
This is illustrated by fig. 2 for one of the dent trials.

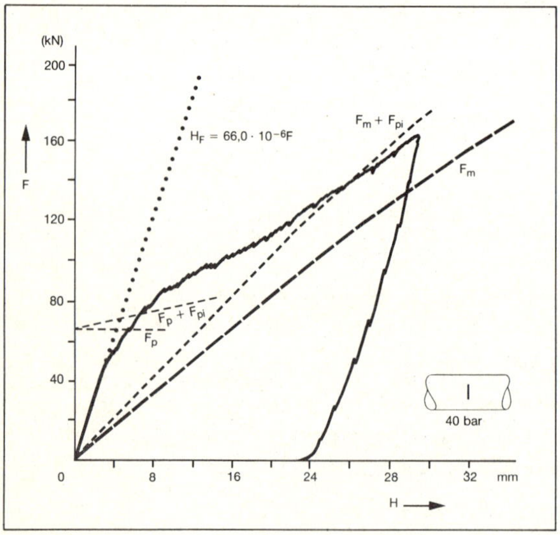

Fig. 2 Load-dent depth diagram; pipe section ∅323,8-7,14 mm.

Photo 1 Test apparatus; pipe diam. 323,8 mm

Photo 1 shows the test apparatus for indentation. Photo 2 shows the pipe after indentation.

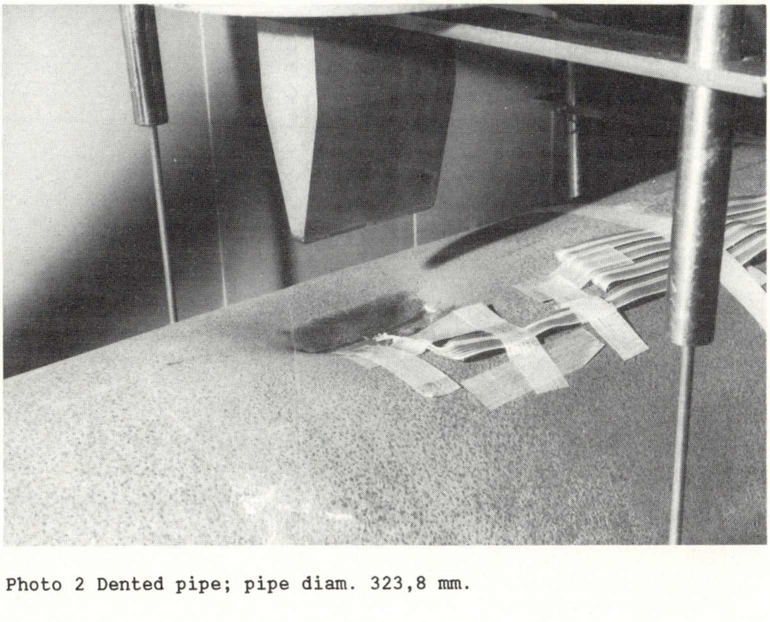

Photo 2 Dented pipe; pipe diam. 323,8 mm.

FINITE ELEMENT CALCULATIONS

For the pipe loaded by a local strip load finite element calculations with elasto-plastic material behaviour, non-conservative pressure loading and large displacement options were carried out using the DIANA* package of IBBC-TNO. DIANA (DIsplacement method ANAlyser) is a comprehensive computer package for stress analysis of a wide variety of structures, and has been developed at the TNO Institute for Building Materials and Structures. A great number of element types are available, such as solid, shell, membrane, plane strain, axisymmetric, spring, beam, truss, plate and reinforcement elements. Also various options for nonlinear analysis such as plasticity (also under cyclic loading), creep, influence of temperature on the material properties, cracking of concrete, large displacements, large strains etc. are available. These phenomena can all be taken into account in static as well as in dynamic analyses.
To simulate the deformation near the local strip load as accurately as possible a refined element mesh had to be employed. An example of an element mesh of the quarter pipe and the specification for the boundary conditions, is shown in figure 3.

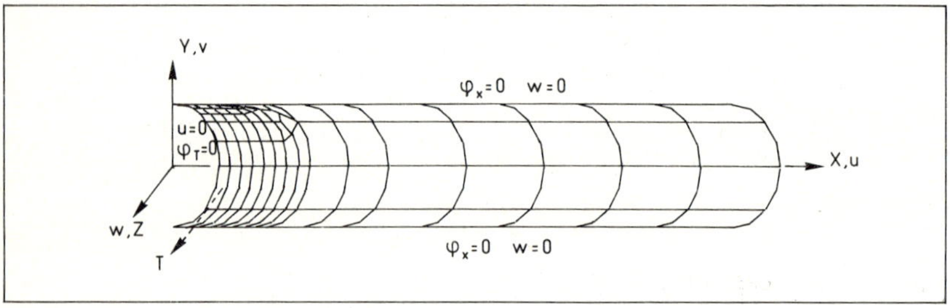

Fig. 3 Element mesh and boundary conditions

Further, by means of finite element calculations the influence of support conditions was established.

The following conditions were examined:

- vertical force with line support;
- two opposite vertical forces;
- vertical force with support at a certain support angle and relieving lateral soil pressure, as in field conditions (modelled by means of non-linear springs).

These calculations showed that support conditions had only little influence on the force/dent depth curve.

The force/dent depth diagram which was obtained by means of finite element calculations closely matched the force/dent depth curve resulting from tests. This is illustrated by fig. 4 for one of the cases. A plot of the deformed configuration in the ultimate loading stage is shown in fig. 5.

* - DIANA is a trademark of TNO

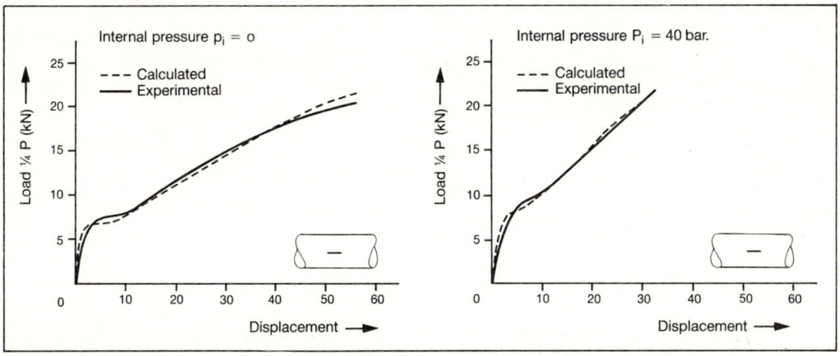

Fig. 4 Comparison of load-dent depth curve according to finite-element calculations with experimental load-dent depth curve; pipe section ∅165,1-4,82 mm; width of indenting body 40 mm.

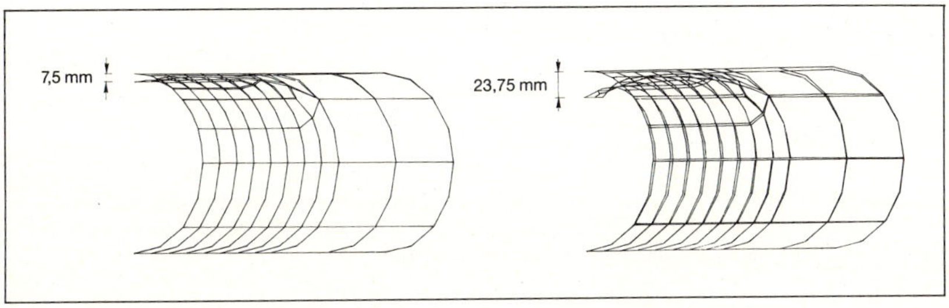

Fig. 5 Indendation of pipe; diam. 165,1 - 4,82 mm; length of strip load 40 mm, internal pressure 40 bar.

CONCLUSION

With the finite element calculations, supported by a few dent tests, the influence of the different support conditions could be investigated. By means of the dent tests and results from finite element calculations semi-empirical formulas could be developed with which the relation between local load and dent depth could be described in a sufficiently accurate manner. This calculation model provides a good insight into the influence of the various parameters on the depth.

Literature

(1) Aas-Jakobsen: "Einzellasten auf Kreiszylinderschalen"; Bauingenieur 22 (1941), P. 343-346.

(2) P.P. Bijlaard:" Stresses from local loadings in cylindrical pressure vessels"; Trans. ASME77 (1955), P. 805-814.

(3) P.P. Bijlaard: "Stresses in spherical vessels from local loads transferred by a pipe + additional data on stresses in cylindrical shells under local loading"; Welding Research Council Bulletin 50, New York, May 1959.

(4) K.R. Wichman, A.G. Hopper, J.L. Mershon: "Local stresses in spherical and cylindrical shells due to external loadings"; Pressure Vessel Research Committee of the Welding Research Council, New York, ± 1960.

(5) K.W. Bieger: "Circular cylindrical shells subjected to concentrated radial loads"; Springer,1976.

(6) W. Flügge: "Statik und Dynamik der Schalen"; Springer, Berlin, 1934.

(7) R. de Borst, G.M.A. Kusters, P. Nauta and F.C. de Witte: "DIANA, a three-dimensional, non-linear finite element package on a micro-computer"; International Conference on Engineering Software for Micro-Computers, Venice, Italy, April 2nd-5th, 1984.

(8) R. de Borst, G.M.A. Kusters, P. Nauta and F.C. de Witte: "DIANA, a comprehensive, but flexible finite element system"; 4th International Seminar on Finite Element Systems, Southampton, England, July 6-8th, 1983.

(9) O.C. Zienkiewicz: "The finite element method", Mc Graw-Hill Book.

(10) J. Spiekhout, A.M. Gresnigt, C. Koning and H. Wildschut: "Calculation models for the evaluation of the resistance against mechanical damage of pipelines", 3R International, Heft (4), April 1986.

LARGE DISPLACEMENT ANALYSIS OF THIN SHELLS

Approximate updated Lagrangian formulation considering both initial stresses and initial displacements

Sun Jingyu, Anders Peterson and Hans Petersson
Division of Structural Mechanics
Lund University
Box 118, S-221 00 LUND, Sweden

1. INTRODUCTION

In solid mechanics Lagrangian motion descriptions are normally applied where a given reference volume is associated with the same set of material particles at all stages of deformation. Two forms of the Lagrangian description, the total Lagrangian description (TLD) and the updated Lagrangian description (ULD), are noted in the literature see for example [1-5].

The TLD and ULD are two examples of the referential description - one of several ways to describe the motion of a continuum according to Truesdell [6,7]. In a referential description the independent variables are the position of a particle in a chosen reference configuration and the time t. In TLD and ULD, respectively the initial undeformed and the current stressed configurations are usually chosen as the reference configurations.

Another possible referential description is where the reference configuration coincides with a translated and rotated region of the undeformed body. This description, named partially updated Lagrangian description (PULD), has been reported in [8] and has been used successfully in beam analysis [9].

The reference to an undeformed element subregion makes the integration easier to carry out than to integrate over a deformed element volume (which is the case in an exact ULD-formulation). This advantage of PULD can be retained without loss of accuracy by adding an extra term in the basic formulation considering the influence of local element deformations. The PULD formulation may thus be considered as a modified or 'approximate' updated Lagrangian formulation, similar to various co-rotational approaches reported in the literature [5,10,11].

In the present study a PULD formulation is used for nonlinear analysis of thin shells. Curved shells are discretized into flat shell elements. Kirchhoff's assumptions of the theory of thin plates [12] are applied. Shell materials are assumed to be isotropic and linear elastic. The strains remain small, but the displacements may be large.

In the first part of this paper the basic formulation is given for general nonlinear FEM analysis and then applied to two types of shell elements. Finally, some numerical examples are presented.

2. GENERAL FEM FORMULATION OF NONLINEAR PROBLEMS

In the Lagrangian description used, a Cartesian coordinate system is chosen. The position of a material point of a body can be described by

$$x_i = x_i(X_1, X_2, X_3, t) \tag{2.1a}$$

or

$$x_i = X_i + u_i \qquad i = 1,2,3 \tag{2.1b}$$

where X_i are the material coordinates, x_i are the spatial coordinates and u_i are the displacements. The equilibrium equations may be written as [6]

$$\frac{\partial}{\partial X_J}(\tilde{T}_{JI} \frac{\partial x_i}{\partial X_I}) + f_i^0 = 0 \tag{2.2}$$

where $\tilde{T}$ is the second Piola-Kirchhoff stress tensor and f_i^0 is a component of the body force. An incremental form of Eq. (2.2) is

$$\frac{\partial}{\partial X_J}(d\tilde{T}_{JI} \frac{\partial x_i}{\partial X_I}) + \frac{\partial}{\partial X_J}(\tilde{T}_{JI} \frac{\partial dx_i}{\partial X_I}) + df_i^0 = 0 \tag{2.3}$$

Multiplication of Eq. (2.3) by the weight functions v_{im} and integration over the initial volume V^0 of the body yield

$$\int_{V^0} v_{im} \frac{\partial}{\partial X_J}(d\tilde{T}_{JI} \frac{\partial x_i}{\partial X_I}) dV^0 + \int_{V^0} v_{im} \frac{\partial}{\partial X_J}(\tilde{T}_{JI} \frac{\partial dx_i}{\partial X_I}) dV^0 + \int_{V^0} v_{im} df_i^0 dV^0 = 0 \qquad (2.4)$$

After integration by parts, Eq. (2.4) results in

$$\int_{S^0} v_{im} \hat{N}_J d\tilde{T}_{JI} \frac{\partial x_i}{\partial X_I} dS^0 - \int_{V^0} \frac{\partial v_{im}}{\partial X_J} d\tilde{T}_{JI} \frac{\partial x_i}{\partial X_I} dV^0 + \int_{S^0} v_{im} \hat{N}_J \tilde{T}_{JI} \frac{\partial dx_i}{\partial X_I} dS^0$$

$$- \int_{V^0} \frac{\partial v_{im}}{\partial X_J} \tilde{T}_{JI} \frac{\partial dx_i}{\partial X_I} dV^0 + \int_{V^0} v_{im} df_i^0 dV^0 = 0 \qquad (2.5)$$

where S^0 is the initial surface of the body. $\hat{\mathbf{N}}$ denotes the outward normal vector of S^0. Using the relationships of the surface traction $\tilde{t}$ and t^0 of the body [6]

$$\hat{N}_J \tilde{T}_{JI} = \tilde{t}_I \qquad (2.6a)$$

and

$$t_i^0 = \tilde{t}_I \frac{\partial x_i}{\partial X_I} \qquad (2.6b)$$

Eq. (2.5) may be written as

$$\int_{V^0} \frac{\partial v_{im}}{\partial X_J} d\tilde{T}_{JI} \frac{\partial x_i}{\partial X_I} dV^0 + \int_{V^0} \frac{\partial v_{im}}{\partial X_J} \tilde{T}_{JI} \frac{\partial dx_i}{\partial X_I} dV^0 =$$

$$= \int_{S^0} v_{im} dt_i^0 dS^0 + \int_{V^0} v_{im} df_i^0 dV^0 \qquad (2.7)$$

An incremental form of the constitutive equation is introduced,

$$d\tilde{T}_{JI} = C_{IJKL} dE_{KL} \qquad (2.8)$$

where $\mathbf{C}$ is a symmetric material tensor and dE is the incremental Green strain tensor

$$dE_{KL} = \frac{1}{2}\left(\frac{\partial dx_i}{\partial X_K}\frac{\partial x_i}{\partial X_L} + \frac{\partial x_i}{\partial X_K}\frac{\partial dx_i}{\partial X_L}\right) \tag{2.9}$$

Derivations of Eq. (2.1) give

$$dx_i = du_i \tag{2.10a}$$

and

$$\frac{\partial x_i}{\partial X_I} = \delta_{iI} + \frac{\partial u_i}{\partial X_I} \tag{2.10b}$$

where δ_{iI} is the Kronecker delta.

Substitution of Eqs. (2.8)-(2.10) into Eq. (2.7) yields

$$\int_{V^0} \frac{\partial v_{im}}{\partial X_J} D_{iJjL} \frac{\partial du_j}{\partial X_L} dV^0 + \int_{V^0} \frac{\partial v_{im}}{\partial X_J} \tilde{T}_{JI} \frac{\partial du_i}{\partial X_I} dV^0 =$$

$$= \int_{S^0} v_{im} dt_i^0 dS^0 + \int_{S^0} v_{im} df_i^0 dV^0 \tag{2.11}$$

where

$$D_{iJjL} = (\delta_{iI} + \frac{\partial u_i}{\partial X_I}) C_{IJKL} (\delta_{jK} + \frac{\partial u_j}{\partial X_K}) \tag{2.12}$$

The studied body is divided into a number of subregions (finite elements). The nodal variables of a typical element are denoted by $\bar{u}$. The interpolation functions **N** are introduced on the form

$$u_i = N_{in} \bar{u}_n \tag{2.13a}$$

and

$$du_i = N_{in} d\bar{u}_n \tag{2.13b}$$

where n is a summation index which normally exceeds 3.

Using the Galerkin method the weight functions are chosen as

$$v_{im} = N_{im} \tag{2.14}$$

Substitution of Eqs. (2.13) and (2.14) into Eq. (2.11) yields

$$K_{mn} d\bar{u}_n = dR_m \tag{2.15}$$

where

$$K_{mn} = \int_{V^0} \frac{\partial N_{im}}{\partial X_J} D_{iJjL} \frac{\partial N_{jn}}{\partial X_L} dV^0 + \int_{V^0} \frac{\partial N_{im}}{\partial X_J} \tilde{T}_{JI} \frac{\partial N_{in}}{\partial X_I} dV^0 \tag{2.16}$$

$$dR_m = \int_{S^0} N_{im} dt_i^0 dS^0 + \int_{V^0} N_{im} df_i^0 dV^0 \tag{2.17}$$

From Eq. (2.12) it can be proved that D_{iJjL} is symmetric since C is supposed to be symmetric. The notation

$$D_{iJjL} = C_{iJjL} + D^u_{iJjL} \tag{2.18}$$

is introduced where

$$D^u_{iJjL} = C_{iJKL} \frac{\partial u_j}{\partial X_K} + \frac{\partial u_i}{\partial X_I} C_{IJjL} + \frac{\partial u_i}{\partial X_I} C_{IJKL} \frac{\partial u_j}{\partial X_K} \tag{2.19}$$

Eq. (2.16) may then be written on the form

$$K_{mn} = K^0_{mn} + K^\sigma_{mn} + K^u_{mn} \tag{2.20}$$

where

$$K^0_{mn} = \int_{V^0} \frac{\partial N_{im}}{\partial X_J} C_{iJjL} \frac{\partial N_{jn}}{\partial X_L} dV^0 \tag{2.21a}$$

$$K^\sigma_{mn} = \int_{V^0} \frac{\partial N_{im}}{\partial X_J} \tilde{T}_{JI} \frac{\partial N_{in}}{\partial X_I} dV^0 \tag{2.21b}$$

$$K^u_{mn} = \int_{V^0} \frac{\partial N_{im}}{\partial X_J} D^u_{iJjL} \frac{\partial N_{jn}}{\partial X_L} dV^0 \tag{2.21c}$$

K^0 is the linear part of the element stiffness matrix. $\mathbf{K}^{\sigma}$ and $\mathbf{K}^{u}$ are the nonlinear contributions to the element stiffness matrix. They are called the initial stress stiffness and the initial displacement stiffness, respectively.

3. DESCRIPTION OF MOTION

As mentioned in Chapter 1, the PULD is used in this study. In Fig. 3.1, the initial and current configurations are denoted by V^0 and V, respectively.

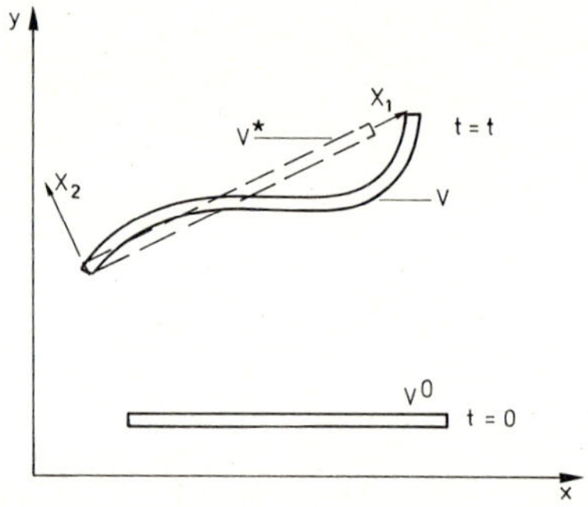

Figure 3.1. Different reference configurations of a studied body

The configuration V^* (dash-lined in the figure) is obtained by a rigid body motion of V^0 to a position close to the current position V. V^* is used as the reference configuration in PULD. A local element coordinate system is established with reference to some characteristic points in V^*. The difference between V^* and V^0 is only a rigid body motion. Such a motion of the reference configuration has no influence on the Green strain tensor and the second Piola-Kirchhoff stress tensor. If the procedure is kept in equilibrium conditions, then the equations of equilibrium remain unchanged. The finite element equations obtained in Chapter 2 are thus applicable also for a PULD formulation if reference is made to V^* and the local element coordinate system.

4. SHELL ELEMENT FORMULATION

Two kinds of finite elements are used in this study. They have the same shape but are of different order. The shape of the elements is a space quadrilateral subdivided into four triangular elements. The stiffness and load matrices of the quadrilateral element are formed from the stiffness and load matrices of the four triangular elements through transformation, assembly and condensation operations.

The two types of triangular elements are called TS1 (12 d.o.f.) and TS2 (27 d.o.f.) respectively, see Fig. 4.1. The X_1-X_2 plane in the figure is in the midsurface of the shell element.

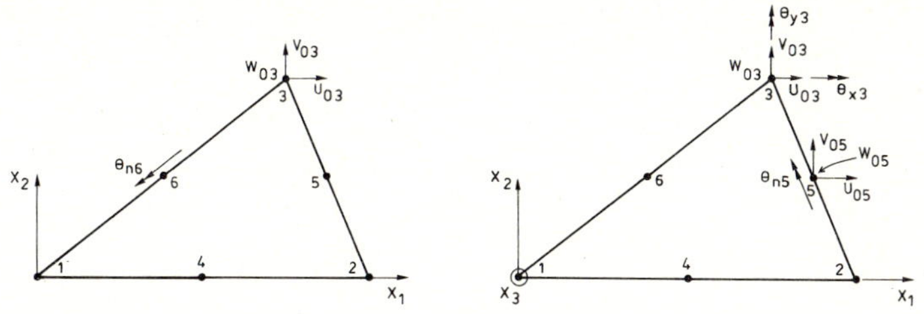

Figure 4.1. Two triangular shell elements: TS1 (left) and TS2 (right)

Inserting Kirchhoff's assumptions and an isotropic material assumption into Eq. (2.21a) yields

$$K^0_{mn} = \int_{V^*} \left[\frac{\partial N_{1m}}{\partial X_1} C_{1111} \frac{\partial N_{1n}}{\partial X_1} + \frac{\partial N_{1m}}{\partial X_1} C_{1122} \frac{\partial N_{2n}}{\partial X_2} + \frac{\partial N_{2m}}{\partial X_2} C_{2211} \frac{\partial N_{1n}}{\partial X_1} + \right.$$

$$+ \frac{\partial N_{2m}}{\partial X_2} C_{2222} \frac{\partial N_{2n}}{\partial X_2} + \frac{\partial N_{1m}}{\partial X_2} C_{1212} \frac{\partial N_{1n}}{\partial X_2} + \frac{\partial N_{2m}}{\partial X_1} C_{2121} \frac{\partial N_{2n}}{\partial X_1} +$$

$$\left. + \frac{\partial N_{1m}}{\partial X_2} C_{1221} \frac{\partial N_{2n}}{\partial X_1} + \frac{\partial N_{2m}}{\partial X_1} C_{2112} \frac{\partial N_{1n}}{\partial X_2} \right] dV^* \qquad (4.1)$$

Introducing

$$[B_{0m}] = \left[\frac{\partial N_{1m}}{\partial X_1} \quad \frac{\partial N_{2m}}{\partial X_2} \quad \frac{\partial N_{1m}}{\partial X_2} + \frac{\partial N_{2m}}{\partial X_1} \right]^T \qquad (4.2)$$

and using the symmetry of **C** yields a compact form of Eq. (4.1)

$$K^{O}_{mn} = \int_{V^*} [B_{Om}]^T [D][B_{On}] dV^* \tag{4.3}$$

where $[D]$ is the material matrix for the plane stress conditions.

The same procedure can be applied to Eqs. (2.21b) and (2.21c). The results are

$$K^{\sigma}_{mn} = \int_{V^*} [B_{\sigma m}]^T \begin{bmatrix} [\sigma] & 0 & 0 \\ 0 & [\sigma] & 0 \\ 0 & 0 & [\sigma] \end{bmatrix} [B_{\sigma n}] dV^* \tag{4.4}$$

$$K^{u}_{mn} = \int_{V^*} \left\{ [B_{Om}]^T [D][B_{Dn}] + [B_{Dm}]^T [D][B_{On}] + [B_{Dm}]^T [D][B_{Dn}] \right\} dV^* \tag{4.5}$$

where

$$[B_{\sigma m}] = \begin{bmatrix} \frac{\partial N_{1m}}{\partial X_1} & \frac{\partial N_{1m}}{\partial X_2} & \frac{\partial N_{2m}}{\partial X_1} & \frac{\partial N_{2m}}{\partial X_2} & \frac{\partial N_{3m}}{\partial X_1} & \frac{\partial N_{3m}}{\partial X_2} \end{bmatrix}^T \tag{4.6}$$

$$[\sigma] = \begin{bmatrix} \tilde{T}_{11} & \tilde{T}_{12} \\ \tilde{T}_{21} & \tilde{T}_{22} \end{bmatrix} \tag{4.7}$$

$$[B_{Dm}] = \begin{bmatrix} \frac{\partial N_{im}}{\partial X_1} \frac{\partial u_i}{\partial X_1} & \frac{\partial N_{im}}{\partial X_2} \frac{\partial u_i}{\partial X_2} & \frac{\partial N_{im}}{\partial X_2} \frac{\partial u_i}{\partial X_1} + \frac{\partial N_{im}}{\partial X_1} \frac{\partial u_i}{\partial X_2} \end{bmatrix}^T \tag{4.8}$$

The derived finite element equations for shell analysis have been implemented in a computer program which is a subset of the CAMFEM-system [13].

5. NUMERICAL EXAMPLES

Some numerical examples are presented in order to verify the results of the derivations of the finite element equations and the computer program.

5.1 An introductory example of large displacement analysis

A cantilever plate beam is chosen as an introductory example, see Fig. 5.1. Concentrated loads are acting at its free end. The beam is divided into five shell elements (TS1). The load step in the analysis is 0.01 EI/L^2.

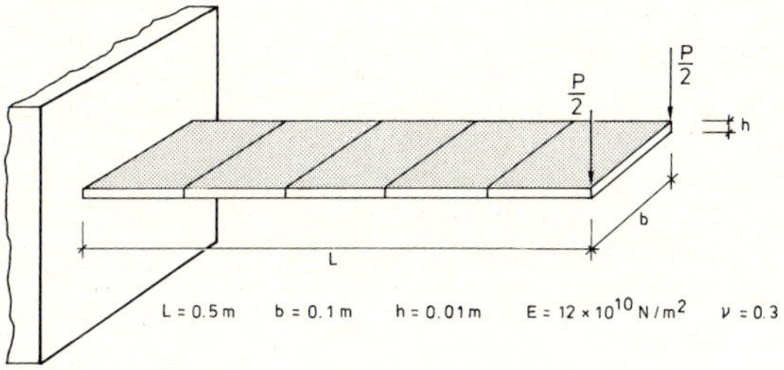

Figure 5.1. A cantilever plate beam subjected to concentrated loads

The resulting and deflection w is given in Fig. 5.2. An elliptic integral solution of the same problem [14] is used as an exact one in Fig. 5.2. The figure shows the importance of the initial displacement stiffness matrix $\mathbf{K}^u$. When this term is considered, the results are improved substantially.

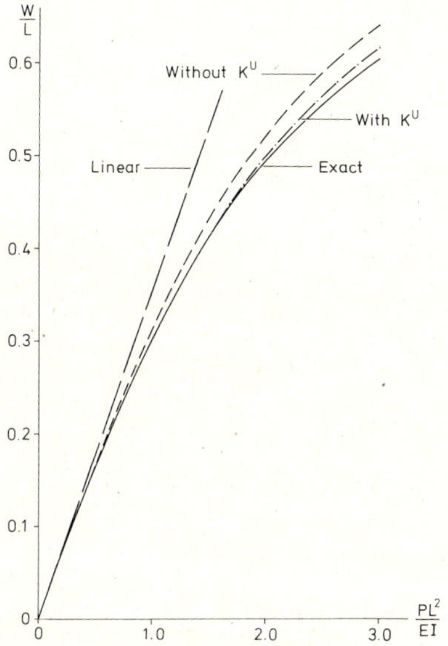

Figure 5.2. Load displacement curve

5.2 Large displacement and post-buckling analysis of a spherical shell

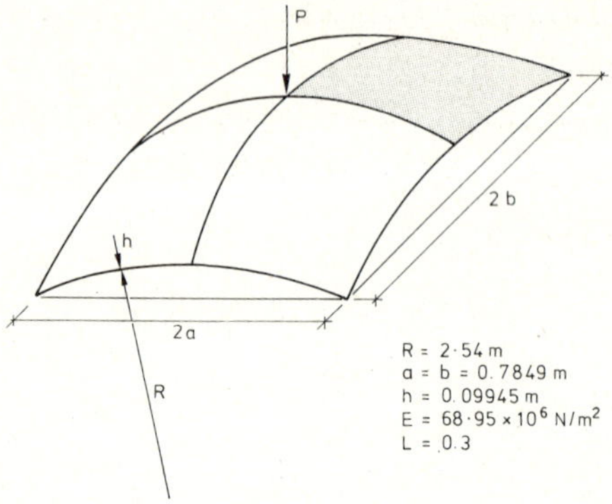

Figure 5.3. A spherical shell subjected to a central load

A spherical shell with data is shown in Fig. 5.3. The shell is subjected to a point force P at its crown. All edges are hinged and immovable. One quarter of the shell is analyzed by using only a 1×1-mesh for TS1 and TS2. The results are shown in Fig. 5.4. A series solution by Leicester [15] and some numerical results by Harte and Eckstein [16] and Thomas and Gallagher [17] are also shown in the figure. From Fig. 5.4, we see that the addition of the term $\mathbf{K}^u$ gives better results for both TS1 and TS2.

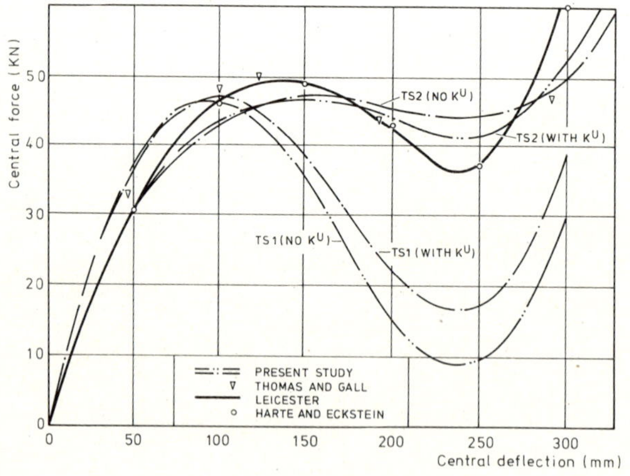

Figure 5.4. Central deflections of a spherical shell

5.3 Large displacement and post-buckling analysis of a cylindrical shell

A cylindrical shell is shown in Fig. 5.5. The curved edges are free whereas the longitudinal boundaries are hinged and immovable. The shell is analysed by using the element TS1 and a 4×4-mesh for one quarter of the shell.

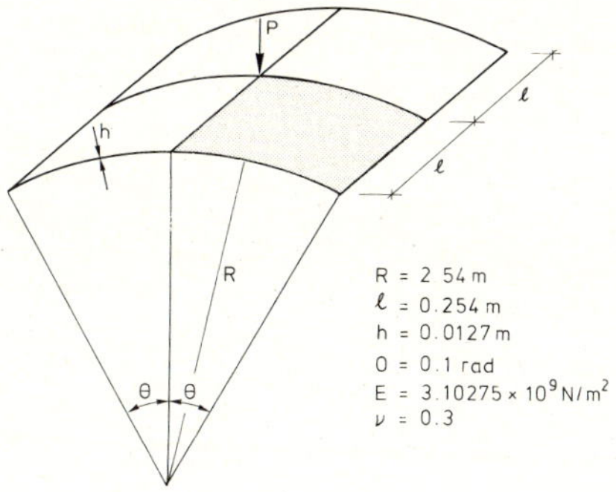

Figure 5.5. A cylindrical shell subjected to a central load

The snap-through behaviour is shown in Fig. 5.6. Some results presented by Horrigmoe [18] and Bergan and Nygård [19] are also shown in Fig. 5.6 for reference.

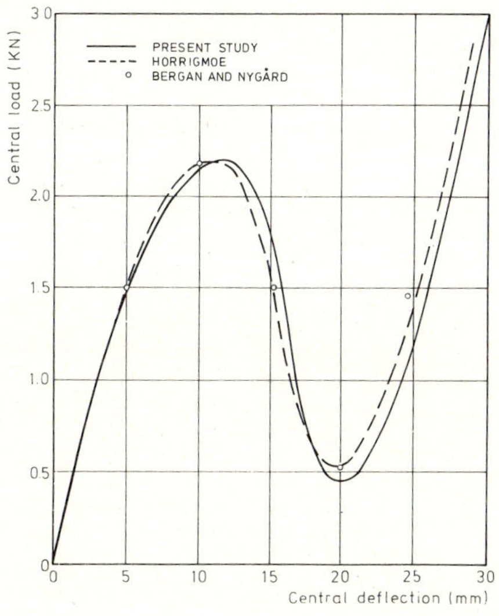

Figure 5.6. Central deflections of a cylindrical shell

6. CONCLUDING REMARKS

The work described in this paper deals with nonlinear shell analysis. The numerical examples show that the suggested PULD formulation can be an attractive alternative to an ULD formulation in shell analysis. The advantage of maintaining the initial shape of the integration volumes of the elements can be obtained without loss of accuracy. The analysis results can be improved significantly by considering the initial displacement term in the basic formulation. By this the influence of local element deformations are taken into account.

7. REFERENCES

[1] BATHE, K-J., RAMM, E. and WILSON, E.L.: Finite element formulations for large deformation dynamic analysis, International Journal of Numerical Methods in Engineering, 9, (1975), pp. 353-386.

[2] CESCOTTO,S., FREY, F. and FONDER, G.: Total and updated Lagrangian descriptions in nonlinear structural analysis: A unified approach, in: Glowinski, R., Rodin, E.Y. and Zienkiewicz, O.C., eds., Energy Methods in Finite Element Analysis (Wiley, Chichester, 1979), pp. 283-296.

[3] GADALA, M.S.,DOKAINISH, M.A. and ORAVAS, G.AE.: Formulation methods of geometric and material nonlinearity problems, International Journal of Numerical Methods in Engineering, 20, (1984), pp. 887-914.

[4] BATHE, K-J.: Finite element procedures in engineering analysis (Prentice-Hall, Englewood Cliffs, New Jersey, 1982).

[5] MATTIASSON, K. and SAMUELSSON, A.: Total and updated Lagrangian forms of the co-rotational finite element formulation in geometrically and materially nonlinear analysis, in: Numerical Methods for Non-linear Problems, Taylor, C., Hinton, D. and Owen, D.R.J., eds., (Pineridge Press, Swansea, 1984).

[6] MALVERN, L.E.: Introductions to the mechanics of a continuous medium, (Prentice-Hall, Englewood Cliffs, New Jersey, 1969).

[7] TRUESDELL, C.: The elements of continuum mechanics (Springer-Verlag, New York, 1965).

[8] PETERSON, A. and PETERSSON, H.: On finite element analysis of geometrically nonlinear problems, Computer Methods in Applied Mechanics and Engineering, 51, (1985), pp. 277-286.

[9] PETERSON, A.: Finite element analysis of structures at high temperatures, Dr. thesis, Publication TVSM-1001, Lund University, Division of Structural Mechanics, Lund, (1984).

[10] ARGYRIS, J.H., BALMER, H., DOLTSINIS, J.St., DUNNE, P.C., HAASE, M., KLEIBER, M., MALEJANNAKIS, G.A., MLEJNEK, H-P., MÜLLER, M. and SCHARPF, D.W.: Finite element method - the natural approach, Computer Methods in Applied Mechanics and Engineering, 17/18, (1979), pp. 1-106.

[11] BELYTSCHKO, T. and HSIEH, B.J.: Nonlinear transient finite element analysis with convected coordinates, International Journal of Numerical Methods in Engineering, 7, (1973), pp. 255-271.

[12] SZILARD, R.: Theory and analysis of plates (Prentice-Hall, Englewood Cliffs, New Jersey, 1974).

[13] DAHLBLOM, O. and PETERSON, A.: CAMFEM - Computer Aided Modelling based on the Finite Element Method, Publication TVSM-3001, Lund University, Division of Structural Mechanics, Lund, (1983).

[14] MATTIASSON, K.: Numerical results from large deflection beam and frame problems analyzed by means of elliptic integrals, International Journal of Numerical Methods in Engineering, 17, (1981), pp. 145-153.

[15] LEICESTER, R.H.: Finite deformations of shallow shells, ASCE, Journal of the Engineering Mechanics Division, EM6, (1968), pp. 1409-1421.

[16] HARTE, R. and ECKSTEIN, U.: Derivation of geometrically nonlinear finite shell elements via tensor notation, International Journal of Numerical Methods in Engineering, 23, (1986), pp. 367-384.

[17] THOMAS, G.R. and GALLAGHER, R.H.: A triangular thin shell finite element: nonlinear analysis, NASA CR-2483, (1975).

[18] HORRIGMOE, G.: Finite element instability analysis of free form shells, Publication No. 77-2, The Norwegian Institute of Technology, Division of Structural Mechanics, The University of Trondheim, Norway, (1977).

[19] BERGAN, P.G. and NYGÅRD, M.K.: Nonlinear shell analysis using free formulation finite elements, Europe-US Symposium Proceedings, (Trondheim, 1985)

Nonlinear Analysis of Tension Fabric Structure

Arata Yoshida
Kajima Corporation, Information Processing Center, Tokyo, Japan
Shigeru Ban
Kajima Corporation, Department of Structural Design, Tokyo, Japan
Haruji Tsubota and Kazuo Kurihara
Kajima Corporation, Institute of Construction Technology, Tokyo, Japan

1. Introduction

Membrane structures possess significant advantages over other conventional structures for enclosing a large volume. Generally speaking, membrane structures are classified into two types, i.e. air-supported type and tensile type.

A new type of tensile fabric structure, as shown in Photo. 1 and Fig. 1, has been designed by the authors. This structure covers an approximately 30 m x 30 m free space and it consists of steel frames (main arches and sub arches), fabric panels and steel cables. Steel cables connect the intersecting points of the main arches and the center points of the fabric panels. One of the most important items for tensile fabric structures is the prestress introducing system. In the case of this structure, prestress is introduced in two stages, as shown in Fig. 2. Firstly, initial prestress is introduced into the fabric panel by tightening clamping nuts which are attatched to clamping bolts. Secondly, additional prestress is induced into the fabric panel by introducing tensile forces into the cables which are attached to the center of the fabric panel. When initial prestress decreases by a long term creep or relaxation of the fabric, prestress can be recovered simply by re-tensioning the cables. This configuration is considered to be the initial state of the structure. Considering the features of this structure, analysis consists of three steps as follows (see Fig.2).

(1) Form finding analysis (when the initial prestress is induced):
In this step the equilibrium shape is calculated when the fabric ends are stretched to the boundary steel frames.

(2) Structural analysis-1 (when the cable tensile forces are induced):
In this step the stress and displacement states are calculated when the tensile forces are induced. This configuration is considered as the initial shape of the structure.

(3) Structural analysis-2 (when snow or wind loads are applied):
In this step the stress and displacement states are calculated when the snow or wind loads are applied.

To investigate the accuracy of the analysis, and to prove the safety of this structure, one unit of the panel was taken out as the specimen at 1/2 scale, and the experiment was carried out. In this paper, the results of the analysis and the experiment are compared, and the structural charactereistics of this structure are discussed.

Photo. 1 Overall View of the Tensile Fabric Structure

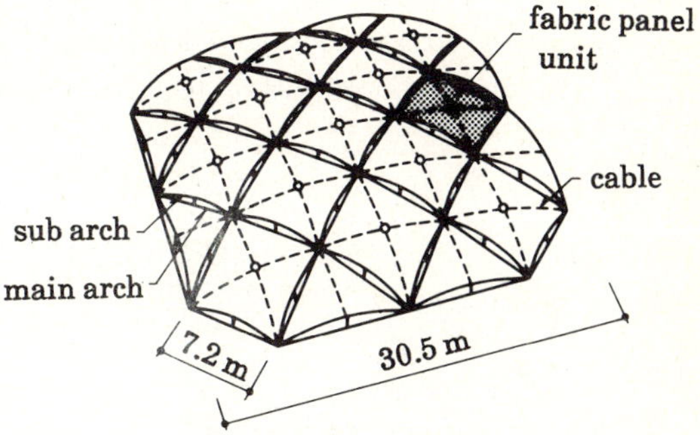

Fig. 1 Structural System

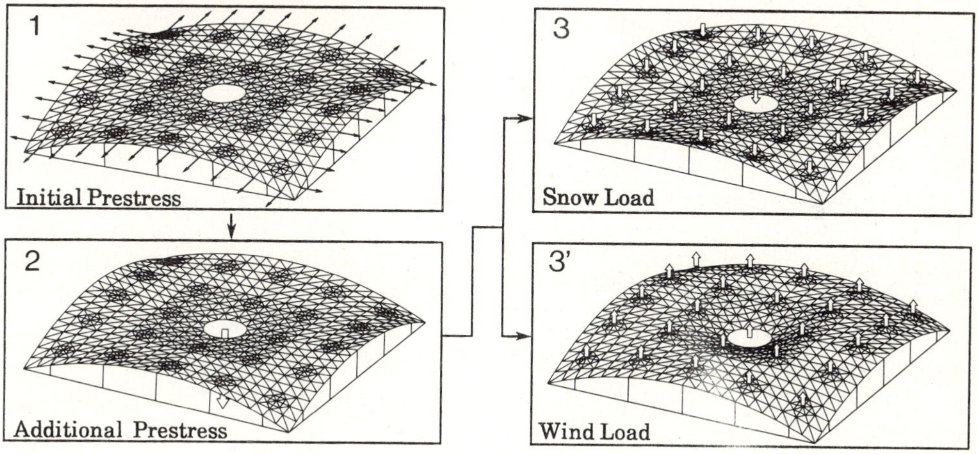

Fig. 2 Prestressing System and Analysis Flow

2. Analysis Method

2.1 Form Finding Analysis

In general, membrane structures are unstable because they do not have flexural stiffness, and the stable configuration is obtained by introducing initial stress or applying inner pressure. In this case, the stress distribution is specified and the equilibrium configuration is treated as an unknown parameter.

Let us consider the triangular membrane element as three bars which form the sides of a triangle and have statically equivalent forces to the element stress. Then the equilibrium equation at a typical free node i, on the surface of the structure that consists of membrane elements, is written as follows (see Fig. 3).

$$\begin{Bmatrix} F_x \\ F_y \\ F_z \end{Bmatrix}_i - \sum_j \begin{Bmatrix} l_{ij} \\ m_{ij} \\ n_{ij} \end{Bmatrix} P_{ij} = \begin{Bmatrix} 0 \end{Bmatrix} \qquad (1)$$

where

F_x, F_y, F_z : loads acting at the node i

P_{ij} : the equivalent force acting on the side ij of the triangular element ijk

l_{ij}, m_{ij}, n_{ij} : the direction cosines of the side ij of the triangular element ijk

Σ : summation of sides of triangular elements converging at the node i.

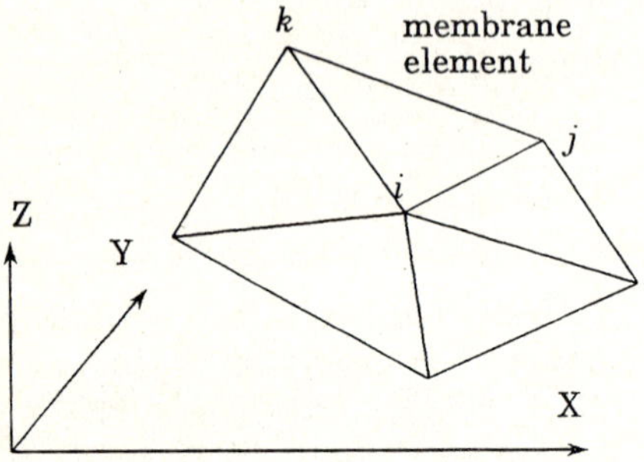

Fig. 3 Membrane Elements Converging at Node i

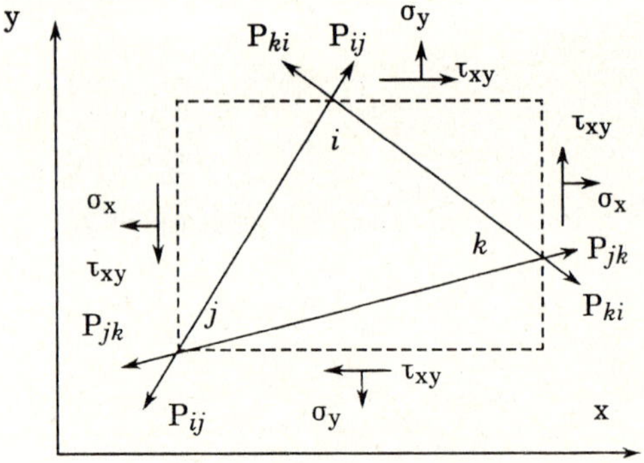

Fig. 4 Membrane Stresses and Equivalent Imaginary Forces

By making equation (1) at all free nodes, we can obtain nonlinear equations of unknown coordinates of each node.

$$g(X) = F_L(X) - L_M(X) P_M(X) = 0 \tag{2}$$

where right subscript L and M represents the term of load, and membrane element respectively. Equation (2) is solved by Newton-Raphson iteration technique. Denoting the n-th approximated solution as X^n, the following expression is obtained.

$$g(X^n + \Delta X^n) = 0 \tag{3}$$

Applying Taylor's expansion to equation (3) with respect to ΔX^n and neglecting over second order terms of ΔX^n, we can get

$$g(X^n) + \Delta X^n g'(X^n) = 0 \tag{4}$$

$$\therefore \quad \Delta X^n = -\left[g'(X^n)\right]^{-1} \cdot g(X^n) \tag{5}$$

where

$$g'(X^n) = \frac{\partial F_L(X^n)}{\partial X} - L_M(X^n)\frac{\partial P_M(X^n)}{\partial X} - P_M(X^n)\frac{\partial L_M(X^n)}{\partial X} \tag{6}$$

As the modified increment of the coordinate is calculated by equation (5), the $n+1$-th approximated value becomes

$$X^{n+1} = X^n + \Delta X^n \tag{7}$$

The calculation is repeated until the out of equivalence force becomes smaller than the allowable level. If the load is independent of the coordinate, the first term of equation (6) can be eliminated. If it is assumed that the state of stress in the triangular element is constant, we can get a linear relation between membrane stresses and equivalent imaginary forces, P_M, as follows[(1)](see Fig.4).

$$\begin{Bmatrix} \sigma_x \\ \sigma_y \\ \tau_{xy} \end{Bmatrix} = \begin{bmatrix} T \end{bmatrix} \cdot \begin{Bmatrix} P_{ij} \\ P_{jk} \\ P_{ki} \end{Bmatrix} \tag{8}$$

where

$$\begin{bmatrix} T \end{bmatrix} = \frac{2}{t} \begin{bmatrix} \lambda_{ij}^2 / h_k & \lambda_{jk}^2 / h_i & \lambda_{ki}^2 / h_j \\ \mu_{ij}^2 / h_k & \mu_{jk}^2 / h_i & \mu_{ki}^2 / h_j \\ \lambda_{ij}\mu_{ij} / h_k & \lambda_{jk}\mu_{jk} / h_i & \lambda_{ki}\mu_{ki} / h_j \end{bmatrix} \tag{9}$$

t : thickness of membrane

h_i, h_j, h_k : the altitude of the triangle measured respectively from the vertex i, j, and k

λ_{ij}, μ_{ij} : the direction cosines of the side ij

It is more natural for the designer to specify the membrane stresses rather than to specify the equivalent imaginary forces. If the designer specifies the membrane stresses (in general case $\sigma_x = \sigma_y$, $\tau_{xy} = 0$), we can get the following equation from equation (8).

$$\begin{Bmatrix} P \end{Bmatrix} = \begin{bmatrix} T \end{bmatrix}^{-1} \cdot \begin{Bmatrix} \sigma \end{Bmatrix} \tag{10}$$

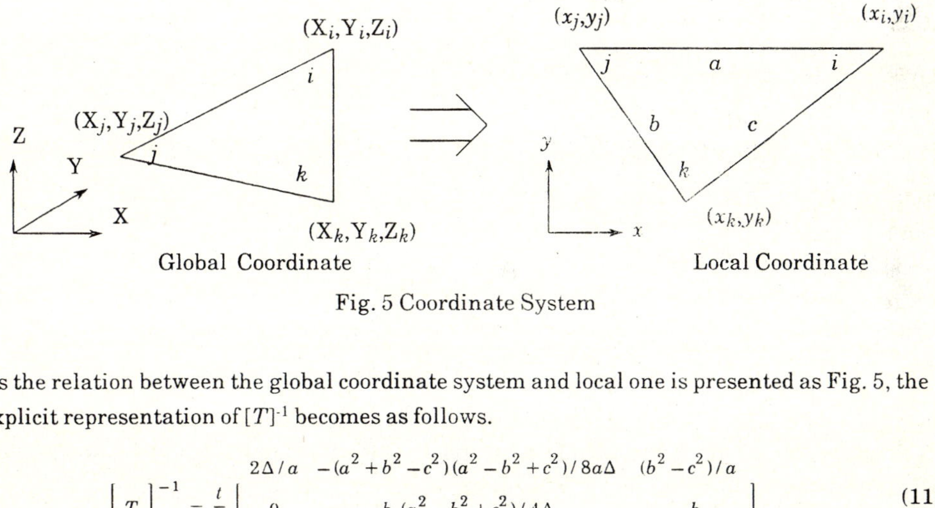

Fig. 5 Coordinate System

As the relation between the global coordinate system and local one is presented as Fig. 5, the explicit representation of $[T]^{-1}$ becomes as follows.

$$\begin{bmatrix} T \end{bmatrix}^{-1} = \frac{t}{2} \begin{bmatrix} 2\Delta / a & -(a^2+b^2-c^2)(a^2-b^2+c^2)/8a\Delta & (b^2-c^2)/a \\ 0 & b\,(a^2-b^2+c^2)/4\Delta & -b \\ 0 & c\,(a^2+b^2-c^2)/4\Delta & c \end{bmatrix} \tag{11}$$

in which

$$a = \sqrt{(x_j-x_i)^2 + (y_j-y_i)^2 + (z_j-z_i)^2} \quad etc. \tag{12.1}$$

$$\Delta = \sqrt{s\,(s-a)\,(s-b)\,(s-c)} \tag{12.2}$$

$$s = 1/2\,(a+b+c) \tag{12.3}$$

The direction cosine of the side ij is

$$l_{ij} = (x_j - x_i) / a \qquad etc. \tag{13}$$

Substituting equations (10) to (13) into equation (6), the tangential stiffness matrix can be estimated.

2.2 Structural Analysis (Nonlinear Finite Element Analysis)

Membrane structures are largely deformed by loads acting on the surface, such as wind and snow loads. This behavior is analysed by the finite element load incremental method using updated Lagrangian description. The fabric is assumed to be an elastic but anisotropic material dependent on the orientation of the fibers. According to the works of K. J. Bathe[2] and R. B. Haber[3], equilibrium equation in finite element method is written as

$$([K_L^n] + [K_G^n]) \{U\} = \{R^{n+1}\} - \{F^n\} \tag{14}$$

in which

$$[K_L^n] = \int_{V^n} [B_L^n]^T [D] [B_L^n] \, dV^n \tag{15}$$

$$[K_G^n] = \int_{V^n} [B_{NL}^n]^T [\tau^n] [B_{NL}^n] \, dV^n \tag{16}$$

$$\{F^n\} = \int_{V^n} [B_L^n]^T \{\tau^n\} \, dV^n \tag{17}$$

$\{R^{n+1}\}$: load vector
$[B^n_L]$: linear strain-displacement matrix
$[B^n_{NL}]$: nonlinear strain-displacement matrix
$[D]$: stress-strain matrix considering anisotropic property
$[\tau^n]$: Cauchy stress matrix

Right superscript n denotes that quantities are estimated at n-th load incrementation. In this paper, the constant strain triangular element and cable element are used.

3. Static Loading Test

One unit of the square fabric panel was taken out as the test specimen at approximately 1/2 scale. The specimen consisted of four flat strips of fabric which were cut in accordance with the cutting pattern, and they were lapped and welded to each other. At the center of the specimen, there was a rigid circlar plate which was connected to two diagonal cables. Dimensions and initial configuration of the test specimen are shown in Fig. 6.

Loading steps were as follows. At first, initial prestress was introduced into the specimen on all four sides by tightening tension rods connecting the fabric end joint and the reaction frame. The amount of initial prestress was measured by load cells attatched to the tension rods and was adjusted to the uniform, biaxial stress state of 5.0 kgf/cm (Loading Condition T1).

Next, additional prestress was introduced into the specimen by pulling the diagonal cables. The amount of tensile force in the cables was also measured by load cells and was adjusted so that their vertical component acting at the center of the panel was equal to 200 kgf (Loading Condition T2).

Finally, vertical external loads equivalent to snow and wind loads were applied to the specimen. These distributed loads were simulated by 25 equivalent concentrated loads. The maximum vertical loads applied in this test were 120 kgf/m^2 (snow load, downward) and 350 kgf/m^2 (wind load, upward) (Loading Condition S and W). The measuring points and loading points are shown in Fig. 7.

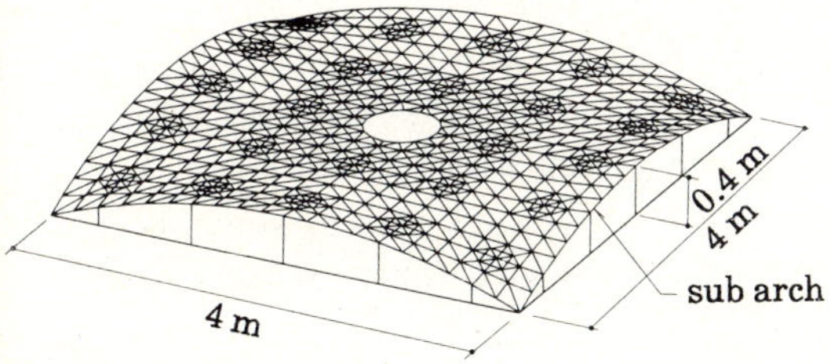

Fig. 6 Configuration of Test Specimen

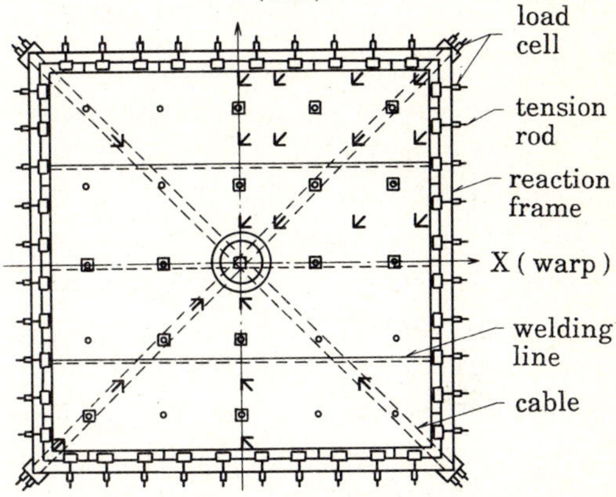

Fig. 7 Measuring Points and Loading Points

4. Comparison Between Analysis and Experiment

4.1 Analytical Condition

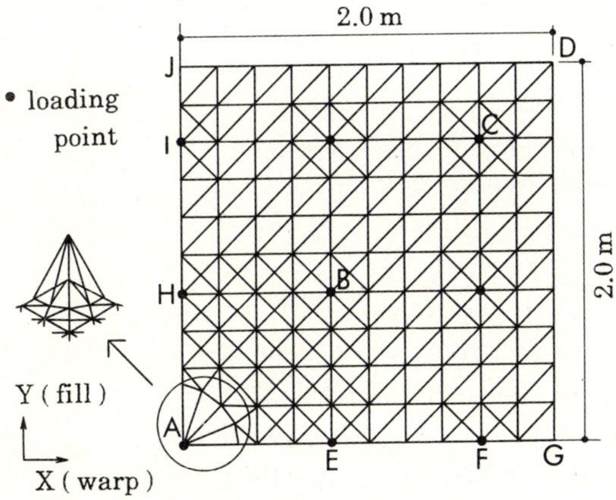

Fig. 8 F.E.M. Model for the Analysis

	Fabric	Cable
Area Thickness	$t=0.08$ cm	1.402 cm^2
Young's Modulus	$E_x \cdot t=658$ kgf/cm $E_y \cdot t=218$ kgf/cm	1.0×10^6 kgf/cm^2
Shear Modulus	$G_{xy} \cdot t=57$ kgf/cm	——
Poisson's Ratio	$\nu_{xy}=0.87$ $\nu_{yx}=0.29$	——

Table. 1 Material Property

The finite element model is shown in Fig. 8. Considering the symmetric condition, a quarter part of the whole structure is taken out. Boundary condition between membrane and reaction frames are assumed to be fixed. The center plate consists of cable elements so as to represent a movable rigid body. Material properties assumed for the analysis are summarized in Table.1[5]. Loading Condition T1 is analysed by Newton - Raphson iteration method described in Section 2.1. The tolerance factor is assumed to be 0.5kgf. Considering the equilibrium form of T1 as the initial state, Loading Condition T2 is analysed by the load incremental method. Then adding a cable element between A-D in Fig. 8, which has the initial axial stress corresponding to the vertical load of the prior load step(T2), Loading Condition S and W are analysed.

4.2 Test Results and Discussions

The measured and calculated results are presented and discussed on the following items.

Initial equilibrium configuration The section of the test specimen of Loading Condition T1 is presented in Fig. 9. The letters A to J correspond to Fig. 8. This configuration is determined only by the shape of the boundary frame and the equilibrium condition of force, so the calculated result is almost in accordance with the measured result.

Tensile Distributions of the Fabric Panel The measured and calculated tensile stress distributions along the edges of the fabric panel are shown in Fig. 10. The measured values of the tensile stress were deduced from the horizontal reaction forces that were measured by load cells attatched to the tension rods. At the initial equilibrium stage (Loading Condition T1), the distributions of tensile stress are uniform (5kgf/cm). However, during Load Condition T2, S and W, the stress state changed from a uniform to an irregular distribution as shown in Fig. 10. At the Loading Condition T2, the increment in tensile stress appeared mainly at the center portion of the panel edges, and the stress state near the corner of the fabric panel was slightly changed. The stress state of Loading Condition S was almost the same as that of T2, as the downward load was sustained mainly in the crosswise direction of the panel (A-G and A-J in Fig. 10). On the other hand, when the wind load was applied to the specimen, the tensile stress increased not only at the center portion but also at the corner of the fabric panel. This can be explained by the fact that upward load is sustained both in the crosswise and diagonal directions (A-D in Fig. 10) of the fabric panel. From the same figure, it becomes clear that external loads are sustained mainly by warp fibers which are stiffer than fill fibers. These tendencies can be seen not only in the calculated results but also in the measured results.

Variations of tensile force in cables Fig. 11 shows the relationship between tensile force in the cables and external loads (snow and wind). In case of the snow load condition, the cable tension decreased linearly as the applied external load increased and finally it became nil. This result indicates that the cables do not work effectively under downward load conditions such as snow load. In case of the upward wind load condition, the cable tension increased almost linearly in proportion to the applied load both in the measured and calculated results. So the effectiveness of the cables against wind load, the most severe load for this structure, is confirmed.

<u>Defomation of the fabric panel</u> Mesured and calculated deformation modes of the fabric panel under snow and wind load conditions are shown in Fig. 12. It should be noted that the displacements presented in Fig. 12 are incremental displacements from Load Condition T2 to S or W. Since the snow and wind loads were applied to the specimen by equivalent concentrated loads instead of ideal distributed loads as mentioned before, displacements near the loading points were considerably larger than those between the loading points. Because of the anisotropic property of the material, displacement along the center line of fill direction (A-J) is larger than that of warp direction (A-G). Fig. 13 shows

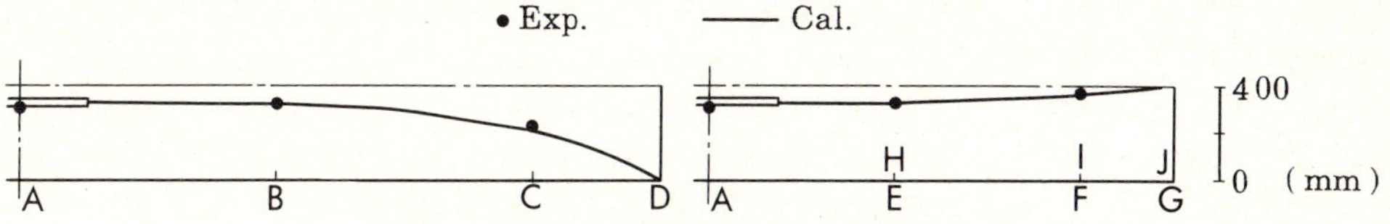

Fig .9 Initial Equilibrium Configuration

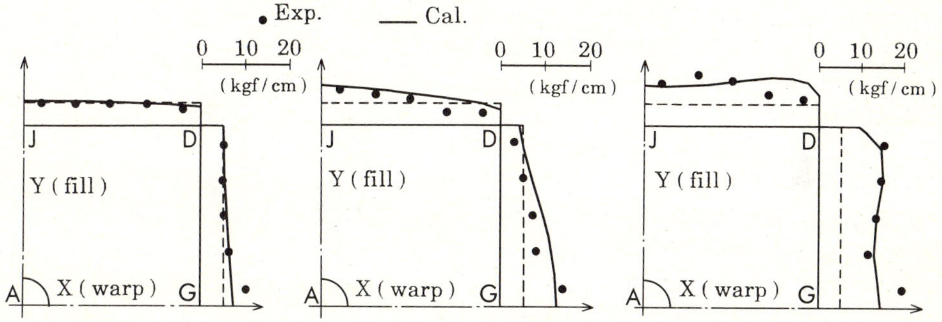

Additional Prestress 200 kgf Snow Load (S) 120 kgf / m2 Wind Load (W) 350 kgf / m2

Fig. 10 Tensile Stress Distribution Along Edges of the Specimen

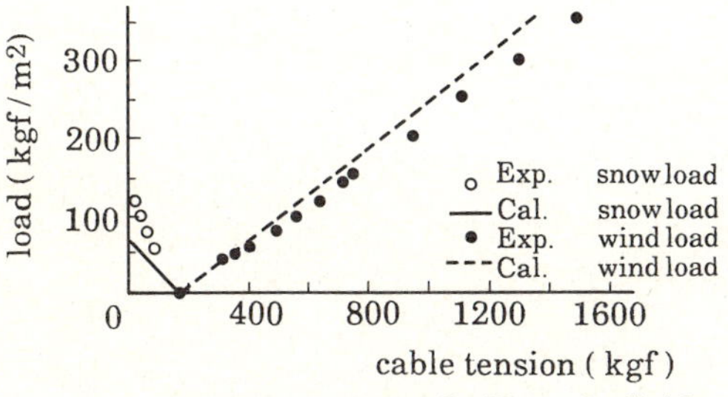

Fig. 11 Variations of Tensile Force in Cable

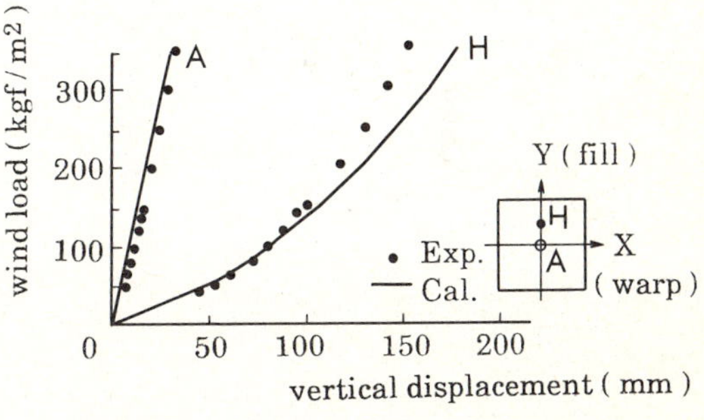

Fig. 13 Load-Displacement Relationship

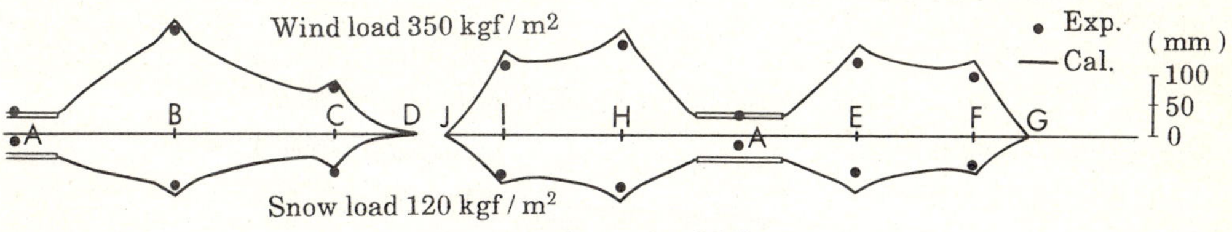

Fig. 12 Deformation Modes

the load-displacement relationships of points A and H. As shown in this figure, a remarkable nonlinearity appears at the membrane point (H), but the relationship is almost linear at the center point (A) both in the measured and calculated results. This proves the effectiveness of diagonal cables. The calculated displacement is very close to that obtained experimentally at point A, where the diagonal cable effect is dominant, but difference between analysis and experiment tends to be large at point H.

Strain distribution in fabric panel Strain distributions in the fabric panel under snow and wind load conditions at 80 kgf/m2 are shown in Fig. 14. These are the incremental strains from Loading Condition T2 to S or W, and strains in X-direction (Warp) and Y-direction (Fill) are denoted by ε_x and ε_y respectively. Under the snow load condition, the increments of strains were very large in the crosswise direction of the panel, but at the points near the corner of the panel strains decreased from Loading Condition T2. On the other hand, under the wind load condition, both ε_x and ε_y increased remarkably at the points near the corner of the fabric panel. The characteristics of these strain distributions are compatible with the stress distributions discussed before.

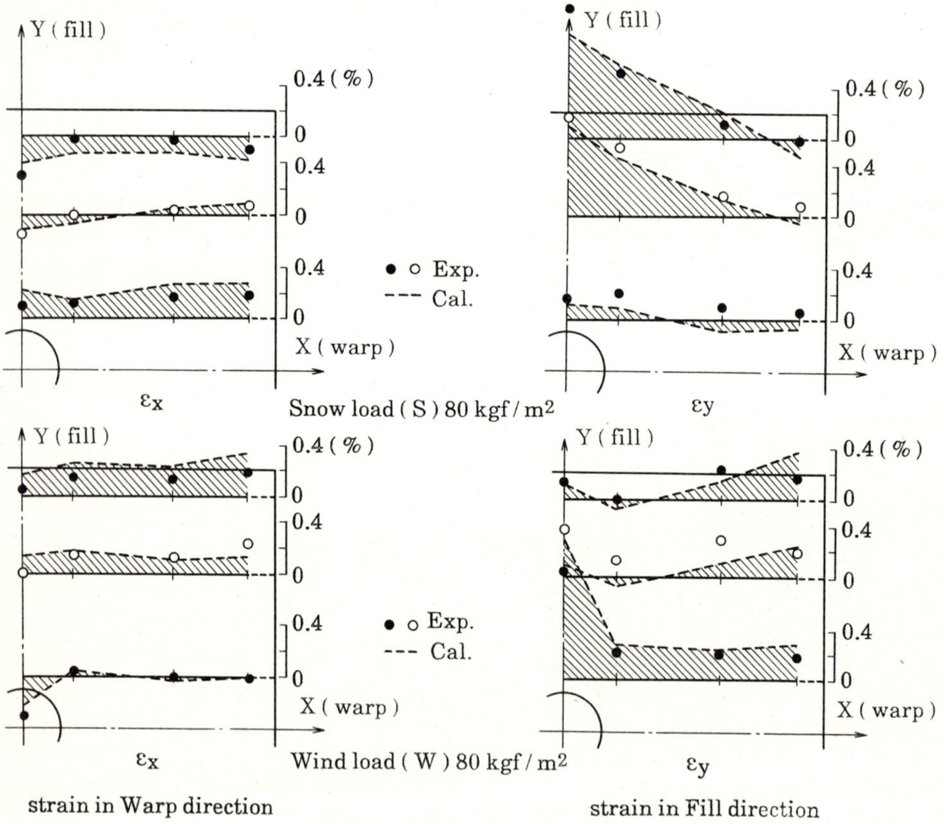

Fig. 14 Strain Distributions

Fig. 15 shows a comparison of the measured and calculated relationships between strains and applied wind load. As shown in this figure, difference between measured and calculated values becomes large as strain increases. The same tendency has already been pointed out in relation to the load-displacement relationship. A major cause of these discrepancies is considered to be a nonlinearity of the fabric material. Therefore, the material nonlinearity should be taken in to account in order to develop a more accurate analytical model.

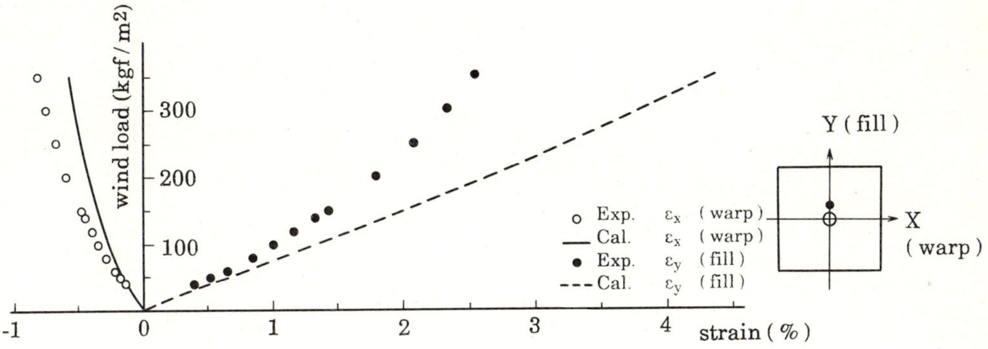

Fig. 15 Load-Strain Relationship

5. Conclusions

Comparing the results of analysis with those of experiment, the following conclusions are made.

(1) The analysis could follow the structual characteristics of the tensile fabric structure.

(2) For more accurate analysis, the nonlinearity of membrane material should be taken into account.

The analytical method presented here was applied to the structural analysis of the actual structure, which was completed in March,1986, in Tokyo.

References

(1) Geiger, D.H., "Large-Span Pneumatic Structures, Reinforced by Steel Cables" Acier. Stahl. Steel, Sept.1977

(2) Bathe, K.J., Ramm, E., and Wilson, E.L., "Finite Element Formulation for Large Deformation Dynamic Analysis" International Journal for Numerical Method in Engineering, Vol. 9, 1975

(3) Haber, R.B., "Computer-Aided Design of Cable Reinforced Membrane Structures" Thesis presented to Cornell University at Ithaca N.Y. 1980

(4) Ishii, K., "Structural Design of Cable-Reinforced Membrane Structure" IASS Internatinal Symposium Shell and Spatial Structure '83 Sept. 1983, Rio de Janeiro

(5) Minami, H., "Biaxial Deformation Property of Coated Plain Fabrics" Summaries of Technical Papers of Annual Meeting, Architectual Institution of Japan, 1984

NON LINEAR ANALYSIS WITH MODEL INVESTIGATION OF BURIED PIPELINES

D.H. JIANG
M.C. CHANG
C.J. HSIAO
C.C. LEE
Y.F. LEE
W.P. LIN

Graduate Institute of Civil & Hydraulic Eng.
Feng Chia University
Taichung, Taiwan, China

Abstract

In this paper our Institute has done the research on the reinferced concrete pipelines. The study includes the review on the development of the theory of thin cylindrical shell and performing the full scale test on pipe up to the collapse with different boundry conditions.
The result of the research done by us has shown that the behavior of buried concrete pipelines under the imposed loading can be better treated by the non-linearity for which the strength of the concrete pipe can have further extension over the yielding point which is still a little bit far from the ultimate strength of the reinforced concrete pipe.

1. Introduction

In Taiwan pipelines has been buried in a large amount during the last few years. These pipelines serve to flow gas, oil, suspended particles, water and chemical products etc. Most of the pipe are made by metal, conctete and plastic materials. Some of them are pressurized vessels.
As a civil engineering institute we have paid more attention on the pipelines made by concrete and plastic materials for drinking water and other public utilities service. Those pipelines are mostly buried along the main street or highway. Because of the poor subsoil happened when pipeline are crossing the river and the additional loading is added by those overweight from the traffic unequal settlement phenomena are expected.
Usually pipeline analysis applies elastic theory only. But elastic theory does not provide the actual resistance which is reserved in the pipe i.e. the effect on the non linear range. Besides, the analysis are more

based on the beam theory. If we take the definition of Roack i.e. if the thickness of pipe is less than one- tenth of the radius then under uniform pressure the membrane stresses are practically uniform througout the thickness of the pipe. It is the only stress of importance present. According to the ACI Code the reinforced concrete can be considered ideally elastic, homogenous and isotropic with identical stress-strain properties in all direction. Based on this principle we have examined the concrete pipe and found out that the thickness are all over the limit of one-tenth of the radius. But if we discount the thickness of cover protection, which is for precast concrete 1/2 in. minimum cover if expose to weather and 3/8 in. minimum cover if it is not exposed to weather, than most of concrete pipe made in Taiwan are able to be considered as thin shell structure.

Thus in this paper we have reviewed the theoretical theory of cylindrical shell and done the experimental analysis on concrete pipe. Though we have also done the test on plastic pipe but in order not to be confuse by those large deformation theory we have limited ourself only on the rigid cylindrical reinforced concrete shell.

2. Theoretical Analysis

The study begins on the elastic theory. Following the famous works of Timoshenko and Gere a formula was derived for a two point loaded ring. The ovalic displacement w in case without internal pressure

$$w = \frac{2PR^3}{\pi EI} \sum_{n=2,4,6..}^{\infty} \frac{(-1)^{n/2} \operatorname{Cos}\phi n}{(n^2-1)^2} \qquad (1)$$

in which

$$EI = \frac{1}{12} Et^3$$

For the bending moment in the ring

$$Mb = \frac{-EI}{R^2}\left(\frac{d^2w}{d\phi^2} + w\right) \qquad (2)$$

In combining the above two equations we have

$$Mb = -\frac{2PR}{\pi} \sum_{n=2,4,6..}^{\infty} \frac{(-1)^{n/2} \operatorname{Cos} n\phi}{(n^2-1)} \qquad (3)$$

The end of the elastic range is reached when the stresses in the cylindrical wall reaches to its yield stress thus:

$$\frac{Mbmax}{1/6\ t^2} = \sigma_e \qquad (4)$$

where:

P: force on pipe

R: Average radius of the pipe

ϕ: Angle

w: displacement

t: thickness of pipe

For the non-linear problem of the Reinforced concrete cylindrical shells it should be pointed out that although there are so well study in the buckling theory but no one can put forth a theory that is applicable to the mathematics and physics of the general buckling problem. Most of the theoretical works on the buckling of the structure assume that buckling occurs in the elastic range. But it is not quite a realistic assumption because of the high stresses are mostly operated. The reinforced concrete shells may be defined as ortho tropic shells since their stiffness and properties are all different in direction and in tension and compression. The different reinforcement in different direction is also an important property in the analysis of the buckling of the R.C. shells. In case of cylindrical shells applied to this study the circumferential stresses are predominanted. For the cylindrical shells of moderately length Donnell's solution is generally recognized as a valid approach to apply on concrete cylinders under hydrostatic loading. The critical stress can be expalined as:

$$\sigma_{cr} = \frac{0.855\ E}{(1-\nu^2)^{3/4}} \left(\frac{t}{R}\right)^{3/2} \frac{R}{L} \eta \qquad (5)$$

where:

E: Young's modulus

R: mean radius

L: length of cylindrical shell

t: wall thickness

ν: Poisson's ratio

η: Plasticity reduction factor

Although expression for η have been developed from the theoretical consideration, these expressions do not appear to apply to concrete structure. Following the study of Chen the plastictity reduction factor is also a function the modulus of elasticity. For design purposes, a conservative curve is selected at $E/fc'=667$. By substituting $E=667fc'$ and $\nu=0.18$ eq. (5) can be reduced as

$$\frac{\sigma_{cr}}{fc'} = 583\left(\frac{t}{R}\right)^{3/2} \frac{R}{L} \eta \qquad (6)$$

Furthermore, when the cylinder wall stress in the hoop direction equals to the ultimate strength of the concrete, material will cause failure. Lamé's solution is then used to predict this pressure. The expression are thus

$$Pcr = 4.0\ \sigma cr \frac{t/R}{(2 + t/R)^2} \qquad (7)$$

and we introduce the end condition β and substituting the Donnell's equation into Lamé's. Then

$$Pcr = 2330 \beta \eta fc' \frac{(t/R)^{5/2} R/L}{(2 + t/R)^2} \qquad (8)$$

where β = 1 for simple ends, β = 1.2 for fixed ends.

3.Experimental analysis

The experimental test are performed by full scale size of R.C. sewage pipe of class I. Its characteristics and dimension are shown in Table 1 and figure 1.

table 1 characteristics of R.C. Pipe specimens (L:2500 mm)

I.dia (D)	NO. of Speciman	Eff. t (te)	Ave. dia (tm)	Reinforcemat Spiral cm2 A			Longitudal		fcr	fc'	I
				cm2 A	BWG NO.	Pitch	BWG NO.	TOTAL NO.			
mm		mm	mm			mm			kg/cm2	kg/cm2	cm4
ф 200	3	7.95	227.0	1.0	12	62	12	10	54	350	104.0
ф 200	3	8.95	278.0	1.0	12	62	12	10	54	350	129.0
ф 300	2	10.95	330.0	1.0	12	62	12	10	54	350	157.0
ф 350	2	12.95	381.5	1.2	12	52	12	12	54	350	187.4
ф 450	3	20.95	490.0	1.5	10	61	10	12	54	350	373.6
ф 500	1	25.95	545.0	1.7	10	53	10	12	54	350	532.0

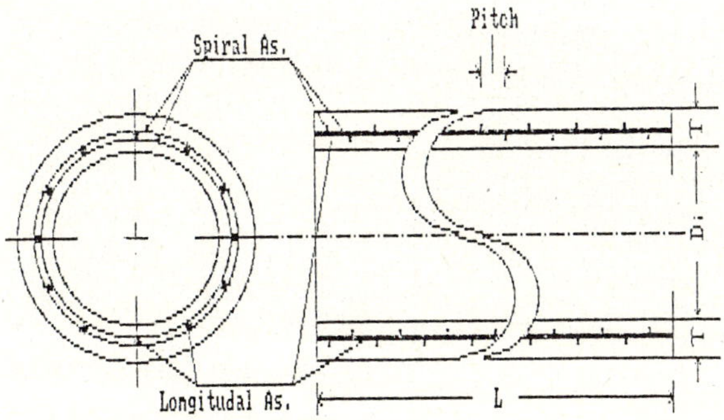

Figure 1

Although we have tested on different size of R.C. pipe in this paper we will only present the results of ϕ_I = 350 m/m pipe to compare with the theoretical study and experimental research contributed by other institutes. The other sizes are no doubt also fulfill the comparison.
The R.C. pipes has been performed by two boundary conditions i.e. linear support fixed underneath and fixed-end support on both sides (Fig.2 & 3).

Fig. 2

Fig. 3

The first examination are comparing the displacement at yield stress by the formula (1) and the test results. Comparing with the test results which is shown in Fig. 4 & 5 we found it can be concided to our result.

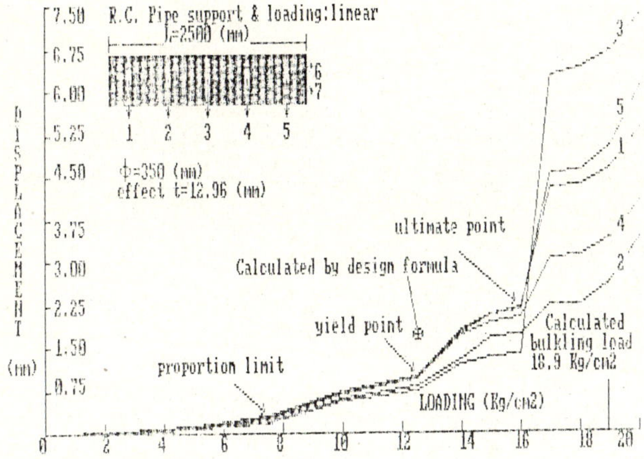

Figure 4

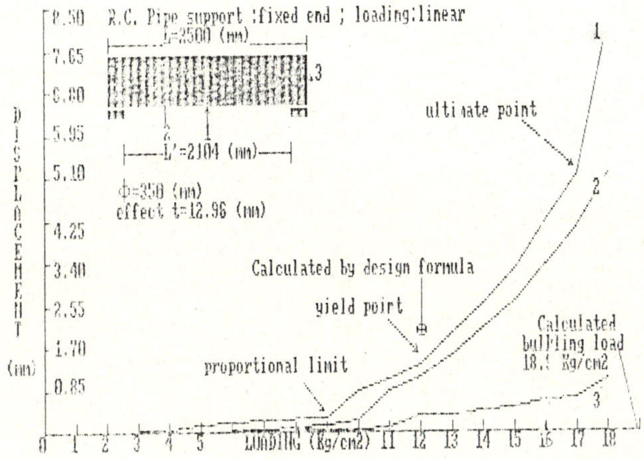

Figure 5

Where in case of line support by eq 1. we have the displacement of 1.726 mm and 2.05mm for fixed-end support. The bending moment found by our calculation are thus 114.15kg-cm/cm for line support and 135.61kg-cm/cm for fixed-end support. By the traditional calculation of the bending moment has the value of

$$Mr = \frac{I \times fcr}{n(T-x)} = \frac{187.42 \times 54}{0.5(3.2-1.32)} = 107.67 \text{ kg-cm/cm}$$

where

T: wall thickness

x: distance to the neutual axis

The results are also quite coincided to our prediction. The buckling load found by the tests are 17kg/cm² for the line and for the fixed-end supports while we have Pcr = 18.9kg/cm² found by eq. (8).

4. Conclusion

During the test on the R.C. pipes we have found that the lateral displacement by the spliting loading is larger than by the fixed-end support test. The critical load are also has the same phenomenon that buckling load by spliting test is smaller than the fixed-end support. Thus in case of buried pipe on the weak soil, pile foundation on the connection of pipe is not only to prevent the leakage but also strengthen the pipe. From the test results we found also the ovalization of the pipe section are very small about 0.002D for elastic range and 0.0035D for plastic range. Thus the effect of ovalization on the section property of the pipe is small can be neglected the effect. According to our measurements buckling occurs at larger curvature on the longitudinal direction and after the measured internal pressure is higher. Therefore buckling occurs after the pipe has partially become plastic. Reinforced concrete has to be considered the cracking phenomena especially in order to prevent the leakage. In our test the visual fissure began right after the yield point and the crack happened on the ultimate strength loading. Crack patterns can be seen in Fig. 2 & 3. The crack propagation stage are also the material charaterisitcs from yield point go up to the ultimate strength. Hence the bond slip and dowel action of the reinforcing steel are actually lead the last stage of failure of the specimen. Crack are also happened on the place where stresses are concentrated, i.e. the place where pipes are largely deformed.

Due to the limit of the equipments for measurement we did not go further to the shear and torsional stresses analysis. It is also recommended to the further study on the non-linearties due to the time-dependent effects and loading rate.

Using the modern computation facilities and programming techniques we have compared with the test results performed by the full scale reinforced concrete pipes. The result are quite coincided. The plasticity behavior of reinforced concrete shells are still a way to develop. The laboratory performance must take into account to the study of plasticity. In fact, plastic range of materials are rather complicated. It is not only performing by the calculation and laboratory test. A well trained engineering judment can be a good help for those merely applying the existed formulas.

5. Acknowledgement

Dept. of Hydraulic and Mechanical Engineerings and Taiwan Provincial Institute of Harbor & Marine Technology provide their facilities to let us performed the test, Hong Chen Cement Production Co. provides their R.C. pipes for to perform the test at their factory are all very much gratitude. Mr. R.C. Yang project manager of Taiwan land development and Trust Corp. sponsors to this study are also very grateful.

Reference

1. A.A. Van Douwen, A.M. Gresnigh & J.W.B. Stark, Plastic design of buried steel pipelines for transport of oil, gas or water, verfied by tests on scale model, TNO-IBBC, Rijswijk, The Netherlands 1974.

2. A.C. Ugural, Stresses in plates and shells, MacGraw-Hills Inc. New York, U.S.A., 1981.

3. A.M. Haas, Thin Concrete shells vol.2., John Wiley & Sons, Inc. New York, U.S.A., 1967.

4. C.L. Dym, The theory of thin shells, Pergamon Press, Oxford, U.K., 1974.

5. D.O. Brush & B.O. Almroth, Buckling of Bars, Hates, and Shells, McGraw-Hills Inc. New York, U.S.A., 1975.

6. E.L. Axelrad & F.A. Emmerling, Flexible shells, Springer-Verlag, Berlin, Germany 1984.

7. F. Mortelmans, Berekening van Kostrukties deel 3 Gewapend beton Acco, Leuven, Belgium. 1981.

8. H. Kraus, Thin elastic shells, John Wiley & Sons, Inc., New York, U.S.A., 1972.

9. K.P. Buchert, Buckling of shell & shell-kike structures, Univ. of

Missouri, U.S.A.,1973.

10. P.C.I. Manual for structural design of architectural and precast concrete, Chicago, U.S.A., 1977.

11. R.J. Roark, Formulas for stress and strain, McGraw-Hills Inc. New York, U.S.A., 1965.

12. W.F. Chen, Plasticity in reinforced concrete, McGraw-Hills Inc. New York, U.S.A., 1982.

13. W. Flügge editor in chief, Handbook of Engineering Mechanics, McGraw-Hills Inc. New York U.S.A., 1962.

14. W. Flügge, Stresses in shells, Springer - Verlag, Berlin, Germany 1968.

15. Y.C. Fung & E.E.Sechler Thin-shell structures, theory, experiment and design, Prentice-Hall Inc. Englewood Cliffs, (N.J.) U.S.A., 1974.

AN UNCONVENTIONAL CLASS OF ELEMENTS FOR NONLINEAR SHELL ANALYSIS

M. K. NYGÅRD and P. G. BERGAN

A.S Veritec and A.S Veritas Research,
P.O. Box 300, 1322 HØVIK, NORWAY

SUMMARY

The paper gives a short introduction to the free formulation theory for finite elements, and indicates how this theory also can be used for structural analysis including geometric and material nonlinearities. Some new flat, triangular shell elements have been developed using this theory. These elements are non-conforming, but satisfy the patch test. A novel feature is that these elements include the normal rotation as well as the conventional three translational and two rotations for flexure at the corner nodes. The drilling freedoms improve the membrane behaviour significantly and are particularly useful in making a complete set of rotational freedoms for shells. The high accuracy of the shell elements are documented through a series of linear and nonlinear examples.

1. INTRODUCTION

Finite elements based on the "Free formulation" by Bergan and Nygård [1] are, in most respects, similar to standard displacement elements. However, they have the added advantage that the displacement patterns need not satisfy interelement compatibility (conformity). The convergence of the free formulation elements is assured through satisfaction of the so-called "individual element test". This test, which was proposed by Bergan and Hanssen [2], is expressed as a set of linear constraints that the element stiffness matrix must satisfy, and it implies intrinsically that the "patch test" by Irons [3] is satisfied.

Up till now, the free formulation theory has been applied for deriving various types of linear membrane and plate elements, [1,4,5,6]. The excellent performance of these elements has motivated the present development of the free formulation theory for application to nonlinear problems with large displacements and inelastic materials. This paper also demonstrates the use of the free formulation theory for thin shell elements. An important aspect of these elements is the introduction of rotational freedoms normal to the element plane ("drilling freedoms"). Bergan and Felippa [4] have shown how these freedoms improve the in-plane bending response substantially over that of similar elements without such freedoms. The drilling freedoms are also convenient in that they eliminate the problem of stiffness singularity for co-planar elements.

2. FREE FORMULATION THEORY

2.1 Linear Formulation

The displacement field $\mathbf{u}$ within an element may be approximated by a set of linearly independent displacement modes

$$\mathbf{u} = \mathbf{N}_{rc}\mathbf{q}_{rc} + \mathbf{N}_h\mathbf{q}_h \tag{1}$$

$\mathbf{N}_{rc}$ is a complete set of rigid/body and constant strain modes (rc-modes) with associated coefficients $\mathbf{q}_{rc}$, and $\mathbf{N}_h$ is a set of "higher order" displacement modes (h-modes) with coefficients $\mathbf{q}_h$. The number of rc-modes plus h-modes must be equal to the number of element freedoms.

The relationship between the generalized coefficients $\mathbf{q}$ and the element freedoms $\mathbf{v}$ is easily obtained by substitution of nodal coordinates into (1) and, for rotational freedoms, into derivatives of $\mathbf{N}$

$$\mathbf{v} = \mathbf{v}_{rc} + \mathbf{v}_h = \mathbf{G}_{rc}\mathbf{q}_{rc} + \mathbf{G}_h\mathbf{q}_h = \mathbf{G}\mathbf{q} \tag{2}$$

The generalized modes must be separable in terms of nodal freedoms, i.e., **G** must be regular and invertible

$$\mathbf{q} = \begin{vmatrix} \mathbf{q}_{rc} \\ \mathbf{q}_h \end{vmatrix} = \mathbf{G}^{-1}\mathbf{v} = \begin{vmatrix} \mathbf{H}_{rc} \\ \mathbf{H}_h \end{vmatrix} \mathbf{v} \tag{3}$$

A standard potential energy or virtual work stiffness formulation based on (1) would give an element that satisfies completeness but not, in most cases, interelement compatibility. Such an element would then not pass the patch test and it would fail to converge. As an alternative, Bergan and Nygård [1] suggested that the element stiffness should be formed from

$$\mathbf{k} = \frac{1}{V}\mathbf{LCL}^T + \mathbf{H}_h^T\mathbf{k}_{qh}\mathbf{H}_h \tag{4}$$

The first contribution to **k** is denoted "basic stiffness" and it is only associated with the constant strain/stress behaviour of the element. **L** is the so-called "lumping matrix" which transfers surface traction from constant stress states to consistent nodal forces. **C** is the constitutive matrix and V is the element volume. The second contribution to **k** is the "higher order stiffness" which is based on the stiffness matrix for the generalized higher order modes

$$\mathbf{k}_{qh} = \int_V \mathbf{B}_{qh}^T\mathbf{C}\mathbf{B}_{qh}dV \tag{5}$$

$\mathbf{B}_{qh}$ is the strain-producing matrix associated with the h-modes. It was demonstrated by the authors [1] that the stiffness in (4) passes the individual element test/the patch test even for incompatible displacement functions. This relaxation of constraints when choosing displacement functions opens the door to a new class of finite elements. The free formulation (4) is particularly useful for C^1 type of elements, such as plate and shell elements.

Equation (4) may also be written on traditional form

$$\mathbf{k} = \int_V \mathbf{B}^T\mathbf{C}\mathbf{B}dV \tag{6}$$

in which the strain-producing matrix is

$$\mathbf{B} = \frac{1}{V}\mathbf{L}^T + \widetilde{\mathbf{B}}_{qh}\mathbf{H}_h \tag{7}$$

2.2. Nonlinear Formulation

The free formulation theory may be extended to nonlinear situations with displacements and strains of arbitrary magnitude, see Nygård [8]. For instance, the lumping matrix in (4) may be found by use of virtual work for "updated", "co-rotated" and "total" Lagrangian descriptions. Unfortunately the scope of the present article does not allow for presentation of this.

The derivation of free formulation elements for nonlinear situation with small deformational strains is rather straight-forward. In cases of fixed or co-rotated coordinate frames the internal, element nodal reaction forces $\mathbf{S}_{int}$ is found from the usual expression

$$\mathbf{S}_{int} = \int_V \mathbf{B}^T\sigma dV \tag{8}$$

B is defined in (7). The Cauchy stress σ may be found from the current strain, or, for inelastic materials, the strain history.

The incremental equilibrium is expressed through the relation

$$\Delta \mathbf{S}_{int} = \mathbf{k}_T \Delta \mathbf{v} = (\mathbf{k}_{mat} + \mathbf{k}_{geom}) \Delta \mathbf{v} \tag{9}$$

The tangential or incremental stiffness $\mathbf{k}_T$ is comprised of material and geometric stiffness contributions. The material stiffness is obtained from the conventional expression

$$\mathbf{k}_{mat} = \int_V \mathbf{B}^T \mathbf{C} \mathbf{B} dV \tag{10}$$

where $\mathbf{C}_T$ is the tangential or incremental material law. $\mathbf{k}_{geom}$ is the usual geometric stiffness matrix; its form depends somewhat on the type of application. It is found from the element displacement field defined by (1) and (3).

3. SHELL ELEMENTS

3.1 Formulation of Plate and Shell Elements

The plate and shell elements are initially flat, and the usual Kirchhoff thin plate assumption are adopted. Introducing an orthogonal x,y,z coordinate system with corresponding displacements u,v,w when x and y follow the plane of the plate, the strain producing displacements are

$$\begin{aligned} u &= u_o - z w_{,x} \\ v &= v_o - z w_{,y} \end{aligned} \tag{11}$$

Index o indicates reference plane value. The in-plane (membrane) displacements $\mathbf{u}_m$ and the out-of-plane displacement w are interpolated between the nodal freedoms in the normal fashion

$$\mathbf{u}_m = \begin{bmatrix} u_o \\ v_o \end{bmatrix} = \begin{bmatrix} \mathbf{N}_u \\ \mathbf{N}_v \end{bmatrix} \mathbf{v}_m = \mathbf{N}_m \mathbf{v}_m \,, \qquad w = \mathbf{N}_b \mathbf{v}_b \tag{12}$$

The interpolation functions $\mathbf{N}_m$ and $\mathbf{N}_b$ will be dealt with later under specific elements. The membrane and bending freedoms for an element with n nodes are

$$\mathbf{v}_m = \begin{bmatrix} \mathbf{v}_{m1} \\ \mathbf{v}_{m2} \\ \vdots \\ \mathbf{v}_{mn} \end{bmatrix} , \qquad \mathbf{v}_b = \begin{bmatrix} \mathbf{v}_{b1} \\ \mathbf{v}_{b2} \\ \vdots \\ \mathbf{v}_{bn} \end{bmatrix} \tag{13}$$

The freedoms at node i are defined by

$$\mathbf{v}_{mi} = \begin{bmatrix} u_o \\ v_o \\ \frac{1}{2}(v_{o,x} - u_{o,y}) \end{bmatrix}_i = \begin{bmatrix} u_o \\ v_o \\ \Theta_z \end{bmatrix}_i , \qquad \mathbf{v}_{bi} = \begin{bmatrix} w \\ w_{,y} \\ -w_{,x} \end{bmatrix}_i = \begin{bmatrix} w \\ \Theta_x \\ \Theta_y \end{bmatrix}_i \tag{14}$$

The rotational freedom Θ_z about the z-axis ("drilling freedom") is defined as the mechanical rotation derived from the in-plane displacement gradients. This quantity is invariant with respect to the reference coordinates. However, there is no unique relationship between the freedom Θ_z and the rotations of the adjacent element sides, see discussion by Bergan and Felippa [4]. This kind of freedom can therefore not be used in connection with the usual displacement formulation which requires C^o interelement continuity for the membrane action. This is, however, no object for the free formulation.

3.2 Triangular Element

Membrane Action

The membrane element suggested by Bergan and Felippa [4] is adopted here. As shown in Fig. 1a, nondimensional coordinates with origin at the centroid (x_c, y_c) is introduced

$$\xi = \lambda(x - x_c) \quad , \qquad \eta = \lambda(y - y_c) \tag{15}$$

where the length scaling factor is

$$\lambda = \frac{1}{\sqrt{A}} \tag{16}$$

The complete set of six rigid-body and linear strain modes may be written

$$\mathbf{N}_{rcm} = \begin{vmatrix} 1 & 0 & -\eta & \xi & 0 & \eta \\ 0 & 1 & \xi & 0 & \eta & \xi \end{vmatrix} \tag{17}$$

According to equations (12), (13) and (14) there are nine freedoms controlling the membrane displacements; three additional higher order modes are thus required. Following [4], the best choice is the in-plane "bending" modes oriented along the median lines of the triangle (see Fig. 1a)

$$\mathbf{N}_{hm,i} = \begin{vmatrix} \bar{\xi}_i \bar{\eta}_i \\ -\frac{1}{2}\bar{\xi}_i^2 \end{vmatrix} \quad , \quad i = 1,2,3 \tag{18}$$

These natural modes are objective with respect to the reference system. They also give a **G** matrix which is regular.

Bending action

The bending modes are expressed most conveniently in terms of area coordinates ζ_i. The complete set of rigid-body and constant curvature modes may be written

$$\mathbf{N}_{rcb} = \begin{vmatrix} \zeta_1 & \zeta_2 & \zeta_3 & \zeta_1\zeta_2 & \zeta_2\zeta_3 & \zeta_3\zeta_1 \end{vmatrix} \tag{19}$$

The three higher order modes suggested in [9] are used. These modes are force orthogonal and have proved to give very good results

$$\mathbf{N}_{hb} = \begin{vmatrix} \zeta_1\zeta_2(\zeta_1 - \zeta_2) & \zeta_2\zeta_3(\zeta_2 - \zeta_3) & \zeta_3\zeta_1(\zeta_3 - \zeta_1) \end{vmatrix} \tag{20}$$

The **H** -matrix for these bending modes has also been found explicitly [8].

Integration

The stiffnesses associated to the membrane and bending actions are completely uncoupled in case of a linear, homogeneous material. However, these stiffnesses become coupled for inelastic materials and can thus only be computed by use of numerical integration. It has been found that a three-station numerical integration based on the mid-side points gives sufficient accuracy. In addition, cross-sectional quantities are computed using either the two point Gaussian or the three to ten point Lobatto integration schemes.

3.3 Quadrilateral Element

Membrane Action

A set of nondimensional coordinates ξ and η with origin at the centroid is introduced, see Fig. 1b. As for the triangular element, ξ and η are defined by (15), however, the scaling factor is here chosen as

$$\lambda = \frac{2}{\sqrt{A}} \tag{21}$$

The basic rc-modes are the same as in (17). The six, additional higher order modes are chosen as

$$\mathbf{N}_{hm} = \begin{vmatrix} \xi\eta & \eta^2 & 0 & 0 & \eta(\xi-\eta)^2+f_{51} & -\eta(\xi+\eta)^2+f_{61} \\ 0 & 0 & \xi\eta & \xi^2 & -\xi(\eta-\xi)^2+f_{52} & \xi(\eta+\xi)^2+f_{61} \end{vmatrix} \tag{22}$$

A main consideration in the selection of these modes is that they do not give a singular **G** regardless of the geometry. It is seen that these h-modes do not correspond to complete polynomials of a certain degree, hence, these modes have a directional dependency in accordance with the orientation of the ξ,η -system. The ξ,η -system is therefore chosen such that it follows "natural" directions of the quadrilateral [8,10], see Fig. 1b. The functions f_{51}, f_{52}, f_{61}, and f_{62} are linear functions giving energy orthogonality [7] for the two last h-modes.

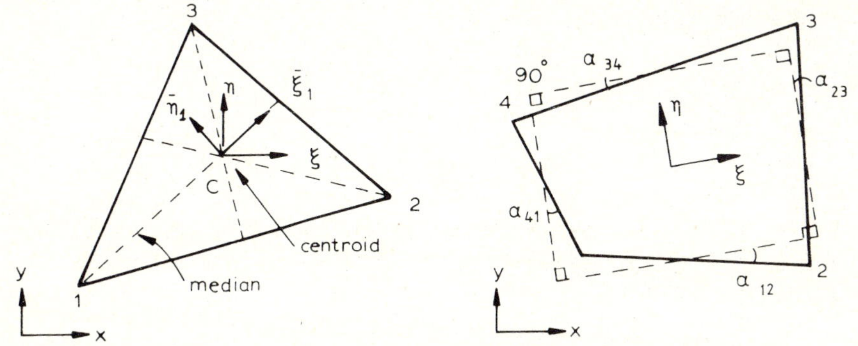

Figure 1. Local coordinate systems for triangular and quadrilateral elements.

Bending action

The rc-modes for the out-of-plane displacement are easily expressed as the six terms of a complete second degree polynomial. The following functions are chosen for the higher order modes [8]

$$\mathbf{N}_{hb} = \left| \xi^3 \quad \xi^2\eta \quad \xi\eta^2 \quad \eta^3 \quad \xi^3\eta - c_2\xi^2 - c_1\xi\eta \quad \xi\eta^3 - c_3\xi\eta - c_2\eta^2 \right| \tag{23}$$

c_1, c_2 and c_3 are constants determined from the energy orthogonality requirement.

Integration

The forces and stiffness matrices may be found using a two by two Gaussian integration over the area of the quadrilateral. This order of integration is sufficient for the bending stiffness, however, it renders the membrane stiffness with a defect of one (one spurious zero energy mode). This mode can result in singularity only for a single element with minimum boundary conditions. However, it cannot propagate through neighbouring elements. Two by two integration is thus safe and sufficient for all practical applications. The defect disappears for three by three integration. As for the triangle, either two Gaussian or three to ten Lobatto points are used for the integration through the thickness.

4. ANALYSIS OF SHELLS

4.1 Ghost reference Description

The large displacement motion of the shell is defined in a Lagrangian description. The deformation of each element is referred to the initial configuration of each element, however, this reference configuration is closely displaced and rotated along with the deformed configuration in order to minimize the rigid-body part of the rotational motion. This method has been used by several investigators, and it is here denoted "ghost reference description". It is applicable to large strains, as well as small deformational strains. This description has the advantage that, when the element coordinate system is moved along with the ghost reference, the components of the Green strain and the symmetric Piola-Kirchhoff stress tensor need not be transformed each time the ghost reference is updated [11].

The establishing of a ghost reference element is based on a procedure which is completely objective with respect to rotations and reference frame, see [8,10,12] Also, a new procedure for treating large rotations in space is applied. Thorough discussion of this may also be found in [8,10].

5. NUMERICAL EXAMPLES

5.1 Computer Programs

The element and methods described herein are implemented in the general purpose nonlinear finite element program FENRIS [13,14]. This program performs all the standard tasks of a nonlinear program, such as assemblage of loads, forces and stiffnesses as well as solution of nonlinear equations. The program is part of the SESAM system [15] and may utilize the interactive pre- and postprocessors available in this system. The FENRIS program has a highly modular structure which allows for combination of substructures of different types of structures.

The free formulation triangular and quadrilateral shell elements described herein are denoted FFTR (or FFT) and FFQ.

5.2 Shear-Loaded Cantilever Beam

The shear-loaded cantilever beam in Fig. 2a is commonly used for evaluating membrane elements, and give a good test on the elements ability to represent in-plane bending modes. The cantilever was analysed using regular mxn meshes as typified for the triangular element in Fig. 2b. The results for the tip deflection, v_c, are tabulated in Tab. 1. This table reveals excellent performance by the free formulation elements. Bergan and Felippa [4] have shown that the accuracy of the FFT element may be increased further by utilizing a scaling factor β for the higher order stiffness, resulting in the FFTR element.

The quadrilateral element is used in three versions, that is 2x2 integration with $\alpha = 1.5$ and 1.0 and 3x3 integration with $\alpha = 1.5$. α is a parameter relating the side rotations to the drilling freedoms [4]. The table shows that all cases give good results, and that the difference between them is rather small. In fact, the difference between the 2x2 and 3x3 integration is negligible, and the former thus appears as a winner because of computational efficiency. As for the FFTR-element [4], an α-value of 1.5 instead of 1.0 slightly increases the elements flexibility.

The other elements in Tab. 1 are the constant strain triangle (denoted CST, see e.g. [16]), and the traditional linear four-noded displacement based quadrilateral, Q4. Both these are completely inferior to the free formulation elements.

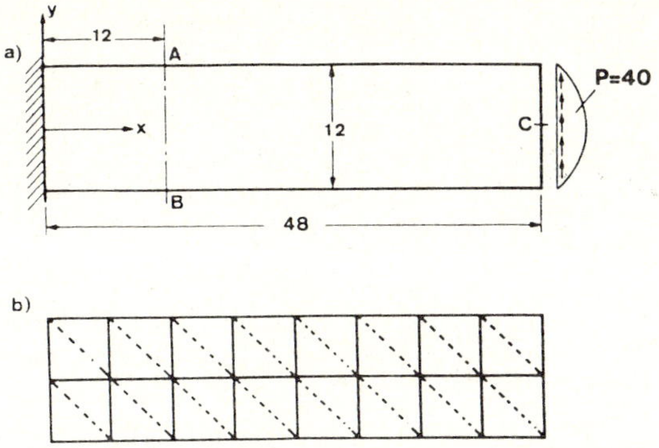

Figure 2. (a) Cantilever beam with load P parabolically distributed along right edge ($E=30000$, $\nu=\frac{1}{4}$); (b) regular 8x2 mesh.

Mesh	CST	FFT $\alpha=1.5$ $\beta=1.$	FFTR $\alpha=1.5$ $\beta=0.5$	FFQ $\alpha=1.5$ $\beta=1.$	FFQ $\alpha=1.$ $\beta=1.$	FFQ 3x3 $\alpha=1.5$ $\beta=1.$	Q4
4x1	25.48	74.24	94.15	92.19	92.19	92.17	68.13
8x2	55.24	91.51	97.72	99.95	98.57	99.69	88.91
16x4	82.66	97.41	99.16	99.88	99.48	99.79	96.91
32x8	94.96	99.28	99.75	99.93	99.85	99.90	99.30
64x16	98.71	99.83	99.96	100.00	99.99	100.00	

Tab. 1 Tip deflection (v_C) for shear-loaded cantilever beam (exact solution equal to 100.).

5.3 Cantilever beam subjected to end moment.

The cantilever in Fig. 3 is used to demonstrate the large rotation and large displacement capabilities. This example has been used by many investigators, but usually the problem is formulated such that bending about only one of the global axes takes place. The cantilever is here oriented "arbitrarily" in space, such that the rotational components about all three global axes become activated. This is a large rotation problem in which the treatment of rotations as non-vectorial quantities are put on a true test. The position of the cantilever may be defined by using and auxiliary $(\bar{x},\bar{y},\bar{z})$-system oriented as

$$\begin{Bmatrix} \bar{x} \\ \bar{y} \\ \bar{z} \end{Bmatrix} = \frac{1}{7} \begin{bmatrix} 6 & -2 & -3 \\ 3 & 6 & 2 \\ 2 & -3 & 6 \end{bmatrix} \begin{Bmatrix} x \\ y \\ z \end{Bmatrix} \tag{24}$$

and letting one side of the cantilever follow the $\bar{y}$-axis. However, this auxiliary system is used for modelling purposes only, all actual computations are performed in the global (x,y,z)-system. The six nodal degrees of freedom are free except at the clamped edge where all freedoms are fixed. The moment is also applied in the global coordinated system with components $M_x=-\frac{6}{7}M$, $M_y=\frac{2}{7}M$ and $M_z=\frac{3}{7}M$.

The size of the load steps was taken as $\Delta M=\frac{\pi}{10}\frac{Ml}{EI}$ for the FFQ element, hence, 20 steps were

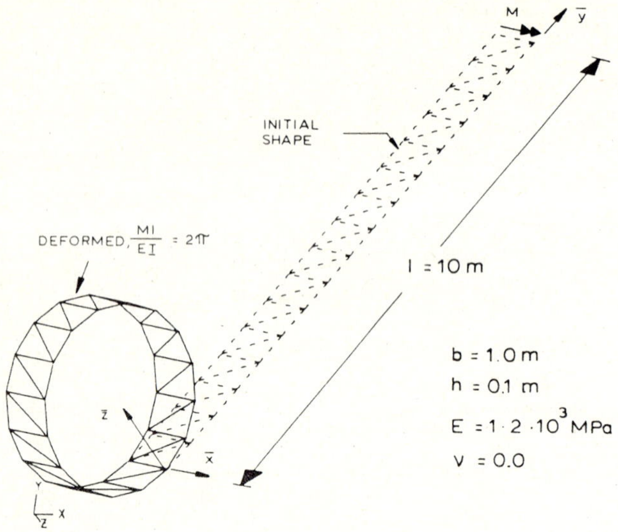

Figure 3. Cantilever beam oriented "arbitrarily" in space subjected to end moment.

required to bend the cantilever 360 degrees, which gives 18 degrees per step. Steps of half that size were used for the FFTR element, since this element does not give full symmetry about the center line of the cantilever. The problem was solved using full Newton-Raphson with equilibrium iterations.

Both elements give very accurate results for this problem, and the free end of the cantilever hits the clamped end within 0.01mm for $\frac{Ml}{EI}=2\pi$. Further discussions of this problem are given in [8].

5.4 Cylindrical Shell Roof Subjected to Gravity Loading

The cylindrical shell roof supported on end diaphragms suggested by Scordelis and Lo [17], has become a classical problem for testing shell elements. The geometry of the shell is given in Fig. 4.

Nonlinear analyses with elastic-perfectly-plastic material with yield limit $\sigma_y=4.2\text{N/mm}^2$ were performed. Von Mises yield criterion was assumed, and ten integration points were used for integration through the thickness of the shell. A uniform gravity force was gradually increased up to a maximum value of $4.0\cdot10^{-3}$ N/mm^2. Fig. 5a and 5b show the vertical deflection at the midpoint of the free edge against the intensity of the vertical load. Plots are given for several uniform nxn meshes for one quarter of the roof. The results are compared to those achieved by Kråkeland [18], who employed a curved 8-node isoparametric element in a 4x4 mesh. It is again seen that both the FFTR and FFQ element give highly accurate results. Only the very coarse 2x2 meshes give results which show significant inaccuracies. Fig. 6 show similar results as in Fig. 5, but comparisons are made with the three node TRUMP element by Argyris et. al. [19] and the four node element by Bathe and Dvorkin [20]. The results for these elements are taken from the mentioned references. The 4x4 mesh for the FFTR element is also plotted. It is clearly observed from Fig. 6, when also considering Fig. 5, that the present free formulation elements are superior to the two other elements for this nonlinear analysis.

5.5 Hinged Cylindrical Shell Under Concentrated Load

The hinged cylindrical shell with a concentrated load shown in Fig. 7 is a good test case for geometric nonlinearities. Assuming symmetrical response only one quarter of the shell is modeled. Thicknesses

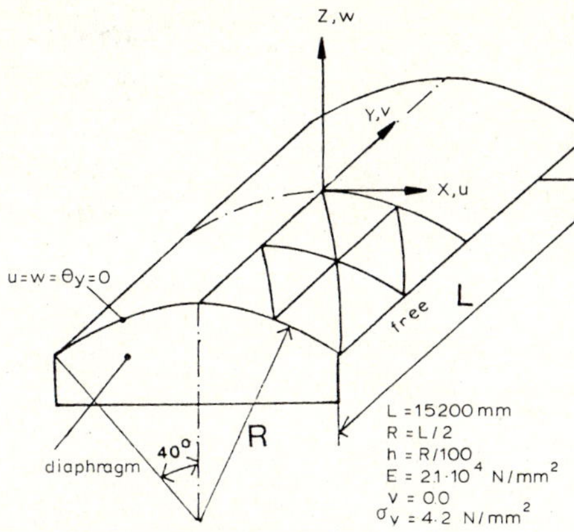

Figure 4. Cylindrical shell roof (Scordelis-Lo shell).

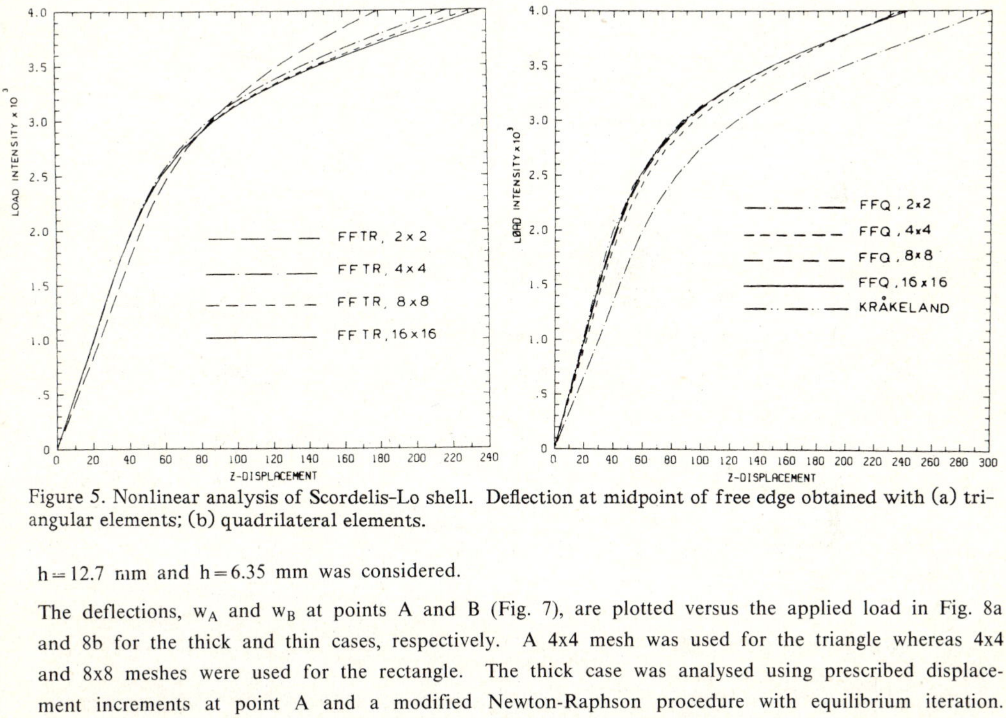

Figure 5. Nonlinear analysis of Scordelis-Lo shell. Deflection at midpoint of free edge obtained with (a) triangular elements; (b) quadrilateral elements.

$h = 12.7$ mm and $h = 6.35$ mm was considered.

The deflections, w_A and w_B at points A and B (Fig. 7), are plotted versus the applied load in Fig. 8a and 8b for the thick and thin cases, respectively. A 4x4 mesh was used for the triangle whereas 4x4 and 8x8 meshes were used for the rectangle. The thick case was analysed using prescribed displacement increments at point A and a modified Newton-Raphson procedure with equilibrium iteration. The results compared are in good agreement with other results reported in the literature (e.g. [21,22]).

A much more severe test case is the snap-through behaviour of the thin shell. It is seen in Fig. 8b that point A moves upwards again after the first instability point of the path has been passed. Such snap-back problems requires more advanced solution algorithms. The present example was solved with automatic load application/reversal in a single solution sequence using the so-called hyperplane displacement control method in [23]. There is some discrepancy between the three cases, particularly near the extremum points. The results obtained with the 4x4 mesh for the FFQ-element are very close to the flat element solution by Horrigmoe and Bergan [21], while the 8x8 FFQ results compare very well with the higher order curved element solution by Oliver and Onate [22].

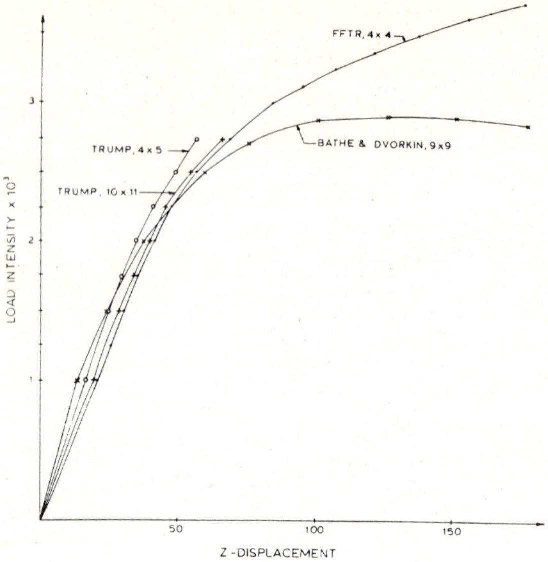

Figure 6. Deflection at midpoint of free edge, various elements.

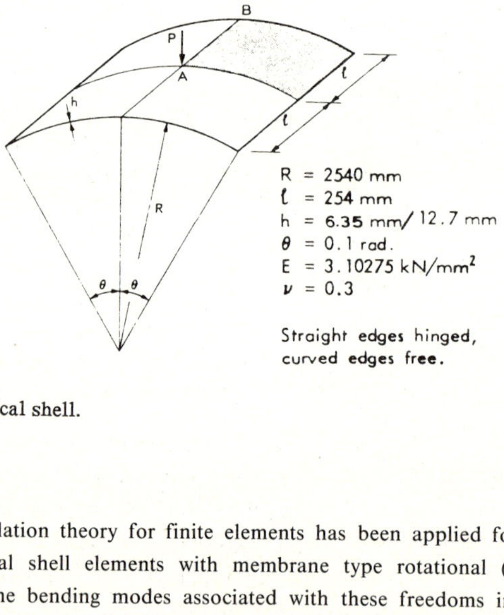

Figure 7. Hinged cylindrical shell.

CONCLUSION

The socalled free formulation theory for finite elements has been applied for developing some new, triangular and quadrilateral shell elements with membrane type rotational ("drilling") freedoms. It is demonstrated that inplane bending modes associated with these freedoms improve the accuracy significantly. These elements are also applicable to situations of very large displacements and rotations as well as material nonlinearities. The elements have been implemented in the nonlinear general purpose program FENRIS. It is demonstrated through a series of numerical examples that the new elements are significantly more accurate than traditional, displacement based elements.

ACKNOWLEDGEMENT

The authors would like to express their appreciation to their colleagues in the FENRIS project for their contribution in developing the computer code that was used.

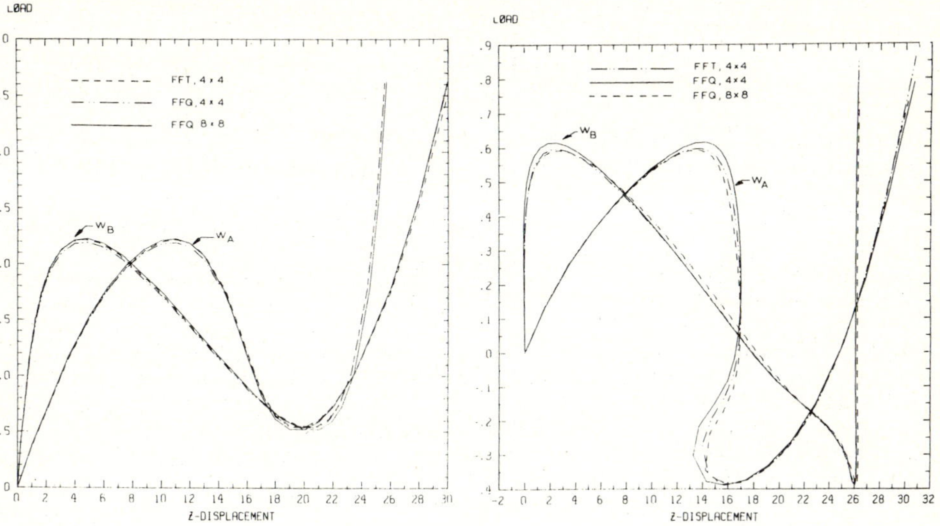

Figure 8. Load-deflection curve for hinged cylindrical shell. (a) thick case (h=12.7mm); (b) thin case (h=6.35mm).

REFERENCES

1 Bergan, P.G. and Nygård, M.K.: Finite elements with increased freedom in choosing shape functions. Int. J. of Num. Meth. in Engng., 20 (1984) 643-664.

2 Bergan, P.G. and Hanssen, L.: A new approach for deriving "good" finite elements. In J.R. Whiteman (ed.), The Mathematics of Finite Elements and Applications. Vol. II. 483-498, London: Academic Press 1976.

3 Irons, B.M. and Razzaque, A.: Experience with the patch test for convergence of finite elements. In K. Aziz (Ed.), Mathematical Foundations of the Finite Element Method with Application to Partial Differential Equations. 557-587, New York, London: Academic Press 1972.

4 Bergan, P.G. and Felippa, C.A.: A triangular membrane element with rotational degrees of freedom., Comput. Meth. in Appl. Mech. and Engng., 50 (1985) 25-69.

5 Bergan, P.G., Nygård, M.K. and Wang, X.: A class of quadrilateral plate bending elements. In H. Guangquian and Y.K. Cheung (Eds.), Proc. Int. Conference on Finite Element Methods. 112-117, Beijing: Science Press 1982.

6 Bergan, P.G. and Wang, X.: Quadrilateral plate bending elements with shear deformations, Comput. and Struct., 19 (1984) 25-34.

7 Bergan, P.G.: Finite elements based on energy orthogonal functions. Int. J. of Num. Meth. in Engng., 15 (1980) 1541-1555.

8 Nygård. M.K.: The Free Formulation for Nonlinear Finite Elements with Applications to Shells. Dr. Ing. Thesis, Div. of Struct. Mech., Norwegian Inst. of Technology, Trondheim, 1986.

9 Hanssen, L., Syvertsen, T.G. and Bergan, P.G.: Stiffness derivation based on element convergence requirements. In J.R. Whiteman (ed.), The Mathematics of Finite Elements and Applications. Vol. III. 83-96, London: Academic Press 1979.

10 Bergan, P.G. and Nygård, M.K.: Nonlinear shell analysis using free formulation finite elements. In P.G. Bergan, K.J. Bathe and W. Wunderlich (Eds.), *Finite Element Methods for Nonlinear Problems.* Springer Verlag, 1985.

11 Bergan, P.G.: Lecture Notes in Nonlinear Finite Element Analysis. Trondheim 1981.

12 Nygård, M.K., Kjeken, J. and Bergan, P.G.: Nonlinear shell analysis based on free formulation elements, In T. Kant (Ed.): *Finite Elements in Computational Mechanics,* Proceedings of FEICOM'85, IIT Bombay, India, Vol. 1, 61-71, Pergamon Press, 1985.

13 Bergan, P.G. and Arnesen, A.: FENRIS - A general purpose nonlinear finite element program. In Proc. 4th Int. Conf. Finite Element Systems. Southampton 1983.

14 Bergan, P.G.: Nonlinear finite element analysis by use of FENRIS. In J.F. Gouldman et al. (Ed.), Proc. 7th Int. Seminar on Computational Aspects of the F. E. Method. Chicago, CAFEM-7 1983.

15 SESAM'80 Description, A.S. Veritec, Høvik, Norway, 1985.

16 Zienkiewicz, O.C.: The Finite Element Method. 3rd edition, London, McGraw-Hill 1977.

17 Scordelis, A.C. and Lo, K.S.: Computer analysis of cylindrical shells. J. Amer. Concr. Inst. 61 (1969) 539-561.

18 Kråkeland, B. : Large Displacement Analysis of Shells Considering Elasto-Plastic and Elasto-Viscoplastic Materials. Report No. 77-6, Div. of Struct. Mech., Norwegian Inst. of Technology, Trondheim (1977).

19 Argyris, J.H., Balmer, H., Kleiber, M. and Riesemann, W.A.: Natural description of large inelastic deformations of shells - application of TRUMP element, *Computer Methods in Applied Mechanics and Engineering,* Vol. 22, 361-389 (1980).

20 Bathe, K.J. and Dvorkin, E.N.: A formulation of general shell elements - the use of mixed interpolation of tensorial components, In J. Middelton, G.N. Pande (Eds.): *Numerical Methods in Engineering: Theory and Applications,* Balkama Press, Rotterdam, 1985.

21 Horrigmoe, G. and Bergan, P.G.: Nonlinear analysis of free-form shells by flat shell elements. Comput. Meth. in Appl. Mech. and Engng., 16 (1978) 11-35.

22 Oliver, J and Onate, E.: A total Lagrangian formulation for the geometrically nonlinear analysis of structures using finite elements. Part I: Two-dimensional problems: Shell and plate structures. Int. J. of Num. Meth. in Engng., 20 (1984) 2253-2281.

23 Simons, J.F., Bergan, P.G. and Nygård, M.K.: Hyperplane displacement control methods in nonlinear analysis. In W.K. Liu, T. Belytschko and K.C. Park (Eds.), Proc. Int. Conf. on Innovative Methods for Nonlinear Problems. 345-364, Pineridge Press, 1984.

SHAKEDOWN AND LIMIT ANALYSIS OF SHELLS - A VARIATIONAL AND NUMERICAL APPROACH

P. Morelle, Boursier I.R.S.I.A.
Ingénieur de Recherches

G. Fonder, Professeur Ordinaire
Université de l'Etat, Liège

1. INTRODUCTION.

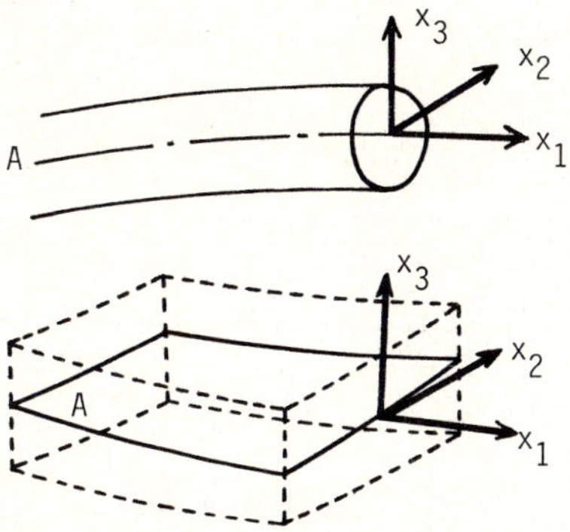

Figure 1

We shall consider in this paper a family of structures which satisfy linear kinematic hypothesis that means for example the Kirchhoff - Love hypothesis for plates and shells, the Bernoulli hypothesis for beams.

The description of the behaviour of these "thin structures" like beams, plates and shells (Fig. 1) in terms of generalized variables is quite usual in the linear elastic range.

A linear distribution of the strains ε_{ij}, along the x_3 axis (Fig. 1), involves a similar distribution for the stresses σ_{ij}. It is then well known that a complete description of the behaviour of the shell can be performed in terms of generalized strains and stresses. A very satisfactory way to reach the whole set of equations which govern the problem in the variational approach, as has been shown by Washizu [1] for example.

However, the situation is much different in the case of an elasto-plastic material : a linear distribution for strains does not automatically generate a linear variation of the stresses along the x_3 axis. So, the generalized variables are not sufficient anymore to describe the complete behaviour of the body.

The solution we have choosen in this paper is to complete the set of generalized variables with a new set of parameters called by J.A. König "pseudo-residual stresses" [2].

Let us see now how they can be introduced.

2. ELASTO-PLASTIC BEHAVIOUR OF BEAMS, PLATES AND SHELLS.

According to Kirchhoff hypothesis, the strain distribution across the thickness of a shell (or a plate) can be written as follows :

$$\varepsilon_{ij} = \lambda_{ij}\,(x_1,\,x_2) + x_3\,\kappa_{ij}\,(x_1,\,x_2) \qquad i,j = 1,2 \qquad (2.1.)$$

$$\varepsilon_{i3} = \varepsilon_{3i} = 0 \qquad (2.2.)$$

where

(x_1,x_2,x_3) coordinates are defined on Fig. 1

λ_{ij} and κ_{ij} are the generalized strains (elongation and curvatures).

We shall symbolically denote relations (2.1.) and (2.2.) by

$$\varepsilon_i = C_{ij}(x_3)\; q_j\;(x_1,x_2) \qquad (2.3.)$$

where ε_i denotes the vector of ε_{ij} components appearing in (2.1.),(2.2.) (we use in this paper Einstein sommation convention over repeated indices). Furthermore, it is generally assumed that

$$\sigma_{33} = 0 \qquad (2.4.)$$

We shall assume, as a rule, that the ε_{ij} components with indices which do not appear in (2.3.) vanish.

The generalized stresses are naturally defined by means of an internal work equation :

$$\int_I \sigma_{ij}\;\varepsilon_{ij}\;dI = g_k\;q_k = \int_I \sigma_k\;\varepsilon_k\;dI \qquad (2.5.)$$

where I is the domain [-h, h] of the x_3 coordinate, h being half the thickness of the shell. We then write :

$$g_k = \hat{C}_k\;[\sigma_{ij}] \qquad (2.6.)$$

where $\hat{C}$ is a linear integral operator defined by :

$$\hat{C}_k\;[\sigma_{ij}] = \int_I \sigma_i\;C_{ik}\;dI \qquad (2.7.)$$

In the case of Kirchhoff-Love hypothesis (2.1.), (2.2.), we obtain in vector notation :

$$\underline{q} = \begin{bmatrix} \lambda_{ij} \\ \kappa_{ij} \end{bmatrix} \quad (2.8.) \qquad\qquad \underline{g} = \begin{bmatrix} N_{ij} \\ M_{ij} \end{bmatrix} \quad (2.9.)\; i,j = 1,2$$

with

$$\left.\begin{aligned} N_{ij} &= \int_I \sigma_{ij}\;dI && (2.10.) \\ M_{ij} &= \int_I \sigma_{ij}\;x_3\;dI && (2.11.) \end{aligned}\right\}\; i,j = 1,2$$

Here, we assume that the beams, plates and shells are made of an elastic-perfectly plastic material. Its constitutive equations are :

$$\begin{cases} \dot{\varepsilon}_{ij} = \dot{\varepsilon}^{E}_{ij} + \dot{\varepsilon}^{p}_{ij} & (2.12.) \\ \varepsilon^{E}_{ij} = H_{ijkl}\,\sigma_{kl} & (2.13.) \\ \dot{\varepsilon}^{p}_{ij} = \dot{\lambda}\,\dfrac{\partial f}{\partial \sigma_{ij}} \quad \begin{cases} \dot{\lambda} = 0 \text{ if } f(\sigma_{ij}) < 0 \\ \dot{\lambda} > 0 \text{ if } f(\sigma_{ij}) = 0 \end{cases} & (2.14.) \end{cases}$$

where :

H_{ijkl} is the linear elastic Hooke tensor

$\dot{\lambda}$ is a plastic multiplier

$f(\sigma_{ij}) = 0$ is the plastic yield surface

the dot "˙" denotes a time derivative.

The equilibrium equations in terms of generalized variables are written

$$L_k\,[g_r\,(x_1,\,x_2)] + p_k = 0 \qquad (2.15.)$$

where p_k are the external load resultants and L_k is a linear operator.

On the other hand, the compatibility equations in terms of generalized strains are

$$q_k(x_1,x_2) = R_k\,[u_r\,(x_1,\,x_2)] \qquad (2.16.)$$

R_k being algebraic or differential linear operators and

u_r being the displacements of the middle mean surface of the shell (we restrict ourselves to small strains, small displacements theory).

To introduce the residual quantities, let us first define the generalized variables g_k^{*} and q_k^{*} corresponding to the same shell problem in the case of an elastic material satisfying :

$$\varepsilon^{*}_{ij} = H_{ijkl}\,\sigma^{*}_{kl} \qquad (2.17.)$$

These generalized variables for this problem obey (2.3.) → (2.7.). The residual quantities are defined as the differences between the elasto-plastic and the elastic solutions :

$$\rho_{ij} = \sigma_{ij} - \sigma^{*}_{ij} \quad (2.18.) \qquad \eta_{ij} = \varepsilon_{ij} - \varepsilon^{*}_{ij} \quad (2.19.)$$

$$v_i = u_i - u_i^{*} \qquad (2.20.)$$

and

$$g_k^{r} = g_k - g_k^{*} \quad (2.21.) \qquad q_k^{r} = q_k - q_k^{*} \quad (2.22.)$$

The residual variables g_k^{r}, q_k^{r} obviously satisfy (2.15.)(2.16.) with zero external conditions.

In the elastic shell problem, it is possible to invert the relation (2.6.) to obtain :

$$\sigma_i^* = G_{ik}(x_3) \quad g_k^*(x_1,x_2) \tag{2.23.}$$

where $G_{ik}(x_3)$ is a linear algebric operator. For example, in the case of Kirchhoff-Love plates or shells, the following relations are obtained

$$\sigma_{ij}^* = \frac{N_{ij}^*}{2h} + \frac{3}{2}\frac{x_3}{h^3} \quad M_{ij}^* \qquad i,j = 1,\ 2 \tag{2.24.}$$

Let us now focus on the elastó-plastic shell.
From (2.18.), we have :

$$\sigma_i = \sigma_i^* + \rho_i = G_{ik}\ g_k^* + \rho_i \tag{2.25.}$$

The residual stress field can also be decomposed as :

$$\rho_i(x_1,x_2,x_3) = G_{ik}(x_3)\ g_k^R(x_1,x_2) + r_i(x_1,x_2,x_3) \tag{2.26.}$$

The first term takes into account the linear part of $\rho_i(x_3)$ and is featuring the residual generalized stress, defined by :

$$g_k^r = \hat{C}_k\ [\rho_{ij}] \tag{2.27.}$$

For what concerns the second term in (2.26.), it can easily be shown that it satisfies :

$$\hat{C}_k\ [r_i] = 0 \tag{2.28.}$$

and so results in vanishing residual stresses.
This is the reason why it has been called by J.A. König [2] "pseudo-residual stress".

We can now proceed to the description of plastic behaviour of shells in terms of generalized quantities.

A very common approach is provided by *limit analysis*, which assumes *rigid-plastic* behaviour ($\dot{\varepsilon}_{ij} \simeq \dot{\varepsilon}_{ij}^p$, $\dot{\varepsilon}_{ij}^E \simeq 0$: see (2.12.)) and unconstrained plastic flow. In this case, it exists a plastic yield surface:

$$F\ (g_k) = 0 \tag{2.29.}$$

with an associated flow rule :

$$\dot{q}_r = \dot{\Lambda}\frac{\partial F}{\partial g_r} \qquad \begin{array}{ll} \dot{\Lambda} > 0 & \text{if } F(g_k) = 0 \\ \dot{\Lambda} = 0 & \text{if } F(g_k) < 0 \end{array} \tag{2.30.}$$

and it is well known that the "Prager reactions" do not appear in (2.29.).
If we return to our elastic-perfectly plastic material, the constitutive equations (2.12.) → (2.14.) are still valid so that, using (2.14.), (2.25.),(2.26.), the plastic yield surface can now be written :

$$f(\sigma_i) = f[\sigma_i^* + \rho_i] = f\ [G_{ik}\ (g_k^* + g_k^r) + r_i] = 0 \qquad (2.31.)$$

For a given $r_i(x_1,x_2,x_3)$, this equation defines, in the space of generalized stresses, a hypersurface called *elastic locus* [3]

$$F\ (g_k^* + g_k^r) = 0 \qquad (2.32.)$$

which is of direct interest in shakedown analysis of plates and shells subjected to variable loading. Indeed, (2.29.),(2.30.) are surely non valid in this case. The surface (2.31.) corresponding to vanishing r_i is called *initial elastic locus*.

3. VARIATIONAL STUDY OF HENCKY AND KIRCHHOFF PLATE.

From hereabove developments, a question is still not very clear : do or do not pseudo-residual stresses have to satisfy equilibrium equations in the sense of a three dimensional continuum ?

The only way to obtain a rigorous answer is to study the elasto-plastic problem from a variational viewpoint.

We are now going to pursue the developments in the case of the Hencky theory of plates. According to this theory, the kinematic hypothesis are :

$$u = x_3\ \kappa_1\ (x_1, x_2) \qquad (3.1.)$$

$$v = x_3\ \kappa_2\ (x_1, x_2) \qquad (3.2.)$$

$$w = w\ (x_1, x_2) \qquad (3.3.)$$

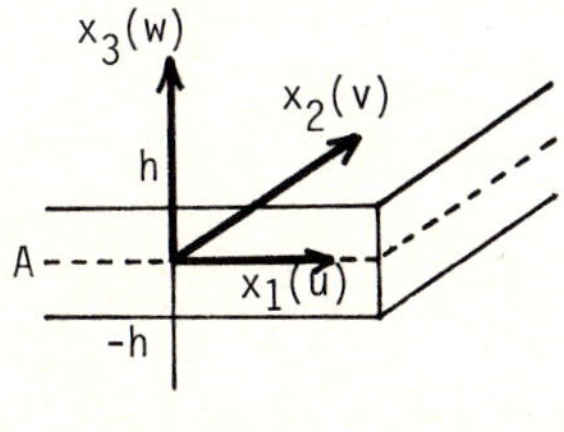

Figure 2

where u, v, w are the displacement components defined on Fig. 2. These kinematic hypotheses are completed by the static hypothesis (2.4.) hereafter recalled :

$$\sigma_{33} = 0 \qquad (3.4.)$$

This theory contains as a particular case the classical Kirchhoff-Love theory provided :

$$\kappa_1 = -\frac{\partial w}{\partial x_1} \quad (3.5.) \qquad \kappa_2 = -\frac{\partial w}{\partial x_2} \quad (3.6.)$$

The set of mixed hypothesis (2.4.), (3.1.) → (3.3.) imposes the choice of a mixed (hybrid) variational principle for elasto-plastic materials.

We are going to use the following-one [4] (analoguous to the Hellinger-Reisner principle in elasticity) :

$$\delta\dot{F} = 0 \qquad (3.7.)$$

with :

$$F(\dot{u}_i, \dot{\sigma}_{ij}, \dot{\lambda}) = \int_V [\dot{\sigma}_{ij}\, \partial_i\, \dot{u}_j - \phi(\dot{\sigma}_{ij}) - \dot{\bar{X}}_i\, \dot{u}_i - \gamma\dot{\lambda} \frac{\partial\delta}{\partial\sigma_{ij}}\, \dot{\sigma}_{ij}]\, dV - \int_{\delta\sigma} \dot{\bar{t}}_i\, \dot{u}_i\, dS_\sigma - \int_{S_u} n_i\, \dot{\sigma}_{ij}\, (\dot{u}_i - \dot{\bar{u}}_i)\, dS_u \qquad (3.8.)$$

where

$$\phi(\dot{\sigma}_{ij}) = \frac{1}{2} \dot{\sigma}_{ij}\, H_{ijkl}\, \dot{\sigma}_{kl} \qquad (3.9.)$$

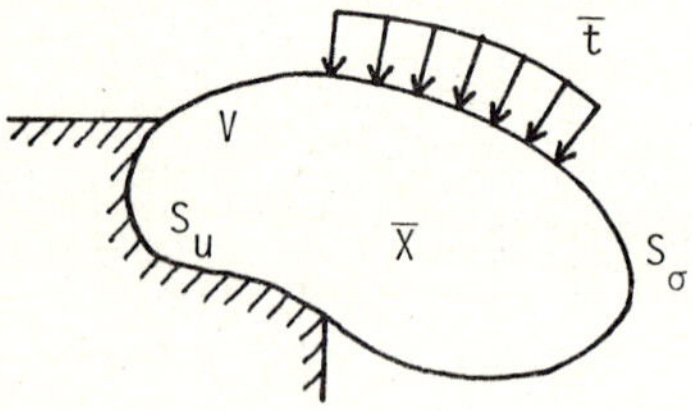

Figure 3

∂_i is the spatial derivative operator

$\dot{\bar{X}}$ are imposed body forces (see Fig 3)

$\dot{\bar{t}}_i$ are the imposed surface forces imposed on part S_σ of the overall surface S

$\dot{\bar{u}}_i$ the imposed displacements (on the remaining part $S_u = S - S_\sigma$)

When applying this principle, we have to satisfy the "a priori" condition (2.14.) written in the following form :

$$\dot{\varepsilon}_{ij} = H_{ijkl}\, \dot{\sigma}_{kl} + \gamma\dot{\lambda} \frac{\partial f}{\partial\sigma_{ij}} \quad \begin{cases} \gamma = 0 \text{ if } f(\sigma_{ij}) < 0 \\ \gamma = 1, \lambda > 0 \text{ if } f(\sigma_{ij}) = 0 \end{cases} \qquad (3.10.)$$

The Euler equations then give:

$$\delta\dot{\sigma}_{ij} : \frac{1}{2} (\partial_i\, \dot{u}_j + \partial_j\, \dot{u}_i) = H_{ijkl}\, \dot{\sigma}_{kl} + \gamma\dot{\lambda} \frac{\partial f}{\partial\sigma_{ij}} \qquad (3.11.)$$

$$\delta\dot{u}_i : \partial_i\, \dot{\sigma}_{ij} + \dot{\bar{X}}_s = 0 \qquad (3.12.)$$

$$\delta\dot{\lambda} : \dot{\sigma}_{ij} \frac{\partial f}{\partial\sigma_{ij}} = 0 \qquad (3.13.)$$

and the boundary conditions :

$$\delta\dot{u}_i : n_i\, \dot{\sigma}_{ij} = \dot{\bar{t}}_j \quad \text{on } S_\sigma \qquad (3.14.)$$

$$\delta\dot{\sigma}_{ij} : \dot{u}_i = \dot{\bar{u}}_i \quad \text{on } S_u \qquad (3.15.)$$

If we restrict ourselves to the Hencky plates, relations (2.23.) become :

$$\sigma_{ij} = \frac{3}{2} \frac{x_3}{h^3} M_{ij}\, (x_1, x_2) + r_{ij}\, (x_1, x_2, x_3) \quad i,j = 1,2 \qquad (3.16.)$$

and we also have :

$$\sigma_{i3} = \frac{Q_i(x_1, x_2)}{2h} + r_{i3}\, (x_1, x_2, x_3) \quad i = 1,2 \qquad (3.17.)$$

where the shear forces Q_i are nothing else but Prager's reactions for our problem.

What we have to do now is to substitute (3.1.),(3.2.),(3.3.) and (3.16.), (3.17.) into the expression (3.8.) of our functional F and try to eliminate the x_3 coordinate. After variation of F with respect to the independent variables. We should then be lead to the complete set of equations governing the generalized quantities as well as the pseudo-residual stresses.

Calculations are detailed in appendix 1. Here we only give the final results, namely the set of equations obtained, after imposing condition (2.28.), which governs the elasto-plastic problem for plates. We have:

a) Equilibrium equations for the generalized stresses (relation(2.15.))

$$\frac{\partial \dot{M}_{11}}{\partial x_1} + \frac{\partial \dot{M}_{12}}{\partial x_1} = \dot{Q}_1 \qquad (3.18.)$$

$$\frac{\partial \dot{M}_{12}}{\partial x_1} + \frac{\partial \dot{M}_{22}}{\partial x_2} = \dot{Q}_2 \qquad (3.19.)$$

$$\frac{\partial \dot{Q}_1}{\partial x_1} + \frac{\partial \dot{Q}_2}{\partial x_2} = p \qquad (3.20.)$$

b) Compatibility equations for the generalized strains (relation (2.16.))

$$\dot{\varepsilon}_{11} = x_3 \frac{\partial \dot{\kappa}_1}{\partial x_1} \qquad (3.21.)$$

$$\dot{\varepsilon}_{22} = x_3 \frac{\partial \dot{\kappa}_2}{\partial x_2} \qquad (3.22.)$$

$$\dot{\varepsilon}_{12} = \frac{x_3}{2} \left(\frac{\partial \dot{\kappa}_1}{\partial x_2} + \frac{\partial \dot{\kappa}_2}{\partial x_1}\right) \qquad (3.23.)$$

$$\dot{\varepsilon}_{13} = \frac{1}{2} \left(\dot{\kappa}_1 + \frac{\partial \dot{w}}{\partial x_1}\right) \qquad (3.24.)$$

$$\dot{\varepsilon}_{23} = \frac{1}{2} \left(\dot{\kappa}_2 + \frac{\partial \dot{w}}{\partial x_2}\right) \qquad (3.25.)$$

c) Constitutive equations.

$$\dot{\varepsilon}_{ij} = \dot{\varepsilon}^E_{ij} + \dot{\varepsilon}^p_{ij} \qquad (3.26.) \quad \{i,j\in[1,2]\} \cup \{i\in[1,2], j = 3\}$$

with :

$$\dot{\varepsilon}^E_{11} = \frac{1}{E} \left[\left(\frac{3}{2} \frac{x_3}{h^3} \dot{M}_{11} + \dot{r}_{11}\right) - \nu\left(\frac{3}{2} \frac{x_3}{h^3} \dot{M}_{22} + \dot{r}_{22}\right)\right] \qquad (3.27.)$$

$$\dot{\varepsilon}^E_{22} = \frac{1}{E} \left[\left(\frac{3}{2} \frac{x_3}{h^3} \dot{M}_{22} + \dot{r}_{22}\right) - \nu\left(\frac{3}{2} \frac{x_3}{h^3} \dot{M}_{11} + \dot{r}_{11}\right)\right] \qquad (3.28.)$$

$$\dot{\varepsilon}^E_{12} = \frac{1}{G} \left(\frac{3}{2} \frac{x_3}{h^3} \dot{M}_{12} + \dot{r}_{12}\right) \qquad (3.29.)$$

$$\dot{\varepsilon}^E_{13} = \frac{1}{G} \left(\frac{1}{2h} \dot{Q}_1 + \dot{r}_{13}\right) \qquad (3.30.)$$

$$\dot{\varepsilon}^E_{23} = \frac{1}{G} \left(\frac{1}{2h} \dot{Q}_2 + \dot{r}_{23}\right) \qquad (3.31.)$$

$$\dot{\varepsilon}^p_{ij} = \gamma\dot{\lambda} \frac{\partial f}{\partial \sigma_{ij}} \qquad (3.32.)$$

together with Prager "consistency" relation :

$$\frac{\partial f}{\partial \sigma_{ij}} \dot{\sigma}_{ij} = \frac{\partial f}{\partial g_{ij}} \dot{g}_{ij} + \frac{\partial f}{\partial r_{ij}} \dot{r}_{ij} = 0 \qquad (3.33.)$$

with

$$g_{ij} = M_{ij} \quad \text{if} \quad i,j \in [1,2] \qquad (3.34.)$$

$$g_{ij} = Q_i \quad \text{if} \quad i \in [1,2] \quad j = 3 \qquad (3.35.)$$

It is necessary underscore the fact that equations of sections a) and b) are "plate type" equations : they have to be satisfied for every point (x_1, x_2) of the middle surface A of the plate.
On the contrary, section c) contains "volume" equations and confirms the well known fact that the elasto-plastic problem for plates is truly three dimensional.
The boundary conditions we have obtained are the following ones :

a) either $\dot{\kappa}_n = \dot{\bar{\kappa}}_n$, or $\dot{M}_n = \dot{\bar{M}}_n$ (3.36.)

depending on the part of C (boundary of the middle surface A): belonging to C_{S_u} (imposed displacements) or C_{S_σ} (imposed forces) ;

b) either $\dot{\kappa}_{nt} = \dot{\bar{\kappa}}_{nt}$ or $\dot{M}_{nt} = \dot{\bar{M}}_{nt}$ (3.37.)

c) either $\dot{w} = \dot{\bar{w}}$ or $\dot{Q}_n = \dot{\bar{Q}}_n$ (3.38.)

In those equations, the subscripts n and t are used to represent the normal and tangential components of the quantities as defined on fig. 4.

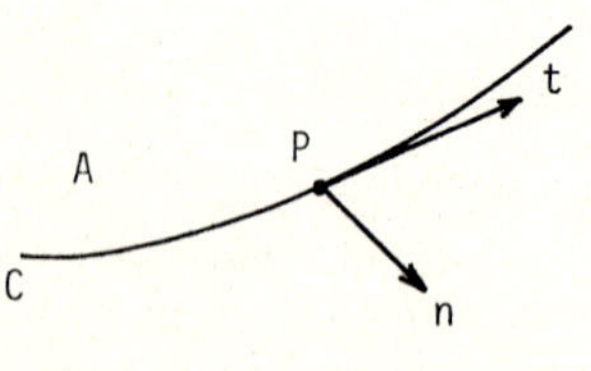

Figure 4

The conclusion of this study is that the pseudoresidual stresses r_{ij} *have not to satisfy any "volume" equilibrium equations* but only conditions (2.28.). It is clear that this result is of general application.

4. SHAKEDOWN ANALYSIS OF PLATES AND SHELLS: GENERAL CONSIDERATIONS.

From a practical point of view, the adaptation of shakedown of a structure can be defined through the unique condition :

$$\dot{\varepsilon}^{p}_{ij} \ (x_1,x_2,x_3,t) = 0 \qquad \forall \ t > t_a \qquad (4.1.)$$

This means that after initial limited plastic deformations occuring during the very first cycles of loading, the behaviour become perfectly elastic and no further plastic flow does occur.

Let us examine what it involves as far as stresses are concerned. From (2.12.) → (2.14.), we have :

$$\dot{\varepsilon}_{ij} = H_{ijkl} \ \dot{\sigma}_{kl} \qquad (4.2.)$$

and thus, $(\dot{\sigma}_{ij}, \dot{\varepsilon}_{ij})$ is the unique solution of an elastic problem the solution of which was defined by $(\dot{\sigma}^{*}_{ij}, \dot{\varepsilon}^{*}_{ij})$; from (2.18.) → (2.20.) it comes that

$$\dot{\rho}_{ij} = \dot{\eta}_{ij} = \dot{v}_i = 0 \qquad (4.3.)$$

From (2.6.) and (2.7.), we have :

$$\dot{g}_k - \dot{g}^{*}_k = \hat{C}_k \ [\sigma_{ij} - \sigma^{*}_{ij}] = 0 \qquad (4.4.)$$

and hence :

$$\dot{g}^{r}_k = 0 \qquad (4.5.)$$

Using (2.23.) (2.25.) (2.26.), we then obtain :

$$\begin{aligned} \rho_i = \dot{\sigma}_i - \dot{\sigma}^{*}_i &= G_{ik} \ \dot{g}_k + G_{ik} \ \dot{g}^{r}_k + \dot{r}_i \\ &- G_{ik} \ \dot{g}^{*}_k \\ &= \dot{r}_i \end{aligned} \qquad (4.6.)$$

and finally shakedown implies :

$$\dot{g}^{r}_k = \dot{r}_k = 0 \qquad (4.7.)$$

Taking into account the definition we have given for the elastic locus it is therefore understood that classical Melan's theorem not recalled here, see for example [5] can be restated in the following way for plates and shells (J.A. König [2]) :

> Theorem 1 : a given structure will shakedown if there exist :
> a) a time-independent generalized residual stress field $g^{r}_k(x_1)$
> b) a time independent field of elastic loci $S(x_1,x_2)$: $(g_k) \leq 0$
> such that, for every t and at every point (x_1,x_2) of the mean surface S, the sum
> $$g^{*}_k \ (x_1,x_2,t) + g^{r}_k \ (x_1,x_2)$$

is contained within the elastic locus $S(x_1, x_2)$

$$f\,(g_k^* + g_k^r) < 0 \qquad (4.8.)$$

Thus, in the generalized stress problems, shakedown analysis requires to construct, in addition to residual generalized stress fields, an appropriate family of elastic loci : that means, a good choice of pseudo-residual stresses.

What we propose in this paper is an automatic method of determination. To this purpose, we use finite element and mathematical programming methods in which not only g_k^* and g_k^r but also r_{ij} are discretized over each element. The plastic condition (4.8.) we wish to satisfy will then become condition (2.32.): this is the plastic condition in local stress space, $f(\sigma_{ij})$ with substitution of (2.23.) (2.25.). That means for our Hencky plates obeying the von Mises condition :

$$\Big[\left(\frac{3x_3}{2h^3} M_{11} + r_{11}\right)^2 + \left(\frac{3x_3}{2h^3} M_{22}+r_{22}\right)^2 + 3\left(\frac{3x_3}{2h^3} M_{12}+r_{12}\right)^2 - \left(\frac{3x_3}{2h^3} M_{11}+r_{11}\right)\left(\frac{3x_3}{2h^3} M_{22}+r_{22}\right) + \left(\frac{Q_1}{2h} + r_{13}\right)^2 + \left(\frac{Q_2}{2h} + r_{23}\right)^2\Big]^{1/2} \leqslant R_e \qquad (4.9.)$$

with

$$\left.\begin{aligned} M_{ij} &= M_{ij}^* + M_{ij}^r \qquad i,j = 1,2 \\ Q_i &= Q_i^* + Q_i^r \qquad i = 1,2 \end{aligned}\right\} \qquad (4.10.)$$

where R_e is the plastic yield stress in pure tension .

Let us define the load domain D_α by :

$$D_\alpha = \{p = \sum_{k=1}^{n} \mu_k p_k \;,\; \mu_k \in \alpha\,[\mu_k^- \;,\; \mu_k^+]\} \qquad (4.11.)$$

The α multiplier (4.11.) is called shakedown load factor. The shakedown problem consists in finding α^1, called shakedown limit factor, namely the biggest α which guarantees adaptation of the structure for any variations of the loads within D_{α^1}.

The following theorem can then easily been demonstrated [5] :

Theorem 2 : If a given structure shakes down in a cyclic loading process which contains all m vertices of the load domain D, then it shakes down in any loading path (even non cyclic) contained within D.

The vertices, $\hat{P}$ of the load domain are the $m = 2^n$ loads of $\hat{D}$ defined by:

$$\hat{D} = \{\hat{P} = \sum_{k=1}^{n} \mu_k \, p_k, \quad \mu_k = \alpha \mu_k^- \text{ or } \alpha \mu_k^+\} \tag{4.12.}$$

According to this theorem, we only have to check satisfaction of the plastic condition (4.8.) for the vertices $\hat{p}$ of the load domain.

For $\alpha > \alpha^1$, two types of collapse are possible :

a) alternating plasticity or low-cycle fatigue
b) ratcheting or incremental collapse, characterized by an unconstrained accumulation of plastic deformation during the load cycles.

It is useful to mention that limit analysis is a particular case of shakedown for which

$$\mu_k^- = \mu_k^+ \qquad \forall \; k = 1, n \tag{4.13.}$$

5. STATIC SHAKEDOWN FORMULATION FOR PLATES AND SHELLS.

We are now going to develop a generalization to plates and shells of the method proposed by J.A. König [6] for frames and arches. Basically, we are going to use statically admissible elastic finite elements to compute the g_k^*; then, after similar discretization of g_k^r, we use a special form of discretized r_{ij}, obeying (2.24.), to express the plastic condition (4.8.). Adding equilibrium relations for g_k^r, we obtain a mathematical programming model.

5.1. Computation of elastic solution.

Let us first briefly recall the statically admissible finite element formulation in terms of generalized variables. In each element, we discretize the generalized elastic stress : $\underline{g}^*$ in vector notation, with (x)

$$\underline{g}^* = S_1 \, \underline{b}_1^* + S_2 \, \underline{b}_2^* \tag{5.1.}$$

where the two terms of right member satisfy equation (2.15.) noted here:

$$L \, (S_1 \, \underline{b}_1^*) = 0 \tag{5.2.}$$

$$L \, (S_2 \, \underline{b}_2^*) = - \underline{p}$$

We have to choose a set of static connectors, called $\underline{g}_c^*$, such that they allow perfect transmission of the generalized surface tensions between the elements.

(x) Einstein's summation convention is dropped throughout in this paragraph.

We have :

$$\underline{g}_c^* = C_1 \, \underline{b}_1^* + C_2 \, \underline{b}_c^* \tag{5.4.}$$

The kinematic connectors $\underline{q}_c^x$ (degrees of freedom) are then defined by :

$$V_{ext} = - \int_{C_u} \underline{t}^T \, \underline{u}^- \, ds = - \underline{q}_c^{*T} \, \underline{g}_c^* \tag{5.5.}$$

Using the elastic complementary energy principle, we obtain :

$$\underline{g}_c^* + \tilde{\underline{g}}_c^* = K \, \underline{q}_c \tag{5.6.}$$

with :

$$K = C_1 \, F_{11}^{-1} \, C_1^T = \text{stiffness matrix} \tag{5.7.}$$

$$\tilde{\underline{g}}_c^* = (C_1 \, F_{11}^{-1} \, F_{12} - C_2) \, \underline{b}_2^* = \text{implicit loads} \tag{5.8.}$$

The flexibility matrix is defined by :

$$F_{11} = \int_A S_1^T \, H \, S_1 \, dA \quad (5.9.) \qquad F_{12} = \int_A S_1^T \, H \, S_2 \, dA \qquad (5.10.)$$

H being the Hooke's tensor such that

$$\underline{q}^* = H \, \underline{g}^* \tag{5.11.}$$

A classical assembly is then performed :

$$(\underline{q}_c)_i = L_i \, \underline{q}_c^s \tag{5.12.}$$

$$(\underline{b}_1)_i = M_i \, (\underline{b}_1)^s \tag{5.13.}$$

where the subscript i indicates the ith element, the subscript the s structural vectors, L_i and M_i are Boolean matrix. We obtain :

$$K^s = \Sigma \, L_i^T \, K_i L_i \tag{5.14.}$$

$$C_1^s = \Sigma \, L_i^T \, (C_1)_i \, M_i \tag{5.15.}$$

5.2. Discretization of the residual generalized stress field.

This field must be self-equilibrated. The simplest choice of discretization is the one we have made in the preceeding section. So, we shall write :

$$\underline{g}^r = S \, \underline{b}^r \tag{5.16.}$$

which satisfies (5.2.). The global self-equilibrium relation will be, using (5.4.)(5.13.)(5.15.) :

$$C_1^s \, (\underline{b}^r)^s = \underline{0} \tag{5.17.}$$

5.3. Discretization of pseudo-residual stress field.

We propose here to test a multi-layer discretization of the pseudo-residual stress field : the structure is divided into n layers of equal thickness. For each-layer, a constant or linear function is used ; namely, we shall choose :

a) either :

$$r_{ij} = C^k_{ij} \;,\; x_3 \in [h-(k-1)\frac{2h}{n} \;;\; h - \frac{2hk}{n}] \qquad (5.18.)$$

$$k = 1,\ldots, n$$

b) or

$$r_{ij} = C^{2k-1}_{ij} + C^{2k}_{ij}\, x_3 \;,\; x_3 \in [h-(k-1)\frac{2h}{n} \;;\; h - \frac{2hk}{n}] \qquad (5.19.)$$

$$k = 1,\ldots,n$$

where the C^k_{ij} coefficients will be functions of x_1, x_2 such that the same dependence is obtained for a r_i component as for the corresponding g^r_i. We shall note :

$$\underline{r} = R_k(x_3)\;\underline{C}^k(x_1,x_2) \qquad k = 1,\, n \qquad (5.20.)$$

R_k being the interpolation matrix valid for :

$$x_3 \in I_k = [h - (k-1)\frac{2h}{n} \;;\; h - 2\frac{hk}{n}] \qquad (5.21.)$$

In addition, conditions (2.28.) have to be imposed, either directly (small number of layers) which allows to eliminate a certain number of C_{ij} parameters, or explicitly in the final mathematical programming form.

5.4. Shakedown problem.

From theorems 1 and 2 and the preceding paragraphs, we can easily see that the shakedown problem can be formulated in the following form :

$$\begin{cases} \max \alpha \\ f(\alpha G\underline{g}^* + G\underline{g}^r + \underline{r}) = f(\alpha G\underline{g}^* + GS\underline{b}^r + R_k\,\underline{C}^k) \leqslant 0 & (5.22.) \\ C^s_1\,(\underline{b}^r)^s = \underline{0} & (5.23.) \\ + \text{ conditions } (2.28.) \end{cases}$$

The inequality (5.22.) will be checked :

a) x_3 axis : at each extreme value of the internal I_k, that means at 2 n points.

b) x_1, x_2 axis : at a finite number, p_z, of points of the mean surface A of the element , depending on the degree of discretization. If p_f is sufficient to ensure that (5.22.) is nowhere violated, then a lower bound can be expected.

6. APPLICATIONS.

Shakedown analysis results for plate or shell problems remain very scarce in the litterature. Only König's statical results for plates [7] allow us to test the present formulation for bidimensional problems. Besides, limit analysis and shakedown of beams and frames furnish a good field of experiment. Solution of the mathematical programme problem is performed using ACDPAC [8].

6.1. Plane frames with rectangular cross-section.

This case corresponds to hypothesis (2.1.)(2.2.) with λ_{ij}, κ_{ij} depending only of x_1 (fig. 5).

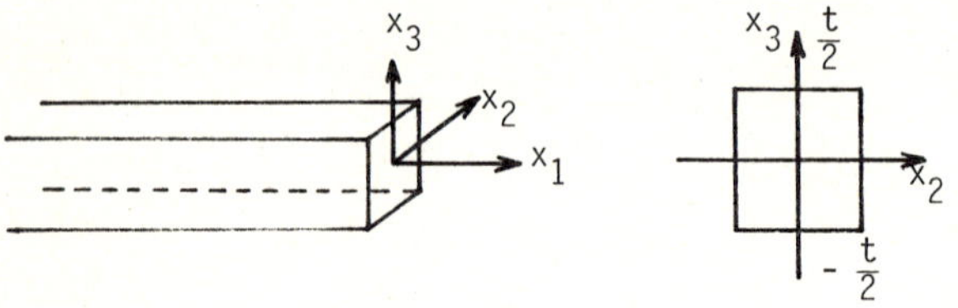

Figure 5

Let us assume that the shear force can be neglected so that, σ_{11} is the sole non-vanishing one stress component. We obtain :

$$\sigma_{11} = \alpha\left[\frac{M^*(x_1)}{I} x_3 + \frac{N^*(x_1)}{A}\right] + \frac{M^r(x_1)}{I} x_3 + \frac{N^r(x_1)}{A} + r_{11}(x_1,x_3) \qquad (6.1.)$$

We shall consider here two discretizations for r_{11} :

a) Constant multi-layer discretization : We choose n = 3 or 4. Using (2.24.), we obtain the following distributions (over the beam's thickness t (fig. 5))

three layers : (Fig. 6)

$$\begin{cases} r_{11} = C_1(x_1) & , x_3 \in [\frac{t}{2} ; \frac{t}{6}] \\ r_{11} = -2\, C_1(x_1) & , x_3 \in [\frac{t}{6} ; -\frac{t}{6}] \\ r_{11} = C_1(x_1) & , x_3 \in [-\frac{t}{6} ; -\frac{t}{2}] \end{cases} \qquad (6.2.)$$

four layers : (Fig. 7)

$$\begin{cases} r_{11} = C_1(x_1) & , x_3 \in [\frac{t}{2} ; \frac{t}{4}] \\ r_{11} = C_2(x_1) & , x_3 \in [\frac{t}{4} ; 0] \\ r_{11} = -3C_1(x_1) - 2C_2(x_1) & , x_3 \in [0 ; -\frac{t}{4}] \\ r_{11} = 2C_1(x_1) + C_2(x_1) & , x_3 \in [-\frac{t}{4} ; -\frac{t}{2}] \end{cases} \qquad (6.3.)$$

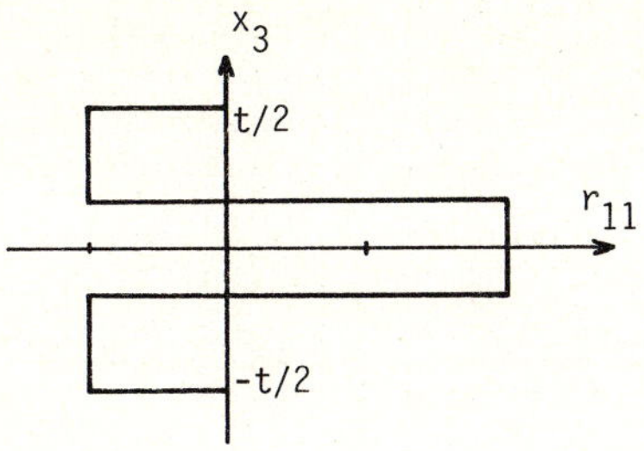

Figure 6

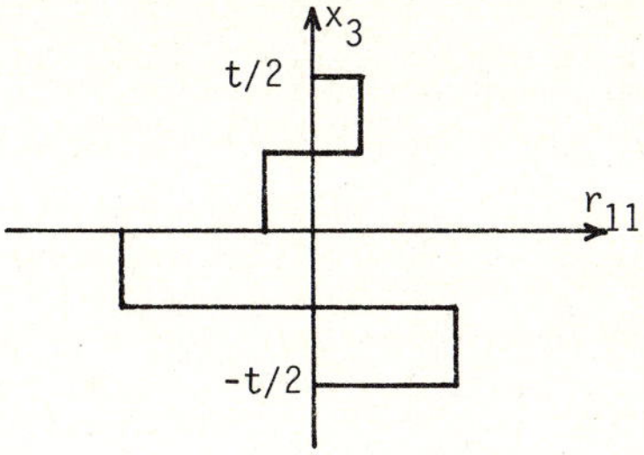

Figure 7

b) Linear multi-layer discretization : We choose here n = 2 or 4. In the first case (n = 2), using (2.24.), we obtain the following distribution (Fig. 8)

$$\begin{cases} r_{11} = C_1(x_1) + C_2(x_1)\, x_3 & , x_3 \in [\frac{t}{2} ; 0] \quad (6.4.) \\ r_{11} = [-7C_1(x_1) - 2C_2(x_1)h] + \frac{x_3}{h}[-24C_1(x_1) - 7C_2(x_1)h] & , x_3 \in [-\frac{t}{2} ; 0] \end{cases}$$

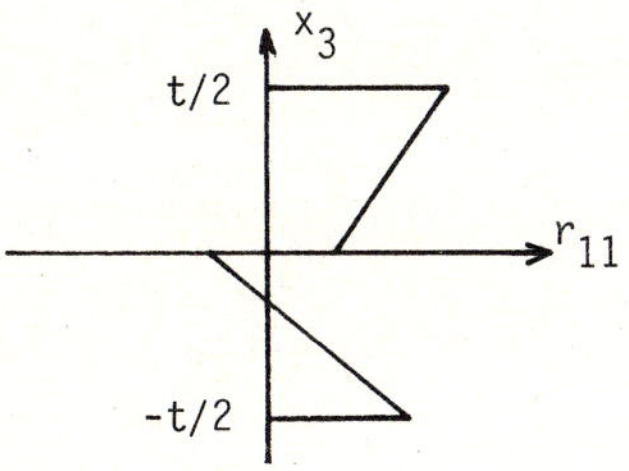

Figure 8

In the second case (n = 4), equation (2.24.) become :

$$8C_1 + 3C_2 t + 8C_3 + C_4 t + 8C_5 - C_6 t + 8C_7 - 3C_8 t = 0 \quad (6.5.)$$

$$18C_1 + 7C_2 t + 6C_3 + C_4 t - 6C_5 + C_6 t - 17C_7 + 7C_8 t = 0 \quad (6.6.)$$

They have to be added to the set of inequalities (5.22.) with r_{11} given by (5.19.) for n = 4.

6.1.1. Beam subjected to M and N.

As a first example, let us examine the case of an infinite beam subjected to a bending moment M and an axial force N.

We define dimensionless variables :

$$m = \frac{M}{M_p} \quad (6.7.) \qquad \text{and} \qquad n = \frac{N}{N_p} \quad (6.8.)$$

with :

$$M_p = 3\, R_e \frac{I}{t} \quad (6.9.) \qquad N_p = R_e\, h\, t \quad (6.10.)$$

where - R_e is the yield stress in pure traction
- h is the beam's thickness
- t is the width of the cross section

Fig. 9 shows the limit analysis interaction curve obtained with a r_{ij} discretization using 2 or 4 linear layers.

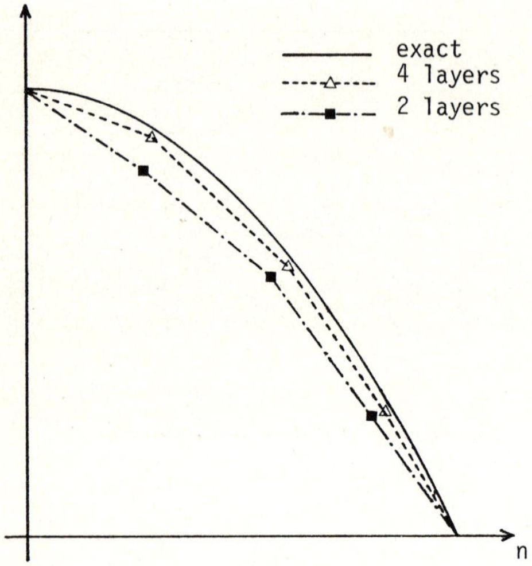

For the case N = O, table 1 gives the results obtained with different r_{ij} discretization while Fig. 10 shows the σ_{11} distributions for the different cases.

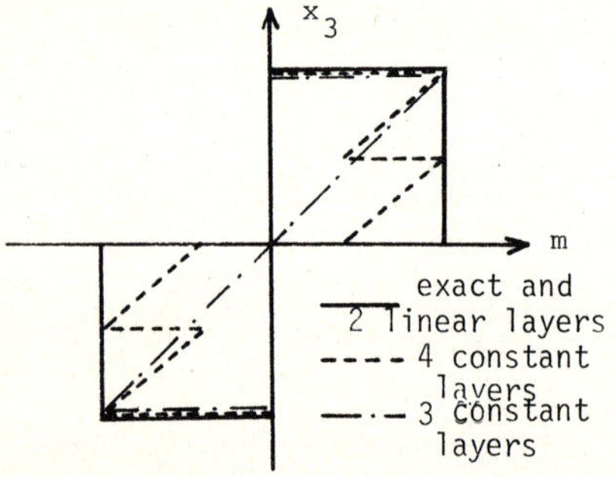

Figure 10

r_{ij}	$(\alpha M)^{calc}/M_p$
constant n = 3	$\frac{2}{3}$
constant n = 4	$\frac{16}{21}$
linear n = 2	1

Table 1

In case of a shakedown problem, Fig. 11 gives the shakedown load factor α corresponding to the load domain

$$D = \{m \in \alpha\, [-k\,\bar{m},\ \bar{m}]\ ,\ n \in \alpha\,[-k\,\bar{n}\ ,\ \bar{n}]\} \qquad (6.11.)$$

with $\bar{m} = \bar{n} = 0{,}5$ and $k \in [0,1]$, when using the linear two layers model.

The results are exact and this is also true in the case of the initial elastic locus corresponding to $r_{11} = 0$, as it is easily understood.

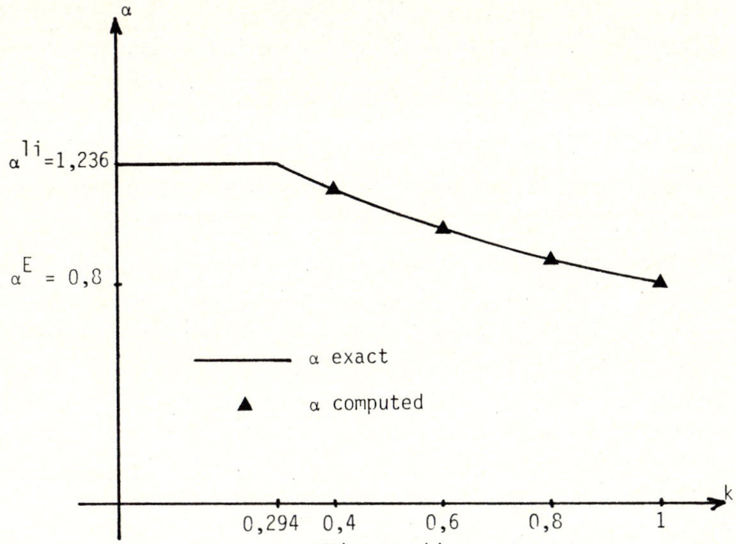

Figure 11

6.1.2. Portal frame.

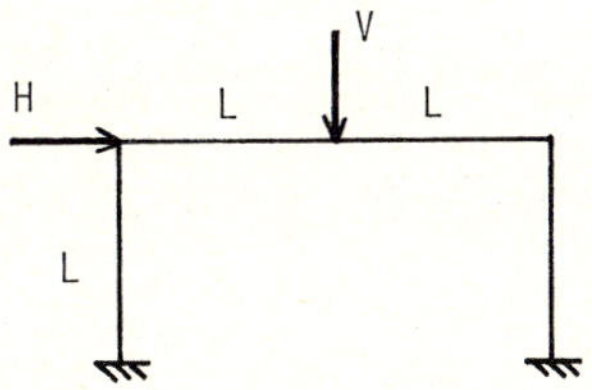

Figure 12

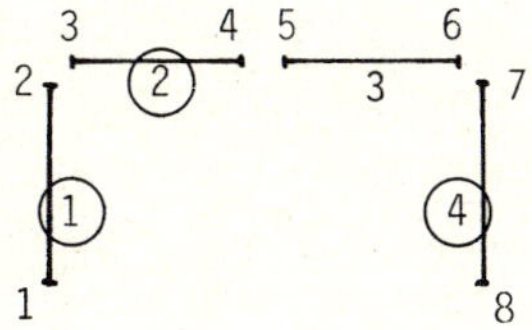

Figure 13

We consider the portal frame of Fig.12, subjected under two types of loading :

a) limit analysis :

$$D=\{H\in\alpha[F,F],\ V\in\alpha\ [F,F]\} \tag{6.12.}$$

b) shakedown analysis :

$$D=\{H\in\alpha[O,F],\ V\in\alpha\ [O,F]\} \tag{6.13.}$$

The discretization is illustrated on Fig. 13 where nodal numerotation is also shown.

The plasticity condition is

$$\left.\begin{array}{l}\sigma_{11} \leq R_e \\ -\sigma_{11} \leq R_e\end{array}\right\} \tag{6.14.}$$

with σ_{11} given by (6.1.). We choose:

- a linear discretization for $M^r(x_1)$, constant for
- a linear two-layers discretization for r_{11}.

The problem is solved for two different values of the geometrical parameter $\bar{t}$ defined by

$$\bar{t} = \frac{t}{4L} \tag{6.15.}$$

that is $\bar{t} = 0,1$ and $\bar{t} = 0,01$. We can compare with the classical solution (neglecting the N force) referred to as the "bending" solution.

a) Limit analysis : Using the dimensionless quantity :

$$\alpha\bar{F} = \frac{\alpha F L}{M_p} \qquad (6.16.)$$

the different values of the limit loads are given at table 2 ; we refer to [9] for the classic "bending" solution.

Approach	$\alpha^1\bar{F}$
Bending	3
Present $\bar{t}$=0,01	2,9631
Present $\bar{t}$=0,1	2,6692

Table 2

The residual bending moments are compared for the three approaches at table 3. The sections are those numbered on Fig. 13.

	Approach $m_r = M^r/M_p$		
Section	Present $\bar{t}$ = 0,1	Present $\bar{t}$ = 0,01	Bending
1	- 0,3722	- 0,3637	- 0,3625
2	0,0078	0,0340	0,0375
3	0,0078	0,0340	0,0375
4	0,0822	0,0979	0,1
5	0,0822	0,0979	0,1
6	0,1566	0,1618	0,1625
7	0,1566	0,1618	0,1625
8	- 0,2234	- 0,2359	- 0,2375

Table 3

b) Shakedown problem : We consider a load domain

$$D = \{H \in \alpha\,[0,\bar{F}]\ ,\ \ V \in \alpha\,[0,\bar{F}]\ \} \qquad (6.17.)$$

and refer to [9] for the classical "bending" solution. The different values for $\alpha^1\bar{F}$ and the residual moments are given in tables 4 and 5.

Approach	$\alpha^1\bar{F}$
Bending	2,857
Present t = 0,01	2,682
Present t = 0,1	2,542

Table 4

	Approach $m^r = M^r/M_p$		
Section	Present $\bar{t}$ = 0,1	Present $\bar{t}$=0,01	Bending
1	- 0,1741	- 0,1207	- 0,108
2	0,0939	0,1598	0,178
3	0,0939	0,1598	0,178
4	0,0984	0,1333	0,143
5	0,0984	0,1333	0,143
6	0,1028	0,1067	0,108
7	0,1028	0,1067	0,108
8	- 0,1652	- 0,1738	- 0,178

Table 5

6.2. Axisymetric plates and shells.

The junior author has already published several papers describing upper and lower bounds for shakedown problems of "sandwich" axisymetric plates and shells [10][11] in which cases statically admissible elastic finite elements have been developed. We shall use here the same elastic elements and complete them with a r_{ij} discretization to be discuss hereafter.

6.2.1. Statically admissible elastic finite elements.

The three elements are illustrated on Fig. 14 with the chosen sets of static connectors. They are :

- a tronconical element (Fig. 14.b)
- a conical hat (Fig. 14.a)
- a cylindrical element (Fig. 14.c).

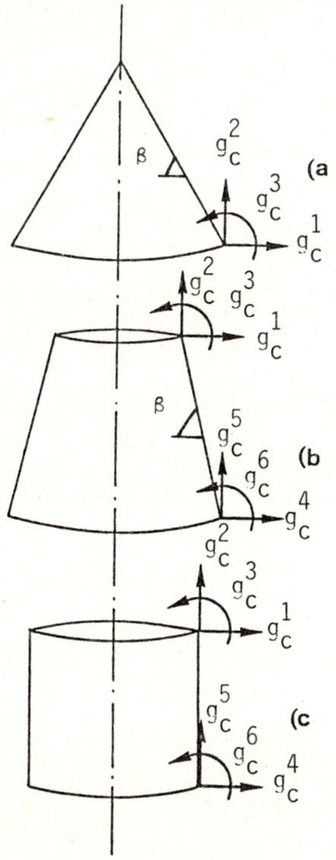

Figure 14

The equilibrium equations are respectively written :

$$\frac{d}{ds}(sn_x) = n_\theta \qquad (6.18.)$$

$$\frac{d}{ds}(sq_x) + n_\theta \ \mathrm{tg}\ \beta + 2\frac{\rho s}{\cos\beta} = 0 \qquad (6.19.)$$

$$\frac{d}{ds}(sm_x) = m_\theta + \frac{sq_x}{h\cos\beta} \qquad (6.20.)$$

for a conical shell and :

$$\frac{d}{ds}(n_x) = 0 \qquad (6.21.)$$

$$\frac{d}{ds}(q_x) + \frac{n_\theta}{\xi} + 2\rho = 0 \qquad (6.22.)$$

$$\frac{d}{ds}(m_x) = \frac{q_x}{h} \qquad (6.23.)$$

for a cylindrical shell; in terms of the dimensionless variables :

$$n_\theta = \frac{N_\theta}{N_p} \ (6.24.) \quad n_x = \frac{N_x}{N_p} (6.25.) \quad q_x = \frac{Q_x}{N_p} \ (6.26.)$$

$$m_\theta = \frac{M_\theta}{M_p} \ (6.27.) \quad m_x = \frac{M_x}{N_p} (6.28.) \quad h = \frac{M_p}{N_p L} \ (6.29.)$$

$$M_p = R_e \frac{t^2}{4} (6.30.) \quad N_p = R_e t (6.31.) \quad \rho = \frac{PL}{2N_p} \ (6.32.)$$

where :
- $t = 2h$ is the shell thickness
- R_e is the tension yield stress
- N_θ, N_x, Q_x, M_θ, M_x are defined on Fig. 15
- L is a reference length.

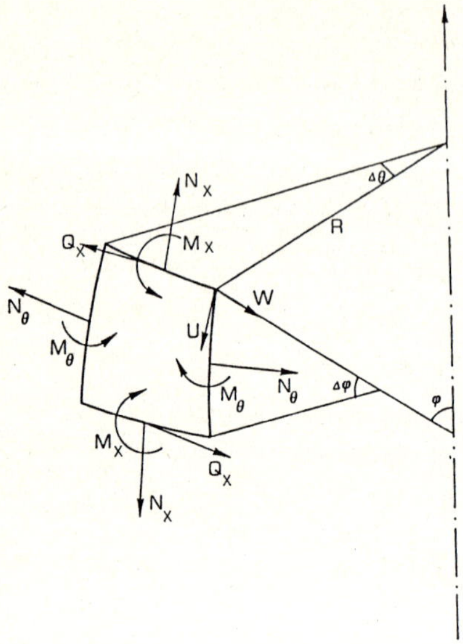

Figure 15

Furthermore, we have defined for the conical shell :

$$s = \frac{r}{L} \qquad (6.33.)$$

β = slope of the shell (see fig. 14.a. and b.)

and for the cylindrical one :

$$s = \frac{z}{L} \quad (6.34.) \qquad \xi = \frac{R}{L} \qquad (6.35.)$$

where R is cylinder's radius.

Let us briefly recall the choice of the discretization for the $\underline{g}^{*}$

- For the conical shell :

$$S_1 = \begin{bmatrix} 1 & 0 & 0 & 0 & 0 \\ 0 & 1 & 0 & 0 & 0 \\ 1 & 0 & 1/s & 0 & 0 \\ -\dfrac{s \sin\beta}{2h\cos^2\beta} & 1 & 0 & -\dfrac{1}{h\cos\beta} & 1/s \end{bmatrix} \qquad (6.36.)$$

$$\underline{g}^{*T} = [n_\theta \,,\, m_\theta \,,\, n_x ,\, m_x] \qquad (6.37.)$$

- For the conical hat :

$$S_1 = \begin{bmatrix} 1 & 0 \\ 0 & 1 \\ 1 & 0 \\ -\dfrac{s \sin\beta}{2h\cos^2\beta} & 1 \end{bmatrix} \qquad (6.38.) \qquad \underline{g}^{*T} = [n_\theta \,,\, m_\theta \,,\, n_x ,\, m_x] \qquad (6.39.)$$

- For the cylindrical shell :

$$S_1 = \begin{bmatrix} \xi & \xi s & 0 & 0 & 0 \\ 0 & 0 & 0 & 0 & 1 \\ -\dfrac{s^2}{2h} & -\dfrac{s^3}{6h} & \dfrac{s}{h} & s & 0 \end{bmatrix} \qquad (6.40.$$

$$\underline{g}^{*T} = [n_\theta \,,\, n_x ,\, m_x] \qquad (6.41.)$$

The reader is invited to check that those discretizations are statically admissible. The S_2 matrices correspond to a linear variation of the external pressure ρ .

6.2.2. Pseudo residual stresses - Discretization.

As already mentioned, we shall discretize the generalized residual stresses with the same S_1 matrices as the elastic ones. For what concerns r_{ij} fields, the only non-vanishing components are such that (Fig. 16).

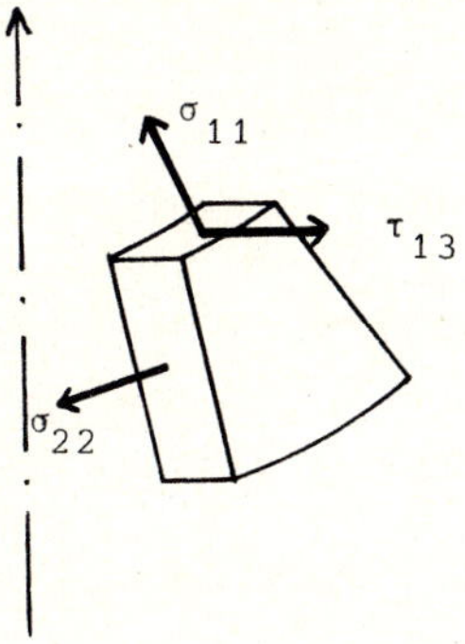

Figure 16

$$\sigma_{11} = \left(\frac{3}{2}\,\frac{M_x\,x_3}{h^3} + \frac{N_x}{2h}\right) + r_{11} \qquad (6.42.)$$

$$\sigma_{22} = \left(\frac{3}{2}\,\frac{M_\theta\,x_3}{h^3} + \frac{N_\theta}{2h}\right) + r_{22} \qquad (6.43.)$$

$$\tau_{13} = \frac{Q_x}{2h} + r_{13} \qquad (6.44.)$$

with :

$$M_i = \alpha M_i^{*} + M_i^{r}, \qquad i = x,\theta \qquad (6.45.)$$

$$Q_x = \alpha Q_x^{*} + Q_x^{r} \qquad (6.46.)$$

We shall choose two $r_{ij}(s)$ discretization :

a) Constant field : With the notations defined in § 5.3. in (5.20.) it leads to :

$$\underline{c}^k(x_1, x_2) = \underline{c}^k(s) = \underline{c}_o^k \qquad (6.47.)$$

b) "Natural" field : By this we mean that we choose for the $r_{ij}(s)$ variation the same function as for the $\underline{g}^r(s)$; it comes then from (6.36.) (6.38.) (6.40.) and the equilibrium equations (6.18.) to (6.23.) that we reach the following discretization :

$$r_{11}(s) = r_{11}^{o} \qquad (6.48.)$$

$$r_{22}(s) = \frac{r_{22}^{-1}}{s} + r_{22}^{o} + r_{22}^{1}\,s \qquad (6.49.)$$

$$r_{13}(s) = \frac{r_{13}^{-1}}{s} + r_{13}^{o} \qquad (6.50.)$$

with r_{ij}^k given by (5.18.) or (5.19.).

6.2.3. Mathematical programming problem.

We have now to decide on what kind of yield surface we are going to whoose. Trying to obtain a linear mathematical programming problem, it is clear from (5.22.) (5.23.) that we have to prefer a linear (or linearized) yield surface. We shall use a linearization, analogous to one we have already adopted in a previous paper [12] ; the yield surface is then :

$$A\,\underline{\sigma} \leqslant \underline{h} \qquad (6.51.)$$

with $\underline{\sigma}$ given by (2.25.) (2.26.) which means in vector notation

$$\underline{\sigma} = \alpha G\underline{g}^{*} + G\underline{g}^{r} + \underline{r} \qquad (6.52.)$$

and

$$\underline{h}^T = [R_e, R_e, R_e, R_e, R_e, R_e, R_e, R_e, R_e, R_e, R_e, R_e, R_e, R_e]$$

$$A^T = \begin{bmatrix} 0 & 1 & 0 & -1 & -1 & 0 & 1 & 0 & -1 & 0 & 1 & 1 & 0 & -1 \\ 0 & -1 & -1 & 0 & 1 & 1 & 0 & 0 & 1 & 1 & 0 & -1 & -1 & 0 \\ 2 & 1 & 1 & 1 & 1 & 1 & 1 & -2 & -1 & -1 & -1 & -1 & -1 & -1 \end{bmatrix}$$

The yield condition should be checked :

i) along the s-coordinate: at the two extreme points of the elements, s_1 and s_2 (Fig.17)

ii) along the x_3-coordinate : at each extreme point of the intervals I_k of (5.21.).

Taking into account the monotone variation of $r_{ij}(s)$ (see (6.47.) to (6.50.)), such a choice guarantees a true lower-bound for the shake-down load factor.

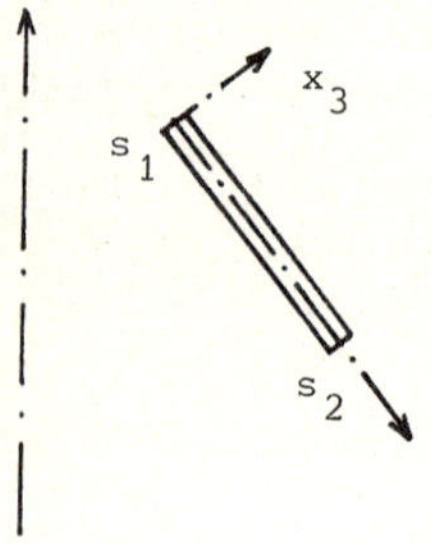

Figure 17

6.2.4. Applications.

An intensive program of applications is actually under development. The authors hope to submit numerous examples particularly in the field of axisymetric thin shells in the near future.

7. CONCLUSIONS.

The authors have presented a general formulation allowing to solve shakedown or limit analysis problems in terms of generalized variables.

A correct determination of the collapse state can be expected, including the very "sensitive" cases of partial collapse over the cross-section (low-cycle fatigue at points located within the shells,...), cases which were not properly detected by the classical "sandwich" formulation. The simple example of the portal frame has demonstrated the practical usefulness of those concepts.

REFERENCES.

[1] K. WASHIZU : Variational Methods in elasticity and plasticity. Pergamon Press.

[2] J.A. KÖNIG : Theory of shakedown of elastic-plastic structures. Archiwum Mechaniki stosowanej 2.18. (1966), pp. 227 - 237.

[3] J.A. KÖNIG : Fundamental theorems of shakedown (in Polish). Mech. Teor.Stos., 8, (1970), 149 - 158.

[4] NGUYEN DANG HUNG : Sur les principes variationnels en plasticité - Collection des publications de la Faculté des Sciences Appliquées, Université de Liège, N° 84 (1980), pp. 61 - 76.

[5] A. BORKOWSKI and M. KLEIBER : On a numerical approach to shakedown analysis of structures. Comp. Math. in Appl. Mech. and Eng., Vol. 22, N° 1, April 1980, pp. 101 - 120.

[6] J.A. KÖNIG : A method of shakedown analysis of frames and arches. Int. J. Solids Structures, 1971, Vol. 7, pp. 327 - 344.

[7] J.A. KÖNIG : Shakedown theory of plates. Archiwum Mechaniki Stosowanej, 5, 21 (1969), pp. 623 - 636.

[8] M.J. BEST and A.T. BOWLER : ACDPAC: a FORTRAN IV routine to solve differentiable mathematical programmes. Department of Combinatories and optimization (1976).

[9] J.B. MARTIN : Plasticity: Fundamentals and general results. MIT Press (1975), pp. 674 - 684.

[10] P. MORELLE and NGUYEN DANG HUNG : Numerical plastic shakedown analysis of plates and shells of revolution by equilibrium finite elements (in French), Journal de Mécanique Théorique et Appliquée, 2, (4) (1983).

[11] NGUYEN DAN HUNG and P. MORELLE : Numerical shakedown analysis of plates and shells of revolution by equilibrium finite elements, Proceedings of New and future developments in commercial finite element methods. Congress, Los Angeles, U.S.A., 1982.

[12] P. MORELLE : Problème de linéarisation optimale du critère de von MISES - Rapport interne N° 145, MSM, Université de Liège (1984).

APPENDIX. VARIATIONAL STUDY OF ELASTO-PLASTIC HENCKY PLATES.

We start from (3.7.)(3.8.)(3.9.) and substitute (3.16.) (3.17.) and (3.1.)to(3.3.). The first term in (3.8.) becomes :

$$\int_V \dot{\sigma}_{ij}\, \partial_i \dot{u}_j \, dV = \dot{H}_1 + \dot{H}_2 \tag{A.1}$$

with

$$\dot{H}_1 = \iint_A [\dot{M}_{11} \frac{\partial \dot{K}_1}{\partial x_1} + \dot{M}_{22} \frac{\partial \dot{K}_2}{\partial x_2} + \dot{M}_{12} (\frac{\partial \dot{K}_1}{\partial x_2} + \frac{\partial \dot{K}_2}{\partial x_1}) \tag{A.2}$$

$$+ \frac{\partial \dot{K}_1}{\partial x_1} \int_{-h}^{h} \dot{r}_{11} x_3 dx_3 + \frac{\partial \dot{K}_2}{\partial x_2} \int_{-h}^{h} \dot{r}_{22} x_3 dx_3 + (\frac{\partial \dot{K}_1}{\partial x_2} + \frac{\partial \dot{K}_2}{\partial x_1}) \int_{-h}^{h} \dot{r}_{12} x_3 dx_3 \,] dA$$

$$\dot{H}_2 = \iint_A [\dot{Q}_1 (\frac{\partial \dot{w}}{\partial x_1} + \dot{K}_1) + \dot{Q}_2 (\frac{\partial \dot{w}}{\partial x_2} + \dot{K}_2) + (\frac{\partial \dot{w}}{\partial x_1} + \dot{K}_1) \int_{-h}^{h} \dot{r}_{13} \, dx_3$$

$$+ (\frac{\partial \dot{w}}{\partial x_2} + \dot{K}_2) \int_{-h}^{h} \dot{r}_{23} \, dx_3 \,] dA \tag{A.3}$$

The second term in (3.8.) is now :

$$\int_V \phi\, (\dot{\sigma}_{ij}) \, dV = \dot{U}_1 + \dot{U}_2 + \dot{U}_3 \tag{A.4}$$

with

$$\dot{U}_1 = \frac{1}{2} \iint_A \{ \frac{3}{2Eh^3} [\dot{M}_{11}^2 + \dot{M}_{22}^2 - 2\nu \dot{M}_{11} \dot{M}_{22} + 2(1+\nu) \dot{M}_{12}^2]$$

$$+ \frac{1}{2Gh} (\dot{Q}_1^2 + \dot{Q}_2^2) \} \, dA \tag{A.5}$$

$$\dot{U}_2 = \frac{1}{2} \iint_A \{ \frac{3}{2Eh^3} [\dot{M}_{11} \int_{-h}^{h} \dot{r}_{11} x_3 dx_3 + \dot{M}_{22} \int_{-h}^{h} \dot{r}_{22} x_3 dx_3 - \nu \dot{M}_{22} \int_{-h}^{h} \dot{r}_{11} x_3 dx_3$$

$$-\nu \dot{M}_{11} \int_{-h}^{h} \dot{r}_{22} x_3 dx_3 + 2(1+\nu) \dot{M}_{12} \int_{-h}^{h} \dot{r}_{12} x_3 dx_3] + \frac{1}{Gh} (\dot{Q}_1 \int_{-h}^{h} \dot{r}_{13} dx_3 + \dot{Q}_2 \int_{-h}^{h} \dot{r}_{23} dx_3) \} dA \tag{A.6}$$

$$\dot{U}_3 = \frac{1}{2} \iiint_V \{ \frac{1}{E} [\dot{r}_{11}^2 + \dot{r}_{22}^2 - 2\nu \dot{r}_{11} \dot{r}_{22} + 2(1+\nu) \dot{r}_{12}^2] + \frac{1}{G} (\dot{r}_{13}^2 + \dot{r}_{23}^2) \} \, dV \tag{A.7}$$

Besides, the fourth term remains volumic :

$$\iiint_V \gamma \dot{\lambda} \frac{\partial f}{\partial \sigma_{ij}} d\sigma_{ij} \, dV = \iiint_V [\gamma \dot{\lambda} (\frac{\partial f}{\partial m_{ij}} d\dot{m}_{ij} + \frac{\partial f}{\partial r_{ij}} d\dot{r}_{ij})] \, dV = \dot{H}_3 \tag{A.8}$$

For what concerns the limit conditions, we obtain :

$$\dot{P} = \iiint_V \bar{\dot{X}}_i \dot{u}_i \, dV + \iint_{S_\sigma} \bar{\dot{t}}_i \dot{u}_i \, dS_\sigma + \iint_{S_u} n_i \dot{\sigma}_{ij} (\dot{u}_i - \bar{\dot{u}}_i) \, dS_u \tag{A.9}$$

$$= \dot{P}_1 + \dot{P}_2 + \dot{P}_3$$

with :

$$\dot{P}_1 = \iint_A p\,\dot{w}(x_1,x_2)\,dx_1dx_2 \tag{A.9}$$

$$\dot{P}_2 = \oint_{C_{S_\sigma}} [\dot{K}_n(\dot{\bar{M}}_n + \int_{-h}^{h} \dot{\bar{r}}_n x_3 dx_3) + \dot{K}_t(\dot{\bar{M}}_{nt} + \int_{-h}^{h} \dot{\bar{r}}_{nt}\, x_3 dx_3) + \dot{w}\,(\dot{\bar{Q}}_n + \int_{-h}^{h} \dot{\bar{r}}_{n3}\, dx_3]\, ds_\sigma] \tag{A.10}$$

$$\dot{P}_3 = \oint_{C_{S_u}} [(\dot{M}_n + \int_{-h}^{h} \dot{r}_n x_3 dx_3)(\dot{\alpha}_n - \dot{\bar{\alpha}}_n) + (\dot{M}_{nt} + \int_{-h}^{h} \dot{r}_{nt}\, x_3 dx_3)\,(\dot{\alpha}_t - \dot{\bar{\alpha}}_t)$$
$$+ (\dot{Q}_n + \int_{-h}^{h} \dot{r}_{n3}\, dx_3)\,(\dot{w} - \dot{\bar{w}})]\; ds_u \tag{A.11}$$

and to which we still have to add the following terms coming from $\dot{H}_1$ and $\dot{H}_2$:

$$\dot{P}_H = \oint_C [\dot{K}_n(\dot{M}_n + \int_{-h}^{h} \dot{r}_n x_3 dx_3) + \dot{K}_t\,(\dot{M}_{nt} + \int_{-h}^{h} \dot{r}_{nt} x_3 dx_3) + \dot{W}(\dot{Q}_n + \int_{-h}^{h} \dot{r}_{n3}\, dx_3)]\; ds$$

After varying with respect to $\dot{K}_i$, $\dot{W}_i$, $\dot{M}_{ij}$, $\dot{Q}_i$, $\dot{r}_{ij}$ and taking into account relations (2.28.), we obtain the set of equations (3.18.) to (3.38.).

DISPLACEMENT FIELDS AND LARGE DEFORMATION ANALYSIS OF 3-D BEAMS

GUO YING QIAO
Mechanical Engineering
Dalian Institute of technology
Dalian
China

JEAN-LOUIS BATOZ
GM/MNM
Université de Technologie
de Compiègne
France

1. INTRODUCTION

The Total and Updated Lagrangian Descriptions (TLD and ULD) are two formulations often used for the finite element non linear analysis of structures. The TLD formulation requires the exact large displacement field for large rotation analysis. But, such expressions are often difficult to obtain and to use. In [1, 2] an exact displacement field is obtained for 2D beam and arch structures, and four different strain expressions with the TLD are used to show the importance of displacement and strain expressions. In [3] a 3D beam displacement field is derived for the finite displacements and moderate rotations using Euler angles. In [4] an another 3D finite displacement field is presented, but it depends on the order of the rotations and is not unique. To the authers' knowledge the exact large displacement field of 3D beam has not been derived and used.

In this paper we will derive the exact field for large displacements and rotations under the Euler-Bernoulli hypothesis. However this field is too complicated to use for a TLD element. Using the UL Description and the small strain hypothesis, the field is greatly simplified. The thin walled section is taken into account and a unified formulation for varied sections is obtained. A finite element procedure based on the ULD has been developped and some results are presented and discussed.

2. LARGE DISPLACEMENT FIELD OF 3D BEAMS

The classical assumptions of Euler-Bernoulli and rigid section contour are retained [5 ,6]. They permit to express the displacements of all particles in terms of the displacements of the centroid locus.

We consider a beam having a superposed gravity and torsion centers which moves from S to S^* without section wraping (Fig.1). $\mathbf{r}$ and $\mathbf{r}^*$ are the position vectors of the centroid P, $\mathbf{R}$ and $\mathbf{R}^*$ are the position vectors of an arbitrary point Q, $\mathbf{u}(u,v,w)$ and $\mathbf{U}(U,V,W)$ are the displacement vectors of the material points P and Q. We have:

$$\overset{*}{r} = r + u = (x+u)\, i + v\, j + w\, k \tag{1}$$

$$\overset{*}{R} = R + U = \overset{*}{r} + \overset{*}{PQ} \tag{2}$$

or

$$\begin{Bmatrix} x + U \\ y + V \\ z + W \end{Bmatrix} = \begin{Bmatrix} x + u \\ v \\ w \end{Bmatrix} + [T]^T \begin{Bmatrix} 0 \\ y \\ z \end{Bmatrix} \tag{3}$$

where [T] is a transformation matrix from the initial reference x y z to the corotational reference $\overset{*}{x}\,\overset{*}{y}\,\overset{*}{z}$.

The tangent direction of the deformed axis is calculated by (1):

$$\overset{*}{x} = \frac{d\overset{*}{r}}{d\overset{*}{s}} = \frac{d\overset{*}{r}}{m\,dx} = a\,i + b\,j + c\,k \tag{4}$$

with

$$a = (1 + u')/m \qquad b = v'/m \qquad c = w'/m$$

$$m = \sqrt{(1 + u')^2 + v'^2 + w'^2} \tag{5}$$

where u', v', w' are derivatives of u, v, w with respect to x.

The matrix [T] is derived by considering: a rotation of biaxial bending from **x** to $\overset{*}{x}$ around the normal of the plane $(x, \overset{*}{x})$ followed by a torsion around the axis $\overset{*}{x}$ (Fig.1). We can demonstrate the matrix [T] is unique in spite of the order of the rotation and the torsion[7]. The matrix [T] depends only on u', v', w', θ and is given by:

$$[T] = \begin{bmatrix} a & b & c \\ -b\cos\theta - c\sin\theta & A\cos\theta + B\sin\theta & B\cos\theta + C\sin\theta \\ b\sin\theta - c\cos\theta & -A\sin\theta + B\cos\theta & -B\sin\theta + C\cos\theta \end{bmatrix} \tag{6}$$

with

$$A = \frac{ab^2 + c^2}{b^2 + c^2} \qquad B = \frac{(a-1)bc}{b^2 + c^2} \qquad C = \frac{ac^2 + b^2}{b^2 + c^2} \tag{7}$$

Hence the exact non linear displacement expression for an arbitrary point (x,y,z) is written as:

$$\begin{Bmatrix} U \\ V \\ W \end{Bmatrix} = \begin{Bmatrix} u \\ v \\ w \end{Bmatrix} + \begin{Bmatrix} y(-b\cos\theta - c\sin\theta) + z(b\sin\theta - c\cos\theta) \\ y(A\cos\theta + B\sin\theta - 1) + z(B\cos\theta + C\sin\theta) \\ y(-A\sin\theta + B\cos\theta) + z(-B\sin\theta + C\cos\theta - 1) \end{Bmatrix} \tag{8}$$

3. SIMPLIFIED SMALL DISPLACEMENT FIELD FOR THE ULD FORMULATION

A TLD element requires the large displacement field and C^1 continuity for u, v, w, but the expression (8) is so complicated that we could not use it into the TLD element. The UL

Description and the small strain hypothesis permit to simplify this field, the displacements and the rotations become small or moderate, we have:

$$m \simeq 1 \qquad a \simeq 1 \qquad b \simeq v' \qquad c \simeq w'$$
$$\cos\theta \simeq 1 \qquad \sin\theta \simeq \theta \qquad A \simeq 1 \qquad B \simeq 0$$
$$C \simeq 1 \tag{9}$$

In this case we obtain:

$$\begin{cases} U = u - y(v' + \underline{w'\theta}) - z(w' - \underline{v'\theta}) \\ V = v - z\theta \\ W = w + y\theta \end{cases} \tag{10}$$

The underlined terms $w'\theta$ and $v'\theta$ can be neglected compared to v' and w' respectively owing to the small or moderate rotations with respect of corotational reference. The following linear field in terms of u, v, w, θ is thus obtained as:

$$\begin{cases} U = u - y\,v' - z\,w' \\ V = v - z\,\theta \\ W = w + y\theta \end{cases} \tag{11}$$

The expression (10) has been consisdered in [8] whereas the expression (11) was adopted in [9] (among others). Both formulas have been used in our calculations leading to very closed results. It is more convenient to use the simpler one (11) for ULD.

When the gravity center O and the torsion center C do not coincide we take O as origin of the axes y z on the section and C as the point P in fig.1. We define u_o, v_c and w_c as the displacements at O and C respectively. θ_x is the torsion of the section around x. If the torsion warping is also included, the displacement expressions are written as:

$$\begin{cases} U = u_0 - y\,v'_c - z\,w'_c + \omega\theta'_x \\ V = v_c - (z - z_c)\,\theta_x \\ W = w_c + (y - y_c)\theta_x \end{cases} \tag{12}$$

where ω is the normalised sectorial area coordinate.

4. DISPLACEMENT FIELD FOR OPEN THIN WALLED BEAM

With the equation (12) we can calculate the torsion shear stress and the torsion rigidity for the solid section:

$$\sigma_T = Ge_T = G\sqrt{(U,_y + V,_x)^2 + (U,_z + W,_x)^2} = G\theta'_x \sqrt{[\omega,_y - (z-z_c)]^2 + [\omega,_z + (y-y_c)]^2} \quad (13)$$

$$GJ = \frac{1}{\theta'^2} \int_A \sigma_T e_T \, dA = G\int_A \{[\omega,_y - (z-z_c)]^2 + [\omega,_z + (y-y_c)]^2\} dA \quad (14)$$

But for the open thin walled beams they are in contradiction with the classical theory: the formula (13) cannot express the correct shear stress distribution which should be proportional to the distance from the mid-surface to the considered point; the formula (14) gives often too great torsion rigidity (ex. $GJ=4GI_z$ for I beam).

To derive a coherent formula, we express the coordinates of all points in terms of the coordinates of the points on the mid-surface and adopt the normal conservation hypothesis for mid-surface (fig.2) [10, 6]:

$$\begin{cases} y = \bar{y} - n \sin\alpha = \bar{y} - mn \\ z = \bar{z} + n \cos\alpha = \bar{z} + ln \end{cases} \quad (15)$$

$$U(x,y,z) = U(x,s) - n \frac{\partial \eta}{\partial x} \quad (16)$$

The equations (15) and (16) permit to get the tangential and the normal displacements ξ, η of the point A:

$$\begin{Bmatrix} \xi \\ \eta \end{Bmatrix} = \begin{bmatrix} l & m \\ -m & l \end{bmatrix} \begin{Bmatrix} V \\ W \end{Bmatrix} = \begin{Bmatrix} lv_c + mw_c + (\rho_t - n)\theta_x \\ -mv_c + lw_c + \rho_n \theta_x \end{Bmatrix} \quad (17)$$

where $\rho_t = m(\bar{y}-y_c) - l(\bar{z}-z_c)$ and $\rho_n = l(\bar{y}-y_c) + m(\bar{z}-z_c)$ (fig.2)

Using the equations (12),(15),(16) and (17), the displacement field for the open thin-walled beam can be now written:

$$\begin{cases} U = u_o - (\bar{y}-mn)v'_c - (\bar{z}+ln)w'_c + (\omega - n\,\rho_n)\,\theta'_x \\ V = v_c - (\bar{z}-z_c+ln)\theta_x \\ W = w_c + (\bar{y}-y_c+mn)\theta_x \end{cases} \quad (18)$$

If we admit that the section curvature radius is large, this formula will give the linear shear stress distribution through the wall thickness and the classical torsion rigidity:

$$\sigma_T = \sigma_{xs} = G(\xi,_x + U,_s) = -2Gn\theta \quad (19)$$

$$GJ = \frac{1}{\theta'^2} \int_A \sigma_T e_T \, dA = G\int_A (-2n)^2 dA = \frac{1}{3} \sum_{i=1}^{m} S_i h_i^3$$

Equations (12) and (18) are equivalent except for the term $-n\,\rho_n\theta'_x$ which gives the correct torsion rigidity for open thin walled section. So in the finite element formulation using the resulting forces we may adopt the equation (12) for varied sections replacing eq.(14) by the classical torsion rigidity.

5. NON LINEAR STRAIN EXPRESSIONS

The non linear Green-Lagrange strains (GL) can be established using eq.(12). With the small strain and moderate rotation hypothesis the relevant strains $\{\varepsilon\}$ are:

$$\begin{Bmatrix} \varepsilon_x \\ \varepsilon_{xy} \\ \varepsilon_{xz} \end{Bmatrix} = \begin{Bmatrix} U,_x + \frac{1}{2}(V,_x^2 + W,_x^2) \\ U,_y + V,_x + W,_x W,_y \\ U,_z + W,_x + V,_x V,_z \end{Bmatrix} \tag{20}$$

Substituting (12) into (20), we obtain:

$$\begin{Bmatrix} \varepsilon_x \\ \varepsilon_{xy} \\ \varepsilon_{xz} \end{Bmatrix} = \begin{Bmatrix} e_x \\ e_{xy} \\ e_{xz} \end{Bmatrix}_1 + \begin{Bmatrix} \eta_x \\ \eta_{xy} \\ \eta_{xz} \end{Bmatrix}_{nl} = \begin{Bmatrix} u' - yv'' - zw'' + \omega\theta' \\ -(z - z_c - \omega,_y)\,\theta' \\ (y - y_c + \omega,_z)\,\theta' \end{Bmatrix}_1 +$$

$$+ \begin{Bmatrix} \frac{1}{2}\{v'^2 + w'^2 + [(y-y_c)^2 + (z-z_c)^2]\,\theta'^2\} + (y-y_c)\,w'\theta' - (z-z_c)\,v'\theta' \\ [w' + (y - y'_c)\,\theta']\,\theta \\ [v' - (z - z_c)\,\theta']\,\theta \end{Bmatrix}_{nl} \tag{21}$$

where u, v, w, θ are functions of x only and represent u_o, v_c, w_c and θ_x for simplicity.

6. UPDATED LAGRANGIEN DESCRIPTION OF 3D BEAM ELEMENT

We consider a straight beam element with 2 nodes and 7 degrees of freedom per node (fig.3):

$$\langle u_n \rangle = \langle u\ v\ w\ \theta_x\ \theta_y\ \theta_z\ \theta'_x \rangle \quad \text{with} \quad \theta_y = -w', \quad \theta_z = v' \tag{22}$$

The correspondent nodal resulting forces are:

$$\langle F_n \rangle = \langle N_x\ T_y\ T_z\ M_x\ M_y\ M_z\ M_\omega \rangle \tag{23}$$

where $M_\omega = \int_A q_x \omega\, dx$ is the warping bimoment.

The interpolation functions are linear for u and cubic (Hermite) for v, w and θ [1,2].

For the ULD the incremental virtual work principle is established on the last known configuration. All variables are mesured with respect of this configuration. We have:

$$\int_V \{\delta e\}^T [D]\{\Delta e\} dv + \int_V \{\sigma\}^T \{\delta\Delta\eta\} dv = \delta W_{ext} - \int \{\sigma\}^T \{\delta e\} dv \tag{24}$$

where $\{e\}$ and $\{\eta\}$ are the linear and non linear parts of $\{\varepsilon\}$, $\{\sigma\}$ is the cauchy stress, $[D]$ is the elasto-plastic constitutive matrix and δW_{ext} is the external virtual work.

After the element discretization and the assemblage step, the solution may be obtained by the Newton-Raphson algorithm as follows:

$$[K_T]^i \{\Delta u\}^i = \{F_{ext}\} - \{F_{int}\}^i \tag{25}$$

with

$$[K_T] = [K_o] + [K_\sigma] = \sum_{e=1}^{n} [T^e]^T ([k_o^e] + [k_\sigma^e])[T^e]$$

$$\{F_{int}\} = \sum_{e=1}^{n} [T^e]^T \{f_{int}^e\}$$

where $[K_T]$, $[K_o]$, $[K_\sigma]$ are the tangent matrix, the linear stiffness matrix and the initial stress matrix, $\{F_{int}\}$, $\{F_{ext}\}$ are the nodal internal and external force vectors and $[T^e]$ is transformation matrix of the element e.

$[k_o^e]$, $[k_\sigma^e]$ and $\{f_{int}^e\}$ are calculated by substituting (21) into (24):

$$[k_o^e] \Leftarrow \int_L [EA\delta u'\Delta u' + EI_z \delta v''\Delta v'' + EI_y \delta w''\Delta w'' + GJ\delta\theta'\Delta\theta' + EI_\omega \delta\theta''\Delta\theta''] dx \tag{26}$$

$$\begin{aligned}[k_\sigma^e] \Leftarrow \int_L [& N_x (\delta v'\Delta v' + \delta w'\Delta w') - (M_y - z_c N_x)(\delta v'\Delta\theta' + \delta\theta'\Delta v') \\ & - (M_z + y_c N_x)(\delta w'\Delta\theta' + \delta\theta'\Delta w') + T_y (\delta w'\Delta\theta + \delta\theta\Delta w') - T_z (\delta v'\Delta\theta + \delta\theta\Delta v') \\ & + M_k \delta\theta'\Delta\theta'] dx \end{aligned} \tag{27}$$

$$\{f_{int}^e\} \Leftarrow \int_L (N_x \delta u' + M_z \delta v'' - M_y \delta w'' + M_x \delta\theta' + M_\omega \delta\theta'') dx \tag{28}$$

where $J = \int_A [(z - z_c - \omega_{,y})^2 + (y - y_c + \omega_{,z})^2] dx$ is the torsion rigidity for solid section,

$J = \frac{1}{3} \sum_{i=1}^{m} S_i h_i^3$ and $J = 4A_o^2 / \oint \frac{dS}{h}$ are for open and closed thin walled sections (A is the area enclosed by the section's mid-line), $I_\omega = \int_A \omega^2 dA$ is the warping rigidity and $M_k = \int_A \sigma_x [(y - y_c)^2 + (z - z_c)^2] dA$ is the Wagner coefficient [9].

We note that: --With the small strain hypothesis, the element geometry (L,A,V) is not modified, --at each iteration the nodal coordinates, the local corotational reference axes and the transformation matrices are updated, --the rigid body displacements are eliminated from the total displacements, --the Cauchy stresses and the small strains are calculated with respect to the last known reference, --for the elasto-plastic analysis the classical Prandtl-Reuss equations with resulting forces and the tangential stiffness method are adopted.

7. SOME NUMERICAL RESULTS

A large number of problems have been solved. Only four examples are shown here:

7.1 Large deformation of an elastic spring (Fig.4)

A spring subjected to concentrated forces is presented in fig.4. The load-displacement curve diverges from the straight line defined by the pure torsion theory due to the influence of the bending moments. Very large 3D displacements and rotations are produced in these examples as shown by the deformed configurations.

7.2 Dome built with open thin-walled beams (fig.5)

This problem has been treated by [8] using eq.(10) as displacement field. Our results with or without warping are presented and compared. The difference using both eq.(10) (11) is small but the torsion warping plays a more important role.

7.3 Elastic and elasto-plastic analysis of 3D frame (fig.6)

The elasto-plastic analysis of this frame has been performed by [11, 9]. Fig.9(a) shows that elastic non linear behavior is consistent with the critical load calculated by the linear buckling program. Torsion warping is not significant in this problem. Our elasto-plastic load - displacement curve is closed to the solutions of [11, 9](Fig.9(b)), the difference is due to the different elasto-plastic yield criteria considered.

7.4 Plastic buckling of a dome (fig.7)

This analysis has been carried out by [12]. His result shows the pre-and post-buckling behaviors. We find that the limit load is reached at the appearance of the second plastic hinge, then the loading capacity decreases. Finally, when the beams like AC become in traction, the loading capacity increases again.

8. Conclusion

In this paper, the exact large displacement field of 3D beam are presented and discussed. It is very difficult to elaborate a 3D thin beam element for large displacements using a Total Lagrangian Formulation. The Updated Incremental Formulation permit to greatly simplify this field. The displacement field for the thin-walled beam was discussed and unified for the solid and thin-walled beam sections usingdifferent torsion rigidity formulas. The formulation and calculation have shown that this 3-D beam element is simple, correct, versatile and efficient.

REFERENCES

[1]BATOZ,J.L and JAMEUX,J.P.,"Post-flambement et grands déplacements d'arcs plans", Journal de Mécanique théorique et appliquée, VOL.5, N°2,1986,P237-257.

[2]BATOZ,J.L. and JAMEUX,J.P.,"Large Displacements and Post-Buckling of plane structures" (European Mechnics Colloquium "Flexible Shells, Theory and Applications", Munich, W-Germany,1983).

[3]ROSEN,A. , FRIEDMANN,P., "The Nonlinear Behavior of Elastic Slender Straight Beams Undergoing Small strains and Moderate Rotations", Journal of Applied Mechanics. 1979, VOL.46, P.161-168.

[4]GRIMALDI,A.,"Flambement élastique des poutres", Le flambement des structures, Editeur SEBTP,1981, PP.73-93.

[5] VASHIZU,K., "Variational Methods in Elasticity and Plasticity", Third edition, Pergamon Press, 1982.

[6] NISHINO,F.,KASEMSET,C. and LEE,S.L., "Variational Formulation of stability Problems for thin-walled Members", Ingenieur-Archiv., 43(1973) P.58-68.

[7] GUO,Y.Q., "Analyse non linéaire des structures formées des poutres tridimensionnelles par la méthode des éléments finis", thèse de doctorat, Université de Compiègne, France, 1986.

[8] PILLER,A., "Spatial Nonlinear Analysis of Steel Beams and Frames", Stability of Metal structures, Paris, 1983, PP. 483-490.

[9] RAMM,E. and OSTERRIEDER,P., "Ultimate Load Analysis of Three-Dimensional Beam Structures with Thin-Walled Cross Sections Using Finite Elements", Stability of Matel Structures, Third International Colloquium, Paris, 1983, PP.201-210 .

[10] GJELSVIV,A., The Theory of Thin walled bars.

[11]ACKERMANN,T., "Traglastberechnung räumlicher Rahmen aus stahl-oder leichtmetallprofilen mit dünnwandigen offenen Querschnitten", Univ, Karlsruhe, Institut für Baustatik und Messtechnik, H.5,1981.

[12]ARGYRIS,J.H.,BONI,B.HINDENLANG, U. and KLEIBER,M.,"Finite Element Analysis of Two-and Three-Dimensional Elasto-Plastic Frames--The National Approach", Comput, Meths. Appl. Mech. Engrg. 35(1982) p.221-248.

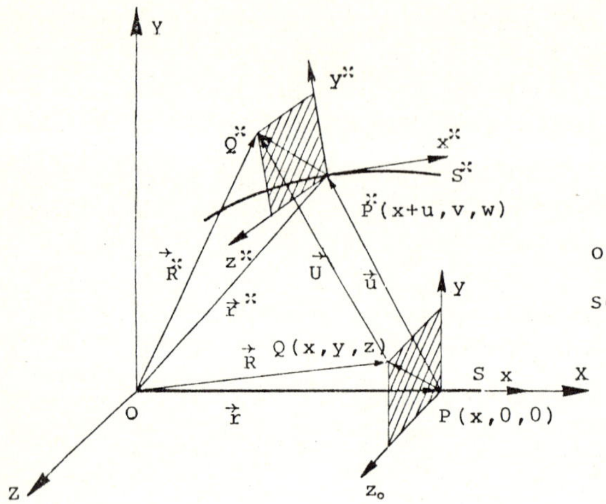

Fig.1 Position vectors of the particles on a section

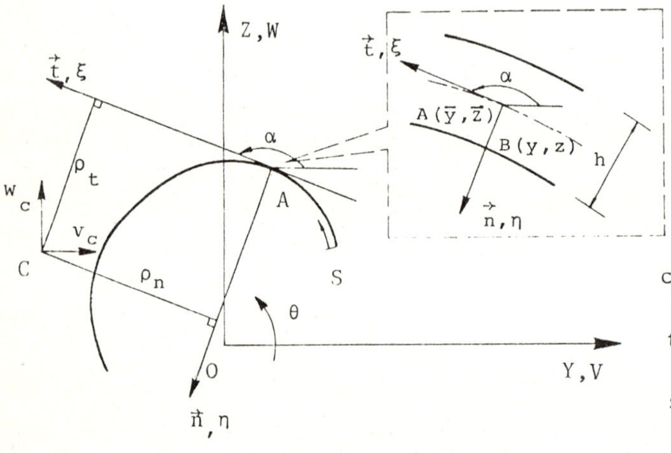

l=cosα
m=sinα

Fig.2 Curved coordinate of the thin walled beam section

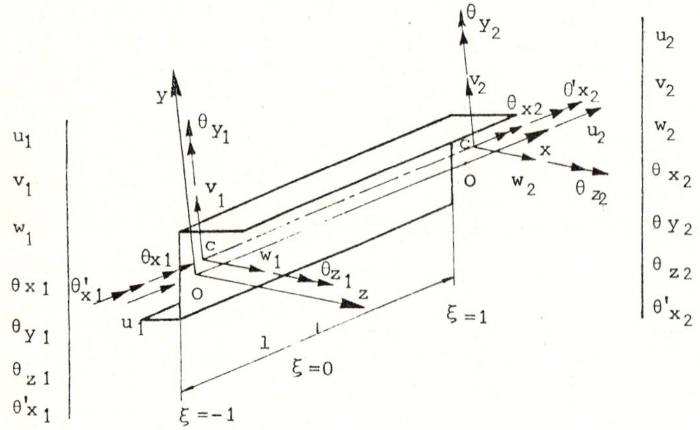

Fig.3 Finite element model

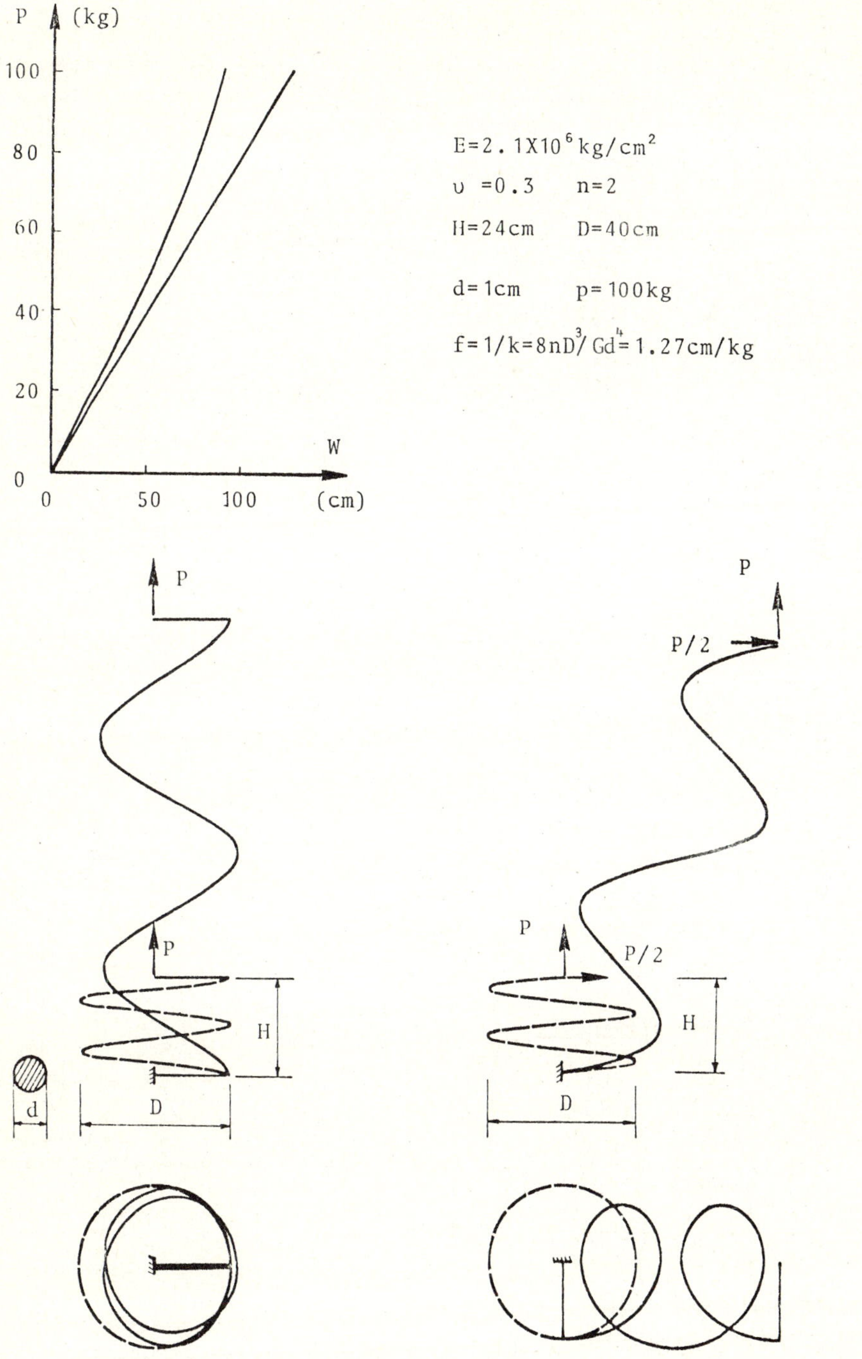

Fig.4 Large deformation of an elastic spring(24 elements)

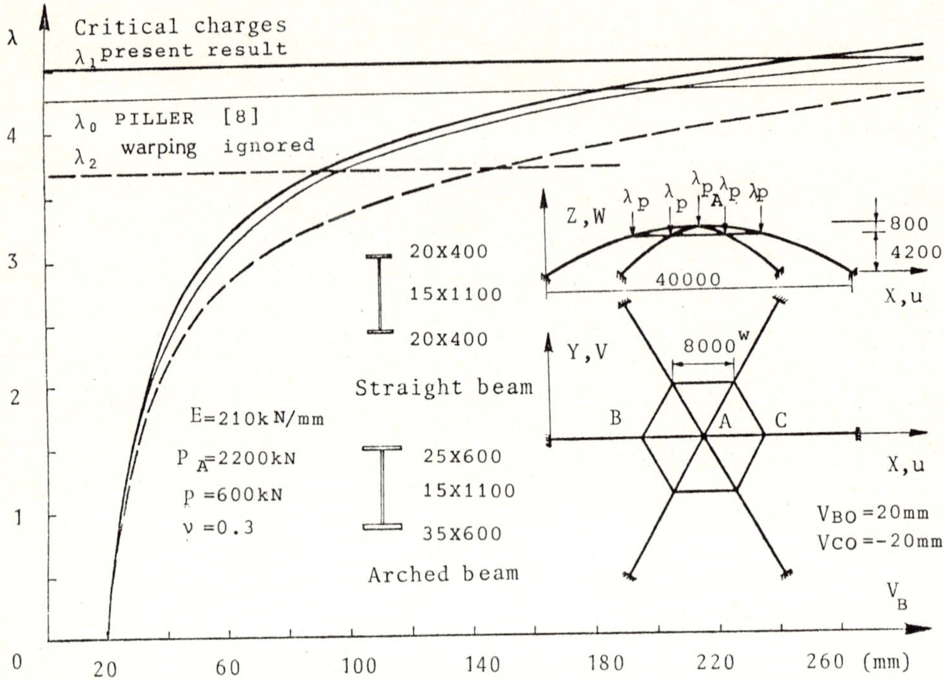

Fig.5 Elastic buckling of a dome

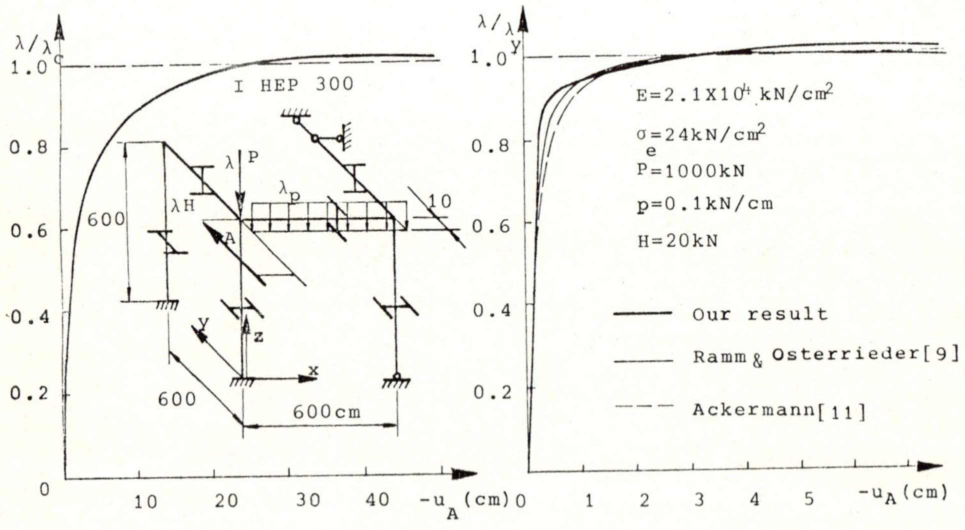

(a) Elastic buckling (λ_c=12.88) (b) Elasto-plastic buckling (λ_y=2.96)

Fig.6 Buckling of a space frame

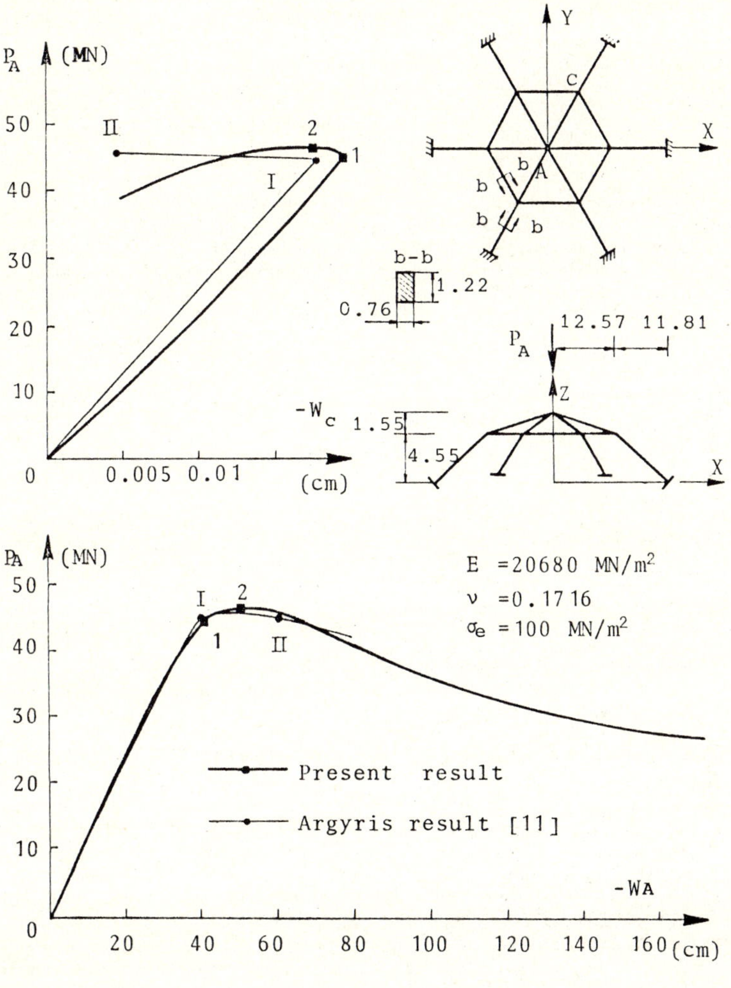

Fig.7. Elasto-plastic buckling of a dome

AN ALGORITHM FOR GRAPHICAL COMPUTER RESULTS IN SHELLS

A. Samartín, J. Cardona
Department of Structural Analysis
University of Santander
39.004-Santander (SPAIN)

SUMMARY

A Finite Element technique to interpolate general data (function values and its derivatives) has been developped. The technique can be considered as a generalized solution of the classical polynomial interpolation, because the condition for the interpolating function to be a polynomial is replaced by a minimizing condition of a given "smoothing" functional. In this way it is possible to find interpolating functions with a given level of continuity according to the class of finite elements used. Examples have been presented in order to assess the accuracy and efficiency of the procedure.

1.- INTRODUCTION

There exists many problems that demand interpolation procedures in order to be solved. Some examples are:

(a) Graphical output in F.E. analysis (stress contour, lines of principal stresses in structural analysis etc.).

(b) Terrain representation in topographical survey.

(c) Design of industrial objects (cars, ships etc) with given conditions of continuity, slope and curvature.

Many techniques dealing with the interpolation problem exist. However most are efficient mainly in 1-D problems (Lagrange, Newton interpolation (1)), others as the splines techniques, (2) are concerned with simple types of data (values of the function) or continuity requirements. The finite element method has been proved to be an excellent tool to deal aproximation problems (3), however its capability to treat general interpolation problems seems to be not completely exploited yet. In fact, since many years considerable efforts have been dedicated to extend the spline ideas to practical surface fitting. Earlier works have dealt with practical aspects (4), (5), but then a more deep theoretical background seemed to be needed in order to tackle efficientely the 2-D problems. The references (6) and (7) represent important contributions on this direction. Unfortunately these developments have not taken place in the area of F.E. therefore many theoretical and practical posibilities of application of the method to these types of interpolation problems have not been completely explored. The objective of this paper is an attempt in this direction. This paper can be considered in some respects a paralell application on the F.E. field of some ideas related with the several variables splines shown in the reference (8).

2.- FORMULATION OF THE INTERPOLATION PROBLEM

The general interpolation problem can be stated as follows:

Given a set of points $P_i \in \bar{R}^2$, $(1 \leq i \leq N)$ and the real numbers $r^i_{l-k;l}$ the objective is to find a function $u \in W^m_2(\Omega)$ defined in a bounded domain Ω in R^2 such that

$$\frac{\partial^l u(P_i)}{\partial x^{l-k} \partial y^k} = r^i_{l-k;l} \qquad (0 \leq l \leq k_i) \qquad (1)$$

where $W^m_2(\Omega)$ is the space of functions $u \in L_2(\Omega)$ and with generalized m-th order derivatives $\leq m$ belonging also to $L_2(\Omega)$.

k_i es the maximum order of derivative associated to the point i and the set of derivatives at a point are not necessarely complete and consecutive.

Besides the function u must minimize the functional

$$Fu = \int_\Omega \{a_{00}u^2 + a_{10}\left(\frac{\partial u}{\partial x}\right)^2 + a_{01}\left(\frac{\partial u}{\partial y}\right)^2 + a_{20}\left(\frac{\partial u^2}{\partial x^2}\right)^2 + a_{11}\left(\frac{\partial^2 u}{\partial x \partial y}\right)^2 + \ldots + a_{0\,m} \cdot \left(\frac{\partial^m u}{\partial y^m}\right)^2\} d\Omega \qquad (2)$$

For Lagrange (only values of the function are gives in all nodes, i.e. $k_i=0$) and Hermite (at each node the set of data is complete, i.e. the function value and its k_i-1 first derivatives are given at each point) interpolation problems with particular points patterns there is an unique polynomial interpolation (9). For the general case theoretical results are scarces.

The Finite Element Method can be used to describe an approximation of the function u (3).

For each element $\Omega_e \subset \Omega$ the vector $\underline{u}^e$ contains the total number n_e of degrees of freedom (dof) , i.e., "dof type a" with values given as data and "dof type b" with values unknown.

This vector $\underline{u}^e$ can be partitioned as follows:

$$\underline{u}^e = \{\underline{u}_1^e, \underline{u}_2^e, \underline{u}_3^e, \ldots, \underline{u}_n^e\} \qquad (3)$$

where n is the number of nodes of the element.

$\underline{u}_i^e$ is a vector containing the values of u and its $n_{e_i}-1$ first derivatives at node i. Obviously $n_e = \sum_{i=1}^{n} n_{e_i}$.

The approximation function u(x) is expressed in the element e by the following equation:

$$u(x) = \underline{N}^e \underline{u}^e = (\underline{N}_1^e \; \underline{N}_2^e \; \ldots \; \underline{N}_n^e) \begin{pmatrix} \underline{u}_1^e \\ \underline{u}_2^2 \\ \vdots \\ \underline{u}_n^e \end{pmatrix} \qquad (4)$$

where $\underline{N}_i^e$ is the shape vector associated to the node i of the element e and it is described as:

$$\underline{N}_i^e = \{N_{i1}(x,y),\; N_{i2}(x,y),\; \ldots,\; N_{in_{e_i}}(x,y)\} \qquad (5)$$

The functional Fu can be expressed as follows:

$$Fu = \int_\Omega (\underline{L}u)^T \underline{D}(\underline{L}u) d\Omega \qquad (6)$$

where

$$\underline{L} = \left(1, \frac{\partial}{\partial x}, \frac{\partial}{\partial y}, \frac{\partial^2}{\partial x^2}, \ldots, \frac{\partial^m}{\partial y^m}\right)^T$$

$$\underline{D} = \mathrm{diag}\{a_{00}, a_{10}, a_{01}, \ldots, a_{0\,m}\}$$

Following the standard FEM steps it is obtained

$$Fu = \sum_{e=1}^{NE} \underline{u}^{eT} \underline{k}^e \underline{u}^e = \underline{U}^T \left(\sum_{e=1}^{NE} \underline{K}^e\right) \underline{U} = \underline{U}^T \underline{K}\, \underline{U} \qquad (7)$$

where $\underline{U}$ is the vector containing the total number of dof, NE is the total number of elements.

The element "stiffness" matrix is given by the expression:

$$\underline{k}_e = \int_\Omega (\underline{L}\underline{N}^e)^T \underline{D}(\underline{L}\underline{N}^e) d\Omega \qquad (8)$$

It is convenient to partitionate the global stifness matrix $\underline{K}$ and the displacement vector $\underline{U}$ according to the two types of dof (a and b, data and unknowns), i.e.:

$$\underline{K} = \begin{vmatrix} \underline{K}_{aa} & \underline{K}_{ab} \\ \underline{K}_{ba} & \underline{K}_{bb} \end{vmatrix} ; \qquad \underline{U} = \begin{vmatrix} \underline{U}_a \\ \underline{U}_b \end{vmatrix}$$

The minimum of (7) is obtained from the condition

$$\frac{\partial Fu}{\partial \underline{U}_b} = \underline{0} \tag{9}$$

i.e.,

$$\underline{K}_{bb}\,\underline{U}_b + \underline{K}_{ba}\,\underline{U}_a = \underline{0} \tag{10}$$

Then the unknowns $\underline{U}_b$ are

$$\underline{U}_b = -\underline{K}_{bb}^{-1}\,\underline{K}_{ba}\,\underline{U}_a$$

The minimum value of the functional (7) is

$$(Fu)_{min} = \underline{U}_b^T\,\underline{K}_{ba}\,\underline{U}_a + \underline{U}_a^T\,\underline{K}_{aa}\,\underline{U}_a = \underline{U}_a^T\,\tilde{\underline{K}}_{aa}\,\underline{U}_a \tag{11}$$

where

$$\tilde{\underline{K}}_{aa} = \underline{K}_{aa} - \underline{K}_{ab}\,\underline{K}_{bb}^{-1}\,\underline{K}_{ba}$$

Once $\underline{U}_b$ is known the interpolant function is given in each element by the formula (4).

3.- COMMENTS

The value of $(Fu)_{min}$ can be used to compare different interpolations. If $\underline{K}_{aa}=0$ the interpolant function becomes a polynomial of (m-1)-th degree.

In order to diminish the number of unknowns it may be interesting to use elements with different dof (n_e) among them and even with different dof (n_{e1}, n_{e2},...n_{en}) betwen the n nodes of one element.

Local discontinuities of the interpolant function can be represented by means of the introduction of "hinges" at the nodes where discontinuities occur. Then, equal unknowns are considered to represent common dof and distint unknowns to simulate the other dof in which the discontinuities appear. This situation can be very useful if Green functions (influence lines in structural analysis) are to be interpolated.

In real applications with a large amount of data to be handled it may be convenient to divide the interpolation procedure in parts with some overlapping elements and reduce in this way the computational effort. The use of this technique presents in some cases, similarities with the spline interpolation.

The use of the interpolation procedure just described in use graphical output of results obtained from F.E. analysis (stress contour, lines of principal streses,..) can be very appealing due to the fact that the same FE mesh can be used in the analysis and the interpolation.

The elements to be used must be of class C^{m-1}. Conforming and non-conforming elements can be considered as well as simple and hyperlements. Several hierarchic families have been reported in the FE literature, see for example (10). In figures 1 and 2 the shape functions of particular hyperelements of class C^0 and C^1 respectively are shown.

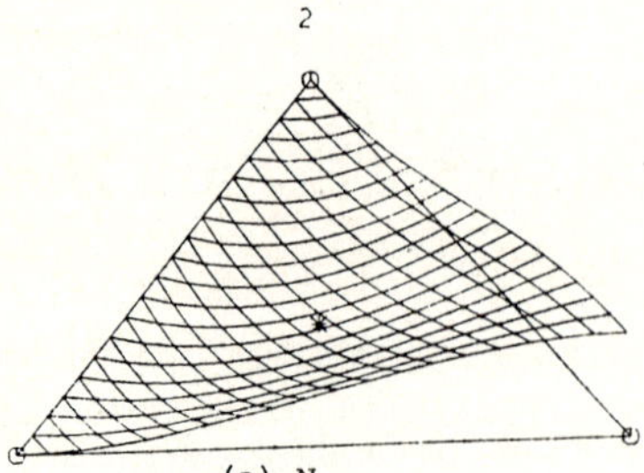

(a) N_1

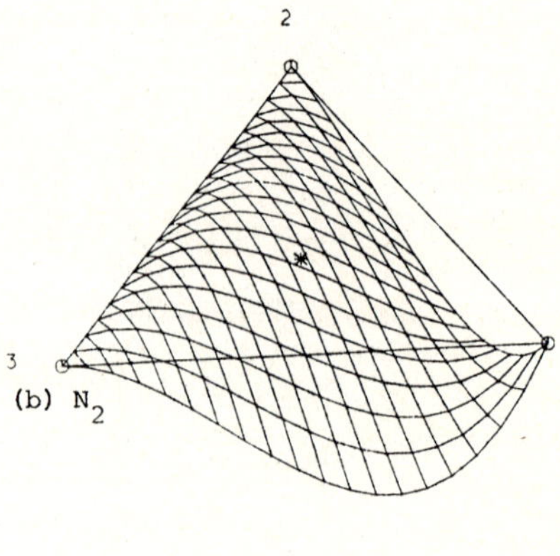

(b) N_2

○ : u, u_x, u_y * : u

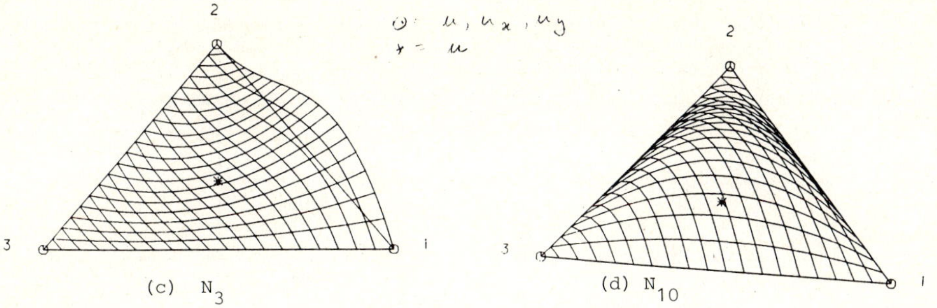

Figure 1.- Element C^0. Shape functions N_i.

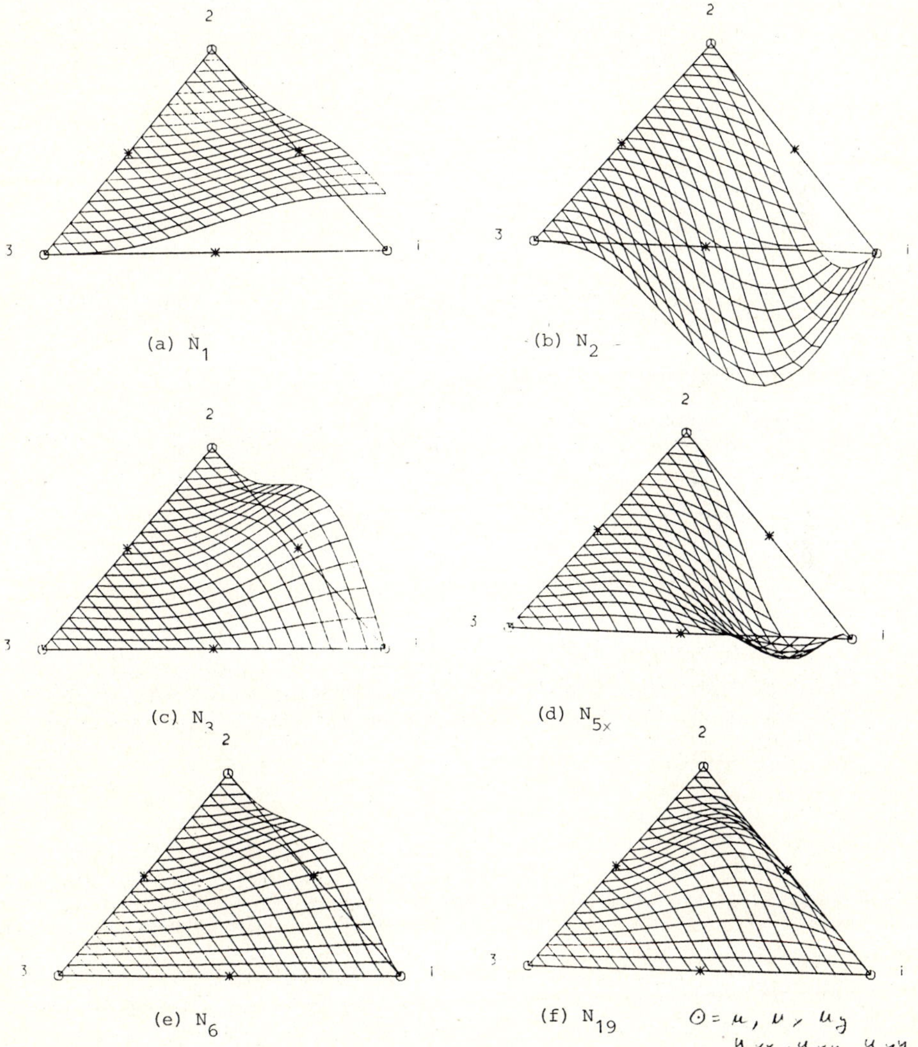

Figure 2.- Element C^1. Shape functions N_i.

Figure 3.- Example of application to a 2-D case. F.E. mesh.

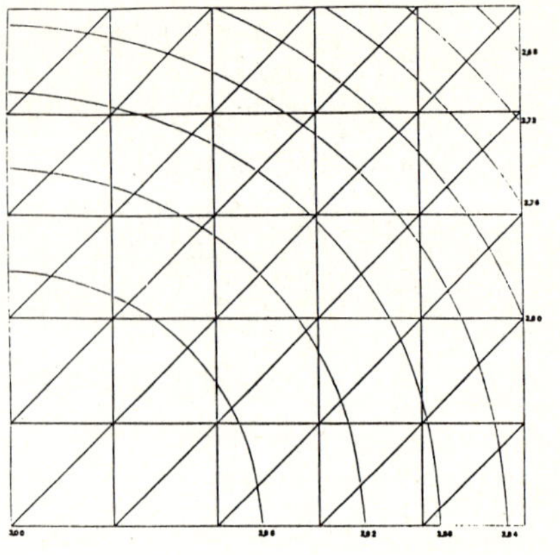

F(UN) = 8.6572
NCARG/NGDL = 0.4319

Figure 4.- Example of application to a 2-D case. Results. (Contour lines between 3.00 - 2.68/0.04).

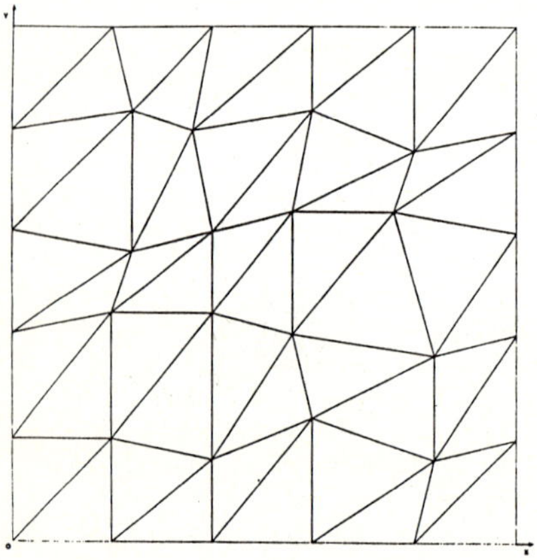

Figure 5.- Example 2-D. Irregular mesh.

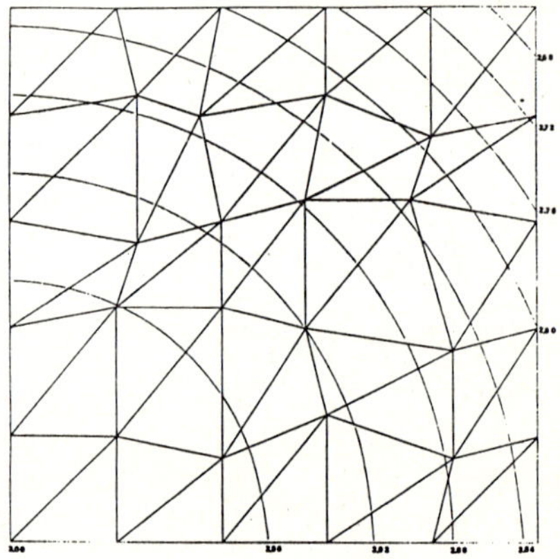

F(UN) = 8.6648
NCAEG/NGDL = 0.4319

Figure 6.- Example 2-D. Results. (Contour lines between 3.00 - 2.68/0.04).

Table I.- Example of application $u=\sqrt{R^2-x^2-y^2}$ (R=3)

FUNCTIONAL	DATA		EFFICIENCY η (%)	NG	ELEMENT CLASS	FUNCTIONAL VALUE		OBS.
	DOF	DISTRIB.				F.E.	EXACT	
$\int_\Omega (u_x^2+u_y^2)\,d\Omega$	u	vertices nodes	22.79 54.43	9 9	C^0	0.08027 0.08122	0.03254	
$\int_\Omega (u^2+u_x^2+u_y^2)\,d\Omega$	u	vertices nodes	22.79 54.43	9 9	C^0	8.39150 8.41328	8.41618	
$\int_\Omega (u_x^2+u_y^2)\,d\Omega$	u u_x u_y	random	22.79 31.65 37.98 50.63	9 9 9 9	C^0	0.05822 0.06251 0.06448 0.06672	0.03254	
$\int_\Omega (u^2+u_x^2+u_y^2)\,d\Omega$	u u_x u_y	random	22.79 31.65 37.98 50.63	9 9 9 9	C^0	8.27121 8.33684 8.34416 8.37105	8.41618	
$\int_\Omega (u_x^2+u_y^2)\,d\Omega$	u	vertices	11.96	12	C^1	0.06965	0.032547	
$\int_\Omega (u^2+u_x^2+u_y^2)\,d\Omega$	u	vertices	11.96	12	C^1	8.33600	8.41618	
$\int_\Omega (u_{xx}^2+u_{xy}^2+u_{yy}^2)\,d\Omega$	u	vertices	11.96	12	C^1	0.22749	0.26232	
$\int_\Omega (u^2+u_x^2+u_y^2+u_{xx}^2+u_{xy}^2+u_{yy}^2)\,d\Omega$	u	vertices	11.96	12	C^1	8.64156	8.67850	
$\int_\Omega (u_x^2+u_y^2)\,d\Omega$	u u_x u_y u_{xx} u_{xy} u_{yy} u_{ni}	random	33.22 43.19 53.49	12 12 12	C^1	0.03946 0.04935 0.05198	0.03254	
$\int_\Omega (u^2+u_x^2+u_y^2)\,d\Omega$	u u_x u_y u_{xx} u_{xy} u_{yy} u_{ni}	random	33.22 43.19 55.48	12 12 12	C^1	8.11713 8.26929 8.30316	8.416182	
$\int_\Omega (u_{xx}^2+u_{xy}^2+u_{yy}^2)\,d\Omega$	u u_x u_y u_{xx} u_{xy} u_{yy} u_{ni}	random	32.22 43.19 55.48	12	C^1	0.22719 0.24257 0.25254	0.26232	
$\int_\Omega (u^2+u_x^2+u_y^2+u_{xx}^2+u_{xy}^2+u_{yy}^2)\,d\Omega$	u u_x u_y u_{xx} u_{xy} u_{yy} u_{ni}	random	32.22 43.18 55.48	12	C^1	8.63902 8.65720 8.66791	8.87850	Fig 4

η = number of data / total number of d.o.f.
NG = number of Gauss integration points

4.- EXAMPLES OF APPLICATIONS

In order to check the efficiency of this interpolation procedure a set of points on a surface u=f(x,y) have been considered. The selected surface is the spherical one, i.e.

$$u = f(x,y) = \sqrt{R^2-x^2-y^2}$$

with radius R=3, defined over the domain (0,1)x(0,1).

Some cases have been studied using different types of data and classes of finite elements. In all cases the same FE mesh has been used and it is represented in figure 3. The functional corresponds to $a_{ij}=1$ for every i and j.

First it was considered to be known as data the values of the function f(x,y) in the vertices of the triangles of the FE mesh, i.e. 36 values. In a second case, the function f(x,y) was known at every node of the Fe mesh i.e. 86 values. The finite element mesh shown in figure 3 has been used in these two interpolation cases, and the stiffness matrix of each elements have been calculated using a 9-point Gauss quadrature. In table 1 the obtained results for two functionals are given.

Two similar problems to the above have been studied but now with data including values of u and its first derivatives u_x, u_y randomly distributed among the total number of dof. Table 1 reproduces the results.

If a different class of triangular elements are used, as the ones of class C^1 of figure 2, the interpolant surface is "smoother" than the previous one. The results obtained for some data distributions are summarized also in the table 1. A plot of the results corresponding to one case is given in figure 4.

Finally in order to show the possibilities of this interpolation procedure even in situations with irregular mesh of data (figure 5), the same problem as the one represented in the figure 4 has been studied and the results have been plotted in the figure 6.

5.- CONCLUSIONS

An interpolation procedure based in the FEM has been developed. The mesh used to obtain the results of a FE shell analysis can be again used in the interpolation procedure. In this way the imput data for the last case is reduced to a minimum. Continuity requirements for the graphical output of the results can be easily fulfilled by increasing the order of the interpolating functional.

6.- REFERENCES

(1) G. Dahlquist and A. Bjorck, Numerical Methods. Prentice Hall, (1974).

(2) T.R.F. Nonweiler, "Computational Methematics", Ellis Horword Ltd. John Wiley and Sons, (1984).

(3) O.C. Zienckiewicz and K. Morgan, Finite Elements and Approximation. John Wiley and Sons, (1983).

(4) M.S. Caceli and W.P. Cachesis, "Fitting curves to Data", Byte, (1964).

(5) R.L. Harder and R.N. Desmarais, "Interpolation Using Surface Splines", Journal of Aircraft, Feb. (1972).

(6) J.W. Jerome and L.L. Schumaker, "On Lg-Splines", Journal of Approximation Theory 2, (1969).

(7) R. Holmes, "R-Splines in Banach Spaces: Interpolation of Linear Manifolds", Journal of Mathematical Analysis and Applications 40, (1972).

(8) J. Duchon "Sur l erreur d'interpolation des functions de plusieurs variables par les D^m-splines", R.A.I.R.O., Numerical Analysis Vol. 12, n.4 (1978).

(9) M. Gasca and J.I. Maeztu, "On Lagrange and Hermite Interpolation in R^k". Numerische Mathematik 39,1-14 (1982).

(10) A. Samartín, "Desarrollo de familias jerárquicas de elementos finitos de clase C^m", VI. C.E.D.Y.A. Congreso de ecuaciones diferenciales y aplicaciones", Jaca (Huesca), 26-30. Sept. (1983).

ANALYSIS OF SHELL STRUCTURES ON MICROCOMPUTERS

J.P. RAMMANT - E. BACKX - L. KNOOPS
SCIA S.V., Herk-de-Stad, Belgium

Symmary

The analysis of shell structures by finite elements in all cases leads to a relatively large size problem. Nevertheless, when the data- and program structure is chosen adequately, micro-computers can solve most of the shell structures at a fraction of the cost of a mainframe.
The program presented has the added advantage that it was developed for a microcomputer and not scaled down from a mainframe system. Therefore, the program is highly user friendly and is embedded in an engineering environment that was developed for engineering offices.

The program structure

The program structure is shown in Figure 1. Boxes in dashed lines indicate optional modules. The main difference compared to conventional finite element programs is in the input : the availibility of a data editor. Where in conventional programs, the input is either given in a file created by a general editor or given by using a command language, the advantage of a data editor incorporated in the program itself is appreciated by the users. The use of a general editor presents an additional difficulty for those people who are not educated with computers. The use of a command language, although a very powerful tool, necessitates the availability and study of manuals.
The data editor is made up as follows.
If a new analysis is started, and the user wants to enter all his data from the keyboard (no preprocessing or library is used), all data items (coordinates, topology, materials ...) will be asked in sequence.
When all data are collected, the user can change or browse selected data items from a menu. The data can also be printed on paper or on the screen or verified by drawings of the structure. A menu on the screen always gives the different possibilities. Before the computation is started, the program will verify if there is sufficient disk space available for the requested calculations.

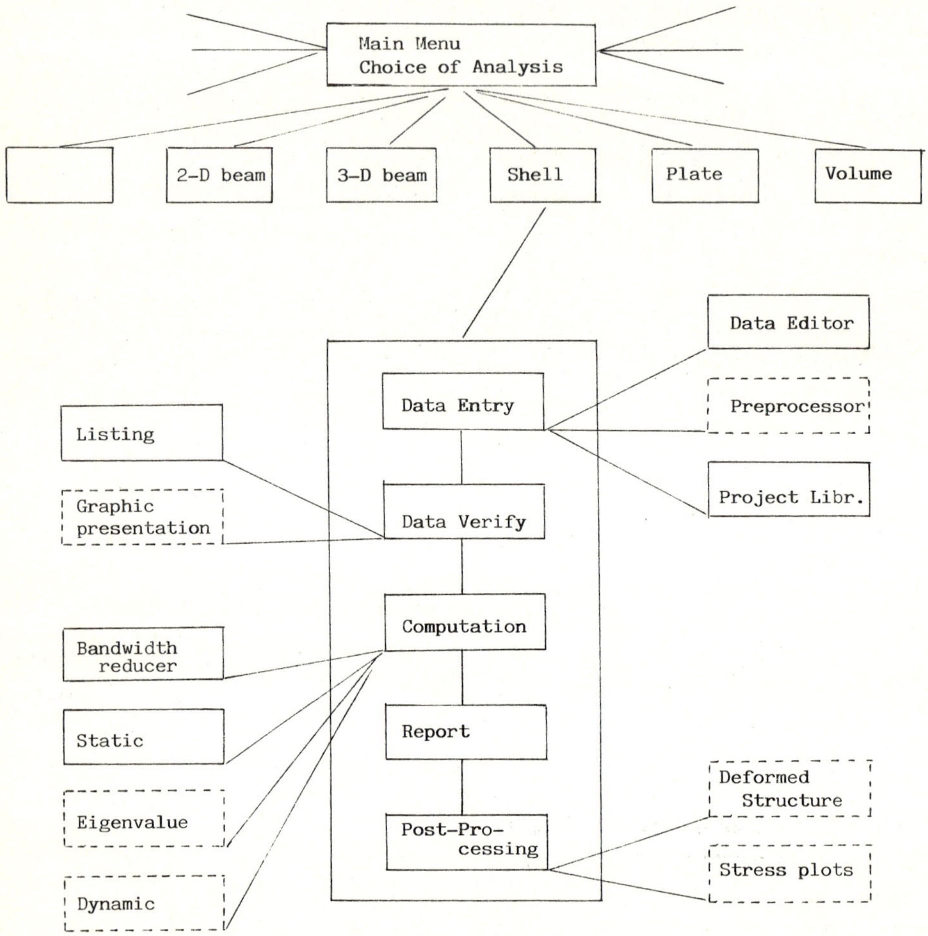

Figure 1 : Program Structure

Afterwards, the screen will display the status of the program while computing the required results. Those results can be listed selectively on the screen or on paper. Plots of the deformed structure and stress distribution can also be generated.

Technical description of shell program.

The elements used in the shell program are :

1. Quadrilateral shell element
2. 3-D beam element

The quadrilateral shell element consists of 4 triangular plate elements as shown in Figure 2, and a quadrilateral membrane element.

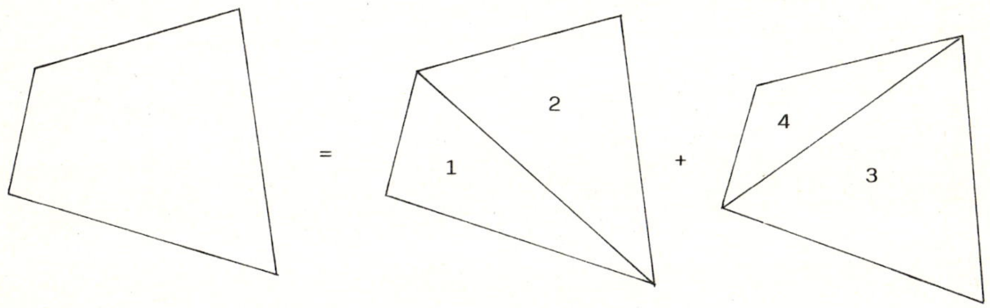

Figure 2 : Shell element

The triangular plate element is based on discrete Kirchoff theory. This element was derived by Batoz (1972) and is shown to perform very good in extreme conditions.

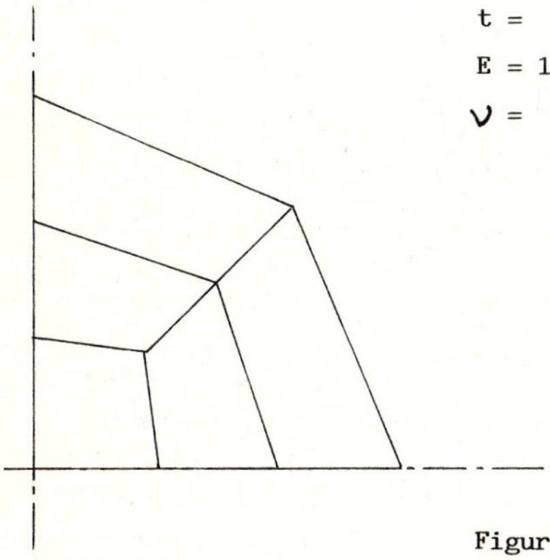

$t = 0.5$, $R = 10$ $w_{th} = 1$ w (computed) $= 0.971$

$E = 10.5$ E 6 $M_{max} = 4079.62$

$\nu = 0.33$ M_{max} (computed) $= 4073.4$

Figure 3 : Plate element example

The quadrilateral membrane element is a 4 noded isoparametric element. This element is well known. It is often critisized for its poor behaviour to represent bending in its plane. Normally, this behaviour is improved by using "incompatible modes" which are connected to extra degrees of freedom. These degrees of freedom are however condensed out at the element stiffness level. In the original SARIV program (1973), this technique is also implemented but its benificant behaviour is restricted to a rectangular element only. Using the solution proposed by Taylor et al (1976), also general quadrilateral elements can be used together with good bending behaviour. This is demonstrated in Figure 4 where a cantilever beam is modelled using membrane elements. Different options are presented in the same figure. It can be seen that the element implemented in the present shell program behaves very well.

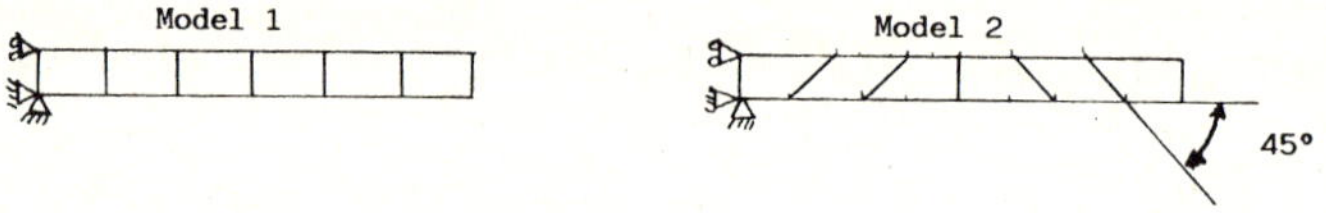

$t = 0.1$
$h = 1.$
$l = 10.$
$E = 10^7$

load case 1 (500, 500)	Model 1	Model 2
Exact	4	4
Program	3.992	3.978

load case 2 (100, 100)	Model 1	Model 2
Exact	0.6	0.6
Program	0.5995	0.6004

Figure 4 : Tip deflection of beam

The shell elements and the beams can be loaded by different forces. It is possible to define many different load cases and combinations of load cases. During postprocessing, the user can define new combinations of load cases.

The number of nodes, elements and load cases is virtually unlimited and only dependent on the available disk space. This is possible because of the use of a data base manager which is described further.

In dynamics, following analyses are possible :

1) Eigenvalue/eigenvector extraction
2) Time integration using modal analysis
3) Spectral analysis
4) Time integration and base acceleration.

Preprocessor

The preprocessor is made to decrease the amount of data necessary to build a complete finite element mesh. It is sufficient to specify the geometric data of the structure, together with the number of subdivisions wanted on each border,used to describe the geometry.
An example is given in Figure 5. This figure gives the input used to generate the mesh in Figure 6. Using the same notation, boundary conditions and edge beams can be generated simultaneously.
All input of geometry and loading is with reference to the preprocessor. The user does not have to interfere with the separate elements and/or nodes of the mesh.

Data verification

It is always very frustrating to receive a message on the screen that the program was stopped because of a singular stiffness matrix or a division by zero or another mathematical problem. Most of these problems can be avoided by a careful data verification. To help the user, adequate means are provided. This starts from a print-out of data in a clear A4-format, coupled with graphical presentations and continues with an automatic procedure that checks each element separately and predicts and verifies the disk space needed to solve the problem and to save all the results which will be generated by the program.

Project library

The users of a finite element program don't have to be computer specialists. Therefore, the program does not require to rename or copy files when a different project is started. Projects are kept as long as the engineer likes in one or more project libraries which are called from the finite element program itself. There is no need to remember the names of system utilities because every action is menu-controlled.

Post-processing

All finite element programs generate masses of output. This output can be printed on A4-size paper, selectively, such that it can be used to fit into a report directly. Plots of deformed structure, stress plots and eigenvectors can be generated. The stresses are given directly in the nodes of the element. The results can be stored in the library together with the input data and recalled at any time for additional information.
It is possible to define new load combinations without any recomputation. Not only membrane stresses and bending moments are computed but also stresses in the extreme fibers and principal stresses and directions.

Data base management

It is possible to handle very large finite element meshes in a micro computer with only restricted memory size and no virtual memory because the data is managed by a data base manager that was developed for this purpose. This data base manager is extremely fast and generates only a very small overhead in code and execution time. The detailed principles of this part of the program have been described in previous papers (Backx, 1985).
The essential feature consists in the management of two pointer lists : one list keeps track of the addresses on disk of the various items (topology, loading, displacements, etc) while the other list has the references of the items that are currently in the main memory. Exchange between disk and main memory is performed automatically.

Report writing

Report generation is an important aspect of finite element analysis because of the very large amount of data and output. All the printing is made on A4 paper-format with the appropriate headings so that it can be used directly in a report. The user can select a complete print-out or a selection of results. The user can browse through the results and the data on the screen prior to printing.
Output levels (with user indicated limits) help to decrease the amount of printed results : only the relevant results of displacements, internal forces, stresses are outputted.

Graphics

Finite elements do also heavily rely on graphical representations : plots of the preprocessor mesh, the generated geometry, the loading, the deformations, the membrane forces, the bending moments, stresses, reactions, etc. are offered on a variety of graphic screens and plotters. As well perspective drawings as projections are used to represent iso-lines and stress patterns. Examples are shown below.

CAD coupling

To extract the finite element model from a CAD system a data transfer link has been realized.
The 2D popular AutoCAD program is used to store its graphical information on a number layer. These data are copied towards a plane of the 3D shell model. The preprocessor input is proceeded to generate the mesh.

Example

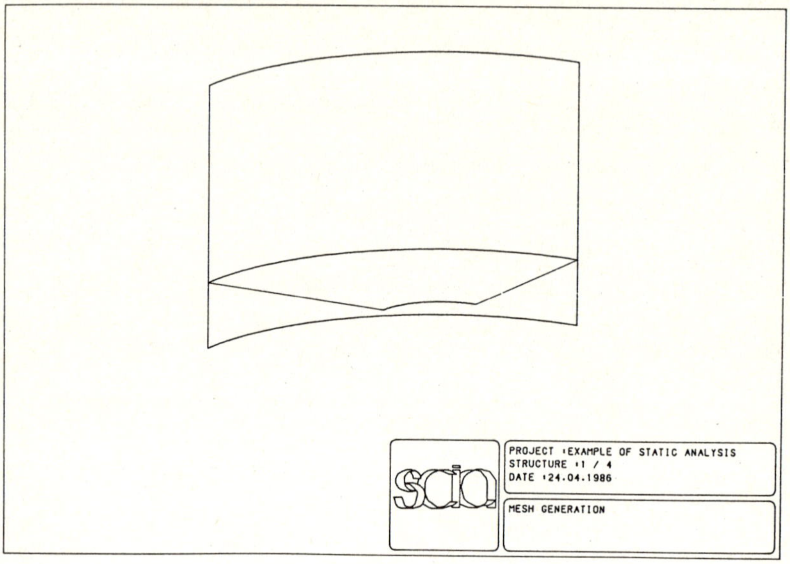

Figure 5

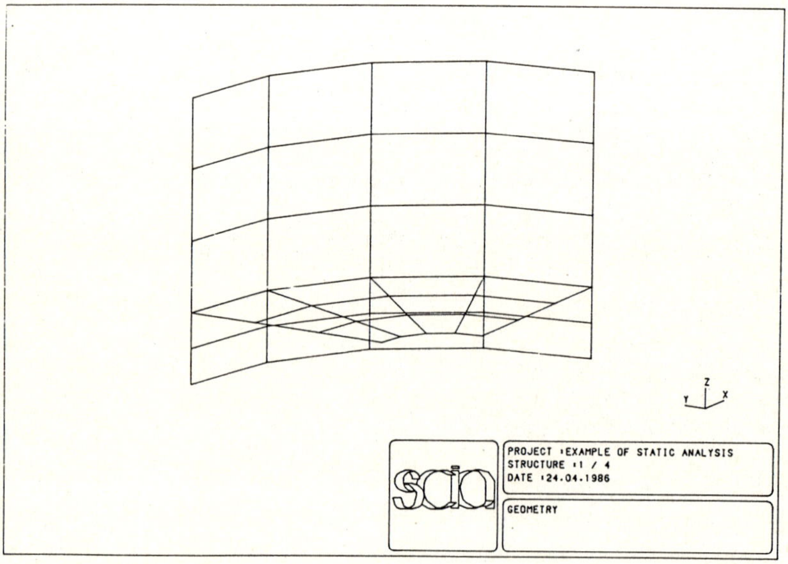

Figure 6

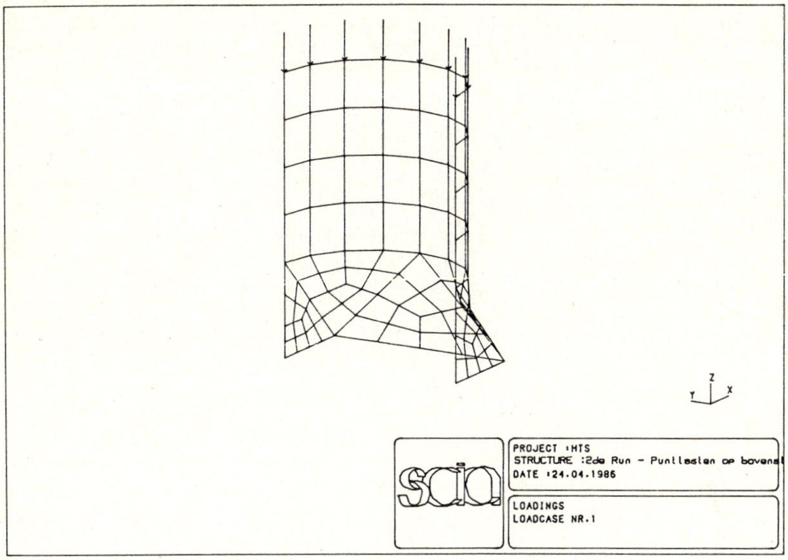

Figure 7

Figure 8

Conclusions

The program presented contains all the features of a finite element program that runs on a large computer. Moreover, because of its development on a microcomputer,the program is much more user-friendly than most other finite element programs. The program was written to be used by engineers such that no system commands have to be renumbered to work with it.
The shell program is only one out of a family of programs, which aims to form a friendly computerized engineering environment.

Batoz J.L., Dhatt G.S. "Development of two simple shell elements", AIAA J. vol.10, No 2, Feb 1972

SAP IV - A structural Analysis Program for static and dynamic response of Linear Systems, Report no. EERC73-11, College of Engineering Univ. of Cal., Berkeley, USA, 1973

Taylor R.L., Beresford P.J., Wilson E.L. "A non-conforming element for stress analysis" Int. J. Num. Math. Eng. 10, 12 1-20,1976

USING MICROCOMPUTERS FOR THE INTERACTIVE GENERATING OF SPATIAL MESHES AND RESULTS EVALUATION

J.F.Stelzer
Nuclear Research Centre (Kernforschungsanlage, KFA) D-5170
Juelich, FR Germany

SUMMARY

Reported is about useful features to facilitate the otherwise tedious task of generating meshes for finite element analysis. Combined herewith is the necessity of effective computer graphics. Microcomputers working with an interpreter language offer with their interactivity possibilities a superior quality in this respect, which is also valid what the postprocessing concerns. The graphics features appropriate for the results interpretation are described, too.

INTRODUCTION

The described pre- and postprocessor are part of a finite element software[1] consequently developed according to principles considered to be decisive for a modern generation of finite element programs. The two steps during a FE analysis which consume the most time of the engineer are the preprocessing (mesh generation, input of properties, boundary conditions and loads) and the postprocessing (results representation and interpretation). It is therefore very necessary to spend thorough considerations how to organize these routines userfriendly and comfortable.

THE PREPROCESSOR

The preprocessor should offer maximum convenience with mesh generation and data input. The described routine has the conception to bring the user on the way by putting questions, comments and menus. It is as well suited for temperature field- and rigidity analysis preprocessing. The high interactivity necessary in this respect is realized by using the interpreter technique of BASIC. If a temperature run shall be coupled with a rigidity analysis it is necessary to go after the temperature field calculation a second time through the preprocessor to adjust the mesh to the rigidity analysis. The arrays are established in the correct size by dynamic dimensioning. The necessary external files are created automatically in the requisite size, too. If the data already exists, at least the topology, a menu is offered presenting e.g. *checking the data, calling the automatic mesh generator, input of boundary conditions (restraints), graphical mesh representation, front width minimisation, mesh transformation, translatory or rotatory, mesh doubling by mirroring, joining of two or more meshes, adding or removing of nodes or elements*. It would exceed the content of this paper to deal with every topic. Because of the innovation potential combined herewith we may confine the theme more to the points related with computer graphics.

Automatic mesh generator. By hand the topology of one or more giant elements must be put in. It is then possible to advise the program to subdivide every element in a wanted amount of elements. This is done according to Hinton and Owen[2] in the natural coordinate system using the shape functions and consquently the subdivision follows all special shapes of the structure. For example, look at Fig.1 showing a machine foundation. To generate this 90-element model only the input of two elements with 8 nodes each is necessary, see Fig.2a. These both elements are automatically subdivided, Fig.2b, and then doubled by mirroring, as Fig.1 showed. The whole process is very short. To find the numbers of the nodes the appropriate regions are zoomed and attached with the node point numbers, see Fig.3.

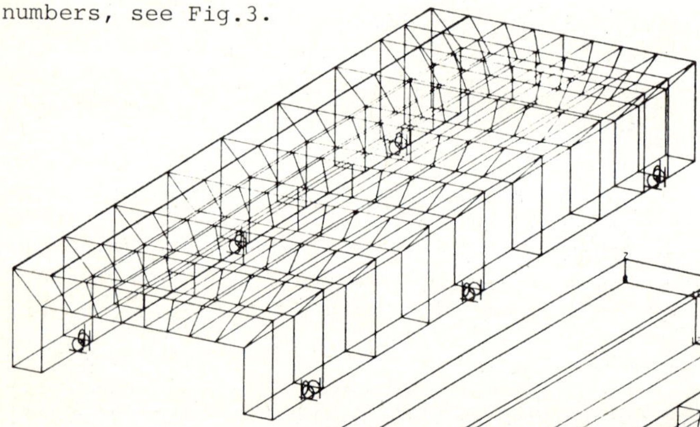

Fig. 1 (left) Mesh of a machine foundation consisting of 90 elements with 20 nodes each.

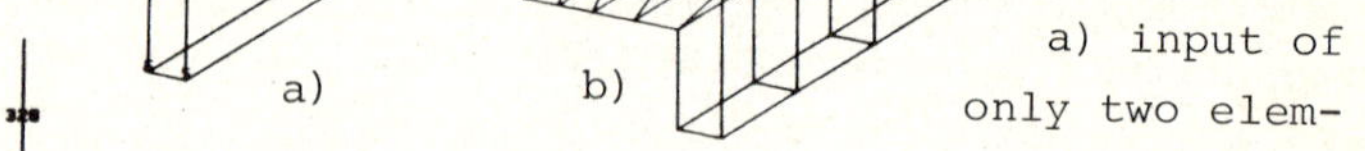

a) b)

Fig.2 a) input of only two elements. b) automatic mesh refinement

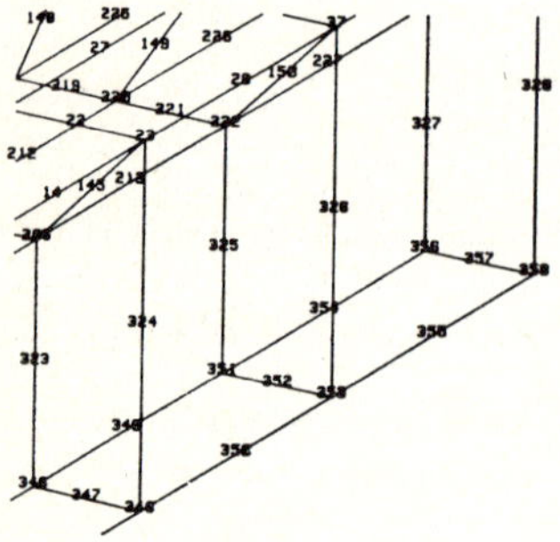

Fig.3 Partial zooming with node numbering to seek the nodes where degrees of freedom must be restrained (left)

Coordinate transformation and joining of two meshes. It may be shown by an example, see Fig.4. The consequences of an impact of a big container crashing on the ground were to be examined. The container is modelled in cylinder coordinates. There is the possibility to change from cylinder coordinates to cartesian. The lower part representing a road section is modelled separately. Finally, the program rotates the container about its centre point at an angle causing the centrepoint to be above the impact point. Both meshes are now combined using this point. The program asks for the nodal points of the separate meshes which must be combined to one and renumbers then nodes and elements.

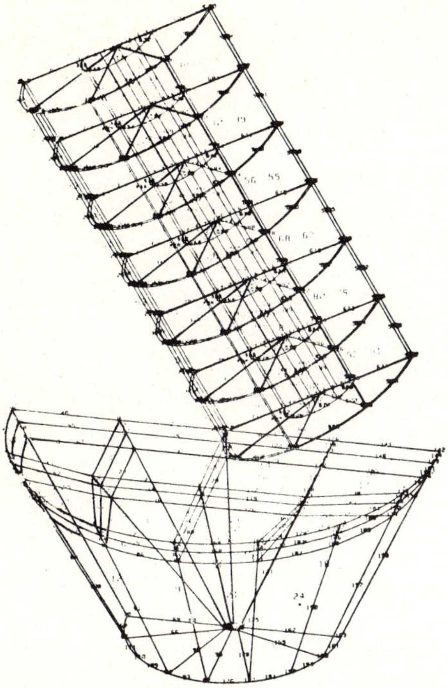

Fig.4 Example for the joining of two meshes

Using different coordinate systems. Look for this case at Fig.5 where a torus part and a tube are combined. The penetration points in the fitting region had to satisfy as well the torus equation as that one of the tube. However, the coarse element meshing in tube and torus were generated in the appropriate coordinates and finally transformed into the cartesian space. In the latter the mesh refinement was executed with the due automatism.

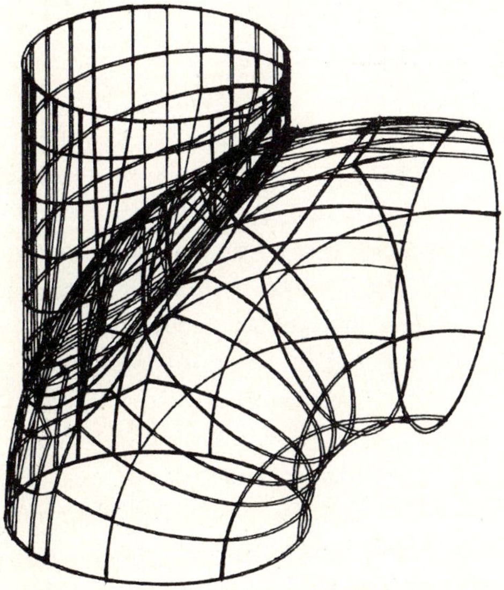

Fig.5 Joined structure combined of a quarter torus and a tube (wireframe)

Graphics representation of 3-D structures. The Figs.1 to 5 were displayed in the wire frame mode which does not give very realistic aspects. Therefore it is necessary to consider the various additional features existing for the presentation of 3-D bodies.

It is not at all sufficient to dispose only of one kind of graphical structure presentation. The usual microcomputer offers principally two ways of graphics: the *vector-* and the *raster (or pixel)* graphics. For producing paper pictures it is necessary to have a plotter (for the vector graphics) and an inkjet printer (for the raster graphics). With four-colour inkjets it is even possible to reproduce colour pictures. The following graphics features are advantageous to represent the meshes generated in the preprocessor:
1) wire frame presentation; 2) hidden line presentation; 3) hidden surface removal. 1) and 2) work in the vector graphics mode, 3) in the raster mode. We may look at these topics in more detail and with examples.

Wire frame presentation. It may be remarked that curved edges are drawn with curved lines. The bendings are piecewise calculated using the shape functions.

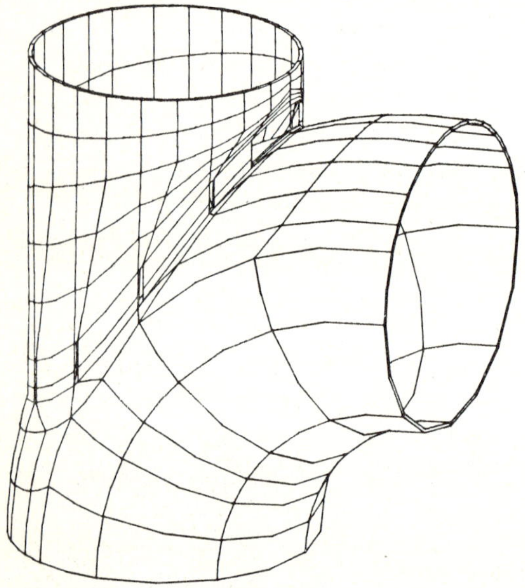

Fig.6 The torus-tube arrangement portrayed in the hidden line mode

Hidden line removal. We employ for the generation of this kind of pictures, see Fig.6, the algorithm created by Janssen[3]. In comparison to the original we made it double as quick by an a priori sorting out those surfaces the normal of which looks off from the eye point. Unfortunately, the edges cannot be drawn in the bent mode.

Hidden surface removal. The principles of this very versatile method were described by Stelzer[4]. In Fig.7 the torus-tube arrangement is drawn in this mode. The finite pixel resolution of the screen is reflected in this picture. Here, the curved edges are drawn with really bent lines. The basic idea of this mode is the drawing of the single surfaces beginning with the farmost one. The later produced surfaces hide the earlier ones. By the subdivision technique applied to every surface the spatial bending can be considered, too, see Stelzer[5]. To hide the farther positioned surfaces every surface must be filled with colour. For a picture like that in Fig.7 the screen colour is used. However, any colour can be used. In this way photograph-alike shaded images can be produced

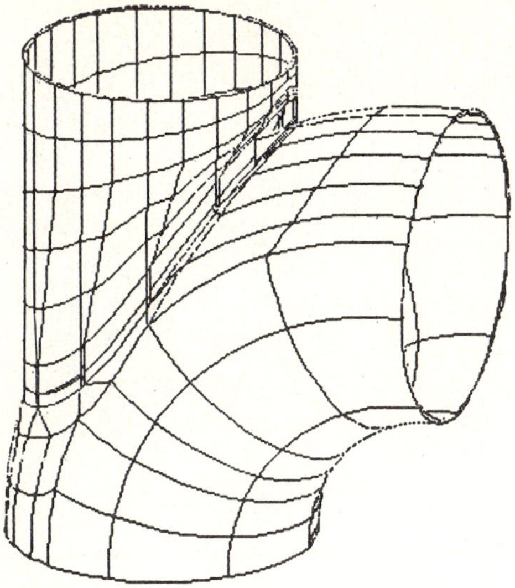

Fig.7 The torus-tube structure displayed in the hidden surface mode

by employing Lambert's cosine law, or presentations of different material zones by different colours.

Generating photo-alike pictures by shading. The spatiality of a body becomes apparently by the effect that the different surface parts reflect the arriving light beams differently according to the distinct angles the surface parts exhibit to these beams. Thus, surfaces positioned normally to the incoming light are bright and others darker, dependent on their inclination (cosine law). This effect can be applied in the hidden surface algorithm. With this feature the Fig.8 was produced. A second example can be seen in Fig.9. Here a model of a magnet is presented. This graphics feature allows to generate pictures resembling to photographs of structures not yet existing in reality.

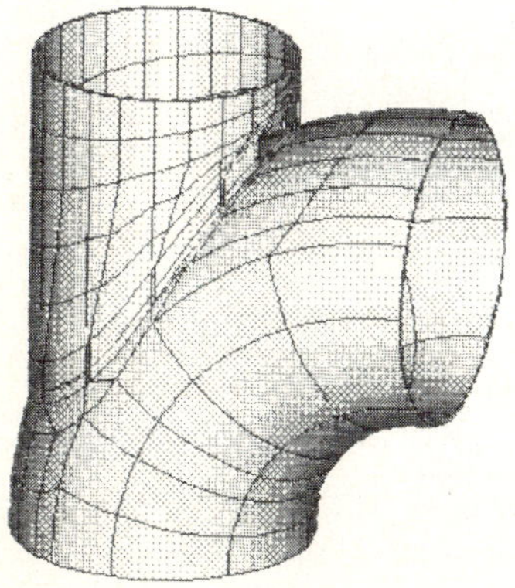

Fig.8 The torus-tube structure in the shaded mode

Surface colouring to show regions of different materials. The magnet of Fig.9 consists of six materials. The graphics routine subdivides the available colour spectrum in six parts and uses the appropriate colours for filling. In this way the Fig.10 was generated. With Fig.11 the appropriate hidden line picture additionally was appended.

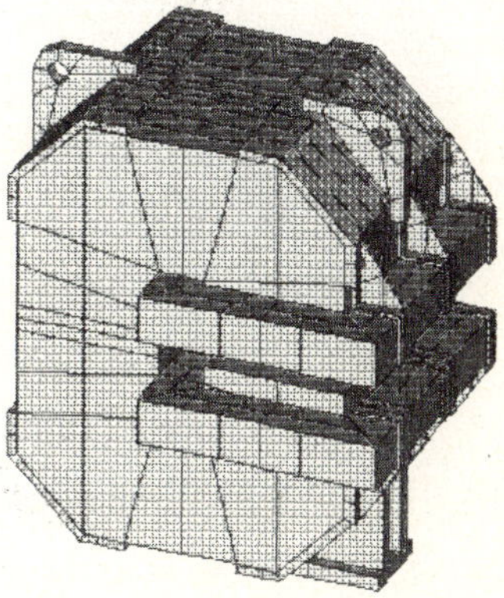

Fig.9 A magnet shown in the shaded hidden surface mode

Displaying subgroups. For the aspect realism it may be necessary to display certain subgroups. A due example is shown in Fig.12.

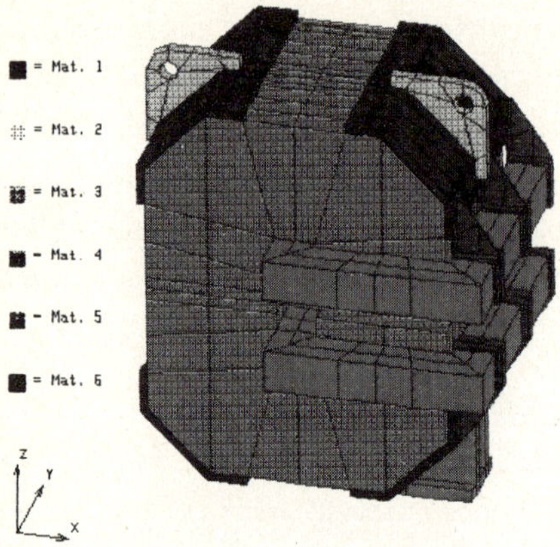

Fig.10 The different material zones of the magnet emphasized by colour

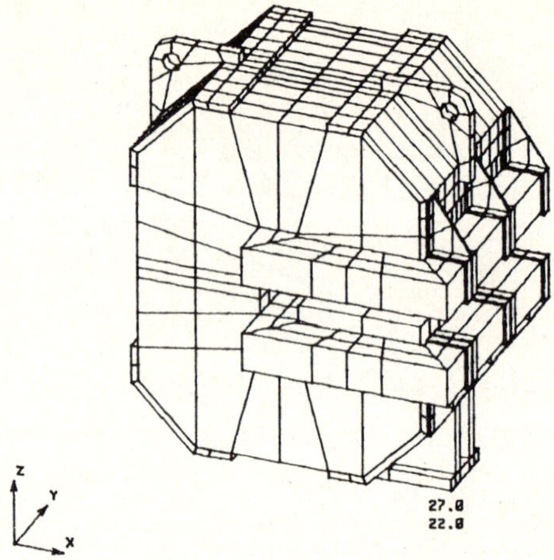

Fig.11 A hidden line plot of the magnet

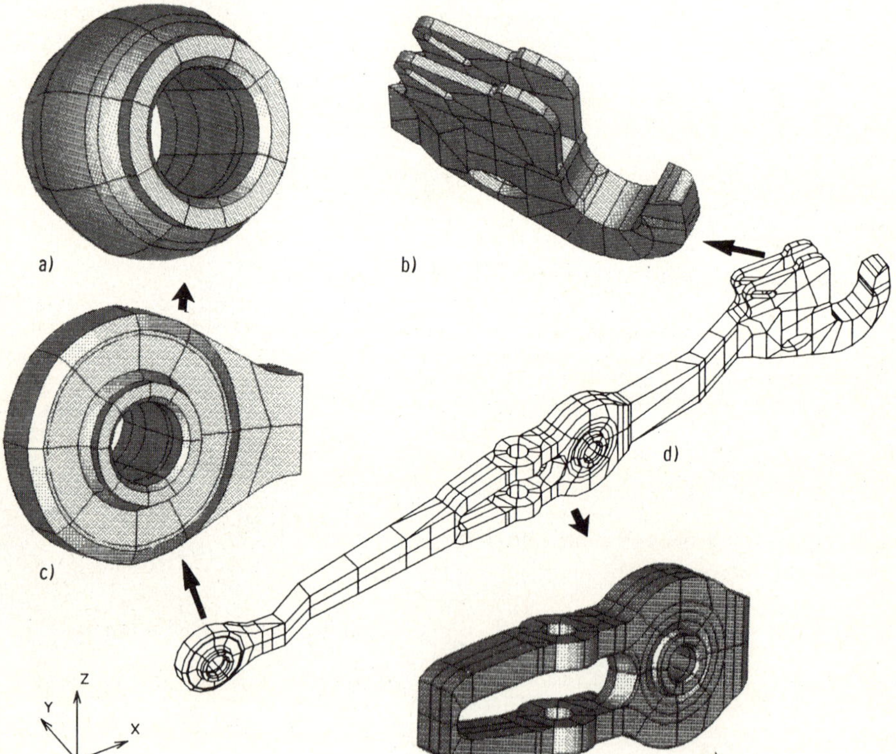

Fig.12 Combination of a hidden line plot and subgroups in the shaded hidden surface plot mode (tractor lever)

THE POSTPROCESSOR FOR THE EVALUATION OF RESULTS

It is no longer timely to deliver structural analysis results in form of staples of number output. Instead, the wise application of result graphics is the up-to-day way to show impressively what is going on in a structure. In our software we offer the following possibilities:

1) *wire frame pictures → a) printing the result values at the nodes; b) deformation plot (poor)*
2) *hidden line display → a) deformation plot; b) with contour lines on the remaining visible surfaces (good)*
3) *hidden surface presentation → a) deformed structure in the shaded mode; b) structure with coloured contour lines on the visible surfaces; c) structure (deformed or undeformed) with smooth colouring of the visible surfaces showing the function values with the spectrum between blue (lowest value) and red (highest value) (excellent).*

Because of its primitivity examples from the 1st category might be omitted. The 2nd category may be represented by Fig.13. The isolines of the reference stresses are plotted on the visible surfaces. To regard also spatially bent surfaces two coordinate transformations are necessary. The contour lines are generated in the natural coordinate system (ξ, η, ζ) by

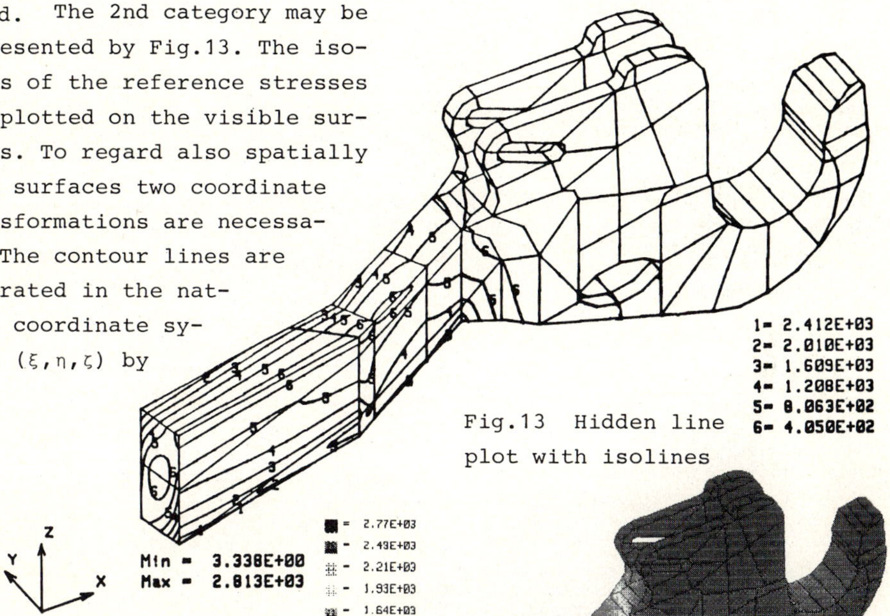

Fig.13 Hidden line plot with isolines

a procedure as described in [4], then transformed into the world coordinate system and, finally, into the screen coordinates. The same content as in Fig.13 is in Fig.14, how-

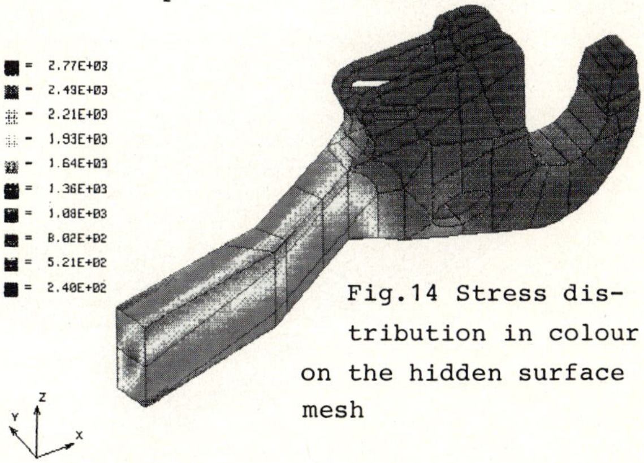

Fig.14 Stress distribution in colour on the hidden surface mesh

ever, now the distribution of the function values is expressed by colours. Undoubtly it is easier to understand where the stress peaks are with the colour picture. With the next series of Figures the coworking of the different display types may be demonstrated with the goal of an optimum result interpretation. There is a part of a lever hous-

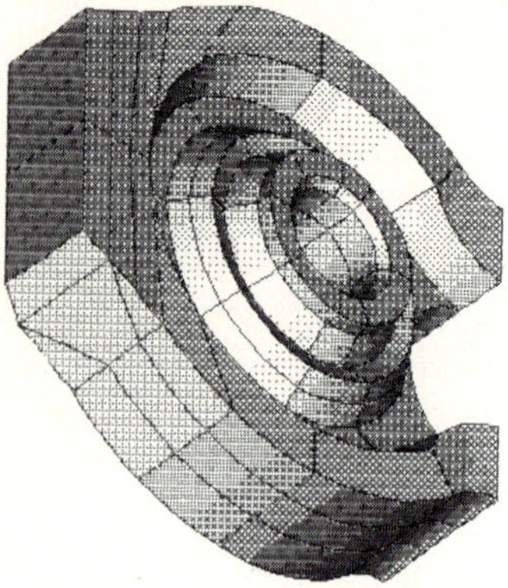

Fig.15 Lever part containing a bearing

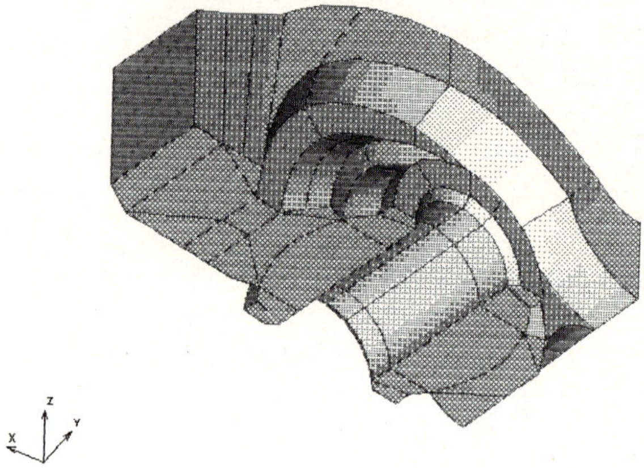

Fig.16 The part of Fig.15 cut through the middle plane

ing a bearing where a shaft is piercing. Fig. 15 shows this part. From an overall stress plot we learned that the highest stresses are to be expected in the internal of the bearing. To arrive at

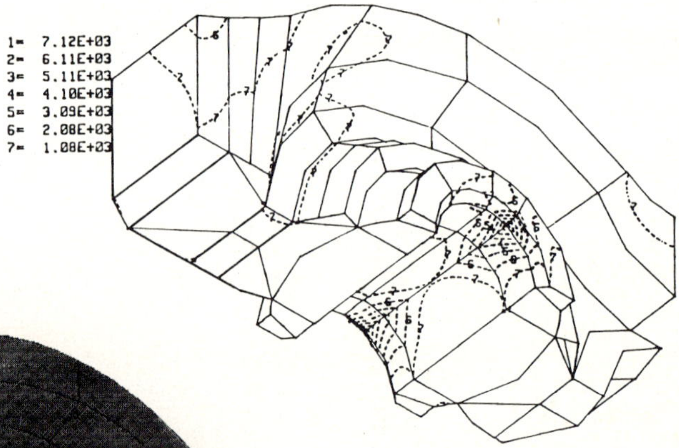

Fig.17 Hidden line plot of Fig.16 with lines of equal reference stress

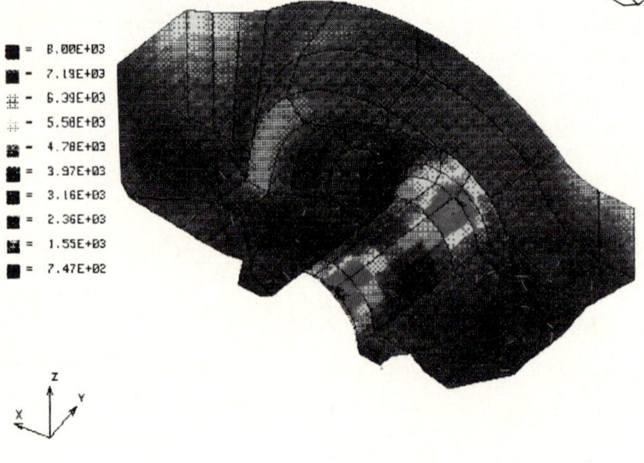

Fig.18 (left) Hidden surface plot with smooth stress colouring

this location it is appropriate to cut this arrangement in the middle plane and to look from below on it, see Fig.16. As a next step we may call the hidden line routine and the isolines feature, Fig.17. We see that the stress peaks occur at the flanks of the bearing as we expect on experience. Finally, we use the hidden surface feature with colour stress display to generate the presentation of Fig.18. The stress maximum is red surrounded by a yellow yard. The stresses decrease where the green colour appears and vanishes gradually showing cyan, light blue and dark blue. The picture shows relationships as we were accustomed to by stress optics. Now, we have a simple, cheep and reliable method to substitute this well-established but expensive means. Of course, the same method is used in the context with finite element analysis of temperature fields. Here, we get pictures in the way as usually with the thermography; of course we have now the advantage to receive a kind of infrared pictures even before the heat conducting structure exists, with all the possibilities of optimisation meaning the saving of money. It might be appropriate to explain a little the technique of the colour graphics.

To the technique of colour graphics. The computer assigns numbers to the single colours. The spectrum begins with red (→0) and ends again with red (→1) so that the following colours belong to the indicated numbers: yellow 0.1667, green 0.333, cyan 0.5, blue 0.6667, magenta 0.833. We apply the spectrum between blue (lowest value) and red (highest one) only and associate the values within this span to the due colour numbers. The colour picture on the screen is mixed from additive colours (e.g. red + green = yellow, red + blue = magenta, green + blue = cyan). It is different with paper pictures which originate from the absorption of coloured light. Here, the subtractive law of the colour mixing is valid basing on the colours yellow, cyan and magenta. The colours green (=yellow+cyan), red (=yellow+magenta) and blue (=cyan+magenta) are now results of mixing. In the to-day computer world it is not the rule that a colour hardcopy can be coupled directly with an just available computer. So it will be necessary that a driver is written which takes the pixel information of the screen and changes it in a way the hardcopy understands and, furthermore, converts the additive in subtractive colours. Moreover, things must be regarded like: the screen has a dark background but the paper a white one. Lines, e.g. the edges of the finite elements, are white on the screen but need be black on the hardcopy. Additionally, the optimisation of the ways of the printhead is advisable to shorten the time necessary for the picture dumping. Our driver, designed by Wimmer[6], couples HP computers of the series 9000 with Tektronix hardcopies.

Application to shell structures. Without difficulties the described graphics features can be employed in the context with shell entities. The Figs.19 and 20 give an example. Here, a complicated shell construction loaded by the external atmospheric pressure is shown.

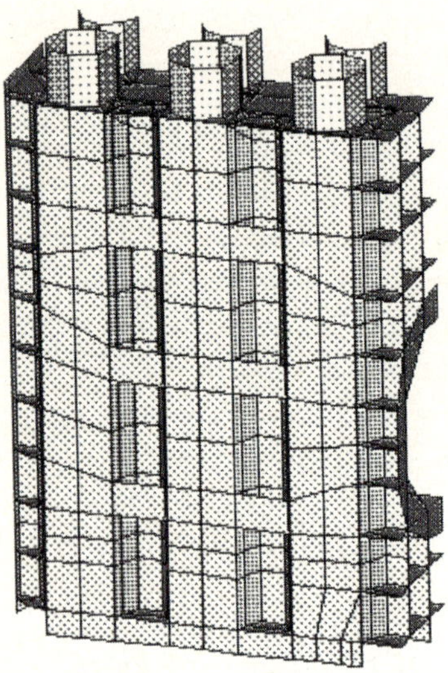

Fig.19 (left) and 20 (right) Structure plot (shaded) and deformation plot of a thin-walled vacuum vessel

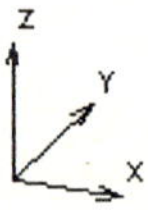

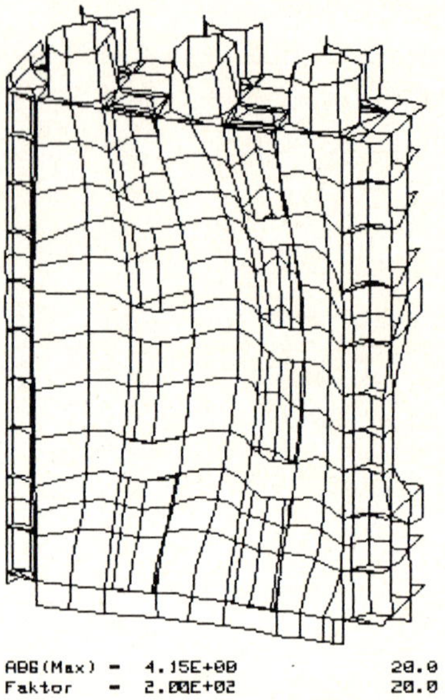

ACKNOWLEDGEMENTS

The described capabilities are part of the finite element package FEMFAM designed for big jobs on small computers. Thanks to A.Sievers, KFA Jülich, H.Stelzer and J.Wimmer, PROFEM GmbH[6], Aachen for assistance.

REFERENCES

1 Stelzer,J.F., Considerations and Strategies in Developing Finite Element Software for Desktop Computers, Eng.Comp. vol.1, no.2, 1984, 106-124

2 Hinton,E., D.R.J.Owen, An Introduction to Finite Element Computations, Pineridge Press, Swansea, 1979

3 Janssen,T.L., A Simple Effective Hidden Line Algorithm, Comp.& Structures, vol.17, no.4, 1983, 563-571

4 Stelzer,J.F., Graphical Postprocessing of Finite Element Results on Microcomputers, in: Microcomputers in Engineering, ed.B.A.Schrefler, R.W.Lewis, Pineridge Press, Swansea, 219-230

5 Stelzer,J.F., A Simple but Effective Method to Produce Colour FEM Results Presentations, Eng.Comp., vol.1, no.3, 1984, 227-231

6 Wimmer, J., Color Dump, internal report of the PROFEM GmbH, Salvatorstr.32, D-5100 Aachen

THE USE OF A CAD-SYSTEM FOR PRE- AND POSTPROCESSING OF FINITE ELEMENT CALCULATIONS

G. De Roeck and C. Vanoverschelde
Department of Civil Engineering
Katholieke Universiteit Leuven

1. INTRODUCTION

It must be clear by now that pre- and postprocessing of finite element programs are essential for engineering-tasks with regard to fast accurate preparation of input and graphical interpretation of results. It is a logical evolution in the horizontal integration of the design process.

In the past years, much effort has been spent building specific pre- and postprocessing integrated into a FE-system (SAP 6/7, SDRC-SUPERB, TITUS,...) [1], with very few possibilities for horizontal integration towards manufacturing, unless you keep on writing your own particular and expensive software.

However, some general purpose pre- and postprocessor programs for FE-applications have been written (FEMGEN/FEMVIEW, PDA-PATRAN/G, SUPERTAB,...) [2], with simple interfaces to different FE-programs.

In our opinion an even better approach is to link FE-programs to powerful existing CAD-systems, equipped with an integrated database, allowing fast development of processing, more interchange-possibilities (IGES,...) and a far better way to expand the horizontal integration, while these systems have in general some modules for manufacturing purposes.

Another improvement would be to use standard graphical software, such as GKS, PHIGS,... instead of writing device dependent graphical software.

2. SHORT PRESENTATION OF CADAM AND SAP4

2.1 CADAM

CADAM stands for Computer Graphics Augmented Design And Manufacturing containing, beside the basic 2.5D-module, several 3D-modules, such as 3D-MESH (a kind of a FE-modeller), 3D-PIPING and 3D-WIREFRAME. Some interfaces are also available, such as the Geometry Interface Module (GIM), providing application programmers with the ability to access and modify the CADAM database. Another important application interface is MeshFile, a software module that transforms database information related to a finite element model into a neutral or universal file, also called a meshfile. This transformation works in inverse direction as well, allowing any neutral file to end up as a drawing on display at the CADAM-scope.

2.2 SAP4

SAP4 is a well known Structural Analysis Program from Berkeley-university, based on the FEM, for static or dynamic analysis of linear structures. It has nine elements in its library of which three are fully supported by the link so far:

- beam element
- thin shell/plate element
- boundary element

3. OVERVIEW OF THE LINK CADAM-SAP4 WITH SURROUNDING PROCESSING [3]

As you can see in figure 3.1 we distinguish three classical parts:

1. preprocessing with 3D-MESH and MACRO-programs
2. SAP4-analysis with input and output-links
3. postprocessing with a batchprogram and MACRO-programs

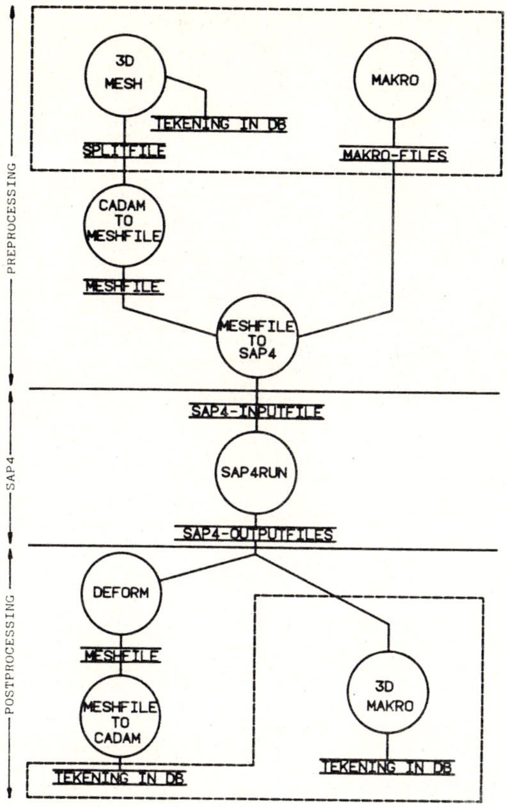

fig 3.1

3.1 Preprocessing

The MESH-module of CADAM permits the user to introduce a finite element model going through the following scheme:

- creation of node- and element-types
- associating attributes to element-types
- meshgeneration
- input of boundary conditions

Finally, all this information can be stored in the database, either as a common drawing or as a splitfile, containing only the relevant information with regard to the finite elements.
The utility Cadam To MeshFile (CTMF) extracts splitfile-information from the database to a meshfile.

On the other hand, some macro-programs have been developed, using a facility of the GIM, namely Cadam Macro Geometry Capability (CADMACGM), allowing interactive execution of user-written FORTRAN-routines through menus in dialogue-mode or parameterlists in display-mode. This interaction is programmed with Assembler language routines, simple enough to use even without familiarity with Assembler language processing. However, this facility was somewhat discouraging, because of the time-consuming procedure of compiling, link-editing and the disappointing restrictions regarding loops, mixing dialogue- and display-mode, selecting nodes or elements on display and control of action- and sub-menus. This is mainly due to the fact that data has to be passed in a one-dimensional vector. Macros are used in preprocessing mainly to ask for non-geometric information, such as material-characteristics and load-conditions. At the end of each macro the gathered data are written to temporary macro-files.
These macro-files, together with the meshfile, provide the input for the SAP4-interface, a user-written link that creates a formatted SAP4-inputfile.

3.2 SAP4-analysis

SAP4 produces a number of outputfiles for nodal displacements and element-stresses. Conversion of numerical results, such as calculation of principal stresses, is mostly a tedious task and moreover no typical engineers-work. It's more important for the designing engineer to interpret results with regard to structural design, in which graphical display plays a revealing role; where are untolerable strains occurring, how to avoid large stress-gradients, etc.? In this way postprocessing is an essential tool for economical design.

3.3 Postprocessing

For postprocessing purposes two possibilities are tested. First of all, we can use the inverse way to access the graphical database by creating a meshfile, which is to be transformed into the database with the utility MeshFile To Cadam (MFTC). This procedure is used successfully to display the deformed structure, simply by copying the old meshfile and adding scaled nodal displacements to the original coordinates. Secondly, we can make better use of the macro-facility, as there is not much control required, by using the Interface to Cadam Database

(CADCD)-facility of the GIM to generate CADAM-elements or even a complete drawing if you like. In this way macro's have been written so far for the:

- visualization as a vectorplot of principal stresses in plate- and shell- structures (in mid- and outer-planes)
- visualization as a contourplot of Von Mises-stresses in plate- and shell-structures
- visualization as a contourplot of moments in 2D-plate-structures (bending-, torsion- or reinforcement-moments)

Some applications will illustrate these possibilities.

4. Applications

4.1 Hypar shell under distributed load

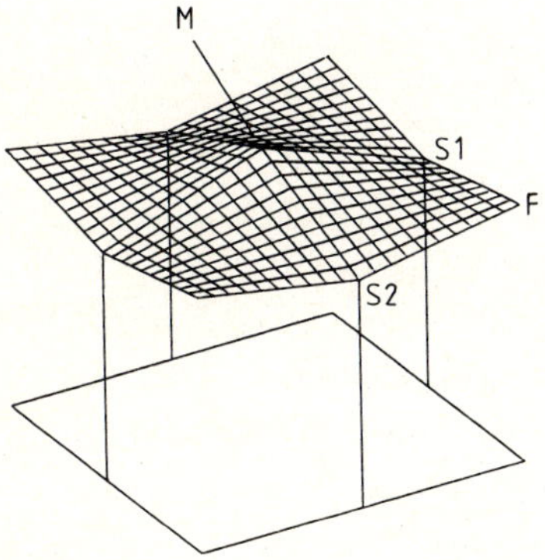

fig 4.1.1

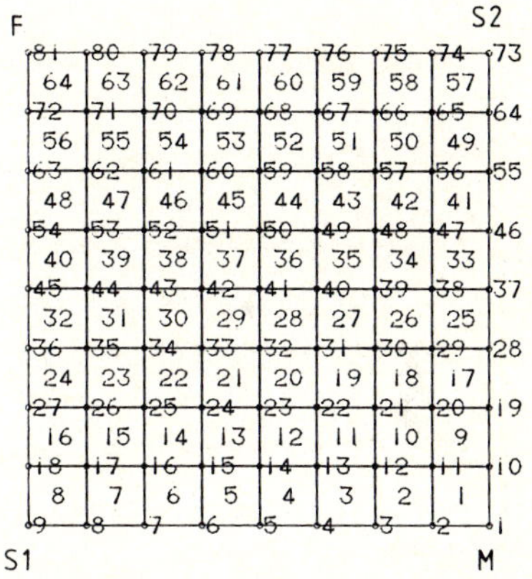

fig 4.1.2

Some research has been done recently, in cooperation with Bekaert nv, concerning the Casapack-system for constructing simple roof-structures for medical 'health-centers' (fig 4.1.1). This system consists of Poutrafils, an orthogonal grid of beams, and Stucanet, a cardboard filled wire mesh as a support for mortar, plaster or even polyurethane foam. To demonstrate some output, one quarter of the roofstructure (fig 4.1.2) has been analyzed of which we present an isometrical Von Mises-stress contourplot (fig 4.1.3) and a vectorplot of the principal membrane stresses (fig 4.1.4).

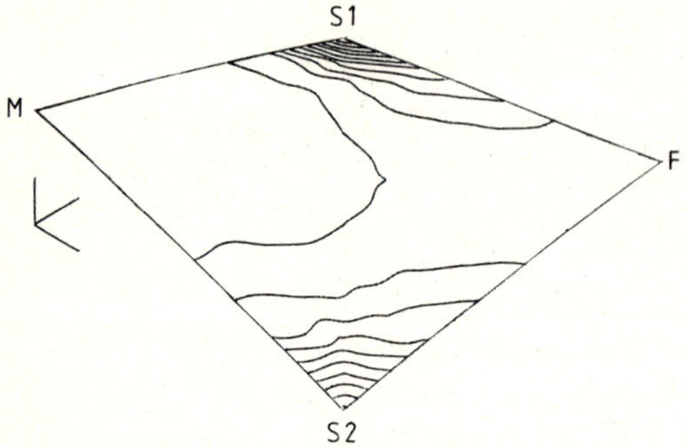

fig 4.1.3

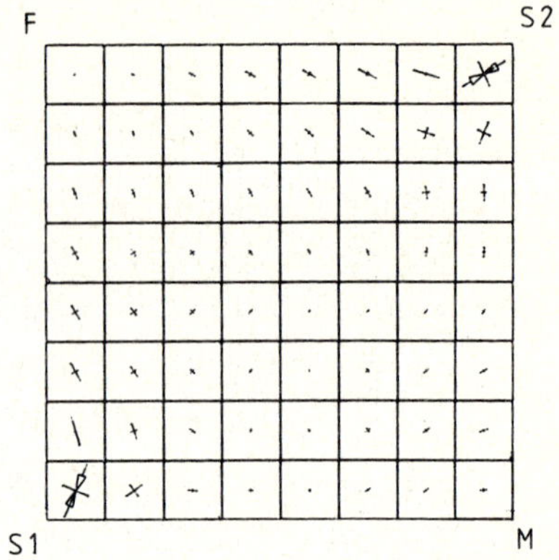

fig 4.1.4

4.2 Clamped square concrete plate

A 7 by 7m square plate with a liftshaft-opening is loaded with 2 symmetric pointloads of 10 kN. Thickness is 0.20m, while modulus of elasticity is 30000 N/mm2 and the coefficient of Poisson is supposed to be 0.

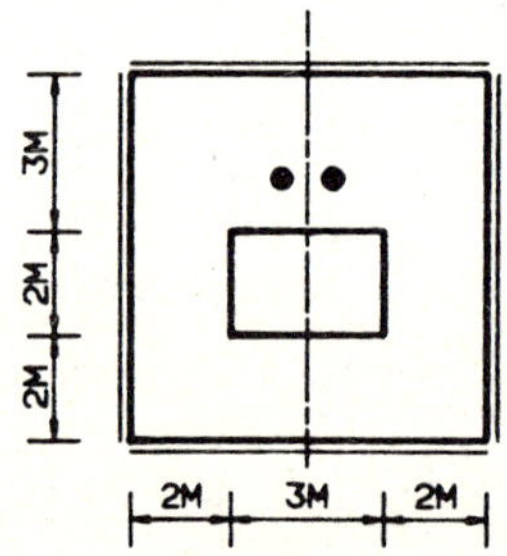

fig 4.2.1

Deformation is shown in fig 4.2.2, while part of a vectorplot of a refined mesh for principal stresses in the lower plane can be seen in fig 4.2.3.

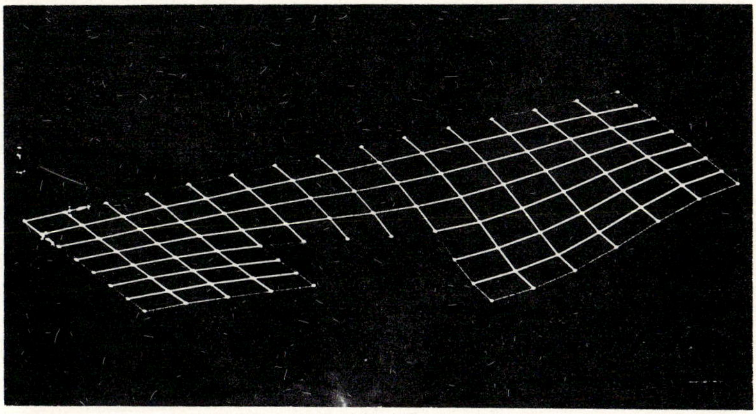

fig 4.2.2

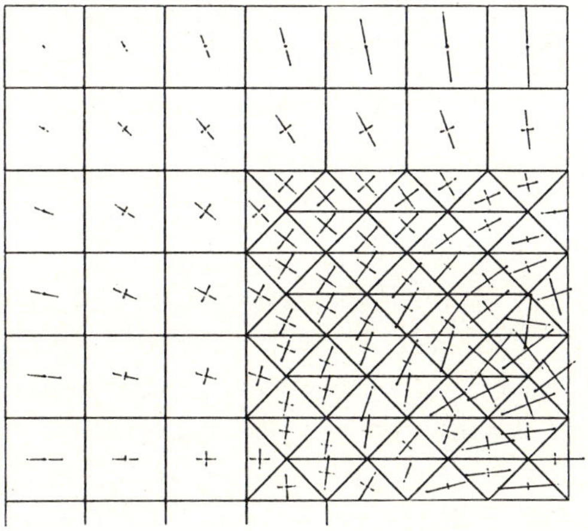

fig 4.2.3

Finally, a contourplot of reinforcement-moments is presented in fig 4.2.4 for reinforcement against MY-moments in upperplane(+Z).

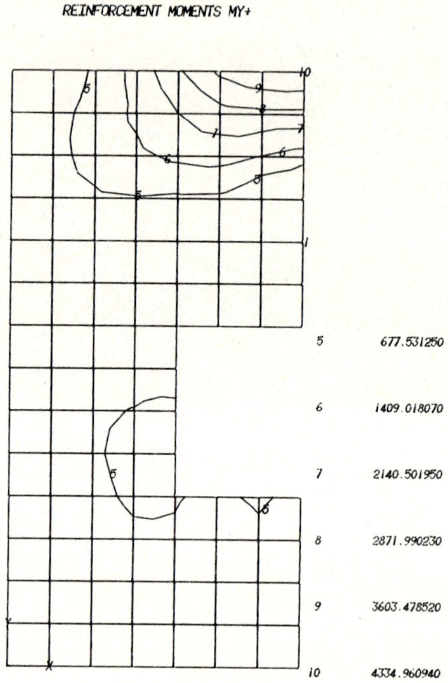

fig 4.2.4

5. Conclusion

As expectations were high concerning full interactive modelling of structures with a CAD-system, we had to experience some nasty drawbacks. Nevertheless, we achieved to link a general CAD-system to a FE-program, allowing designers in general and students in particular to introduce a structure to be analyzed in a economical way. Graphical evaluation of results and alternatives increases structural insight enormously, while further expansion towards integration of manufacturing-modules lies still open, what can be problematical with 'closed' FE-packages.

6. References

[1]_MACKERLE J., Review of pre- and postprocessor programs included in the major commercial general purpose finite element packages, Adv. Eng. Software, Vol. 5, No 1, 1983

[2]_MACKERLE J., Review of general purpose pre- and postprocessor programs for the finite element applications, Adv. Eng. Software, Vol. 5, No 3, 1983

[3]_DEGRANDE G., VERBRUGGEN W., Gebruik van een CAD systeem voor pre- en postprocessing van eindige elementenanalyses, Eindwerk K.U.Leuven, Dept. Bouwkunde, 1986

MATHEMATICAL MODEL FOR CALCULATION OF REBUILD OF R/C MUSHROOM SLAB

D. Van Gemert, C. Vanoverschelde and M. Vanden Bosch
K.U.Leuven, Reyntjens Laboratory
de Croylaan 2
3030 Heverlee - BELGIUM

ABSTRACT

The influence of changes of a reinforced concrete mushroom floor on the distribution of internal forces in it is studied, in view of a strengthening of the plate by means of externally bonded steel strips. The finite element modelling of the plate is discussed, and a comparison of finite element and finite difference results is made. An evaluation of the obtained results, from the design engineers viewpoint, is given.

1. INTRODUCTION

The load bearing structure of a factory building at Amersfoort (The Netherlands) consisted of a three storey reinforced concrete mushroom slab. A longitudinal cross-section of the building is shown in figure 1.

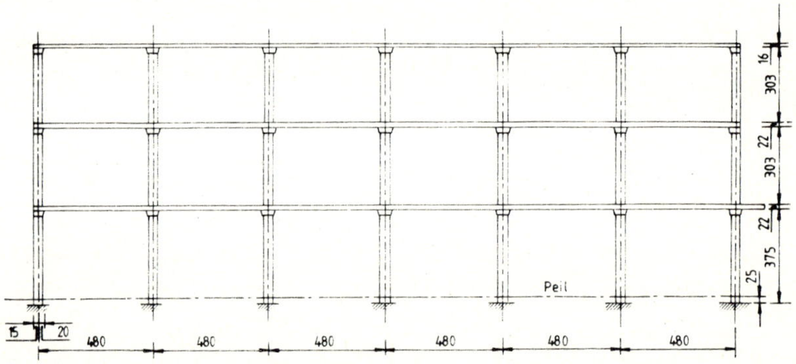

Fig. 1 : Longitudinal cross-section

The plan of the slab is shown in figure 2.

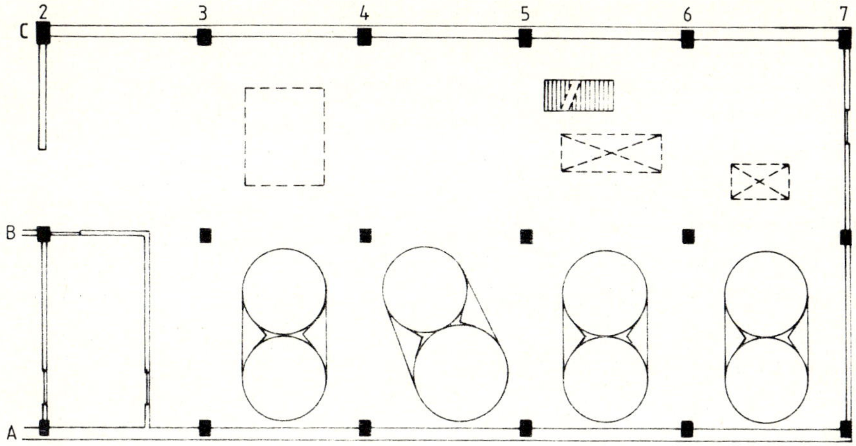

Fig. 2 : Plan of mushroom slab (dotted lines = existing openings)

Some openings were present in this slab. Due to a need for changing production processes, some openings had to be closed, and four big additional openings had to be cut in the slab. The magnitude of these additional openings is also shown in Fig. 2. The existing openings had to be closed.
Normally the force distribution in a mushroom slab is little influenced by openings, as long as they are not made in the immediate vicinity of the column heads. In this case, however, a more detailed analysis was necessary because of the importance of the openings. The only possible methods of analysis are the finite difference method and the finite element method. The method used was the finite element method. The mathematical modelling was realised with the facilities, provided by the finite element program SAP IV, produced at the University of Berkeley U.S.A. A comparison is given with classical numerical techniques : frame-analysis, executed by n.v. Rescona, and finite difference analysis, executed by b.v. Logisterion.

2. DESIGN SPECIFICATIONS

The thickness of the plate is 220 mm. The plate was originally designed for an overall live load of 7 kN/m^2. In the new situation after the adaptations the live loads are 7 kN/m^2 in the zone BC, and 2,5 kN/m^2 in the zone A - B (see Fig. 2).
The analysis of the plate was executed with a total load of 7,5 kN/m^2 in the zone A - B,

and a total load of 12 kN/m^2 in the zone B - C. The existing openings, which had to be closed, were covered with reinforced concrete plates with a thickness of 120 mm. The weight of these plates and the normal live load on them is uniformly distributed over the edges of the openings. In the analysis the openings were kept, because one could not assume that the covering plates would cooperate with the existing structure. Due to the asymmetrical composition of the plate after the introduction of the openings, the influence of the bending stiffness of the columns had to be introduced. This was done in a simplified way, Fig. 3.

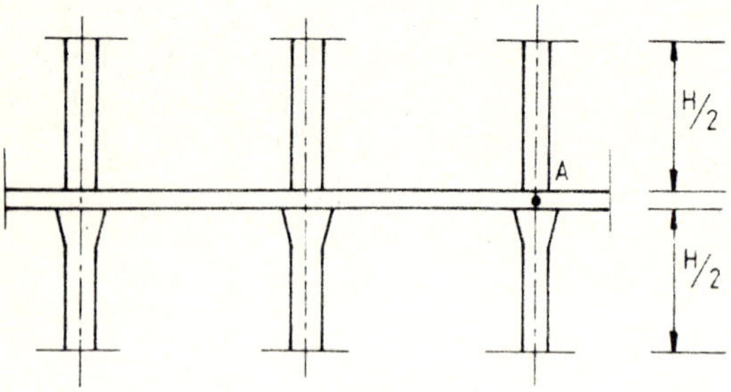

Fig. 3 a

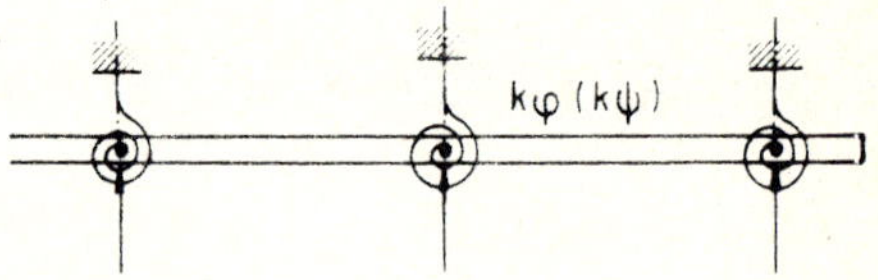

Fig. 3b

The moment distribution in the columns was supposed to have a zero value in the middle of the height. The rotation of the column head A was calculated under the action of a unit moment. The relation between unit moment and rotation gave the spring constants k_φ or k_ψ, which were than introduced as boundary elements in the finite element model, Fig. 3b.

3. FINITE ELEMENT MODEL

The finite element distribution used in the analysis is shown in Fig. 4.
This finite element distribution was chosen to facilitate the comparison of the calculated results, with the results of finite difference calculations. The mesh was refined in the column regions, as well in the horizontal as in the vertical direction. This procedure gave a concentration of small elements around the column heads, were the internal forces are the most important. The total number of elements was 656.

4. RESULTS OF ANALYSIS

The evolution of the mean values of the m_x- and m_y-moments in the column and central strips (Fig. 5) was calculated. The results have been compared with the results of a finite difference calculation. The comparison is made in Table 1 for the moments in the column strips. The moment values in the original plate, calculated by the equivalent frame method, are also indicated. The notations are clarified in Fig. 2.

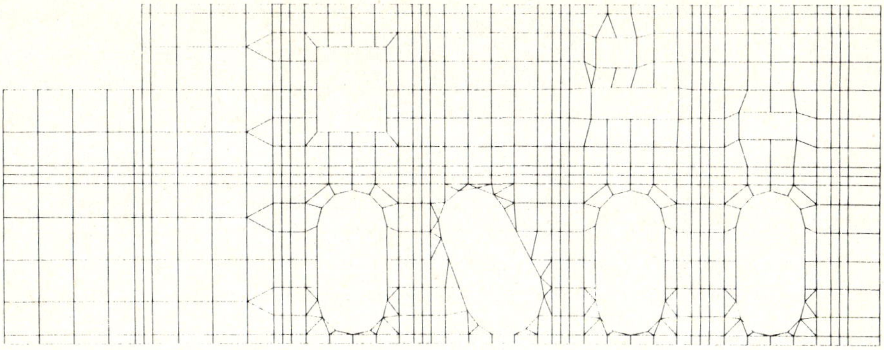

Fig. 4 : F.E. distribution over mushroom slab

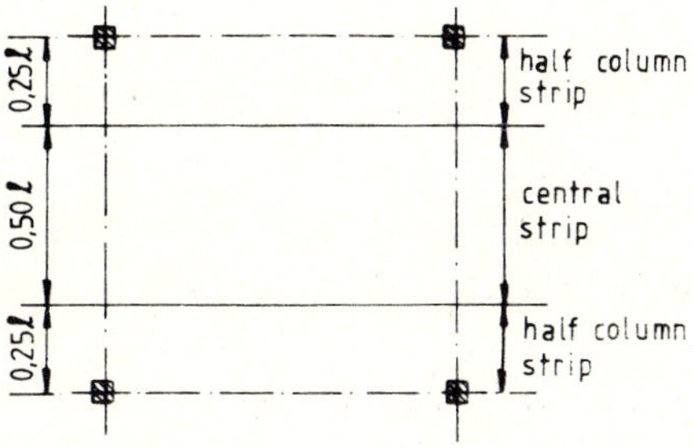

Fig. 5 : Definition of column and central strips in mushroom plate

A, B and C determine the horizontal row of columns, whereas the numbers 2 to 7 refer to the position of the vertical axis.
The general picture of finite difference and finite element calculations looks the same. However, individual moment values can differ considerably. Especially in our case of a special mushroom plate, it was not possible to find a general explanation for these differences, because in certain places the F.E. analysis gives higher values, in other places it is the finite difference method which gives the higher values, and this phenomenon is found as well for positive as for negative moments. In such a case the design engineer is happy to work with reinforced concrete, because this material has a great possibility of plastic deformation. The difficulty of explanation of differences is also due to the fact that the big openings are weakening the plate,

		-37300	20200	-55200	20200	-37300
epuiv. frame		-37300	20200	-55200	20200	-37300
A_2C_2	Dif.	-12548	12068	-39917	26617	-23782
	F.E.	-16929	19939	-42580	9796	-28695
	Max.	-47110	23970	-65720	16500	-30750
A_3C_3	Dif.	-10600	10128	-48441	28275	-25270
	F.E.	-11960	18385	-30254	10457	-7375
	Max.	-35195	20385	-48205	12545	-22470
A_6C_6	Dif.	-7581	6135	-41599	28767	-25601
	F.E.	-8970	14066	-27325	13543	-8845
	Max.	-27365	16750	-41105	14150	-27380

Table 1 : Comparison of m_x- moment (Nm/m)

(Dif. : Finite Difference, mean value

F.E. : Finite Element, mean value

Max. : Finite Element, max. value over strip).

but the loads are also decreased considerably.

The comparison of the moments in the adapted plate with the moments in the original plate, revealed that the moments in the zone A - B, where the big openings are made, are decreasing. The moments in the field B - C are increasing. This is quite normal because in the zone of the openings the loads are diminished from 7 to 2,5 kN/m^2, and because the continuity over the column has dropped due to the fact that the plate in the zone A - B becomes more flexible by the presence of the openings. The same effects are responsable for the fact that the negative moments at the column heads are not increasing. Only the positive moments are increasing.

These results are very favourable for a strengthening procedure by means of epoxy bonded external steel strips, because now all the necessary strengthening works can be done at the bottom side of the plate.

A scheme of the calculated epoxy bonded steel strips is given in Fig. 6.

The design of these reinforcements is discussed in ref. 1 and 2.

The strengthening works are executed in 1985.

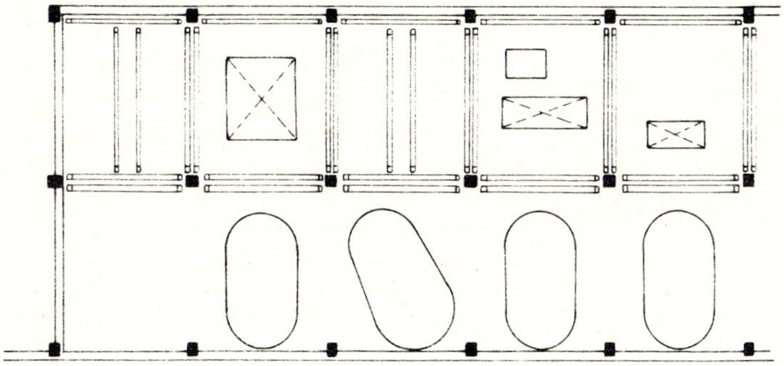

Fig. 6 : Plan of epoxy bonded reinforcements

5. CONCLUSION

The analysis of plates with complex geometrical shapes can easily be done by the finite element technique. However, a considerable amount of experience is needed to find the most appropriate finite element distribution, in order to make the interpretation of the results accessible to the design engineer. In many cases it is not necessary to strive for a minimum number of elements, because the savings in numerical calculation will be more than compensated by the increase in interpretation costs.

6. REFERENCES

1. Van Gemert, D., Vanden Bosch M., : "Design of externally bonded steel reinforcements in bending" (in Dutch). Tijdschrift der Openbare Werken van België, 1982, nr. 1.
2. Van Gemert, D., Vanoverschelde, C., Norrée, F., Janssens, H. : "Praktische berekeningswijze voor orthotrope platen". Tijdschrift der Openbare Werken van België, 1986, nr. 1.

DESIGN AND ANALYSIS OF THE WINTER GARDEN IN NEW YORK CITY

Paul Lew, Abraham Gutman,
Leonid Zborovsky & Lee Petrella

Lev Zetlin Associates, Inc.
641 Avenue of the Americas
New York, New York 10011

I. INTRODUCTION

The Winter Garden is the centerpiece of the World Financial Center (See Fig.1 & 2). Its structure consists of a complex assemblage of structural elements that must interact to achieve the total strength of the structure. These structural elements include a telescoping barrel vault roof, a half dome, a large arch folded plate front wall, and a "U"-shaped stiffening ring. To analyze all these structural elements at once would create an enormous computer analysis. The meaning of the computer analysis would be obscured by the complexity of the results.

Instead of doing one large complex analysis, the structural designers elected to subdivide the structure into basic elements and analyze the interaction of these elements by controlled forces and displacements. This method allowed for a greater understanding of the structure and the interaction of the structural elements.

II. DESCRIPTION OF THE STRUCTURAL ELEMENTS

The roof plan (See Figure 3) provides an overview of the structure. The Telescoping barrel vaults consist of five-feet-deep curved arch trusses spaced at 15'-0" on center (See Figure 4). These trusses span a maximum of 110 feet telescoping down to 90 feet and 70 feet. At each step in span, there is a double truss system - one for the larger span and one for the smaller span. The thrusts of the arches are taken by a "U"-shaped stiffening ring at the base of the arches.

Perpendicular to the arch, lateral forces were taken by in plane bracing within the circular vault surface. To limit the relative movement of the telescoping arches lateral out of plane movement, bridging and intertruss struts were used.

At the front of the building is a 100'-0" wide by 115'-0" high semi-circular glass wall. To span this expanse, the front wall was shaped into a folded plate with in-plane trusses (See Figure 5). The front wall is dependent on the roof for its lateral support. At the other end of the structure the roof terminated in a half dome, which has its main axis horizontal (See Figure 6).

The dome vaults were dependent on the "U"-shaped stiffening ring built into the adjoining roof for its strength (See Figure 7). This assemblage of structural systems could be analyzed together or analyzed separately as long as the influence of each element on the other was determined. It was elected to analyze each element separately so that a better understanding of the basic systems and the interaction of the different systems could be developed.

III. ANALYSED OF THE MAIN STRUCTURAL SYSTEMS

A. The Arch Vaults -

1. Main direction of Span -

The main direction of span was done by five feet deep trusses spaced 15'-0" on center. The analysis was done by modeling the truss with a linear axial straight line truss chords. Two issues had to be investigated that were not included in the basic truss analyses. These were :

a. The effect of the displacement of the stiffening ring at the arch supports.

b. The effect of the top and bottom chords of the truss being continuous and curved rather than straight and hinged at the joints.

The effect of displacement of the stiffening ring was studied by displacing the support of the arch trusses by the maximum amount the stiffening ring would be allowed to move (2"). The effect of displacing the arch truss support 2" laterally was found to be negligible.

The effect of the continuous curved top and bottom chord was studied in one computer run that was modeled as a frame with a curved sections between truss panel points. The result was to develop a ratio between the arches axial forces and bending moment in the top and bottom chords. This ratio was then applied to all the other arch trusses to account for bending due to the continuous curvature of the top and bottom chords.

2. Perpendicular to the main span -

Perpendicular to the main direction of span the arch vaults depended on in-plane bracing following the cylindrical vault for lateral resistance. This in-plane bracing diaphragm has in addition to the in-plane diaphragm forces that would occur if the diaphragm was flat, an overturning effect due to the rise of the vault (See Figure 8).

To properly model this, the vaulted in-plane diaphragm truss was modeled. To take into account the out of plane resistance of the main arch trusses, the top, bottom and intermediate chords of the diaphragm truss which were the top chords of the main arch trusses had an equivalent stiffness included equal to the effective moment of inertia of the main trusses. Each telescoping vault diaphragm truss was designed independently but it was necessary to limit the relative movement of each telescoping vault relative to each other. To accomplish this it was realized that the end dome was rigid relative to the arch vaults and a frame system of purlins, spanning between the main trusses and down the steps in the telescoping vaults could provide this method of limiting deflection. By creating this frame force in one vault diaphragm could be redistributed to the other vault diaphragms. In the analysis of the purlin frames the arch vault diaphragms were linked together (See Figure 8) and the dome was treated as a support.

B. The End Dome -

The end dome was complex in that the axis of dome was horizontal and that the end hub had to be modeled by finite element (See Figure 9). The edge of the dome was supported by a double vault truss at the step between the last vault and the end dome.

C. The Front Wall -

The front wall folded plate is one of the largest glass walls ever built (See Figure 10). The dimensions of the front wall is over 100 feet wide and 115 feet high. The folded plate shape of the front wall gave it its strength. The incline faces had built-in trusses which span to a vierendeel horizontal truss about 70 feet above the base. The front wall was laterally supported by the diaphragm truss of the arch vault roof. The effect of displacements of the arch vault roof on the front wall had to be determined. To determine the effect, the supports of the front wall that would belong to the arch roof were displaced the amount they would be displaced from the roof. The forces and moments that resulted in the front wall from these displacements were added to the primary forces and moments from loads.

D. Stiffening Ring -

The basic strength of a long span roof lies in its boundary conditions. In the case of the Winter Garden, the "U"-shaped stiffening ring on the adjoining flat roofs at the arch vault's base was the key element whose stiffness allowed the other structural elements to be light. The stiffening ring was studied from two aspects; strength and stiffness. The strength was developed by an in-plane truss in the flat roof. This allowed the development of effective connections for the known truss diaphragm forces. The truss by itself did not have adequate stiffness without including the stiffness of the concrete roof deck on top of the stiffening truss. To measure the effectiveness of the roof deck, a second run was done by adding a in-plane infill finite element to the trusses of the thickness of the roof deck. It was found that the roof slab decreased the deflection by a third and the stresses induced in the deck were within range of a concrete diaphragm.

CONCLUSION

The Winter Garden structure was a complex interaction of various structural systems. Each structural system in itself had its own degree of complexity. By separating each structural system and studying the influence of structural distortions of one system on another, it was possible to develop a greater understanding of the primary forces on the structure and the interaction of the various structural systems. These effects would have been lost if the structure had been analyzed in one massive computer run. The final structure will be one of the key indoor spaces in New York City.

Figure 1

Interior View

Figure 2

Rendering

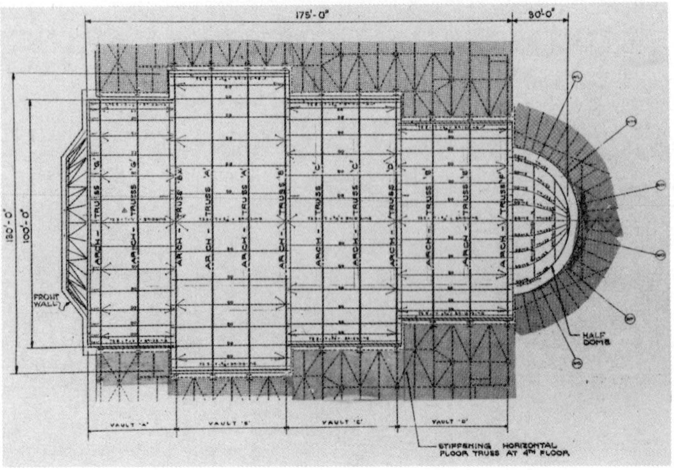

Figure 3

Roof Plan

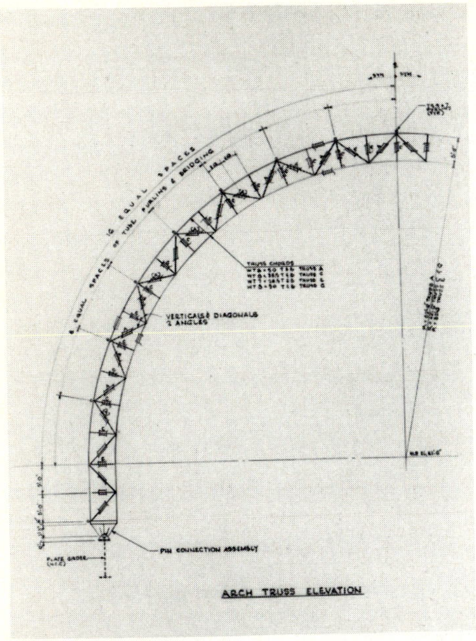

Figure 4 - Arch Truss

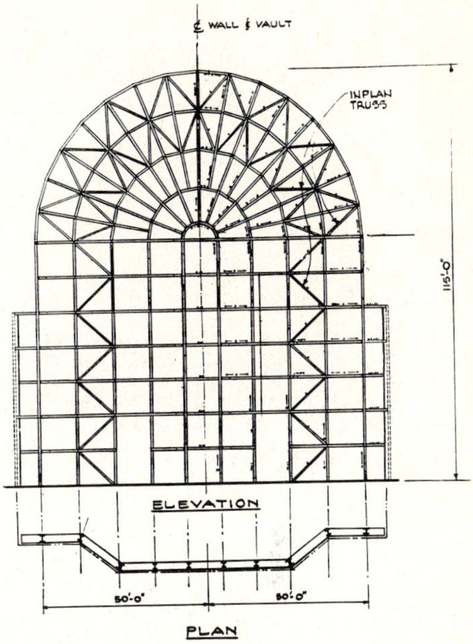

Figure 5 - Front Wall

Figure 6 - Half Dome

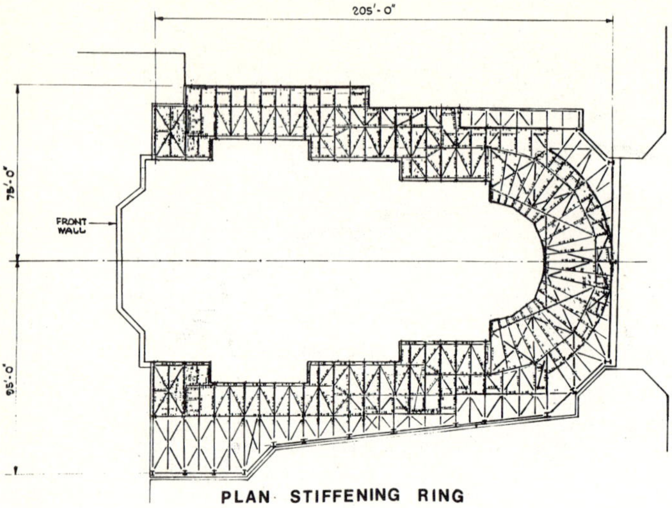

Figure 7

Plan Stiffening Ring

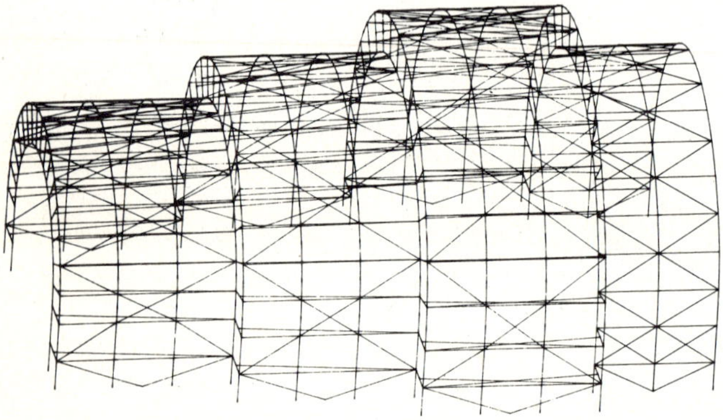

Figure 8

Arch Bracing

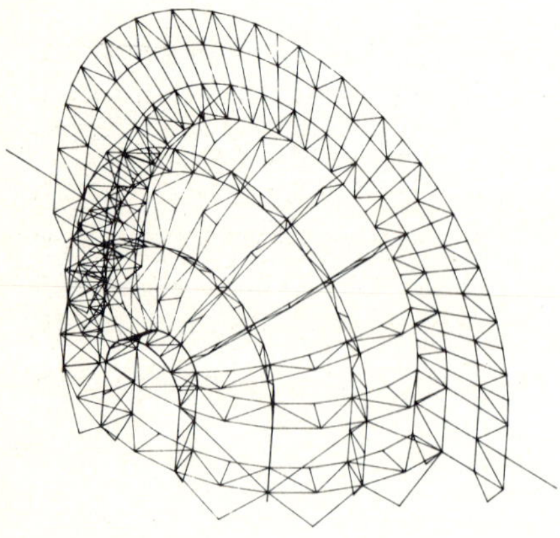

Figure 9 - Half Dome

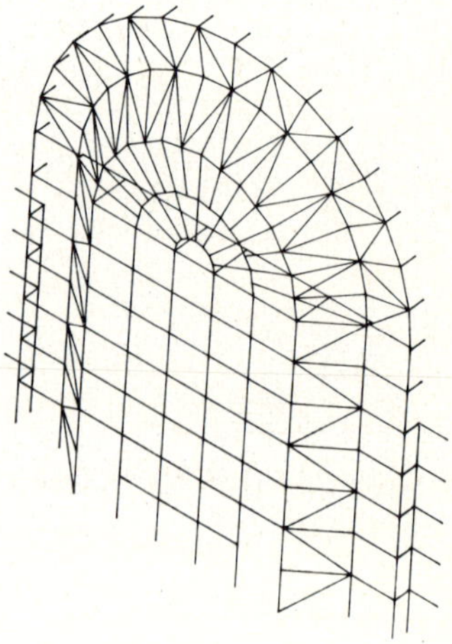

Figure 10 - Front Wall

EXPERIMENTAL AND NUMERICAL VERIFICATION OF THE STRUCTURAL BEHAVIOUR OF A FOAMED HYPERBOLIC PARABOLOID SHELL

G. De Roeck and C. Vanoverschelde
Department of Civil Engineering
3030 Leuven/Belgium

1. Introduction

Casapack is a building system of Bekaert International Trade, specially developed in order to keep housing simple.

The frame consists of lattice girders made of galvanized steel ("Poutrafils") and cardboard filled galvanized wire mesh ("Stucanet") as a mortar/plaster support.

Other fastening and assembling accessories are clips, clamps and rods. Walls can be filled with loosely (cement) bound gravel, local aggregate, insulating material, etc. depending on local availability and climate conditions (Fig.1).

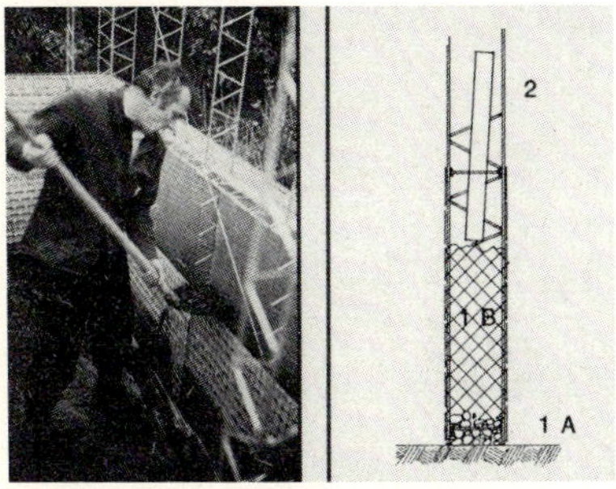

Fig.1 Erection of walls

Some of the advantages of this building system are:

- maximal economical use of (simple) equipment, (unskilled) labour and (local) building materials;
- excellent adaptability to local needs (thermal comfort, resistance to seismic shocks or hurricanes, local building culture, ...);
- favourable quality-to-price ratio.

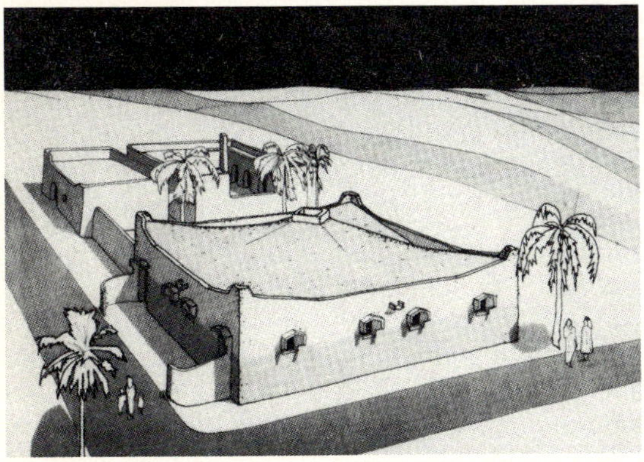

Fig.2

Light-weight roofstructures can also be built with the casapack - system. In this case a grid of poutrafils serves as the prior bearing structure and the base for fixing the Stucanet. Against the Stucanet, polyurethane is sprayed for waterproof covering and thermal insulation. In this way flat roofs, folded plate structures, pyramidal roofs, hyperbolic paraboloids can be erected (Fig.2).

In Fig.3 the construction principle of a hyperbolic paraboloid is illustrated:

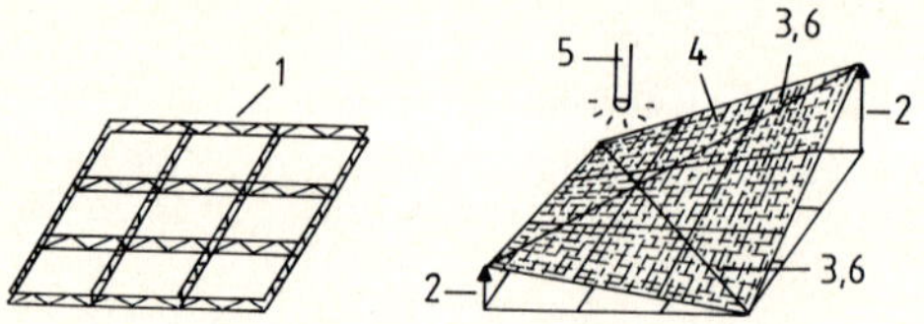

1: construction of a horizontal grid structure
2: elevation of two opposite corners
3: fixing the form by two tensioners
4: application of the Stucanet
5: spraying of the polyurethane foam
6: removal of the tensioners

Fig.3 Construction of a hyperbolic paraboloid with the Casapack system

At the Laboratory Reyntjens of the Department of Civil Engineering two scale models have been made and tested till failure [1]. This contribution treats the mechanical behaviour of hyperbolic paraboloids [1]. Thermal behaviour is verified by a number of numerical simulations in reference [2]. The durability and ultra-violet protection of the polyurethane foam was also taken into consideration.

2. Materialproperties

For modelling purposes the elastic characteristics are to be determined.

2.1. Polyurethane

The mean density of the foam was 62 kg/m^3. Samples with a cross section of 70 mm x 70 mm were tested under different load conditions:

	$\dot{\varepsilon}$ ($\times 10^{-6}/s$)	$E(\sigma)^{(1)}$ (N/mm^2)	$\sigma_{max}^{(3)}$ (N/mm^2)
Tension	167	15,5 (0,20)	0,48
Compression	67	11,6 (0,10)	0,30
Bending	$221^{(2)}$	14,2 (0,35)	-
	$\dot{\gamma}$ ($\times 10^{-6}/s$)	$G(\tau)$ (N/mm^2)	τ_{max} (N/mm^2)
Torsion	$221^{(2)}$	6,5	0,47

Table 1

Notes: (1): A secant modulus is used. The corresponding stress level is indicated.
(2): The strain is non uniform. Maximum strain rates are given.
(3): This value is only indicative because of clamping or buckling effects.

The obtained values correspond very well with other tests of the producer. For the numerical model the values $E = 14\ N/mm^2$ en $\nu = 0$ are retained.

2.2. Poutrafils

The guarantied yield strength of the steel wire is 450 N/mm^2. From the Bekaert product catalogue we retrieve the following bending properties:

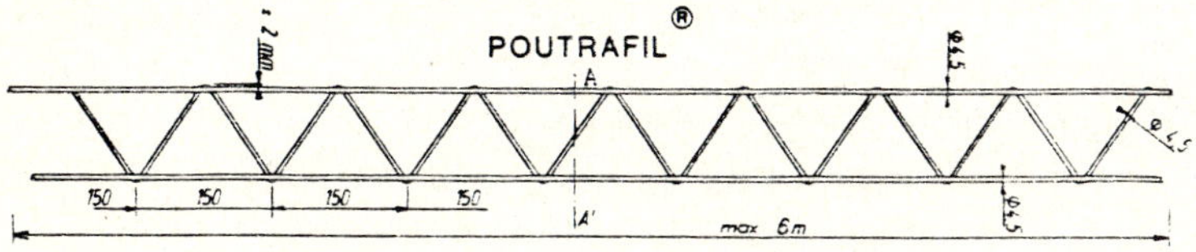

FLEXURAL STRENGTH CHARACTERISTICS

Cut AA'

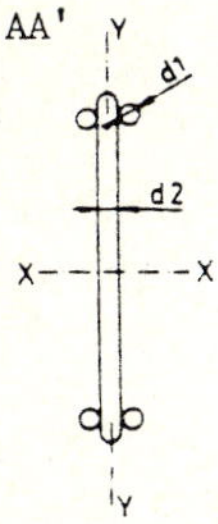

WIRE		PF../4.5	PF../6
d1 (d2)	(mm)	4.5 (4.5)	6.0 (4.5)
Ω I I/v i	(mm^2) (mm^4) (mm^3) (mm)	15.9 24.1 8.95 1.12	28.3 63.3 21.2 1.50
p	(kg/m')	0.125	0.222

Ω = Cross section
I = Moment of inertia
W = Sectional Modula
i = $\sqrt{\frac{I}{\Omega}}$ = Radius of gyration
p = Weight

POUTRAFIL®	PF 40/4.5	PF 50/4.5	PF 90/4.5	PF 100/4.5	PF 140/4.5	PF 150/4.5	PF 190/4.5	PF 200/4.5	PF 100/6	PF 150/6	PT 200/6
Ω tot (cm^2)	0.64		0.64		0.64		0.64		1.13	1.13	1.13
W_x Ix (cm^4) Ix/v (cm^3) ix (cm')	2.00 1.00 1.77	3.29 1.32 2.27	11.63 2.58 4.26	14.51 2.90 4.76	29.20 4.17 6.75	33.67 4.49 7.25	51.51 5.42 8.97	57.40 5.74 9.47	25.00 5.00 4.70	58.60 7.81 7.20	100.00 20.00 9.41
W_y Iy (cm^4) = Iy/v' (cm^3) iy (cm')	0.090 0.165 0.375							-	0.2544 0.283 0.450		
p (kg/m')	0.620	0.661	0.698	0.703	0.775	0.800	0.890	0.910	1.163	1.260	1.130

Table 2

In this table only the longitudinal wires are taken into account for evaluation of the flexure strength and stiffness. A number of bending tests on simply supported beams were performed to verify these values [1]. The obtained experimental stiffness was lower. Numerical simulation of the poutrafil as a plane frame confirmed this result: the difference became smaller with greater span. The conclusion was that the lattice girder can be modelled approximately as a massive beam with the indicated simplified stiffness (Table 2), provided the length of the beam is long enough.

Table 3 contains the experimental torsional stiffness of the poutrafils:

	PF 40/4,5	PF 50/4,5
Torsional stiffness (mm^4)	190	220

Table 3

There was a good agreement with numerical simulation as a grid, loaded by a torque [1].

3. Test on hypar shell models

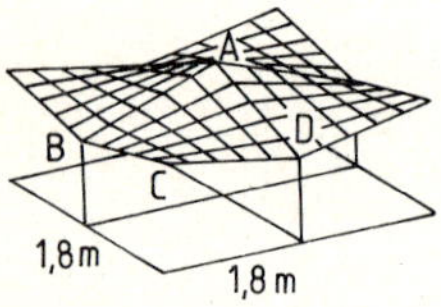

Fig.4 shows the investigated structure, composed by 4 hypar shells and supported by 4 columns. If the loading is symmetrical, the analysis can be restricted to a quarter of the shell (part ABCD).

Fig.4 The investigated structure

3.1. Test frame (Fig.5)

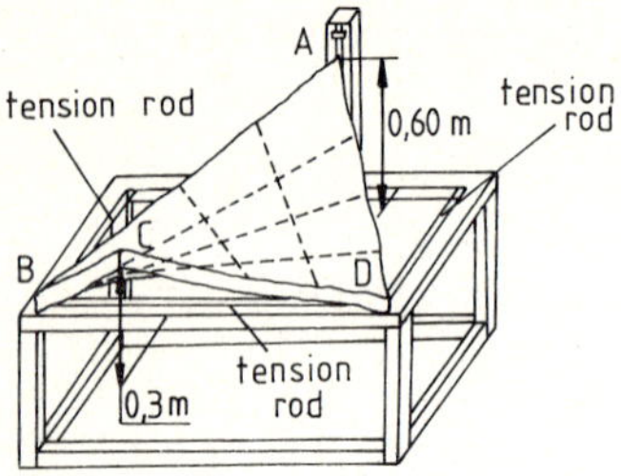

Fig.5 Test frame

The 2 columns are simulated by ball supports (B and D). Tensioners must take up the thrust forces. Point C is free to move and point A is confined to a strictly vertical movement, as it is in the complete structure (Fig.4). The same boundary conditions are also valid for the sides AB, BC, CD and DA. For keeping the test apparatus simply these sides are left free. Calculations have shown that this has a minor effect on the results.

3.2. Construction of the scale models

A first model (I) starts from a grid of poutrafils, every 0,45 m (see also Fig.3). In the second (II) the distance between poutrafils is enlarged to 0,60 m. The mean thickness of the polyurethane foam is 95 mm.

3.3. Loading

In very node of the poutrafil-grid a point load is applied.

3.4. Measurements

A number of vertical displacements and strains in the poutrafils were measured.

3.5. Results

The tests have revealed that there was only a small difference in the structural behaviour of the two models (I and II).

This means that the spacing of the poutrafils is of little consequence. Fig.6 shows some results for the second model.

The maximum load, a concentrated force of 380 N in all free nodes, corresponds with a uniformly distributed load of $\pm$ 1700 N/m^2.

After unloading the shell nearly gets back its original shape.

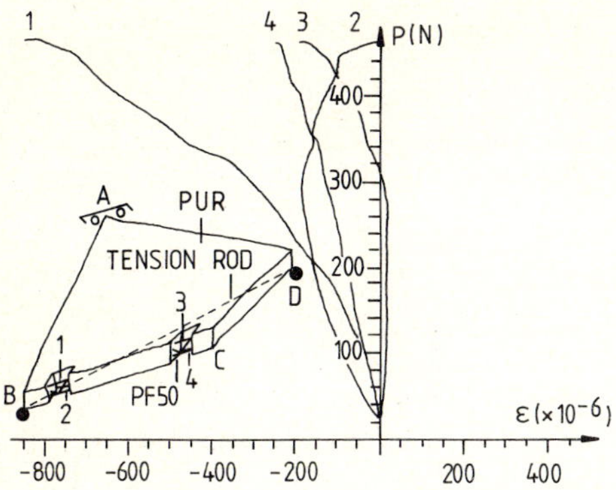

a) Strains in the edge beam

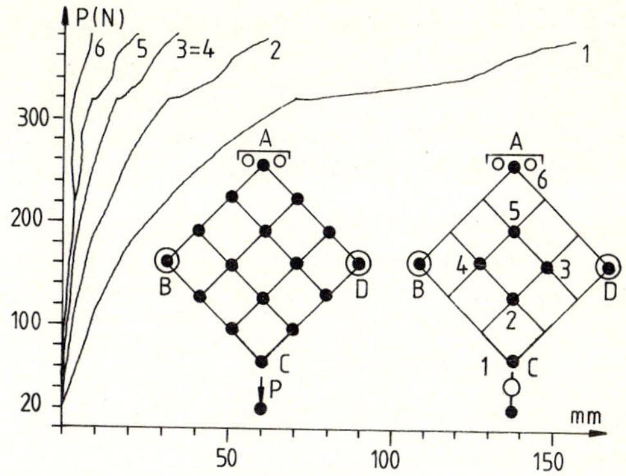

b) Vertical Displacements

Fig.6 Measurements on the scale model II

4. Numerical verification and parameter study

4.1. Linear analysis

The triangular thin shell element of the finite element program TITUS [3] is used (Fig.7). This program has excellent pre- and postprocessing capabilities.

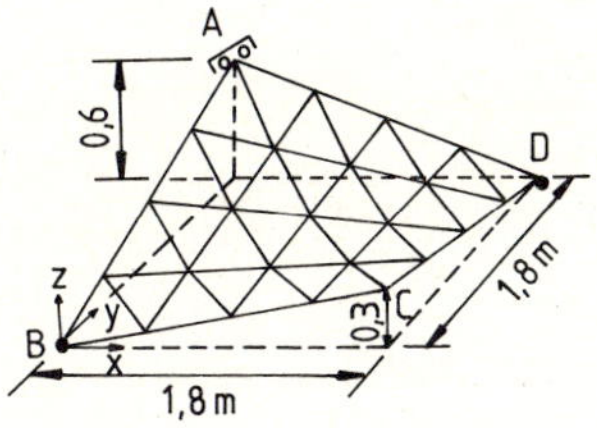

Fig. 7 The element mesh

Different calculations are performed to answer the following questions:

- What is the influence of the thickness of the polyurethane foam?
- Which component is predominant in the load-carrying-capacity: the grid of poutrafils or the polyurethane?
- Is the spacing of the poutrafils important?
- What is the effect of reinforcing the edge beams?
- How favourable is an extra support at the free end C?
- Is it important to use tension rods in the case of weak columns?

Table 4 sums up the different alternatives:

CASE	PUR		POUTRAFILS SPACING	BOUNDARY CONDITIONS			REMARKS
	elem.	thickness (mm)	(mm)	B,D	A	C	
1*	32	95	-	ball	slider	free	lin./non-lin (1')
2	32	60	-	ball	slider	free	
3	32	95	-	weak columns	slider	free	
4	32	95	beam(1800)	ball	slider	free	
5	16	95	beam(600)	ball	slider	free	
6	32	95	beam(450)	ball	slider	free	
7	32	95	2beams(1800)	ball	slider	free	
8	32	95	-	ball	slider	ball	

* Reference case

Table 4

A spacing of 1800 mm means that poutrafils are used exclusively as edge beams (no intermediate poutrafils).

The tabulated results of the finite element calculations for a 100 N load in every node are reproduced in Table 5. In case 5 it is necessary to multiply the values by 25/16 to obtain the same total load.

CASE	(1)	(2) N/mm^2	(3)	(4) N/mm^2	(5)	(6) N/mm^2	(7)
1	71 (1,00)	0,13	Fig. 8				
2	168 (2,37)	0,23	Fig. 8				
3	127 (1,79)	0,22	Fig. 8				
4	30 (0,42)	0,05	Fig. 9	35	Fig.11a	27	Fig.11b
5	16 (0,23)	0,03	Fig.10	22	Fig.11a	24	Fig.11b
6	20 (0,28)	0,05	Fig.10	35	Fig.11a	21	Fig.11b
7	28 (0,39)	0,05	Fig. 9	18	Fig.11a	14	Fig.11b
8	0 (0,00)	0,11	Fig. 8				

(1) = deflection in point C

(2) = maximum effective stress (Van Mises) in the polyurethane

(3) = position of maximum stress

(4) = maximum normal stress in the poutrafils

(5) = position of maximum normal stress

(6) = maximum bending stress in the poutrafils

(7) = position of maximum bending stress

Table 5

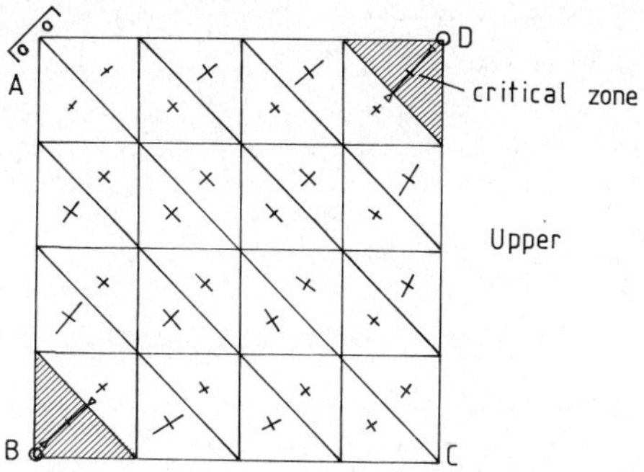

Fig.8 Principal stresses in the polyurethane foam: case 1,2,3,8

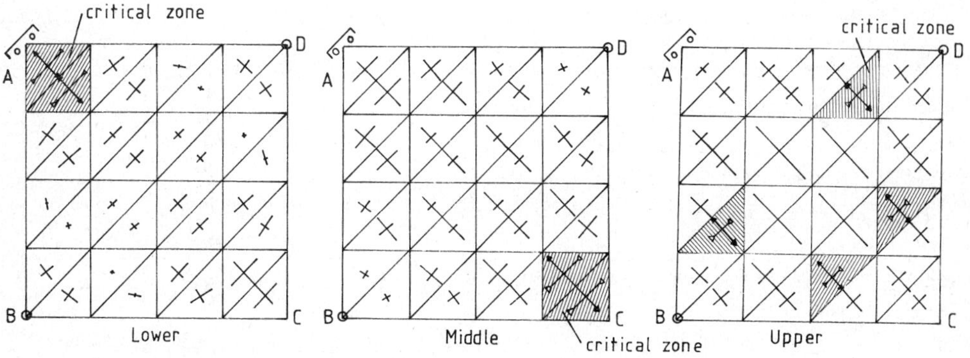

Fig.9 Principal stresses in the polyurethane foam: case 4,7

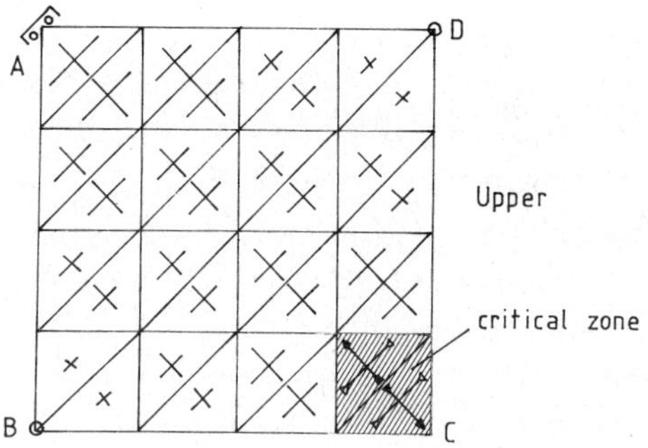

Fig.10 Principal stresses in the polyurethane foam: case 5,6

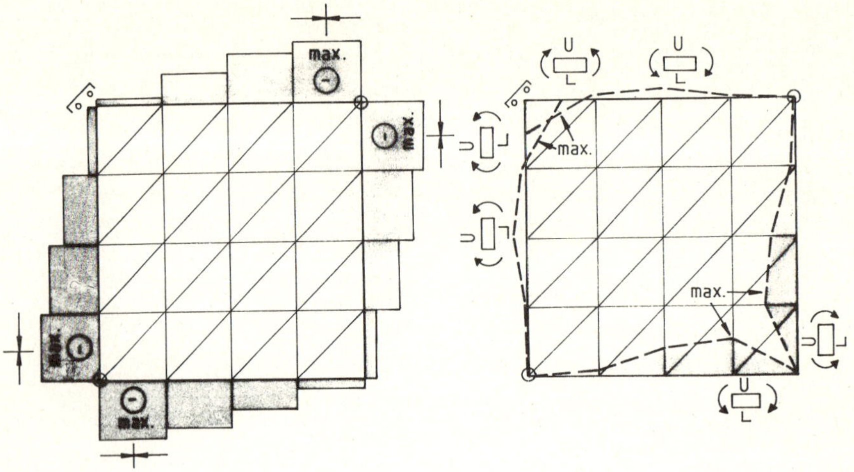

a) Axial forces b) Bending moments

Fig.11 Internal forces in the poutrafils: case 4,5,6,7

We draw attention to the following results:

- The membrane theory gives only an estimation of the stresses in the mid plane of the central part (effective stress ≅ 0,02 N/mm^2). The maximum effective stresses in the cases without poutrafils are at least 5 times higher. The deficiency of the membrane theory is well known in literature [4].
- There is a good agreement between the calculated axial forces in the edge beams (poutrafils) and the values predicted by membrane theory. This is also stated in [5].
- Important bending of the poutrafils occurs in the neighbourhood of the free corner C and the top point A.
- Increasing the thickness of the polyurethane has a great influence on the maximum deflection and the maximum stress (Case 1 ⇔ 2). This influence is greater than predicted by membrane theory.
- Reduction of the stiffness of the supporting columns causes considerable greater stresses and deflections (Case 3). It is worth while to provide tension rods between the (weak) columns.
- On its own the grid of poutrafils has only a very small bearing capacity. Nevertheless addition of boundary poutrafils to the polyurethane shell results in considerably lower stresses in the polyurethane and smaller deflections (factor 2,5) (Case 1 ⇔ 4,7).
- Strenghtening of the boundary (2 poutrafils) causes evidently a force reduction in this boundary but has a minor influence on deflections and polyurethane stresses (Case 4 ⇔ 7).
- The inclusion of poutrafils has the consequence that the critical zone of maximum stress is moving away from the supports (Fig.8 ⇔ Fig.9). A distribution of poutra-

fils moves the critical zone towards the free corner. (Fig.10)

- The greatest bending moment in the boundary poutrafils is found near the free corner C and the top point A. The positions of maximum normal force and bending moment doesn't coincide.
- The intermediate poutrafils have little structural interest: the forces in the boundary poutrafils are nearly the same and so are the stresses in the polyurethane (Case 4 ⇔ 5,6). These poutrafils are necessary for fixing the form of the hypar shell and for providing the punchresistance.
- If the free corner C is supplementary supported, the maximum effective stress in the polyurethane is almost unchanged but the maximum deflection decreases drastically from 71 mm to 17 mm (Case 1 ⇔ 8).

4.2. Non-linear analysis

The question arises whether the observated deformation behaviour can be explained as a geometrical non-linear effect. With TITUS a geometrical non-linear analysis of the reference case 1 is performed, based on a updated Lagrangian formulation. A load step of 50 N is used. Fig.12 and 13 show some results:

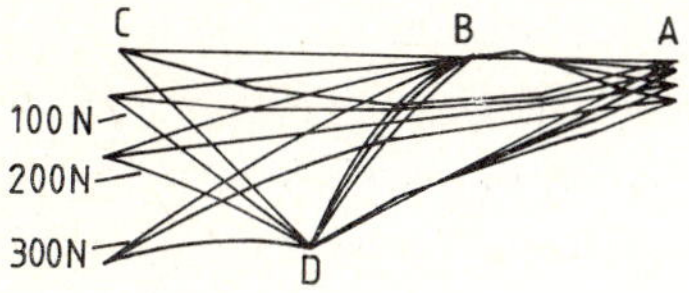

a) Spatial view

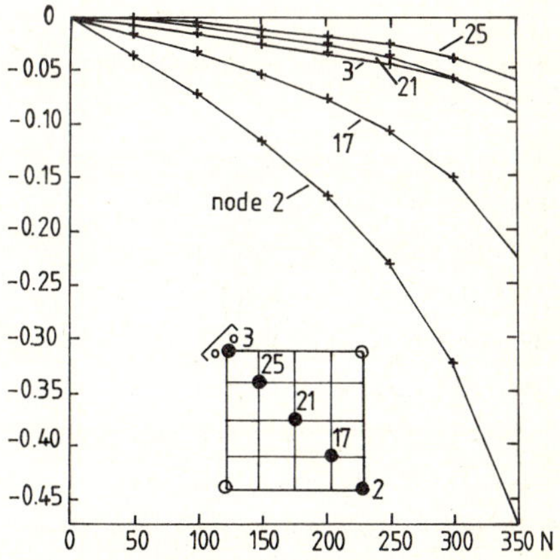

Fig.12 Deflections (Non-linear analysis)

b) Function of the applied load

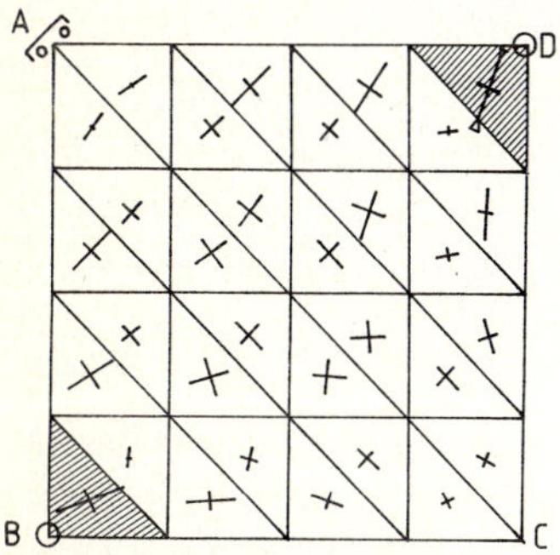

a) Overall view

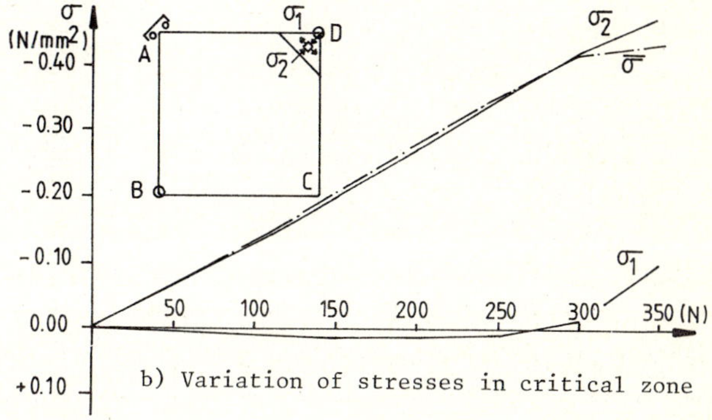

b) Variation of stresses in critical zone

Fig.13 Principal stresses (Non-linear analysis)

It can be seen from these figures that

. the non-linear behaviour is most pronounced in the region of great displacements (Fig.12).
. the principal directions may vary considerably, without significant change in the value of the principal stresses (compare Fig.13a with Fig.8).

5. Confrontation of the experimental results with the results of numerical analysis

5.1. Deflections (Fig.14)

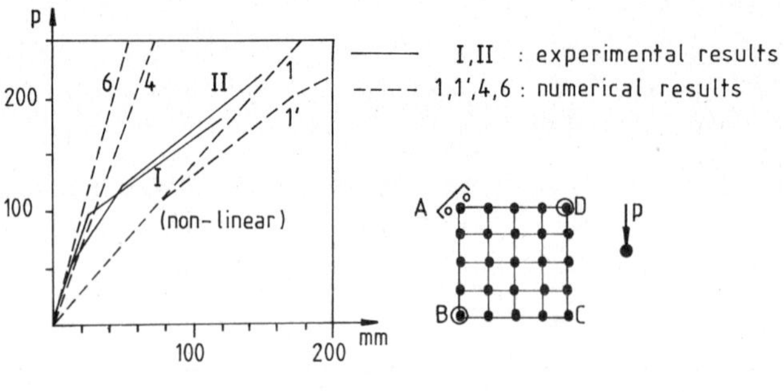

Fig.14 Deflection in the free point C

In the lower load range the deformation behaviour can be predicted by the combined model polyurethane and poutrafils (case 4,6). Beginning from 120 N there is clearly a non-linear effect. The somewhat different experimental behaviour can be caused by some shortcomings in the assumptions of the calculation model: e.g. the influence of the internal stresses, as a result of the construction procedure (Fig.3), is neglected, the link between two crossing poutrafils is assumed to be infinitely rigid.

5.2. Strains (Fig.15)

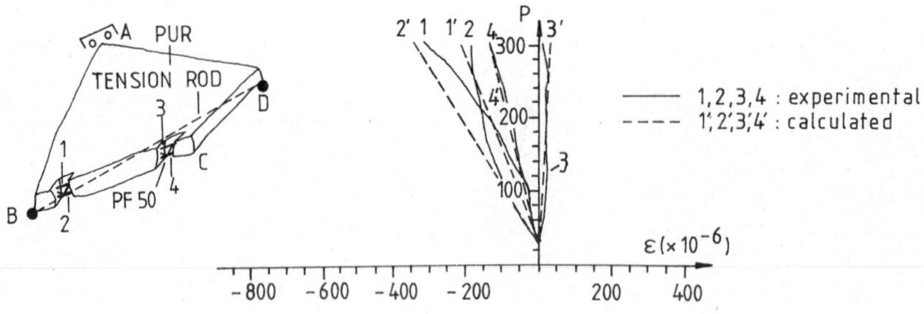

Fig.15 Strains in the edge beam (model II)

There is a reasonable agreement for loads up to 120 N. For higher loads the bending moment in the edge beam near the ball support B takes the other sign during the test.

5.3. Ultimate load

The observed maximum load is 1700 N/m^2. According to case 5, the maximum stresses are 0,11 N/mm^2 in the polyurethane and 110 N/mm^2 in the poutrafils. Although a non-linear calculation will give higher values, it is our conviction that the load-carrying-capacity is not limited by materialstresses but by excessive deformation (buckling, snap through behaviour).

6. Conclusions

The Casapack low-cost building system is used succesfully for the erection of walls and can be extended to the construction of roof structures. The combination polyurethane foam and poutrafils has the necessary load-carrying-capacity, if the following requirements are satisfied:

- Try to use the polyurethane as a structural element by constructing a curved, faceted or folded surface so that geometry activates mainly axial forces. A hyperbolic paraboloid is treated in this contribution.
- The great flexibility of the combined system obliges to control the deformations. It is important to supply the structure with sufficient (stiff) support. The use of tensioner rods can be recommended. In the case of moderate deformations, a linear model can fairly predict the structural behaviour.
- The most advantageous placing of the poutrafils is at the boundary. The intermediate poutrafils serve only as a base for fixing the stucanet.

References

1. Bettens C., Van Valckenborgh D., Studie van een hyparschaal opgebouwd volgens het Casapack-systeem, Thesis, Department of Civil Engineering, Univ. of Leuven, 1986.

2. Senave E., Studie van het Casapack-bouwsysteem, Thesis, Department of Architecture, University of Leuven, 1986.

3. Titus System: User's Manual, Framatome, 1978.

4. Ringleben W., Ein Beitrag zum Tragverhalten van hyperbolischen Paraboloidschalen, Bauingenieur 60, 1985.

5. Schnobrich W.C., Thin Shell Structures, Handbook of Concrete Engineering, Chapter 14, Van Nostrand Reinhold Company, New York, 1985.

6. Casapack makes housing simple / Bekaert Health Centers, Bekaert International Trade, Zwevegem, Belgium, 1985.

NUMERICAL ANALYSIS OF TENSILE STRUCTURES WITH MINIMUM SURFACE BASED ON SURVEY OF SOAP FILM MODELS

G.VALENTE (*)   M.MOSCETTI (**)

(*) Associate Professor, Istituto di Scienza delle Costruzioni, Facoltà di Ingegneria, Università de l'Aquila, Italy.

(**) Dr.Architect.

ABSTRACT

Tensile structures have modified some basic aspects of architectural and structural design. Structural shapes are, in fact, usually defined on the basis of architectural and functional criteria, and structural safety is tested subsequentially. But this procedure cannot be followed in the case of tensile structures, for which only the boundary conditions can be assigned; these conditions are then tested to ascertain that they are structurally valid. A scientific method which can determine the shape of tensile structures and strengths not only represents a new technical achievement, but also opens up a whole new and extremely intriguing problem in the structural mechanics field. This paper proposes to set forth such a new method, using as its starting point the shape derived from a soap film model. Using this method, certain initial hypothesis may be tested and problems of morphology and statics may be solved using mathematical programming techniques.

1. Introduction

The design of tensile structures can be verified immediatly using soap film models obtained on a special frame which reproduces the contour of the structure, in scale. The soap film surface is a minimal area surface.

Generally speaking, it is relatively simple to create soap film models; but they are difficult to measure both because of the films' instability in time and because of its transparency, which makes it almost impossible to photograph.

The survey method originally proposed by Architect M.Moscetti [2] was outlined in an earlier paper [3] .

The maximum error in the results of this procedure -the total of model and photographic errors- is about 10 cm in the surface height.

To obtain more accurate values for the elevation of the minimal surface, we can start from the values obtained by the photographic survey; using a numerical method of mathematical programming, we can then obtain very accurate values for this elevation.

2. Calculation of geometric characteristics.

Let us consider the tensile structure subdivided into m oblique quadrlaterals either by the actual cables or by the planes of the photographic survey on the model.

For a given quadrilateral IJKL in the space XYZ as in the following Fig. 1

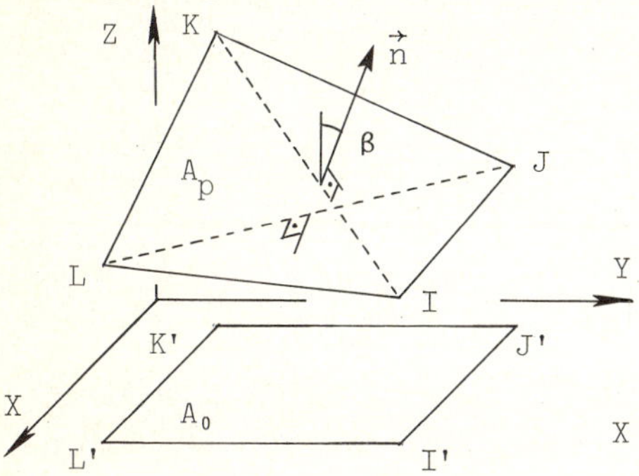

Fig.1. Oblique quadrilateral.

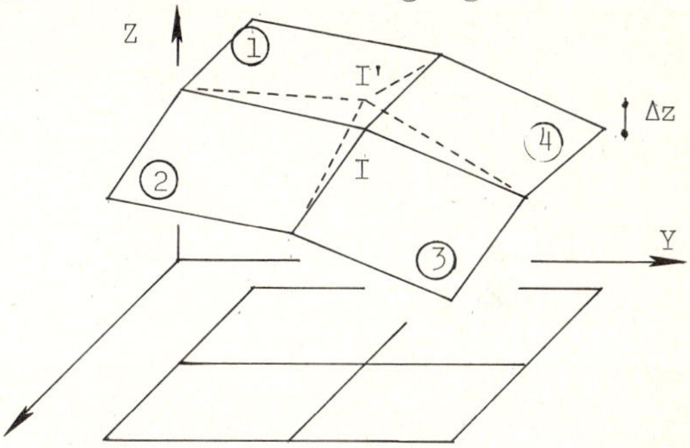

Fig.2. Quadrilateral containing node I.

we compute the values:

A_0 area of the quadrilateral I'J'K'L' projection of IJKL on the plane XY;

β angle between the normal tensor to diagonals IK and JL, and Z axis;

$A_p = A_0 \cos\beta$ the area of this oblique quadrilateral IJKL.

The total surface of the tensile structure is a function of the abscissae z_i of the free nodes:

$$A_{tot} = f(z_i) = \Sigma_p A_p$$

Let us consider all the quadrilaterals containing a certain node I as in Fig. 2. The first partial derivative of the total surface $f(z_i)$ with respect to this node I is numerically computed by varying the z_i abscissa of node I with an assigned value $\Delta z_i = 10^{-5}$ m and by computing the resulting variation for the total surface area. The variation is thus calculated considering only the quadrilaterals (the maximum number of these quadrilaterals is four) that contain this node I; that is:

$$g_i = \delta f(z_i)/\delta z_i = \Delta A_{1234}/\delta z$$

3. Function minimization by conjugate gradients.

The original process proposed by Fletcher and Reeves [1] is used herein.

In particular, we consider a function of n variables z_i whose value f(z) and gradient vector g(z) can be calculated at any point z_i.

We assume that in a neighbourhood of the required minimum h, the function may be expanded in the form

$$f(z) = f(h) + (z-h)' A (z-h)/2 + \text{higer terms} \quad (1)$$

where A, the matrix of second-order partial derivatives, is symmetric and positive definite.

By starting from a position z_i, let z_{i+1} be the position of the minimum of f(z) with respect to variations along the line through z_i in some specified direction p_i.

Setting $g(z_i) = g_i$ for each i, the step from $z_i$to z_{i+1} is determined by the relation

$$g'_{i+1} p_i = 0 \quad (2)$$

where

$$z_{i+1} = z_i + \alpha_i p_i \quad (3)$$

for some scalar parameter α_i.

4. Methods using conjugate directions.

Let us consider the minimization by successive linear searches of the quadratic function

$$f(z) = f(h) + (z-h)' A (z-h)/2 \quad (4)$$

for which the gradient is

$$g(z) = A (z-h) \quad (5)$$

By repeated use of equation (3), we have

$$z_n = z_{j+1} + \Sigma_i \alpha_i p_i \quad (6)$$

for any j in $0 \leqslant j \leqslant n-1$. It then follows from equation (5) that

$$g_n = g_{j+1} + \Sigma_i \alpha_i A p_i \quad (7)$$

and therefore, using equation (2), that

$$g'_n p_j = \Sigma_i \alpha_i p'_i A p_j \quad (8)$$

Now if the vectors p_0 , p_1 , p_2 ,...p_{n-1} are A-conjugate, satysfying

$$p'_i A p_j = 0 \quad \text{for } i \neq j \text{ , then} \quad (9)$$

$$g'_n p_j = 0 \quad (10)$$

and therefore, since p_0 , p_1 , p_2 ,...p_{n-1} form a basis,

$$g_n = 0 \quad (11)$$

whence, by equation (5)

$$z_n = h \quad (12)$$

This demonstrates that the method of successive linear searches is quadratically convergent when using any set of A-conjugate directions.

5. The method of conjugate gradients.

This method is an elegant n-step procedure for solving a set of simultaneous linear equations having a symmetric positive definite matrix of coefficients.

The equivalence of that problem and the minimization of a quadrative function $f(z)$ is clear from equations (4) and (5). The condition for the gradient to vanish is seen to be

$$A\,z = b \qquad \text{where} \qquad b = A\,h \tag{13}$$

In the solution of these equations, directions p_0, p_1, ... are generated such that p_{i-1} is a linear combination of $-g_{i+1}$ and p_0, p_1, ...p_i such that the A-ortogonality conditions (9) are satisfied.

In the event many of the coefficients are zero and the following simple form results

$$p_{i+1} = -g_{i+1} + \beta_i\, p_i \qquad \text{where} \tag{14}$$

$$\beta_i = g_{i+1}^2 / g_i^2 \tag{15}$$

This leads to the following general minimization algorithm

$$\left.\begin{array}{l} z_0 = \text{arbitrary} \\ g_0 = g(z_0)\,, \quad p_0 = -g_0 \\ z_{i+1} = \text{position of minimum of } f(z) \text{ on the line through } z_i \text{ in direction } p_i \\ g_{i+1} = g(z_{i+1}) \\ \beta_i = g_{i+1}^2 \,/\, g_i^2 \\ p_{i+1} = -g_{i+1} + \beta_i\, p_i \end{array}\right\} \tag{16}$$

This process is guaranteed, apart from rounding errors, to locate the minimum of any quadratic function of n arguments in at most n iterations.

For functions which are not quadratic the process is iterative rather than n-step, and a test for convergence is required.

6. The choice of z_0.

For quadratic functions any choice of starting point is in principle equally satisfactory.

For general functions the best can reasonably be expected is that the minimization process will lead as quickly as possible to the bottom of whatever valley it starts in. In some applications it is possible to detect when convergence to an unwanted minimum has occurred, and some form of extrication process is then desirable.

To avoid problems we start from the numerical values obtained from the photographic survey of the soap film shown in the paper [3] .

7. The linear search.

Essentially the linear search problem is to determine t_m such that

$y'(t_m) = 0$ where (17)

$y(t) = f(z_i + t\, p_i)$ (18)

and therefore

$y'(t) = p_i^! \, g(z_i + t\, p_i)$ (19)

Thus $y(t)$ and $y'(t)$ are calculable for any t and, in particular, $y(0) = f_i$ and $y'(0) = p_i^! \, g_i \leqslant 0$ are already available.

The calculation is in three stages; the first estimates the order of magnitude of t_m, the second estabilishes bounds on it, and the third interpolates its value.

There is no comparable feature of the present method, and so we have arbitrarly chosen a unit for t which corresponds to a displacement along p_i of unit length in the z-space.

On the supposition that this is a correct estimate, that the unconstrained minimum lies on the line $z_i + t\, p_i$, and that $f(z)$ is quadratic, we would obtain the value k for t_m where

$k = 2\,(est - f_i)\, /\, p_i^! \, g_i$ (20)

We have therefore taken as a tentative step length

$h = k \quad \text{if} \quad 0 < k < (p_i^2)^{-1/2} \;, \quad = (p_i^2)^{-1/2} \quad \text{otherwise}$ (21)

In the second stage y' is examined at the points t = 0, h, 2h, 4h,...,a ,b, where t is doubled each time, and where b is the first of these values at which either y' is non-negative or y has not decreased. It then follows that $a < t_m \leqslant b$.

The third stage uses the classical cubic interpolation on the line.

8. The convergence chriterion.

Clearly if any g_i^2 vanishes the iterations must stop, both to avoid division by zero in the next iteration and because this is the formal requirement for z_i to be at a minimum.

This condition is unlikely to be realized in practice because of round-off errors.

We therefore continue the iterations until a complete cycle of (n+1) iterations, starting from a steepest descent search, produces no reduction in the value of the function.

9. Hyperbolic paraboloid.

The surface shown in the earlier paper [3] corresponds to the hyperbolic paraboloid defined by the following equation

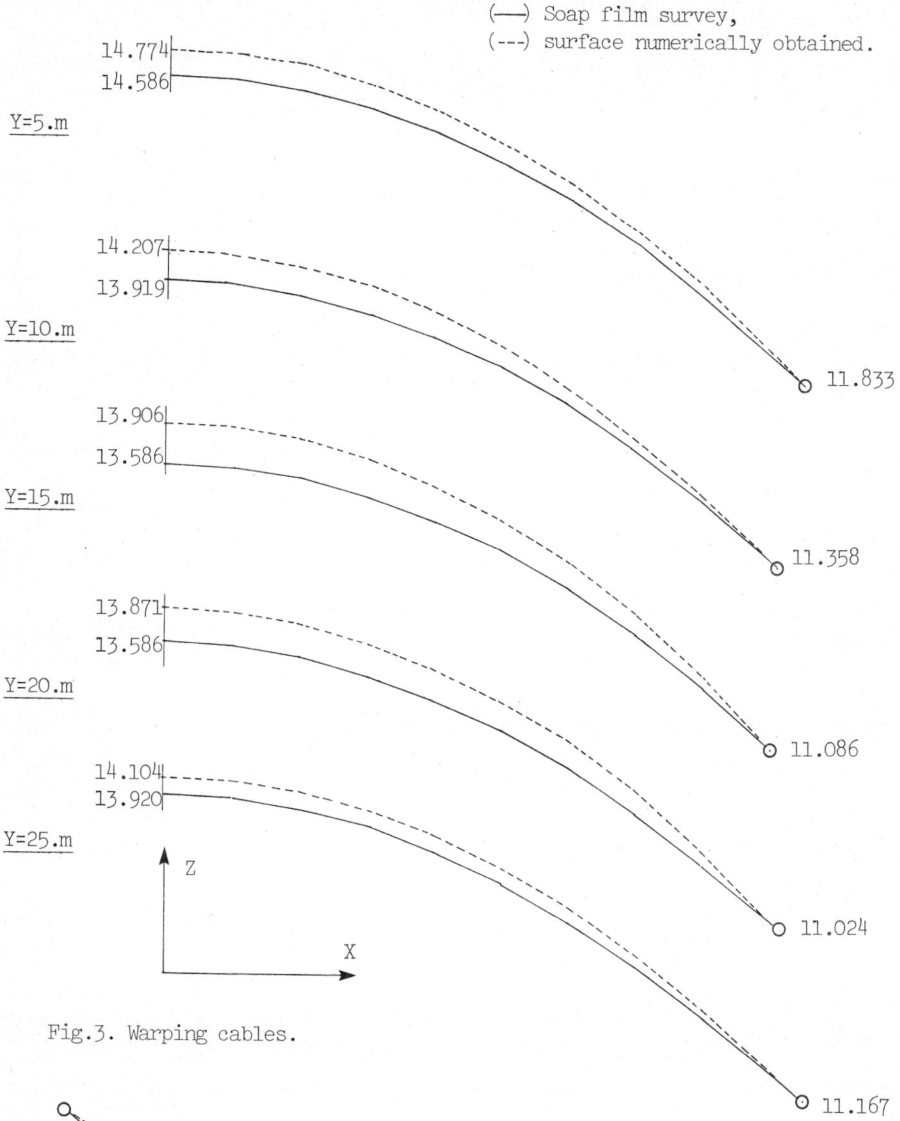

Fig.3. Warping cables.

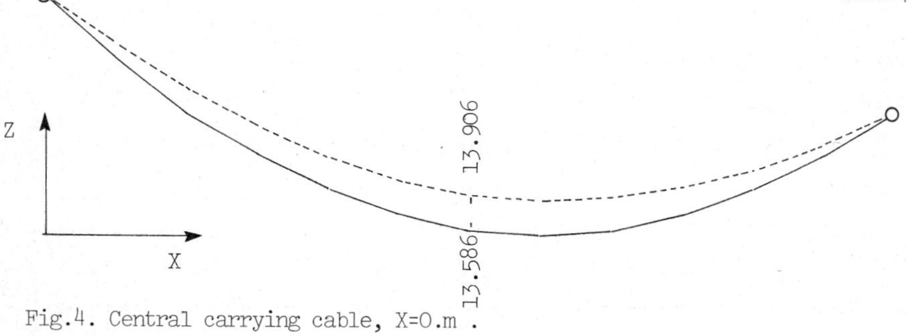

Fig.4. Central carrying cable, X=0.m .

$$z = 15.63 - .2\overline{3}\ y + .00\overline{6}\ y^2 - x^2/\ 202.5$$

We apply the proposed numerical method to this surface by using as input the elevation obtained from the photographic survey of the soap film.

We consider the two resulting meshes, the sides of whose squares are respectively 5. m and 2.50 m as appears in Fig.5.

The minimal surfaces for both pratically coincide, as appears in Figs. 3 and 4.

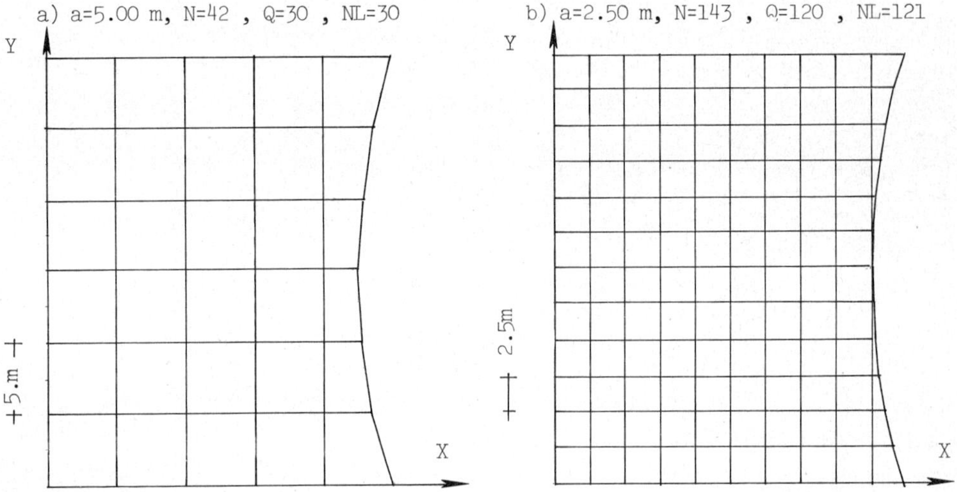

Fig. 5. Subdivision into quadrilaterals.

10. Second surface tested.

This is a general surface that cannot be defined by an equation, as in the first example.

Fig. 6 shows the mesh numerically tested

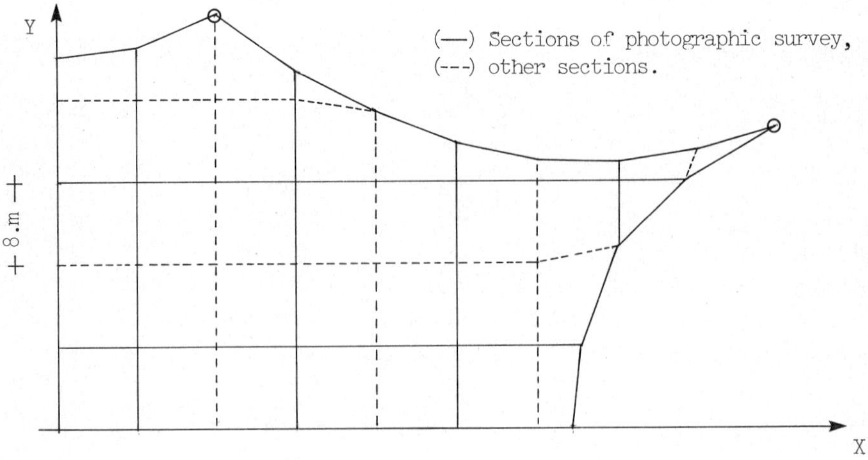

Fig.6. Numerically tested mesh.

The mesh was obtained by taking into account that the planes X=0 and Y=0 are symmetrical planes for the elevations.

Fig.7 is the photo of the model plan.

In the model, the boundary cables are represented by silk threads which are drawn in by the surface tension of the soap film.

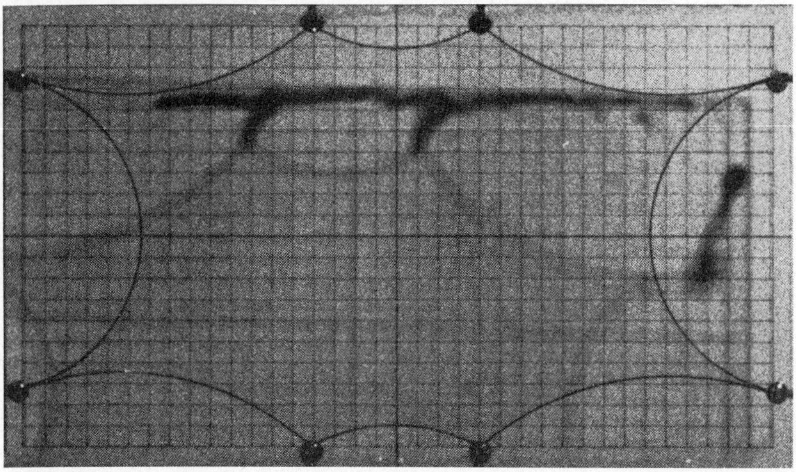

Fig.7. Photo of the model plan.

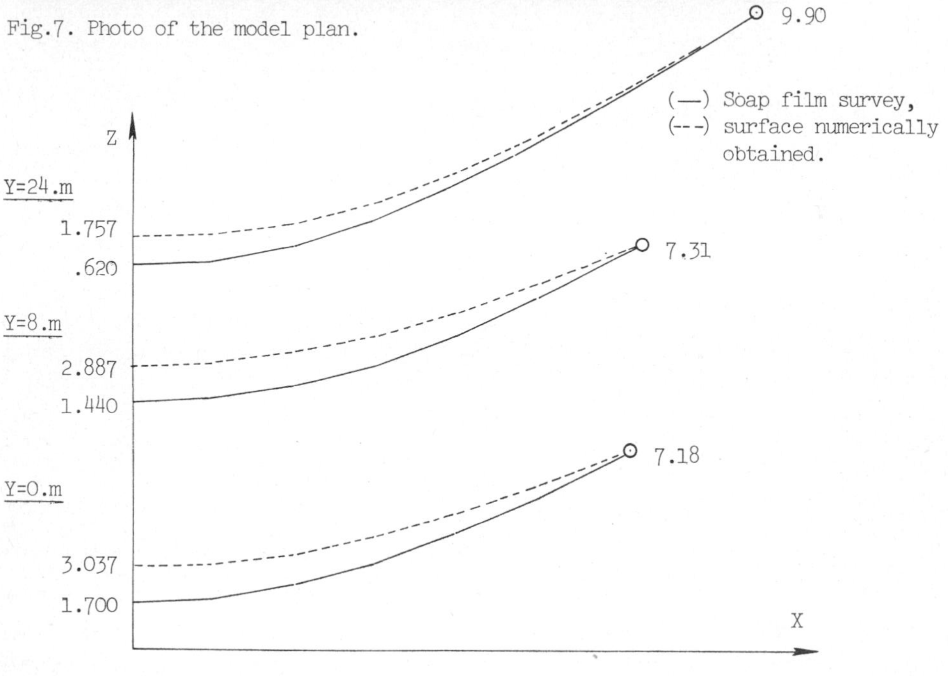

Fig.8. Carrying cables.

We thus attain the elevations shown in the Fig.8.

11. Third surface tested.

Also this is a general surface; like the second surface shown previously, it cannot be defined by an equation. The mesh shown in Fig.9 is obtained taking into account that, for the elevations, X=0 and Y=0 are symmetrical planes and X=Y is an unsymmetrical plane.

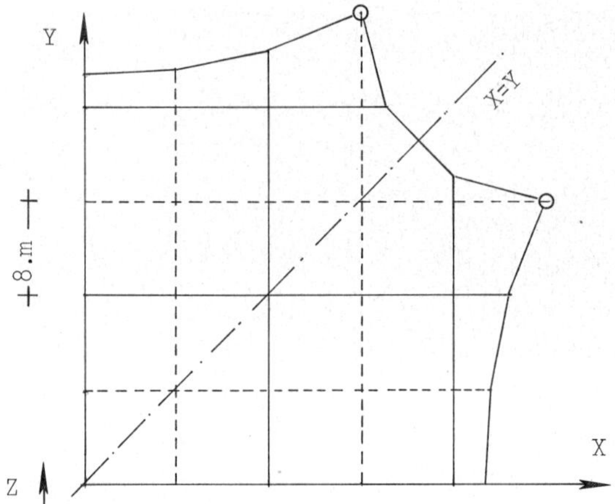

Fig.9. Numerically tested mesh.

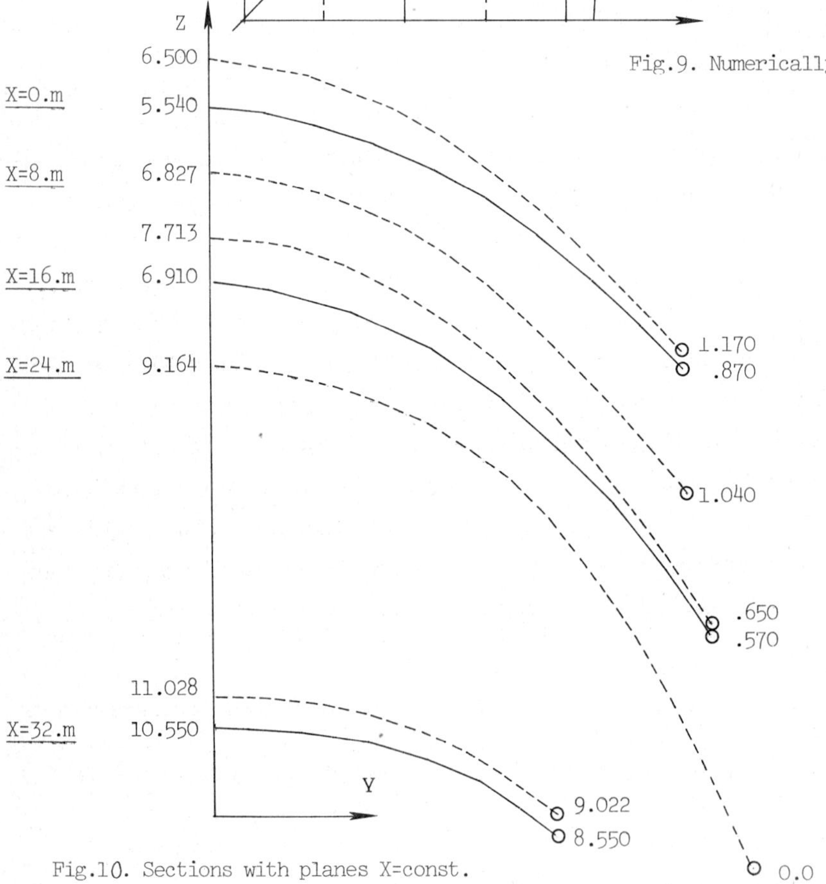

Fig.10. Sections with planes X=const.

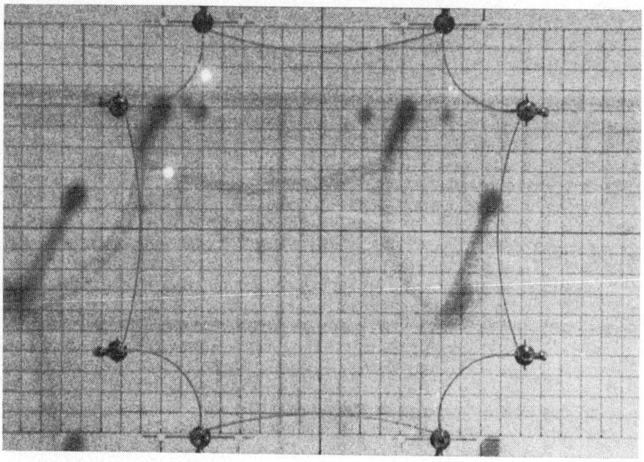

Fig.11. Photo of the model plan.

12. Conclusions

The scale chosen for the model is such that the weight of the soap film doesn't significantly affect the results. The actual effect may be observed by comparing the elevations of the model right side up with those of the model placed upside down. However, despite the precautions used in chosing the scale, in all hte three examples just shown, the elevations obtained by mathematical programming were higher than those resulting from the model. In any case, using the survey of the soap film model proposed in the earlier paper [3] in conjunction with the present numerical analysis,we attain to a powerful tool for the solution of the new problem raised by tensile structures. In this paper we have considered only the variability of elevation; this problem is solved by locating an unconstrained minimum of a function of several variables z_i. It is possible to solve this problem with other boundary conditions, such as a boundary cable of prassigned length, assuming as other variables the abscissae X and Y and expressing several of them as dependant on others.

REFERENCES

1. Fletcher, R.; Reeves, C.M. :"Function minimization by conjugate gradients", Computer Journal, Vol.7, pp. 149-154, 1964 .
2. Moscetti, M. :"I modelli nella ottimizzazione delle tensostrutture", Abstract of Doctoral Thesis, Acciaio, December 1982, pp. 562-565.
3. Valente, G. ; Moscetti, M. :"Tensile structures with minimum surface-Note II", International Congress I.A.S.S., USSR, Moscow, September 1985, pp. 153-166.

Lecture Notes in Engineering

Edited by C.A. Brebbia and S.A. Orszag

Vol. 1: J. C. F. Telles,
The Boundary Element Method
Applied to Inelastic Problems
IX, 243 pages. 1983.

Vol. 2: Bernard Amadei,
Rock Anisotropy and
the Theory of Stress Measurements
XVIII, 479 pages. 1983.

Vol. 3: Computational Aspects of
Penetration Mechanics
Proceedings of the Army Research
Office Workshop on Computational
Aspects of Penetration Mechanics
held at the Ballistic Research Laboratory
at Aberdeen Proving Ground, Maryland,
27–29 April, 1982
Edited by J. Chandra and J.E. Flaherty
VII, 221 pages. 1983.

Vol. 4: W.S. Venturini
Boundary Element Method in Geomechanics
VIII, 246 pages. 1983.

Vol. 5: Madassar Manzoor
Heat Flow Through Extended
Surface Heat Exchangers
VII, 286 pages. 1984.

Vol. 6: Myron B. Allen III
Collocation Techniques for Modeling
Compositional Flows in Oil Reservoirs
VI, 210 pages. 1984.

Vol. 7: Derek B. Ingham,
Mark A. Kelmanson
Boundary Integral Equation
Analyses of Singular, Potential,
and Biharmonic Problems
IV, 173 pages. 1984.

Vol. 8: Linda M. Abriola
Multiphase Migration of Organic
Compounds in a Porous Medium
A Mathematical Model
VIII, 232 pages. 1984.

Vol. 9: Theodore V. Hromadka II
The Complex Variable Boundary
Element Method
XI, 243 pages. 1984.

Vol. 10: C. A. Brebbia, H. Tottenham,
G. B. Warburton, J. M. Wilson, R. R. Wilson
Vibrations of Engineering Structures
VI, 300 pages. 1985.

Vol. 11: M. B. Beck
Water Quality Management:
A Review of the Development and
Application of Mathematical Models
VIII, 108 pages. 1985.

Vol. 12: G. Walker, J. R. Senft
Free Piston Stirling Engines
XIV, 286 pages. 1985.

Vol. 13: Nonlinear Dynamics
of Transcritical Flows
Proceedings of a DFVLR International
Colloquium, Bonn, Germany, March 26, 1984
VI, 203 pages. 1985.

Vol. 14: A. A. Bakr
The Boundary Integral
Equation Method in Axisymmetric
Stress Analysis Problems
XI, 213 pages. 1986.

Vol. 15: I. Kinnmark
The Shallow Water Wave
Equation: Formulation,
Analysis and Application
XXIII, 187 pages, 1986.

Vol. 16: G. J. Creus
Viscoelasticity – Basic
Theory and Applications
to Concrete Structures
VII, 161 pages. 1986.

Vol. 17: S. M. Baxter
C. L. Morfey
Angular Distribution
Analysis in Acoustics
VII, 202 pages. 1986.

Vol. 18: N. C. Markatos,
D. G. Tatchell, M. Cross, N. Rhodes
Numerical Simulation of Fluid Flow
and Heat/Mass Tranfer Processes
VIII, 482 pages. 1986.

Vol. 19: Finite Rotations
in Structural Mechanics
Proceedings of the Euromech
Colloquium 197, Jablonna 1985
VII, 385 pages. 1986.

Vol. 20: S. M. Niku
Finite Element Analysis
of Hyperbolic Cooling Towers
VIII , 216 pages. 1986.